INTEGRATED ANALOG-TO-DIGITAL AND DIGITAL-TO-ANALOG CONVERTERS

INTEGRATED ANALOG-TO-DIGITAL AND DIGITAL-TO-ANALOG CONVERTERS

by

RUDY VAN DE PLASSCHE
Philips Research Laboratories, Eindhoven, The Netherlands

SPRINGER-SCIENCE+BUSINESS MEDIA, B.V.

A C.I.P. Catalogue record for this book is available from the Library of Congress.

ISBN 978-1-4613-6186-2   ISBN 978-1-4615-2748-0 (eBook)
DOI 10.1007/978-1-4615-2748-0

Printed on acid-free paper

Contents

List of Figures

List of Tables

List of Symbols

Symbol	Description	Unit
A	Amplitude	1
ΔA	Amplitude deviation	1
A/D	Analog-to-Digital	
$A_{foldback}$	Ratio between amplifier bandwidth and system bandwidth	1
A_b	Amplitude of square wave	1
A_c	Amplitude of sine wave	1
A_d	Attenuation component: index d	1
A_e	Amplitude of error signal	1
A_f	Amplitude of fundamental	1
A_j	Amplitude level: index j	1
A_l	Limiter constant	1
A_{rms}	Root Mean Square amplitude value	1
A_s	Amplitude of sine wave	1
$A_{stopband}$	Stopband attenuation	dB
ATE	Automatic Test Equipment	
A_{pp}	Peak-to-peak amplitude	1
A_n	Operational Amplifier: index n	1
A_0	Output value of 1-bit quantizer	1
$A_{openloop}$	Open-loop amplification	1
B	Full-Scale Value	1
B_{in}	Input number of bits	1
B_0	Most Significant Bit Value	1
B_m	m_{th} Bit Value	1
B_{n-1}	Least Significant Bit Value	1

B_{out}	Output number of bits	1
BER	Bit Error Rate	1
C	Constant	1
C_{bc}	Base-Collector capacitance	F
C_{be}	Base-Emitter capacitance	F
C_{bex}	Extra Base-Emitter capacitance	F
C_d	Diode capacitance	F
C_g	Comparator gain index: g	F
$C_{H(old)}$	Hold capacitor	F
C_{in}	Input capacitance	F
CML	Current Mode Logic	
C_n	Capacitor: index n	F
$CMOS$	Complementary Metal Oxide Semiconductor	
C_m	Digital code: index m	F
C_n	Capacitor: index n	F
D/A	Digital-to-Analog	
DC	Direct Current	A
D_n	Diode: index n	
DNL	Differential Non-Linearity	1
D_{out}	Digital output data	1
$E(.)$	Statistical expectation	1
ECL	Emitter Coupled Logic	
ENB	Effective Number of Bits	1
EPS	Errors Per Second	1
ERB	Effective Resolution Bandwidth	Hz
$EXOR$	Exclusive OR function	
E_{glitch}	Glitch error	Vs
E_{LSB}	LSB energy	Vs
E_{noise}	Quantization noise voltage	V
E_{total}	Total noise voltage	V
E_{qns}	Quantization noise voltage	V
$E_{qns}^2(f)$	Quantization noise density	V^2/Hz
f	Frequency	Hz
f_b	-3 dB bandwidth	Hz
f_c	Clock frequency	Hz

f_{comp}	Comparator -3 dB bandwidth	Hz
f_{in}	Input signal frequency	Hz
f_{osc}	Oscillation frequency	Hz
f_{qns}	Quantization noise bandwidth	Hz
f_s	Sample frequency	Hz
f_{sig}	Signal bandwidth	Hz
f_{UGB}	Unity Gain Bandwidth	Hz
FSR	Full-Scale Range	1
G_A	Amplifier gain	1
$G(z)$	Transfer function	1
g_m	Mutual transconductance	A/V
$H(z)$	Transfer function	1
$H(\omega)$	System transfer function	1
$HS3$	High-speed bipolar oxide isolated process	
f_1	Unity gain bandwidth	Hz
f_t	Transition frequency	Hz
I	Current	A
ΔI	Current deviation	A
i	As index	1
IC	Integrated Circuit	
I_{comp}	Compensation current	A
$I_{leakage}$	Leakage current	A
I_{out}	Output current	A
I_{Ref}	Reference current	A
$IEEE$	Institute of Electrical and Electronic Engineers	
INL	Integral Non-Linearity	1
i_0	Base-Emitter reverse current	A
I_0	Bias current index: 0	A
I^2S	Inter IC Signal Standard	
I_T	Temperature-dependent current	A/degree C
I_{in}	Input current	A
$J - FET$	Junction Field Effect Transistor	
$J(z)$	Transfer function	1
k	Boltzmann's constant 1.38×10^{-23} J/K As index	1
$K(z)$	Transfer function	1

V_B	Battery voltage	V
V_{be}	AC Base-emitter voltage	V
V_{BE}	DC Base-emitter voltage	V
V_D	Diode voltage	V
$V_{difference}$	Difference voltage	V
V_{droop}	Droop voltage	V
V_{fs}	Full-scale voltage value	V
V_g	Bandgap voltage of silicon	1.208 V
V_{id}	Idle noise voltage	V
V_{in}	Input voltage	V
V_{lr}	Linear voltage range	V
V_{max}	Maximum voltage	V
V_n	Voltage at node index: n	V
V_{out}	Output voltage	V
V_p	Peak voltage	V
V_{ref}	Accurate reference voltage	V
V_{rn}	Reference voltage at node index: n	V
V_{tap}	Reference tap voltage	V
V_{th}	Threshold voltage	V
z	Complex frequency for discrete signals	
β	Transistor current gain	1
β_{square}	Unit area MOS gain factor	$\mu A/V^2 \mu$
ϵ	Amplitude deviation	1
ϵ_m	Error index: m	1
Δ	Matching deviation	1
π	Angular constant	3.14159
λ_n	Integration constant index: n	1
λ	Global transfer	1
λ_b	Global transfer of square wave	1
λ_l	Global transfer of limiter	1
λ_m	Maximum value of λ	1
λ_s	Global transfer of sine wave	1
Λ	Real value	1
$\sigma(P)$	Standard deviation of P	1

τ_d	Delay time	s
Θ_n	Complex phase	rad
ω	Angular frequency	rad/s
Φ	Phase angle	rad

TDA1540	Type indication of 14-bit D/A converter [1]
TDA1541	Type indication of dual 16-bit D/A converter [2]
TDA1534	Type indication of 14-bit A/D converter [3]
TDA1535	Type indication of sample-and-hold amplifier [4]
TDA8716	Type indication of folding A/D converter [5]
TDA8718	Type indication of folding A/D converter [6]

Preface

In this introduction an overview of the contents of this book will be given. Analog-to-digital (A/D) and digital-to-analog (D/A) converters provide the link between the analog world of transducers and the digital world of signal processing, computing, and other digital data collection or data processing systems. Numerous types of converters have been designed that use the best technology available at the time a design is made. High-performance bipolar and MOS technologies result in high-resolution or high-speed A/D and D/A converters that can be applied in digital audio and digital video systems. Furthermore, the high-speed bipolar technologies show an increase in conversion speed into the Giga Hertz range. Applications in these areas are, for example, in high-definition digital television and digital oscilloscopes. The availability of high-speed memory chips results in so-called "one-shot" memory applications in these oscilloscopes. In this book different techniques to improve the accuracy in high-resolution A/D and D/A converters will be discussed. Also, special techniques to reduce the number of elements in high-speed A/D converters by a repetitive use of comparators will be described.

In Chapter one the application of converters in systems will be discussed. If analog-to-digital and digital-to-analog converters are applied in discrete-time systems it is important to perform these operations on frequency-band-limited signals. In most cases filters are needed to limit the input and output spectrum of the analog signals. If no band limitation is performed, then aliasing of the analog signals into the signal band of interest can occur. General criteria that determine the overall system performance in the case of ideal converters are introduced and defined. Combinations of analog and digital filtering operations result in linear phase filtering over the band of interest. Such a linear phase filtering is very important in digital audio systems.

Performance definitions of converters are defined in Chapter two. The performance definitions must be unique for a specific parameter. Good parameter definitions of converters are very important in determining the final performance of a discrete-time system. Furthermore, these definitions can be used to compare the performance of different brands of converters. In particular, a good definition of the dynamic parameters of converters is needed. The application of converters in digital audio and digital video systems, for example, requires these dynamic specifications. Many converters that originally were designed for high-accuracy measurement system applications are not optimized for dynamic operations. In digital audio, for example, many specifications, which are important in instrumentation (such as offset, full-scale accuracy, temperature drift) are of minor value. The specific dynamic parameters must therefore be defined and related to important design parameters. Glitches in D/A converters introduce distortion in digital audio systems, while in video display systems fuzzy images are obtained.

After defining the important specifications of converters, good measurement set-ups and definitions are needed. In Chapter three attention is paid to measuring the static and dynamic performance of converters. Usually DC parameters can be measured with automatic test equipment and are well defined. Much attention has been paid to obtaining measurements and measurement set-ups which give the required parameter in a fairly simple way and use a well-defined test condition. One of the most important parameters to be determined in this way is the dynamic performance of a system. The dynamic range is determined by measuring the signal-to-noise ratio of a converter over half the sampling frequency. This signal-to-noise ratio must, in the case of a well-designed converter, be close to the theoretical value, which is defined over a bandwidth equal to half the sampling frequency. When large discrepancies are found, then this converter is not designed to operate under dynamic conditions. Sometimes it is possible to overcome some of the problems by adding external circuitry, which, for example, can be a deglitcher circuit to reduce the glitch of the converter or by adding data latches which perform the switching of the bit weights in a digital-to-analog converter at the same time moment.

A high-performance sample-and-hold amplifier in front of a parallel-type A/D converter can improve at high frequencies the dynamic performance of this converter. The sampling of the analog signal in this way is performed by the sample-and-hold amplifier. The following A/D converter only needs to have a good settling performance to the applied input signal step from the sample-and-hold amplifier. Distortion introduced by timing uncertainties at high frequencies in the flash converter can be avoided in this way.

In general it can be said that the number of systems and circuits that can perform analog-to-digital conversion is much larger than the number of structures and basic solutions to digital-to-analog conversion. In Chapter four examples of high-speed analog-to-digital converters are discussed. Up until now the full-flash converter was considered the fastest converter type that can be designed. This, however, is only partly true. Due to the use of a large amount of components for example, in an 8-bit converter 255 comparators are needed to detect every code level the size of such a converter becomes large with respect to the time a signal needs to travel over the interconnection lines on a chip. In high-speed converters the time difference for signals traveling at the top of the converter structure with respect to signals traveling in the middle of the structure may become in the order of pico seconds. At high-input frequencies these time differences in signal transfer introduce distortion. Depending on input frequency and resolution of a converter, timing differences may not exceed pico second values. When taking into account that the transmission speed over the interconnection lines in an integrated circuit is about two-thirds of the speed of light, then 1 psec equals about 200 microns of interconnect. When the size of an analog-to-digital converter chip without sample-and-hold amplifier increases, it is practically impossible to match clock- and signal-line delays within the required timing accuracy. Therefore, improved converter systems with a reduced chip size using a continuous input-signal folding architecture overcome these problems. In a final system implementation of such a system, only zero crossings of the analog signal are important for the converter accuracy. To reduce the number of input amplifier-comparator stages in such a system, an interpolation of zero crossings is used. This system results in a compact very high-performance analog-to-digital converter structure. In an MOS technology switches and capacitors are the main design elements. The input signal can be stored on multiple capacitors and then a comparison is made to perform an A/D conversion. Even two-step system solutions can be easily implemented in this structure.

High-speed converter systems depend on the availability of high-performance sample-and-hold amplifiers. Such systems are difficult to design, while furthermore the maximum sampling frequency and the accuracy of the system are determined by the performance of the sample-and-hold amplifier. In digitally sampled (eg, full flash and folding A/D structures) systems mostly a much higher sampling clock can be applied without disturbing the analog signal path. In this way additional resolution and/or accuracy can be obtained by using oversampling, filtering, and subsampling techniques.

In high-speed analog-to-digital converters that are based on the full-flash or folding principles, the question about the relation between maximum analog input frequency and the bandwidth of the comparator-amplifier stages arises. In Chapter five an analysis that determines the relation between the maximum analog input frequency and the bandwidth of the comparator-amplifier stages is performed. Furthermore it will be shown from this analysis that a limited analog bandwidth of the system results in third-order signal distortion. This distortion can fold signals back into the baseband of the converter and results in a reduction of the signal-to-noise ratio of the system. The maximum analog input frequency will be defined as that input frequency for which the signal-to-noise ratio is reduced by a value equal to half the least significant bit. The number of effective bits as a function of input frequency will be used as an accurate measure for the performance of high-speed analog-to-digital converters. A second analysis about the decision failure of comparators in flash-type converters is performed. This analysis gives insight in the relation between the bit error rate (BER) of a comparator as a function of the gain-bandwidth product of a technology. Comparator stages can be optimized for a minimum error.

In high-accuracy analog-to-digital and digital-to-analog converters the accuracy with which the binary weighting of the bit weights is performed is an important design criterion. In Chapters six and seven generally applicable methods to obtain high-accuracy converters will be introduced. These methods will mainly use a combination of accuracy introduced by matched elements and a dynamic method to improve the limited passive accuracy. Resistor or capacitor matching in integrated circuits limits the resolution of converters based on these elements to about ten to twelve bits. Trimming methods can be used to overcome this problem. These trimming methods, however, are expensive, while in addition changes of the trimmed elements due to time or temperature variations destroy the accurate trimming of the converter. In MOS technologies accuracy is obtained by matching binary weighted capacitor banks. Accuracies of ten to twelve bits are possible without trimming.

When monotonicity of a converter is the most important design criterion then special system configurations are possible. Solutions based on current swapping or voltage division will be described. These systems are inherently monotonic while overall linearity is limited.

When the absolute accuracy in a converter is required, then special systems are needed. Systems which convert the digital value into an accurate time need a limited amount of accurate elements. However, speed is limited, while systems with higher sampling rates suffer from extremely high clock frequencies which result in high-frequency signal radiation.

To overcome these problems a combination of passive and active matching of components will be used. The first method combines accurate passive division with a time interchanging concept to obtain a very high weighting accuracy without using accurate elements. Furthermore, this method is independent of element aging and remains accurate over a large temperature range. Examples of 14- and 16-bit digital-to-analog and analog-to-digital converters will be given that use this special method. The system is not limited to bipolar technologies, although at this moment only examples of bipolar implementations exist. In an MOS technology calibration of current sources can be used to obtain a segmented converter structure. Such a structure is less sensitive to element matching, while the calibration of the individual current sources makes the system independent of element aging and less temperature-sensitive. The gate capacitance of an MOS device is used to store the error signal information that is needed to calibrate the bit current sources.

In high-accuracy analog-to-digital converters the analog input signal must be sampled and kept constant during the time the conversion takes place. High-resolution and high-accuracy sample-and-hold amplifiers are a key element for analog-to-digital conversion in, for example, digital audio. Examples of high-speed sample-and-hold amplifiers using a bipolar or CMOS technology will be described in Chapter eight. These systems are very suitable for medium-resolution high-speed A/D converters using a two-step conversion method.

For the most part open-loop systems are used in high-speed sample-and-hold amplifiers to avoid stability problems and obtain the maximum bandwidth of such a system.

Furthermore, a high-resolution design using a bipolar technology and employing an overall feedback will be described. This sample-and-hold amplifier uses high-performance operational amplifiers with improved frequency compensation techniques. A specially designed all-NPN class-B output stage with switching capability completes the design.

Low-noise high-stability reference sources are a basic element in converters. Different methods to obtain accurate and temperature insensitive reference sources will be given. These circuits use special techniques to fabricate zener diodes with low-temperature coefficients and a good long-term stability. In integrated circuits reference sources are designed that are based on the bandgap voltage of silicon. The temperature dependence of the bandgap voltage reference sources is small, but in the case of 16-bit converters this temperature stability is not good enough. Therefore, systems that use a second-order temperature compensation will be discussed in Chapter nine. The noise analysis of these systems is performed, and data will be given on the signal-to-noise ratio. Methods to reduce the noise as a function of frequency will be discussed.

In Chapters ten and eleven examples of noise-shaping techniques to improve the dynamic range of a system are described. Such techniques are very useful when speed can be exchanged with accuracy or with the wordlength used in a system.

An ultimate of the noise-shaping techniques are the well-known sigma-delta A/D and D/A converters which basically use a noise-shaping filter in co-operation with a quantizer and a 1-bit D/A or A/D converter stage. Such a 1-bit converter is extremely linear, which results in a very good differential linearity of such a converter. The most important design criteria and limitations will be given. At the moment the dynamic range of a system must be increased but the maximum clock speed in the system cannot be enlarged, then a multi-bit D/A converter can be used in the feedback loop. At that moment, however, the linearity of the D/A converter determines the overall linearity of the converter. Therefore, dynamic element matching or self-calibration techniques are required to obtain this very high linearity. The root locus method can be used successfully to determine the stability of noise-shaping coders. Especially when a higher-order noise-shaping filter is used, the stability analysis gives an insight in the operation of the system when large input signals are applied. A noise-shaping coder is said to be stable when an idle pattern at half the sampling frequency is present under nearly all input signal conditions. Different architectures will be described which introduce a high-order noise-shaping filter architecture without running into a stability problem. A special system that uses a signal-level-dependent filtering order introduces a kind of feed-forward coupling in the filtering function. At low-level input signals a high-order filtering function is present, but at the moment the signal in the loop increases in amplitude, the filter order is reduced. In this way a very good optimum between filter order and stability of operation is obtained. Examples of designed 16- to 20-bit D/A converters using 1-bit or multi-bit D/A converters will be described. Finally an example of a 5-digit digital voltmeter system to complete this Chapter will be given.

Many circuits given in this book use bipolar technologies. Most of the designs were implemented in a standard 10 micron minimum-size double-metal process. The transition frequency of the transistors in this process is between 350 MHz and 400 MHz. In the high-speed folding and interpolation analog-to-digital converter, an oxide-isolated bipolar process with a transition frequency between 7 and 8 GHz is used. The minimum emitter size of the transistors in this technology is 2 x 3 micron. The MOS circuits are implemented in 1- to 2-μ gate-length complementary MOS devices. Mostly such a technology is optimized for digital applications. Therefore, the systems must be designed to be rather insensitive to MOS parameter variations and to be incorporated onto a large digital signal-processing chip. In that case special attention must be paid to crosstalk problems between the digital system part and the mostly sensitive analog part of the A/D and D/A converter.

Components in an integrated circuit are best matched when they are placed close to each other. Furthermore, these devices must be placed on chip using an isothermal boundary. When more devices such as transistors or resistors must have a very good matching, care must be taken that neighboring, not active elements are placed at the edges of the layout of such a system. In this way an improvement in matching between the individual elements of the array is obtained. Furthermore, the non-linearity of, for example, diffused base resistors can be compensated for using a proper island connection or splitting-up the resistor into parts from which each part is compensated by connecting the island to every individual resistor. These technology-oriented subjects are mostly shortly discussed. More information can be obtained from proper literature.

During the writing of this book an attempt has been made to discuss the most important parameters and design criteria and giving worked-out examples of practical circuits and systems. On the other hand, no attempt has been made to be complete and cover all subjects in detail. By including practically worked-out examples, which in a number of cases are products supplied by semiconductor manufacturers, the reader gets familiar with design problems and practical system implementations under real conditions. For the most part, these implementations differ quite a lot from the basic system solution.

Chapter 1

The converter as a black box

1.1   Introduction

A/D and D/A converters are the link between the analog world of transducers and the digital world of signal processing and data handling. In an analog system bandwidth is limited by device and element performance and by the parasitics introduced. Thermal noise generated in active and passive components limits the dynamic range of an analog system. The ratio between the maximum allowable analog signal and the noise determines the dynamic rangedynamic range of the system. The signal-to-noise (S/N) ratio is a measure of the maximum dynamic range.

In a digital system the amplitude is quantized into discrete steps, and at the same time the signal is sampled at discrete time intervals. When the sampling time moments at quantization differ from the sampling time moments at the time the signal is reconstructed into an analog signal again, a signal distortion is introduced. This phenomenon is very important in discrete-time systems and finds its origin in "time jitter" or time uncertainty of the sampling clock. In particular, input signals at the high-frequency end of the signal band show a great sensitivity to a sampling time uncertainty. Furthermore, the sampling operation of analog signals introduces a repetition of input signal spectra at the sampling frequency and multiples of the sampling frequency. If the input signal bandwidth is larger than half the sampling frequency, aliasing of spectra occurs. In that case frequencies around the sampling frequencies and its multiples are folded back into the baseband of the system. Usually this is an unwanted phenomenon. To avoid aliasing of the signal, the input bandwidth must be limited to not more than half the sampling frequency (Nyquist criterion see [7]). This filtering must be performed by continuous-time filters.

The quantization of analog signals into a number of amplitude-discrete levels places limitations on the accuracy with which signals can be reproduced. This quantization error is often called "quantization noise" to indicate that the errors have a random amplitude distribution and in this way have a noise-like frequency spectrum. This random character, however, implies that under no circumstances should the analog input signal and the sampling clock have any correlation [8]. If a correlation exists then the quantization errors appear at well-defined points in the frequency spectrum which are generally multiples of the signal frequency. The ratio between the frequency of the input signal and the sampling frequency should preferably be an irrational number to avoid this correlation. The dynamic range of a digital system is determined by the number of quantization levels as will be shown in this chapter. By definition the noise is measured over a bandwidth equal to half the sampling frequency of the system.

In this chapter the different criteria mentioned will be described in more detail. The converter will be treated as an ideal black box. Specifications for input and output circuitry connected to the ideal converter will be derived under the condition that the overall system performance must be close to the ideal converter performance.

1.2   Basic D/A and A/D converter function

In Figure 1.1 a block diagram of a D/A converter is shown. Digital signals are applied to the converter as parallel signals. Suppose we have a binary-weighted converter, then the digital input value is converted into an analog value using the following equation:

$$V_{out} = \sum_{m=0}^{n-1} B_m 2^m R_{ref}. \tag{1.1}$$

In this equation V_{out} represents the analog output value and R_{ref} is a reference value. (R_{ref} may be a reference voltage, current or charge).

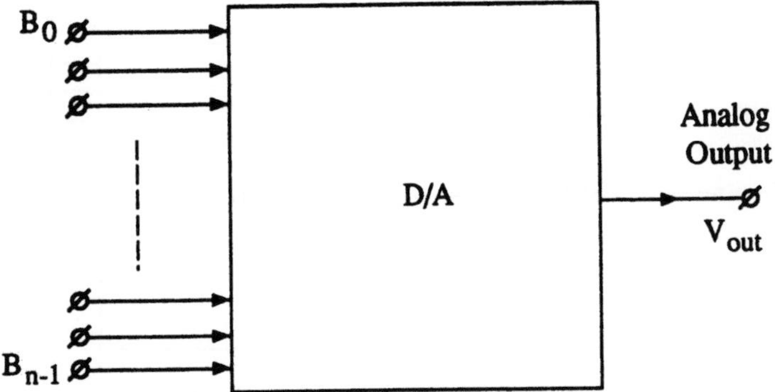

Figure 1.1 : Block diagram of a D/A converter

A reference current source or a reference voltage source is mostly used in practical implementations.

Note that equation 1.1 represents an n-bit converter. B_{n-1} represents the Most Significant Bit (MSB) of the converter and B_0 is the Least Significant Bit (LSB). The factor 2^m indicates the binary weighting of the bit values as a function of the variable m.

Furthermore, the analog output value can only change with a minimum step size q_s equal to R_{ref} because of the quantization process.

At the moment the Most Significant Bit is coupled with the reference R_{ref} then equation 1.1 changes into:

$$V_{out} = \sum_{m=0}^{n-1} B_m 2^{-m} R_{ref}. \tag{1.2}$$

By using the definition given in 1.2 the quantization step size q_s becomes:

$$q_s = 2^{-n+1} R_{ref}. \tag{1.3}$$

In the succeeding part of the book only the definition according to 1.1 will be used. As *quantization step* q_s will be introduced.

A block diagram of an A/D converter is shown in Figure 1.2. A sample-and-hold amplifier is added to sample the input signal and hold the signal information at the sampled value during the time in which the conversion

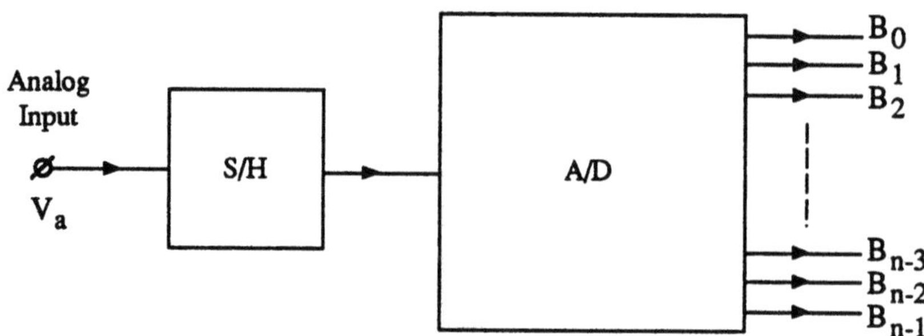

Figure 1.2 : Block diagram of an A/D converter

into a digital number is performed. In an A/D converter the equation 1.1 changes into:

$$\frac{V_a}{R_{ref}} = D_{out} + q_e = \sum_{m=0}^{n-1} B_m 2^m + q_e. \tag{1.4}$$

This can be partly rewritten as:

$$D_{out} = \sum_{m=0}^{n-1} B_m 2^m. \tag{1.5}$$

In this equation D_{out} represents the digitized value of the analog input signal V_a and q_e represents the quantization error. The quantization error represents the difference between the analog input signal V_a divided by R_{ref} and the quantized digital signal D_{out} when a finite number of quantization levels n is used.

1.3   Classification of signals

1.3.1   Analog signals

Analog signals are continuous-time and continuous-amplitude signals. Basically, there is no limitation in bandwidth and amplitude. When analog signals are processed by systems, however, frequency limitations are introduced and noise generated by active or passive components is added.

The dynamic range of a system is determined by the signal-to-noise ratio (S/N) of that system. An ideal test signal is supposed to be applied to that system for the determination of the signal-to-noise ratio. Maximum signal is determined by the distortion components generated in the system. A 10 % distortion level is adopted, for example, in power amplifiers to define the maximum signal level.

1.3.2 Discrete-time signals

Discrete-time signals are generated by sampling an analog signal at discrete time intervals without quantizing the amplitude of this signal. In most systems equal time intervals are used as the sampling clock. The sampling operation, however, introduces replica of the input frequency spectrum around the sampling frequency and multiples of the sampling frequency. To avoid aliasing of frequency spectra, the input signal bandwidth must be limited to half the sampling frequency as implied by the Nyquist criterion [7]. When an analog signal has to be limited in bandwidth, then a continuous-time filter instead of a switched capacitor filter is needed to avoid a repetition of frequency spectra in this filter. Such filters, however, can introduce phase distortion, which may be audible in high-performance digital audio systems.

One example of a discrete-time system is a sample-and-hold amplifier. In such an amplifier analog signals are sampled and stored on a capacitor during the hold time. Samples are taken at well-defined discrete time intervals without quantizing the signal amplitude.

1.3.3 Amplitude-discrete signals

In a continuous-time system the amplitude can be quantized into discrete amplitude levels, resulting in an amplitude-discrete signal. This operation can be performed to maintain well-defined amplitude levels when signals pass through several processing stages. This amplitude quantization introduces quantization errors that limit the accuracy and dynamic range of a system. To obtain step-like output signals of the quantizer, a high gain and a large comparator bandwidth are needed.

1.3.4 Digital signals

Digital signals are obtained if a signal is sampled at discrete time intervals and the amplitude is quantized in discrete amplitude levels. The amplitude quantization introduces quantization errors which limit the accuracy of the system. Sampling at discrete equal time intervals, requires a limited input spectrum to avoid aliasing. Variations in sampling time moments of fast-changing analog signals result in amplitude quantization errors, compared with signals sampled at equal well-defined time intervals. To avoid such errors, the sampling time uncertainty must be small. Especially at the high end of the input frequency spectrum these errors are significant.

Digital values represent well-defined levels. In a binary coded system a "0" or ZERO represents an "off" or "false" state, while a "1" or ONE represents an "on" or "true" state. These well-defined levels are called logic levels which may be TTL, ECL, or CMOS levels in a binary-coded system.

A/D and D/A converters perform digitization or reconstruction of analog signals. The performance of such a system can be measured by measuring the signal-to-noise ratio over a well-defined system bandwidth. To simplify comparison of converters a measuring bandwidth equal to half the sampling frequency is used. In the following paragraphs these definitions will be explained in more detail.

1.4   Sampling time uncertainty

Sampling time uncertainty introduces additional errors when analog signals are sampled at equal time intervals and reconstructed at time intervals that show a timing uncertainty or vice versa. To obtain quantitative insight into the problem of sampling time uncertainty, suppose that a sine wave close to half the sampling frequency is applied to an A/D converter. The maximum slope of the input signal occurs at the zero crossing of the signal as shown in Figure 1.3. If the sampling time moment varies between t and $t + \Delta t$, then an amplitude variation between A and $A + \Delta A$ is obtained. The peak-to-peak amplitude variation ΔA must be at maximum equal to the quantization step q_s of the converter to avoid a significant loss in quantization resolution of the converter. Note that the quantization step q_s is equal to the Least Significant Bit (LSB) value of the converter.

Suppose furthermore that the converter has a binary weighting with n bits, then the number of quantization levels is equal to $2^n - 1$. When $n \gg 1$, the number of levels can be approximated by 2^n.

The sampling time uncertainty Δt must be so small that the LSB amplitude

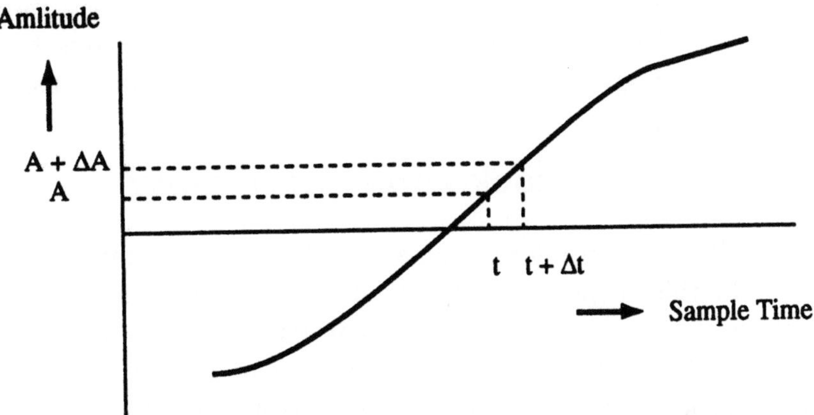

Figure 1.3 : Sampling time uncertainty calculation model

level is not exceeded for signals with a bandwidth equal to half the sampling frequency .

With a sine wave input of $V_{in} = A \sin \omega t$ we obtain for the sampling time uncertainty:

$$\Delta t = \frac{\Delta A}{A \omega \cos \omega t}, \tag{1.6}$$

and with

$$\Delta A = \frac{2A}{2^n}, \tag{1.7}$$

the result becomes:

$$\Delta t = \frac{2^{-n}}{\pi f_{in} \cos 2\pi f_{in} t}. \tag{1.8}$$

Here $2A$ = peak-to-peak amplitude of the signal, $\omega = 2\pi f_{in}$ and f_{in} is the input signal frequency.

Formula 1.8 shows that the sampling time uncertainty depends on the moment the input signal is sampled (t is still in the equation!). The tightest specification for the sampling time uncertainty is obtained when $t = 0$.

Inserting $t = 0$ into formula 1.8 results in:

$$\Delta t_{max} = \frac{2^{-n}}{\pi f_{in}}. \tag{1.9}$$

As an example the sampling time uncertainty is calculated for a 16-bit digital audio system with an input frequency f_{in} of 20 kHz. Inserting these values into equation 1.9 gives a time uncertainty (Δt) of less than $\frac{1}{4}$ nsec.

1.5   Sampling clock time uncertainty

It is difficult to obtain information about the time uncertainty of clock generators. In this section a simple model will be introduced to obtain an estimation of the pulse-to-pulse time uncertainty of a sampling clock generator. In general it can be said that a sine wave that is generated by a highly frequency selective element shows a high stability. Suppose now that a nearly ideal sine wave $V_{cl} = A \sin \omega t$ is generated. When a highly frequency selective element is used then the (white) noise generation added to the sine wave is very small. Mostly a square wave is required as a clock signal in sampled data system. This square wave generation is obtained by applying the sine wave signal V_{cl} to a squaring circuit. Such a squaring circuit, however, generates white noise with a bandwidth equal to the squaring amplifier bandwidth f_b. The addition of the wide-band noise results in a pulse-to-pulse time uncertainty which is determined by the noise amplitude and the slope of the sine wave input signal. In Figure 1.4 the simplified calculation model is shown. In Figure 1.4 e_n represents the (rms) white noise generated

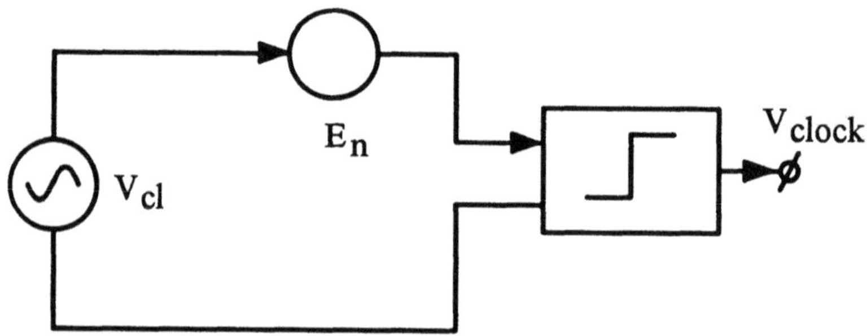

Figure 1.4 : Clock time uncertainty model

by the squaring system. Mostly e_n is thermal noise and can be expressed as:

$$e_n^2 = 4kT R_n \Delta f. \tag{1.10}$$

R_n represents the equivalent noise resistance of the squaring circuit.

The time uncertainty can be calculated. Differentiating V_{cl} we obtain for the time uncertainty:

$$\Delta t = \frac{e_n}{2\pi f_{cl} A \cos 2\pi f_{cl} t}.$$ (1.11)

The maximum sensitivity of the squaring circuit is found at the zero crossing of the sine wave which corresponds to $\cos 2\pi f_{cl} t = 1$. Furthermore, noise equation 1.10 gives the rms value corresponding to 1σ. To obtain the total 1σ time uncertainty the result of equation 1.11 must be doubled. We obtain for the clock time uncertainty Δt_{cl} :

$$\Delta t_{cl} = \frac{e_n}{\pi f_{cl} A}.$$ (1.12)

Equation 1.12 shows that with a constant value of noise e_n the time uncertainty will decrease with increasing clock frequency f_{cl}. When we suppose that the squaring circuit has a single time constant with a bandwidth f_b then 1.10 can be inserted in 1.12, resulting in:

$$\Delta t_{cl} = \frac{\sqrt{2\pi kT R_n f_b}}{\pi f_{cl} A}.$$ (1.13)

Note that the noise-bandwidth of a first order system is equal to:

$$f_{noise} = \frac{\pi}{2} f_b.$$ (1.14)

In 1.14 f_{noise} is the noise bandwidth of the system.

Furthermore, in amplifier systems a general relation exists between the rise time t_r of the output pulse and the bandwidth f_b:

$$f_b \times t_r \approx 0.35.$$ (1.15)

Inserting 1.15 into 1.13 results in:

$$\Delta t_{cl} = \frac{\sqrt{0.7 kT R_n}}{f_{cl} A \sqrt{\pi t_r}}.$$ (1.16)

Suppose we have an ideal 50 Ohm system and we want to generate a clock signal with a rise time of 250 psec and a repetition rate of 10 MHz. With a sine wave amplitude of 0.5 V an rms clock time uncertainty of 2.7 psec is obtained.

1.6   Quantization errors

The quantization process introduces an irreversible error (see [9] [10]). Therefore we want to calculate the relation between the quantization step q_s and the input signal V_{in}. The quantization step is determined by the number of steps a signal is quantized into. This number of quantization steps is expressed in a number of (binary-weighted) bits n. In Figure 1.5 the quanti-

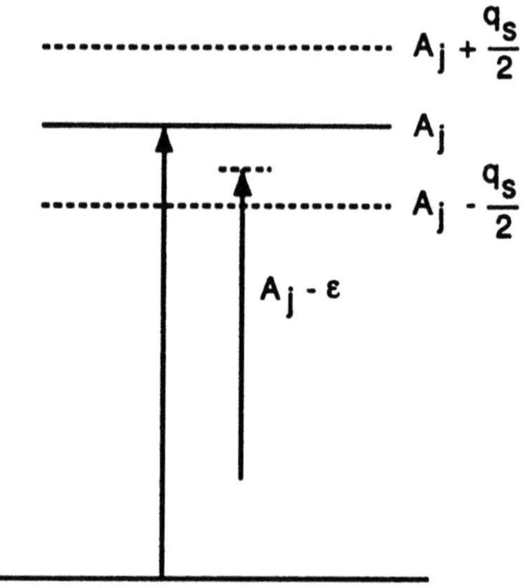

Figure 1.5 : Quantization of a signal at level A_j

zation of a signal at the amplitude level A_j is shown. A signal $A_j + \varepsilon$ is ideally quantized into level A_j as long as the value of ε is between $-\frac{q_s}{2} < \varepsilon \leq \frac{q_s}{2}$. From this example it can be seen that the quantization error basically never exceeds an amplitude level equal to $\pm \frac{q_s}{2}$. Signals that are somewhat larger than $A_j + \frac{q_s}{2}$ are quantized to the next quantization level A_{j+1}.

The mean-square-error due to quantization can now be calculated. We assume that over a long period of time all levels of uncertainty within the quantizing region $A_j \pm \frac{q_s}{2}$ appear the same number of times. A uniform probability density function over the interval $-\frac{q_s}{2}$ to $+\frac{q_s}{2}$ is defined in this way. On this assumption the mean-squared value of ε will then be:

$$E(\varepsilon^2) = \frac{1}{q_s} \int_{-\frac{q_s}{2}}^{\frac{q_s}{2}} \varepsilon^2 d\varepsilon. \tag{1.17}$$

The symbol $E(.)$ represents the statistical expectation. The average value of the error is zero with the assumptions made.

Solving the integral, the quantization error can be expressed as a quantization error voltage e_{qns}^2. The rms quantization error voltage $e_{qns}^2 = E(\varepsilon^2)$ can be represented by:

$$e_{qns}^2 = \frac{1}{12}q_s^2. \tag{1.18}$$

In systems with a small number of quantization steps the quantization error is correlated with the applied signal. To obtain practical usable values the correlation of the quantization errors is recalculated as an amplitude correction. The amplitude correction is based on correlated errors such as distortion of the signal. This distortion component will be used to determine the correction factor for a 1-bit quantizer. In an amplitude quantized system using n binary weighted bits the number of quantization levels (decision levels) equals $2^n - 1$. The maximum peak-to-peak amplitude in this system equals $2^n q_s$ before the next quantization level is reached. However, this maximum peak-to-peak level cannot always be obtained. As a result, the maximum peak-to-peak $sine - wave$ signal amplitude is between:

$$(2^n - 1)q_s \leq A_{pp} \leq 2^n q_s. \tag{1.19}$$

Here A_{pp} is the peak-to-peak amplitude of the input or output signal used in the system.

When a converter with a large number of bits is used ($n \geq 5$), the peak-to-peak signal amplitude can be approximated by:

$$A_{pp} = 2^n q_s. \tag{1.20}$$

It is easy to calculate the rms signal value A_{rms} of the signal A_{pp}:

$$A_{rms} = \frac{q_s \times 2^n}{2\sqrt{2}}. \tag{1.21}$$

In this calculation a *sine wave* signal is adopted.

The signal-to-noise ratio (S/N) can be calculated using equation 1.21 divided by the square root of equation 1.18, resulting in:

$$S/N = 2^n \sqrt{1.5}. \tag{1.22}$$

Converting equation 1.22 into decibels results in:

$$S/N = n \times 6.02 + 1.76 \ \text{dB}. \tag{1.23}$$

An attempt will now be made to come to an estimate for the S/N ratio of a system with a low number of bits ($n \geq 1$). To overcome the problem with the maximum signal definition uncertainty a correction factor S_{cor} will be introduced with: $0 \leq S_{cor} \leq 1$.

The peak-to-peak signal amplitude A_{pp} in a binary weighted n-bits amplitude quantized system then becomes:

$$A_{pp} = \left(2^n - S_{cor}\right) \times q_s. \tag{1.24}$$

S_{cor} is determined by the amplitude of the fundamental of the signal frequency f_{sig} in the quantized system.

This correction factor can be accurately calculated for $n = 1$. At that moment a square wave is generated at the output of the converter. After a series expansion of the square wave, the peak to peak amplitude of the fundamental becomes: $\frac{4}{\pi}q_s$. Using this result to determine the value of S_{cor} from equation 1.24 the following is obtained:

$$\frac{4}{\pi} = 2 - S_{cor}. \tag{1.25}$$

Rearranging equation 1.25 results for the correction factor:

$$S_{cor} = 2 - \frac{4}{\pi} \approx 0.73. \tag{1.26}$$

Although the value of equation 1.26 is exactly valid for a 1-bit quantizer, this value will be inserted into equation 1.24 to give the best approximation for converters having a low resolution.

Inserting 1.26 into equation 1.24 results in:

$$A_{pp} = \left(2^n - 2 + \frac{4}{\pi}\right) \times q_s. \tag{1.27}$$

It is easy to calculate the rms signal value A_{rms} of the signal A_{pp}:

$$A_{rms} = \frac{q_s \times \left(2^n - 2 + \frac{4}{\pi}\right)}{2\sqrt{2}}. \tag{1.28}$$

number of bits n	S/N ($S_{cor} = 1$) dB	S/N (1.29) dB	($S_{cor} = 0$) dB
1	1.76	3.86	7.78
2	11.30	12.06	13.80
3	18.66	19.00	19.82
4	25.28	25.44	25.84
5	31.59	31.66	31.86
6	37.75	37.79	37.88
8	49.89	49.90	49.92
10	61.95	61.95	61.96

Table 1.1 : S/N as a function of the number of bits n

In this calculation a *sine wave* signal is adopted. The signal-to-noise ratio (S/N) can be calculated using equation 1.28 divided by the square root of equation 1.18, resulting in:

$$S/N = (2^n - 2 + \frac{4}{\pi})\sqrt{1.5}. \qquad (1.29)$$

In general equations 1.22 and 1.23 are nearly always used to calculate the dynamic range of a converter.

It must be noted that in systems with a small number of quantization levels (or bits), that the quantization error will show a rather strong correlation with the signal in the system.

In table 1.1 the results of equation 1.29 for $S_{cor} = 1$, $S_{cor} = -2 + \frac{4}{\pi}$ and $S_{cor} = 0$ are shown as a function of the number of bits n. The output of formula 1.29 is converted into decibels to ease comparison of the results.

The error in formula 1.29 is within the accuracy with which the dynamic range of a system can be measured.

When $n = 4$ the error with formula 1.22 is 0.4 dB. This error decreases to below 0.01 dB for $n = 10$.

Formula 1.23 shows that the dynamic range of a system increases by 6 dB when an extra bit is added.

By definition: The dynamic range of a system is equal to the signal-to-noise ratio of that system measured over a bandwidth equal to half the sampling frequency.

The dynamic range of a 16-bit digital audio system can be found by inserting $n = 16$ into formula 1.23. We obtain S/N = 98.1 dB.

Because the quantization error can be modeled by a random process we can compare this error with noise. By this modeling the quantization error is sometimes called quantization noise. It is therefore very useful to express formula 1.18 as a noise density per unit bandwidth ($e_{qns}^2(f)$):

$$e_{qns}^2(f) = \frac{q_s^2}{12 f_{qns}} = \frac{q_s^2}{6 f_s}. \tag{1.30}$$

Here f_{qns} is the quantization noise bandwidth and f_s is the sample frequency.

Because the signal-to-noise ratio in a system is calculated over a bandwidth equal to half the sampling frequency, the following relation is used: $f_{qns} = \frac{1}{2} f_s$.

The signal-to-noise ratio as density derived from 1.22 becomes:

$$S/N(f) = 2^{n-1}\sqrt{3}\sqrt{f_s}. \tag{1.31}$$

The signal-to-noise ratio of a system with a bandwidth equal to f_{sig} is found by dividing equation 1.31 by $\sqrt{f_{sig}}$. The result becomes:

$$S/N_{system} = 2^{n-1}\sqrt{3}\sqrt{\frac{f_s}{f_{sig}}}. \tag{1.32}$$

It is very advantageous to use formula 1.32 for dynamic range calculations of systems that do not use Nyquist sampling.

1.7   Oversampling of converters

When in a system the sampling frequency is made much higher than the maximum signal frequency, this operation is called "oversampling." The quantization error in this case is randomized over a larger frequency band. As a result the quantization noise density is reduced and the effective resolution of the system increases when the system bandwidth is kept equal to the bandwidth of a Nyquist sampled system using the lower sampling frequency.

Using formula 1.32 we have an expression for the signal-to-noise ratio of the oversampled system:

$$S/N_{system} = 2^{n-1}\sqrt{3}\sqrt{\frac{f_s}{f_{sig}}}, \qquad (1.33)$$

or in decibels we get

$$S/N = n \times 6.02 - 1.25 + 10\log\frac{f_s}{f_{sig}} \ \text{dB}. \qquad (1.34)$$

As an example the dynamic range of a four times oversampled system will be calculated. By inserting $\frac{f_s}{f_{sig}} = 8$ into formula 1.34, the result becomes:

$$S/N = (n+1) \times 6.02 + 1.76 \ \text{dB}. \qquad (1.35)$$

Comparing this result with formula 1.23, it can be seen that the dynamic range increases by 6 dB or 1 bit. In Figure 1.6 a graphical approach to the oversampling of converters is shown. Note the reduction in quantization noise density of the oversampled system. The total quantization error shown as shaded areas is equal for both systems because the same number of bits is used. In the dashed areas signals appear that are introduced by the sampling process. For simplicity only the frequency range from zero to f_s or $4f_s$ is shown in the figure. The sampling operation introduces replica of the frequency spectra at multiples of the sampling frequency, too. Usually these replica do not add additional information to the picture and are therefore omitted. This omission is introduced in the following figures as well. When the signal bandwidth is kept at $\frac{1}{2}f_s$, the total quantization error is reduced with a factor four in example b), and the increase in resolution with 1 bit is obtained.

Furthermore, the distortion in an oversampled system using the same signal bandwidth is reduced. This can be explained by supposing that a signal frequency close to the highest signal frequency is applied to the converter. Then in a Nyquist sampled system large quantization steps occur. Such

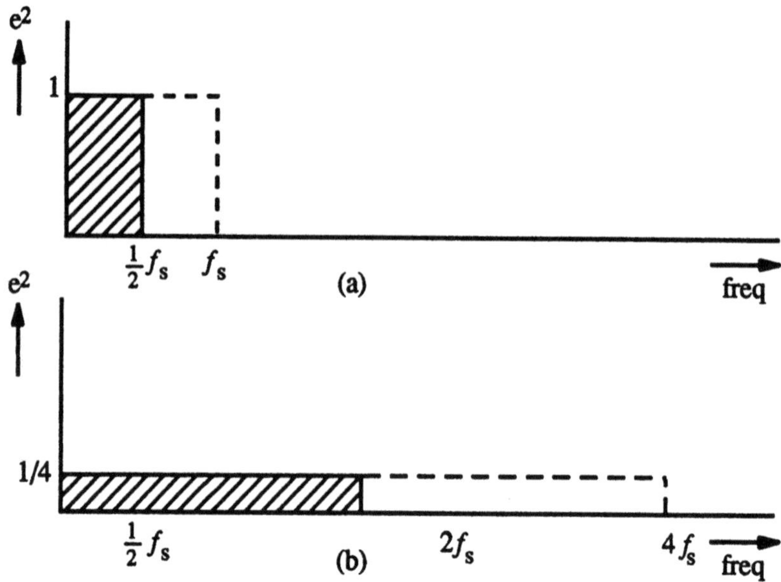

Figure 1.6 : (a) Quantization error of a Nyquist sampled converter system (b) Quantization error of a four times oversampled converter system

large steps result in, for example, slewing of the output amplifier in a D/A converter system or slewing of the sample-and-hold amplifier in an A/D converter system. This slewing results in a nonlinear operation of a part of the circuit that results in distortion. When a large oversampling is used, then the quantization steps are reduced. Mostly slewing of amplifiers can be avoided. As a result, a lower distortion in the system is obtained.

1.8   Quantizer model

To simplify calculations on systems that are using quantizers such as converters, a generalized model of such an quantizer will be given. In Figure 1.7 a block diagram of a quantizer is shown. The system consists of an adder to introduce the quantization error ε onto the signal. At the input an analog or digital signal V_{in} is applied, and at the output a digital output signal B_{out} is obtained. The block with gain $c_g = 1$ is introduced to simplify specific calculations. Applications are found in noise-shaping coders and sigma-delta converters.

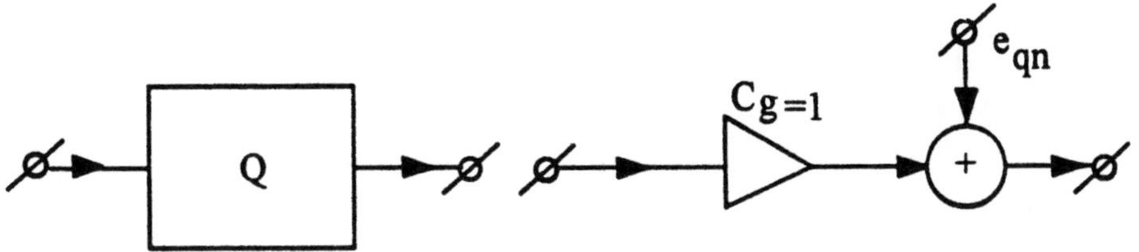

Figure 1.7 : Block diagram of a quantizer

The following simplified equation can be used for an analog quantizer:

$$B_{out} = V_{in} + \varepsilon. \tag{1.36}$$

The quantization error ε which is added to the signal at the input of the quantizer is determined by the number of quantization levels that are used in the quantizer. At the moment this quantizer is a simple zero comparator, only one quantization level is present. To ease the understanding of the system in this case, the zero level is equal to the decision level of the system. When the maximum input signal level is determined by $\pm V_p$, then the quantization step q_s is equal to $2V_p$. The signal-to-noise ratio of this system is now determined by equation 1.29. The quantization error then becomes:

$$q^2_{qns} = \frac{V_p^2}{3}. \tag{1.37}$$

At the moment more than *one* quantization step is used, then the quantization error ε is reduced by the number of quantizations steps in the system. An increase in signal-to-noise of this system is obtained.

1.8.1 Quantizer phase uncertainty

When at the input of a $1 - bit$ analog quantizer a sine wave is applied, then the phase shift between the digital output signal and the analog input signal is not accurately determined. As can be seen in Figure 1.8, a phase uncertainty between two clock moments for an analog input frequency at half the sampling frequency is equal to $\pm\frac{\pi}{2}$. The sign of the comparator output in this case does not change. This can be expressed as:

$$Sign(\cos(n2\pi\frac{f_{sig}}{f_s})) = Sign(\cos(n2\pi\frac{f_{sig}}{f_s} + \Delta\Phi)). \tag{1.38}$$

$\Delta\Phi$ represents the phase uncertainty.

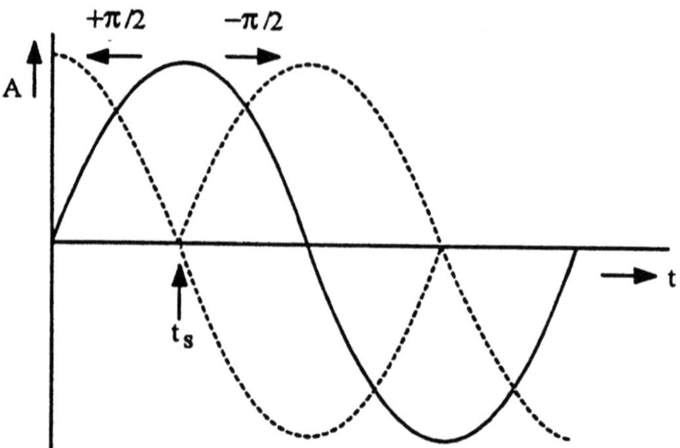

Figure 1.8 : Phase uncertainty of an analog quantizer/comparator stage at $\frac{f_s}{2}$

From Figure 1.8 it is easy to see that a decision "1" can be obtained as long as the sine wave signal has a positive amplitude. As a result, the phase uncertainty of $\pm\frac{\pi}{2}$ at half the sampling frequency is found. In Figure 1.9 the phase uncertainty as a function of frequency is shown. When this quantizer

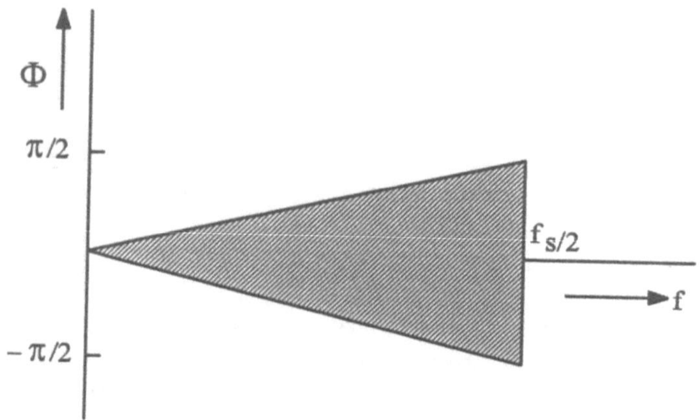

Figure 1.9 : Phase uncertainty as a function of frequency

is used in a noise-shaping coder application, the phase shift introduced by this quantizer must added to the phase shift of the noise-shaping filter introduced in a feedback loop. In Chapter 10 an explanation of noise-shaping coders is given.

If the number of quantization steps is increased, then the phase uncertainty changes. In Figure 1.10 an example of the phase uncertainty in a *three − decision − level* quantizer is shown. From Figure 1.10 it can be seen that for

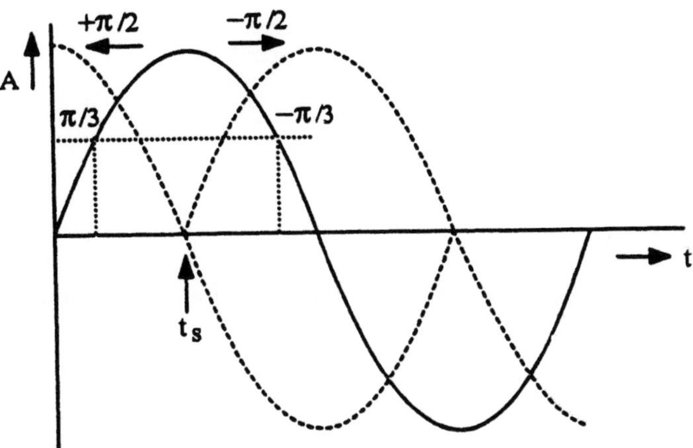

Figure 1.10 : Phase uncertainty of a three-decision-level analog quantizer

the first quantization level the phase uncertainty of $\pm\frac{\pi}{2}$ is maintained. At the moment the next decision level is reached, then the phase uncertainty is reduced to zero when a minimum trespassing of the level is obtained. This phase uncertainty increases to $\pm\frac{\pi}{3}$ when a maximum input level is applied. The $\pm\frac{\pi}{3}$ is determined by $\arccos\frac{1}{2}$. In Figure 1.11 the phase uncertainty of a 3 level quantizer as a function of frequency is shown. The result of this operation has implications on the stability of a system. Especially when a noise-shaping system is used, this phase uncertainty can result in a switching to the next quantization level of the system at a much larger input signal level than is expected. The phase characteristic of the quantizer together with the phase response of the loop filter applied in a noise-shaping analog-to-digital converter can be used as a design criterion for a stable noise-shaping coder. In Chapter 10 more attention is paid to such converter implementations.

1.8.2 Quantizer describing function model

In noise-shaping coders and sigma-delta converters a 1-bit quantizer is used. During stability analysis it is important to know the maximum or minimum gain of such a nonlinear system. Mostly a *describing function method* is used to obtain a first-order approximation of the gain as a function of the

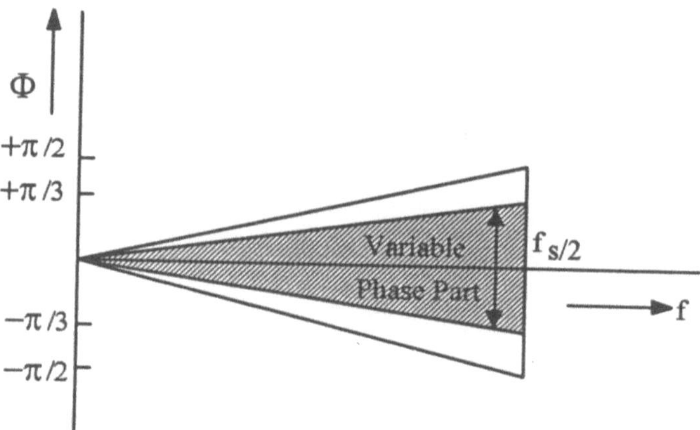

Figure 1.11 : Phase uncertainty versus frequency of a 3 level analog quantizer

input signal [11].

In Figure 1.12 the nonlinear part of a quantizer including a limiter function is shown. The quantizer input signal is S_x, the output signal is S_0 and the error signal $-e_q = -(S_x - S_0)$. The limiter function is added to the system

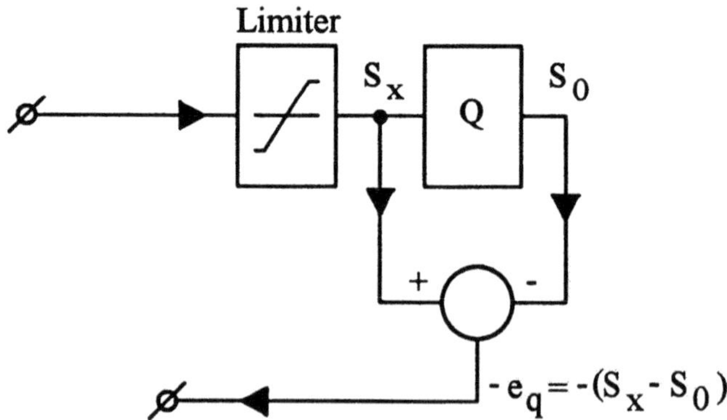

Figure 1.12 : Nonlinear part of a quantizer function

to determine the maximal gain of the 1-bit quantizer, as will be shown in this section.

In a 1-bit coder two output levels $+A_0$ and $-A_0$ are available. For a small input value, the quantizing error exceeds the input signal value, whereas for a large input value the limiter restricts the modulus of the quantizing error to $A_l - A_0$ in which with $A_l > A_0$ is the value at which overload starts. In Figure 1.13 the transfer function is shown. The global transfer λ of the

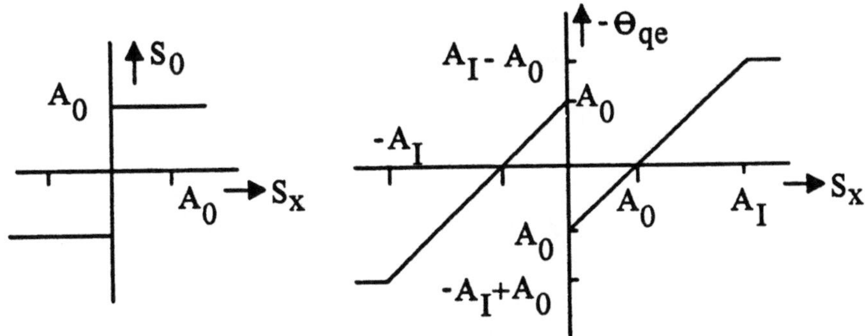

Figure 1.13 : Input versus output of the nonlinear part of a 1-bit quantizer

nonlinear part of this system can be determined. A describing function of a non sampled system will be calculated.

Suppose at the input a square wave with amplitude A_b $(A_b > 0)$ is present, then the error signal A_e also represents a square wave and A_e is obeying the following relations:

$$A_e = A_b - A_0 (A_b \leq A_l) \tag{1.39}$$

$$A_e = A_l - A_0 (A_b > A_l). \tag{1.40}$$

The global transfer λ_b in case of a square wave becomes:

$$\lambda_b = \frac{(A_b - A_0)}{A_b} (A_b \leq A_l) \tag{1.41}$$

$$\lambda_b = \frac{(A_l - A_0)}{A_b} (A_b > A_l). \tag{1.42}$$

The maximum value λ_m is obtained in case $A_b = A_l$.

In Figure 1.14 the global transfer λ as a function of input amplitude is shown. If a sine wave with an amplitude A_s $(A_s > 0)$ is applied at the input, then the error signal is the sum of a square wave and a sine wave as shown in Figure 1.15. If $A_s \leq A_l$, then the amplitude of the fundamental A_f in the input signal is given by:

$$A_f = A_s - \frac{4}{\pi} A_0. \tag{1.43}$$

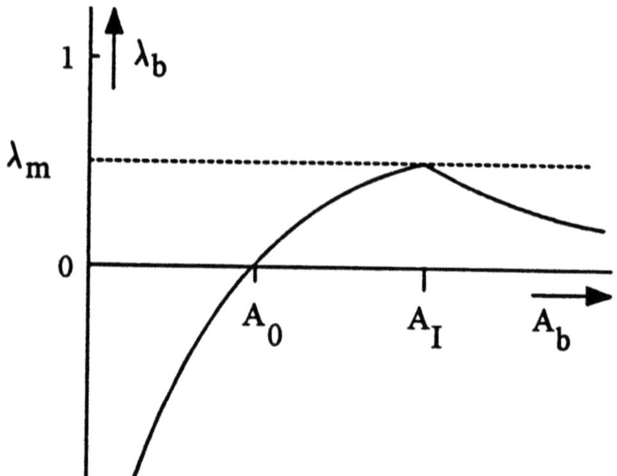

Figure 1.14 : Global transfer λ as a function of the input amplitude A_b

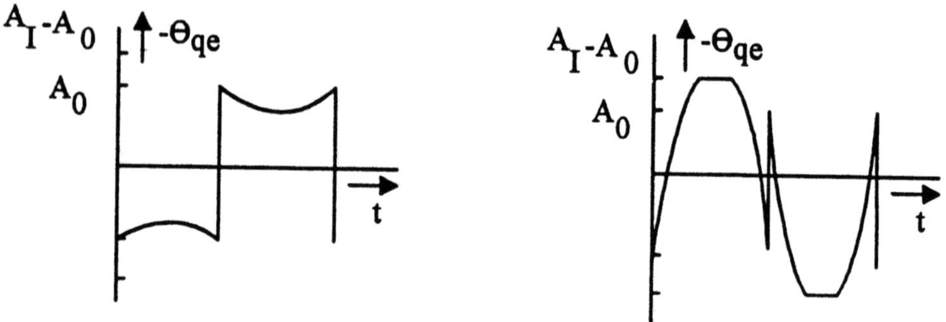

Figure 1.15 : Quantizer error signal with a sine-wave input signal

At the moment the input amplitude increases $A_s > A_l$, the limiter involves the addition to A_f of a term:

$$- A_s + \frac{2}{\pi} A_s \arcsin\left(\frac{A_l}{A_s}\right) + \frac{2}{\pi}\frac{A_l}{A_s}\sqrt{A_s^2 - A_l^2}. \qquad (1.44)$$

The maximal value of the global transfer in case of a sine wave λ_s is slightly larger than the value of λ_s for $A_s = A_l$. Putting $A_s = A_l$ in 1.44 results in the approximation of λ_m in:

$$\lambda_m = 1 - \frac{4}{\pi}\frac{A_0}{A_l}. \qquad (1.45)$$

In Figure 1.16 λ_s as a function of the input sine wave A_s is shown. From Figure 1.16 it can be estimated that the approximated maximal global gain

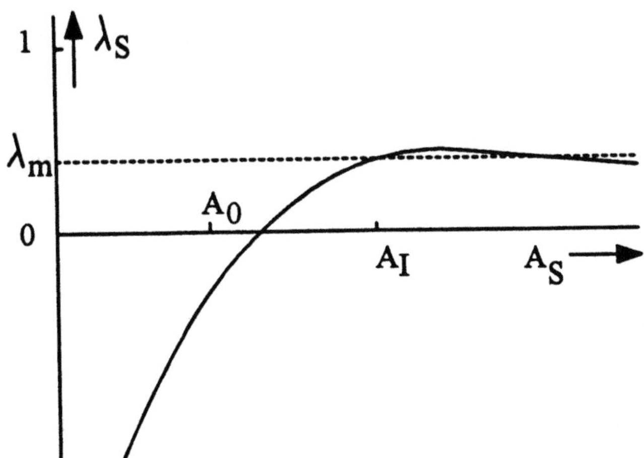

Figure 1.16 : Global gain λ_s as a function of the input sine wave A_s

λ_s is about 7.5 % too small at the moment $A_l = 2A_0$. With increasing A_l the error decreases showing that 1.45 is a sufficient approximation for gain and stability calculation purposes. The values of λ_b or λ_s range from $-\infty$ for small input values to a positive maximum λ_m in the proximity of A_l. The value of λ_m depends on the input waveform and is a function of the maximal input signal amplitude A_l and the output amplitude A_0. In Figure 1.17 the comparison of λ_m for the cases of a sine wave and a square wave as a function of the ratio $\frac{A_l}{A_0}$ show that the square wave gain is slightly larger than the sine wave gain. In a worst-case stability calculation, therefore, the square wave gain must be used. Furthermore, from Figure 1.17 the maximal gain as a function of the input limiter value and the output amplitude A_0 gives an important design parameter for noise-shaping coders (Chapter 10) or sigma-delta converters (Chapter 11). Using properly adjusted input limiter values A_l and output quantizer levels A_0, the maximal loop gain is fixed under nearly all input signal conditions.

1.9   Filtering

As stated before, the signal bandwidth before sampling must be limited to maximally half the sampling frequency. In an A/D converter system an analog filter, therefore, precedes the converter. In a high-resolution system the attenuation of this filter must be large, while the transition band is usually small. Steep filters are needed because of the narrow transition band. These

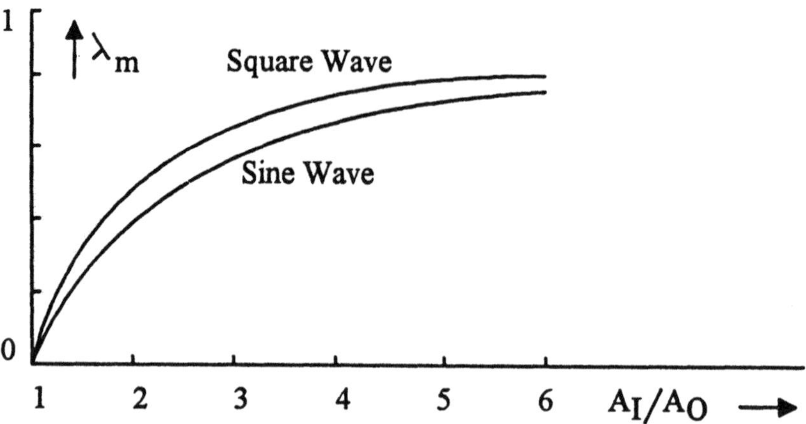

Figure 1.17 : Comparison of sine wave and square wave maximal gain as a function of $\frac{A_I}{A_0}$

filters might be of the elliptical type. Such analog filters have a serious phase distortion. In digital audio systems, for example, this phase distortion can be audible. In stereo systems phase distortion affects the directional information of the sound.

When constructing $L - C$-type low-pass filters for measuring purpose, care must be taken in choosing the components used. Some type of practical capacitors show a nonlinear signal behavior resulting in distortion. High-performance capacitors are needed in low-distortion systems to avoid this problem. Furthermore, the size of the pot cores used to construct the inductors must be large to avoid nonlinear effects due to magnetic saturation in these cores. An additional problem may exist by mounting the pot cores too close to each other. In that way the magnetic fields generated by the coils are transferred through the filter. This results in a limited stop band attenuation. By using adequate elements, low-pass measuring filters can be designed for digital audio systems up to a dynamic range of 20-bits.

To overcome these problems, oversampling of converters is used so a much simpler analog pre-filter can be applied. Such a filter can have much relaxed requirements and can be designed with a smaller phase distortion. The final channel filtering is performed by a digital filter with a linear phase characteristic (see [13]).

In a D/A converter system low-pass filtering is needed to reject the frequency spectra above the band of interest. The reconstruction filter takes care of this process. Depending on the width of the transition band, a high-order analog filter might be required. By using oversampling together with a digital filtering operation, a high-performance low-complexity analog reconstruction filter can be obtained. In the following paragraphs a more detailed analysis of these filtering operations will be given.

1.9.1 Anti-alias filtering in A/D converter systems

In Figure 1.18(a) an A/D converter system with only analog filtering is shown. The shaded area (Figure 1.18(b)) shows the frequency response $H_a(f)$ of an ideal low-pass filter. The drawn line close to half the sampling frequency shows the transition band of a more practical analog filter. The dashed area shows the passband region of the input filter. The frequency response of such an ideal low-pass filter is equal to:

$$H_a(f) \;=\; 1 \text{ for } |f| < \frac{f_s}{2} \tag{1.46}$$

$$\;=\; 0 \text{ for } |f| > \frac{f_s}{2}. \tag{1.47}$$

In systems with a minimum sampling rate f_s compared to the analog bandwidth, a zero transition band of the filter is required. Analog filters with such a steep characteristic mostly have a nonlinear phase characteristic. Furthermore, the stop band rejection must be related to the number of bits in the system. Aliasing of these stop band signals should result in errors which are below the quantization noise level. A system which uses a combination of an analog pre filtering, a digital post filtering, and decimation is shown in Figure 1.19(a). In this system a four times oversampled A/D converter allows the use of a simpler analog nearly linear-phase pre filter to reject the signal band around the high sampling frequency $4f_s$. This simple analog filter is followed by a sample-and-hold amplifier, an analog-to-digital converter and a linear-phase digital filter performing the steep filtering characteristic. In Figure 1.19(b) the frequency characteristic of the analog filter is shown. The digital filter performs the steep filtering at $\frac{f_s}{2}$. Input signals present in the frequency range between $3\frac{1}{2}f_s$ and $4\frac{1}{2}f_s$ have to be filtered out by the analog pre filter. The attenuation of the analog input filter in this frequency band must be so large, that signals which are aliased by the sampling operation of the sample-and-hold amplifier and appear in the input signal band are below the level of the quantization error. The digital post filter then

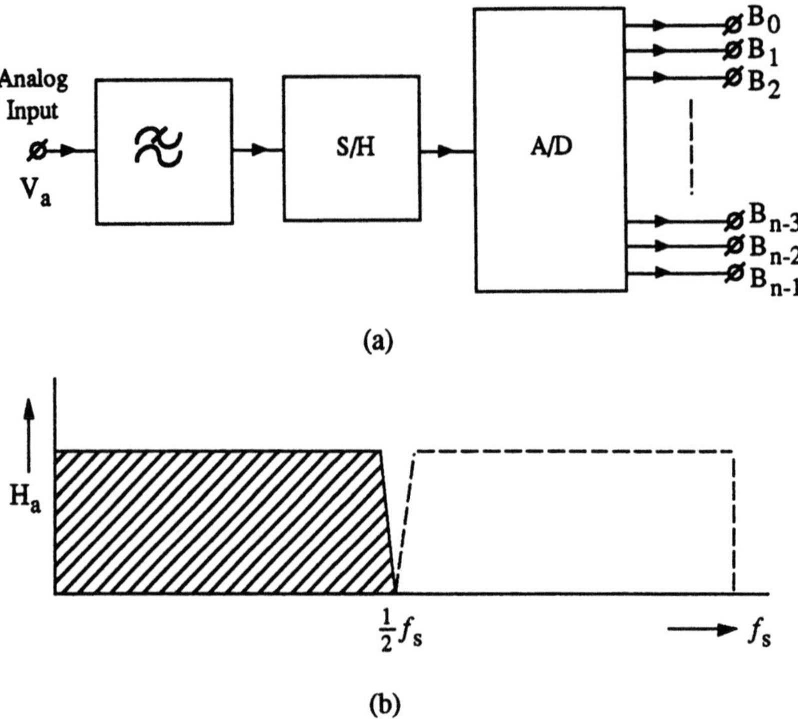

Figure 1.18 : (a) A/D converter system with analog filtering (b) Analog filter response

performs the required steep filtering in between $\frac{f_s}{2}$ and $2f_s$. The final result of the analog pre filtering and digital post filtering operation is shown in Figure 1.19(c). The stop band rejection of the combined analog and digital filter must be made so high that sub sampling can be applied without aliasing the stop band signal components into the baseband. In that case digital output words at the lower sampling rate f_s can be supplied to the following digital signal processing circuitry. In some cases it can be advantageous to apply oversampling and decimation to increase the resolution of a system. (For example, distortion products at the high-frequency band edge are removed by the digital filter.) In Figure 1.19(a) the sub sampling or decimation operation with a factor 4 is shown as the box called R. The final result of this whole operation is an accurate filtering with an almost linear phase characteristic.

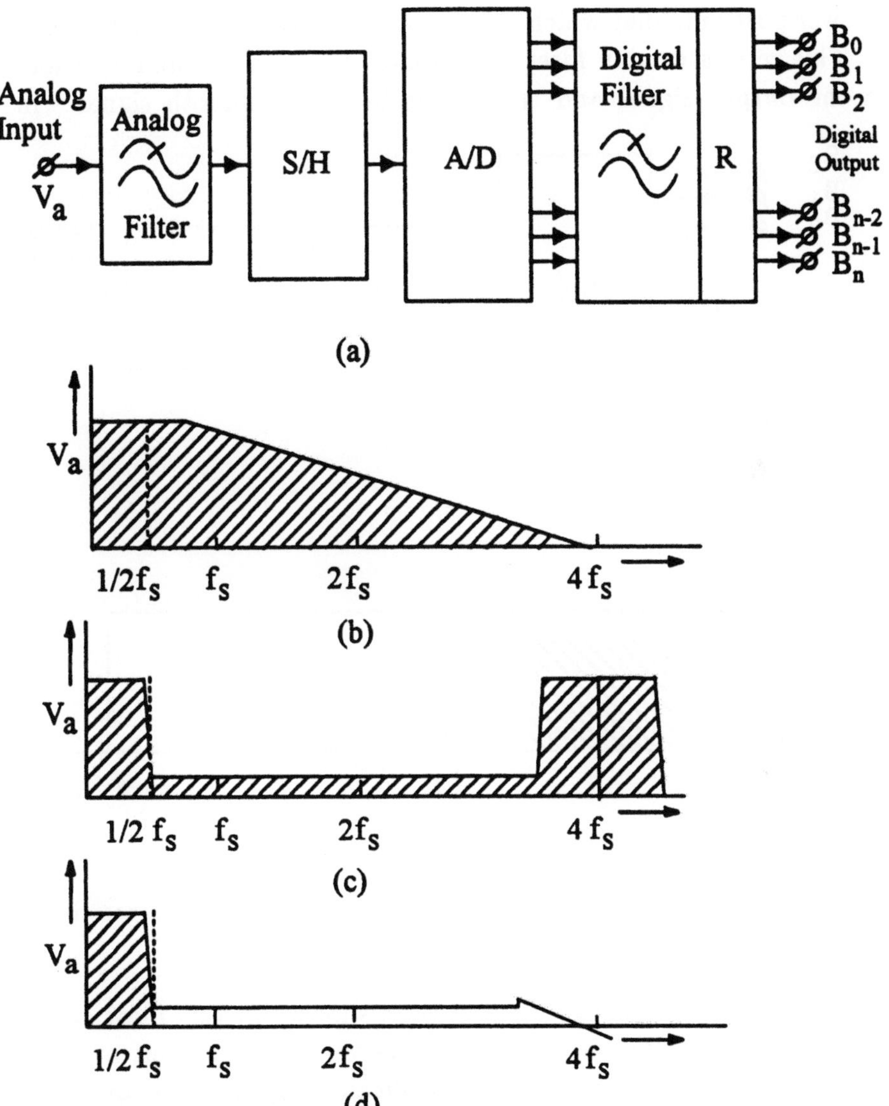

Figure 1.19 : (a) A/D converter system using combined analog and digital filtering (b) Analog filter response (c) Digital filter response (d) Total filter response

1.9.2 Output filtering in D/A converter systems

In D/A converter systems low-pass filtering is needed to reject the repeated spectra around the sampling frequency and multiples of the sampling frequency. Only the baseband is of interest, while the high-frequency components of the output signal of the D/A converter can, for example, overload the power post amplifier which drives the loudspeaker in a digital audio system. This reconstruction filter interpolates between the quantization steps from the output signal and transforms it into a smooth signal. In Fig-

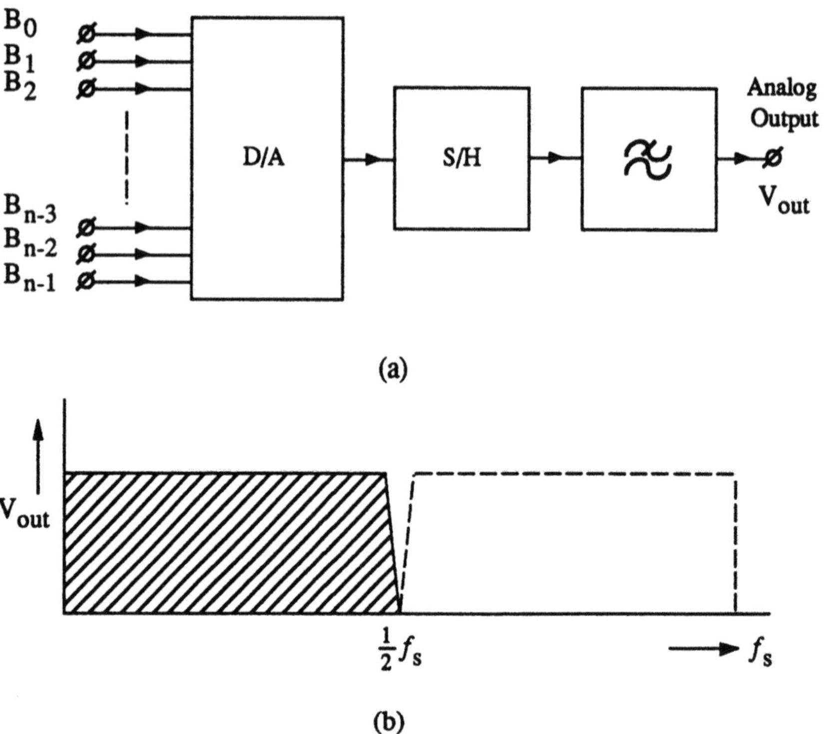

(a)

(b)

Figure 1.20 : (a) D/A converter system (b) Ideal amplitude characteristic of the total system

ure 1.20 an ideal D/A converter system is shown. In this ideal system the sample-and-hold (S/H) amplifier transmits very short output pulses to the reconstruction filter. In this way no significant amplitude distortion at output frequencies close to half the sampling frequency is introduced. Basically this system is the inverse function of the A/D converter system shown in Figure 1.18.

However, there is one big difference. At the moment the length of the output pulses of the sample-and-hold amplifier increase to about the sampling time, an amplitude distortion is found. D/A converters act as a sample-and-hold function when during reproduction of the digital signal into an analog value the output signal is maintained. This "zero-order" hold operation introduces the amplitude distortion. A circuit block called sample-and-hold (S/H) is added for this purpose in Figure 1.20 to make this clear, while Figure 1.21 shows the operation of a zero-order hold. During the hold time t_h of the D/A converter the analog output signal remains constant. The low-pass filter "averages" this signal resulting in the dashed output signal shown in the figure. The frequency response of the zero-order hold can be calculated.

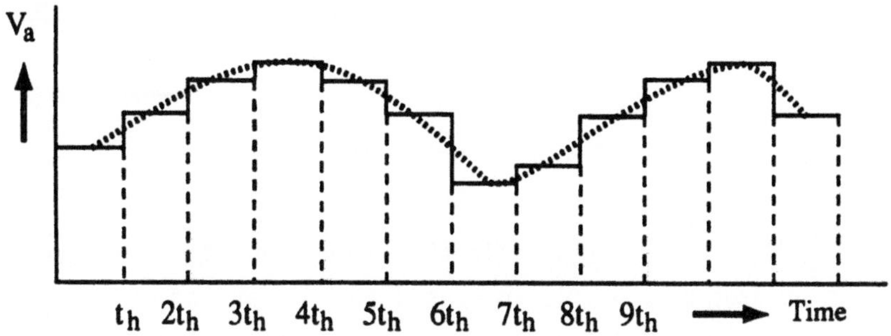

Figure 1.21 : Zero-order hold operation

Suppose t_h is the time during that the circuit is in the hold mode, then the frequency response becomes:

$$H(j\omega) = \frac{1 - e^{-j\omega t_h}}{j\omega}. \tag{1.48}$$

After rearranging we obtain for the amplitude characteristic of the system:

$$\mid H(\omega) \mid = t_h \frac{\sin \omega \frac{t_h}{2}}{\omega \frac{t_h}{2}}. \tag{1.49}$$

This is a well-known amplitude reduction called $\frac{\sin x}{x}$ distortion. In a normal D/A converter as shown in Figure 1.20, the hold time is equal to the sampling time so: $t_h = \frac{1}{f_s}$. Every clock cycle the input data can change and remains constant over that clock cycle. With $\omega = 2\pi f_{in}$ formula 1.49 can be rewritten into:

$$\mid H(2\pi f_{in}) \mid = \frac{1}{f_s} \frac{\sin \frac{\pi f_{in}}{f_s}}{\pi \frac{f_{in}}{f_s}}. \tag{1.50}$$

In a D/A converter system that is sampled at the Nyquist rate, the maximum input frequency is almost equal to half the sampling frequency. Inserting $f_{in} = \frac{1}{2}f_s$ into equation 1.50 results in an amplitude reduction of $\frac{2}{\pi}$ or 3.92 dB. The amplitude response of the $\frac{\sin x}{x}$ operation is shown in Figure 1.22 for

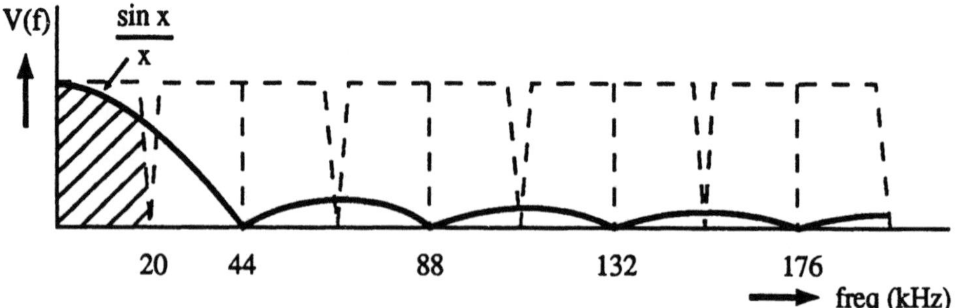

Figure 1.22 : Zero-order hold amplitude reduction

a system using Nyquist sampling. On the other hand, the zero-order hold operation can be used as a simple low-pass filter. When narrow band signals with respect to the sampling frequency are reproduced, then the $\frac{\sin x}{x}$ filtering operation can already introduce sufficient suppression of the repeated signals. In Figure 1.22 the dashed area shows the narrow-band signal in comparison with the sampling frequency. At f_s the sampled signals are significantly reduced in amplitude compared to the base-band.

In some cases it is possible to design an analog low-pass output filter in such a way that the $\frac{\sin x}{x}$ amplitude reduction is compensated for by an increase in the amplitude characteristic of this filter at the high-frequency band edge.

A second possibility to overcome this problem is obtained by a decrease of the hold time t_h. The hold-time reduction at a constant sampling frequency f_s can be obtained by introducing a switch at the output of the digital-to-analog converter. During the hold time t_h the output signal of the digital-to-analog converter is applied to the low-pass output filter and during the time $t_s - t_h$ a zero output signal is applied to the low-pass output filter. (Note: $t_s = \frac{1}{f_s}$.)

Suppose that t_h is a fraction k of t_s, so $t_h = kt_s$; then equation 1.50 can be rewritten as:

$$\mid H(2\pi f_{in}) \mid = \frac{k}{f_s} \frac{\sin k\pi \frac{f_{in}}{f_s}}{k\pi \frac{f_{in}}{f_s}}. \tag{1.51}$$

Note that $0 \leq k \leq 1$.

A decrease of the hold time t_h, for example, by a factor equal to four reduces the amplitude reduction from 3.92 dB to 0.22 dB at $f_{in} = \frac{1}{2}f_s$.

At the moment that k is close to 0, then the limit of the value of $\frac{\sin x}{x}$ is nearly one. No amplitude reduction over a large bandwidth is obtained at the cost of an overall output signal reduction. The disadvantage of this operation is the necessity to use an output amplifier to compensate for the amplitude reduction. At a certain moment the output amplitude can be so small that the noise of the output amplifier is larger than the quantization noise of the digital-to-analog converter. A second problem in practice is found in the accuracy with which the switching of the signal must be performed. Furthermore, large signal pulses are applied to the output filter. When an active solution for such a filter is used, then slew rate errors can be introduced, resulting in distortion of the output signal.

A third possibility to avoid the amplitude reduction is obtained by increasing the sampling frequency of the converter. With the same input frequency f_{in} the ratio $\frac{f_{in}}{f_s}$ decreases, giving the desired result. Such an operation can be performed by an oversampling D/A converter. In Figure 1.23(a) a D/A converter that combines a digital oversampling annex low-pass filtering function is shown. With the box called R the oversampling is introduced. In this specific example a four times oversampling is used. In reference [13] a detailed description of such a system is given. The digital filtering performs the steep filtering around $\frac{f_s}{2}$. The analog post filter takes care of the signal band around $4f_s$. This is shown in Figure 1.23(c) using the dashed line. The analog filter can be designed having a nearly linear phase characteristic. As a result, the amplitude reduction due to the $\frac{\sin x}{x}$ signal reduction is reduced to 0.22 dB, which can be compensated for in the analog or digital filter. Although the total filter exhibits a narrow transition band, a nearly linear overall phase characteristic is obtained.

Nowadays oversampling ratios of 8 or 16 in 16-bit digital audio D/A converter systems is not unusual. Dynamic range of these systems increases in this way to above 98 dB, which is more than the coding used with the compact disk. The increase in resolution of the converter can be successfully applied when a digital processing algorithm increases the number of significant bits. An example of such a processing can be tone control. In such an application a 12 dB bass or treble boost increases the dynamic range needed in the converter to 18 bits. Furthermore, the somewhat larger dynamic range can improve the linearity around zero when small input signals have to be

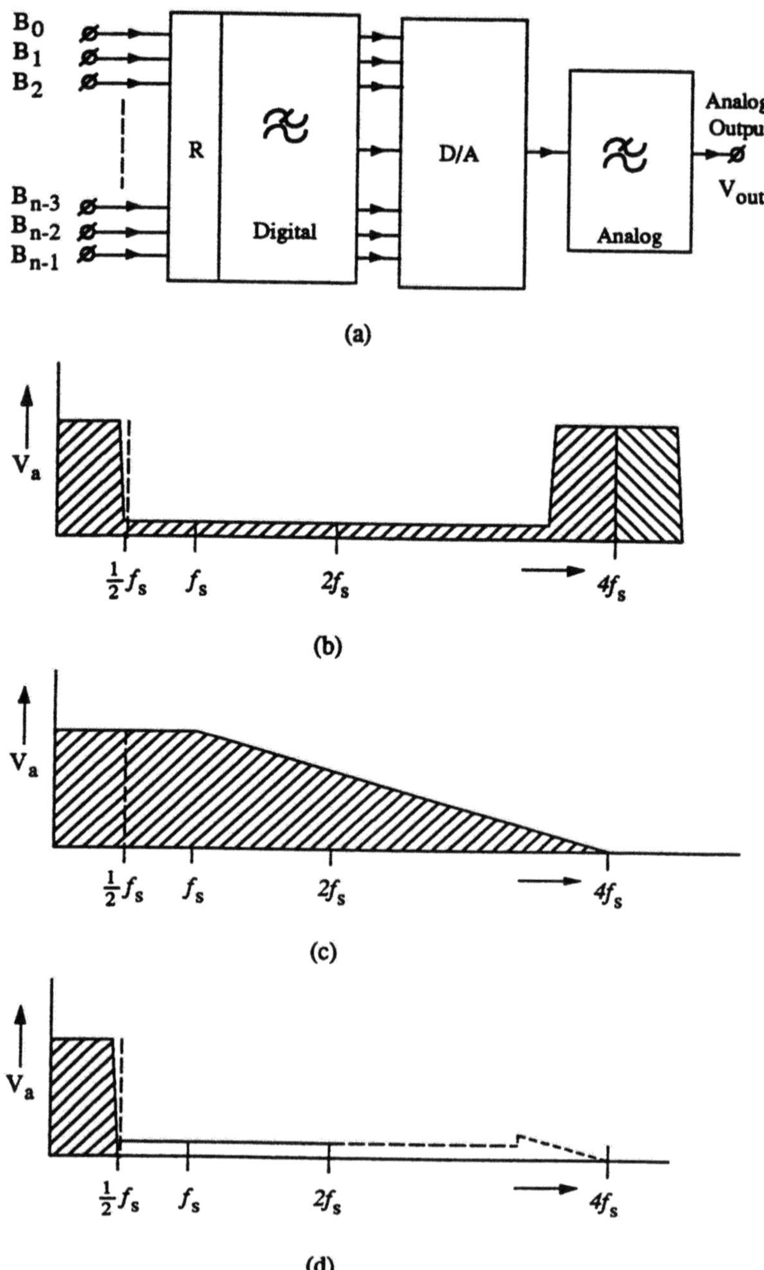

Figure 1.23 : (a) D/A converter system using combined digital-analog low-pass filter (b) Amplitude characteristic of the digital filter (c) Amplitude characteristic of the analog post filter (d) Amplitude characteristic of the total system

reconstructed. Special noise-shaping techniques can be used to increase the dynamic range of a system. In Chapter 10 these systems will be described.

Although an A/D converter is mostly preceded by a sample-and-hold amplifier, the transfer function of this sample-and-hold unit does not introduce the well-known $\frac{\sin x}{x}$ amplitude reduction. A sample-and-hold amplifier tracks the input signal during the sampling mode. At the moment the amplifier is switched from the sample into the hold mode, the momentary signal is applied to the hold amplifier. This means that at the sample moments the exact signal value is sampled and converted into a digital value. No holding or averaging operation is performed on the signal. The extra hold time introduced by the hold amplifier is only needed to allow the A/D converter to perform the conversion from an analog discrete-time signal into a digital output signal.

1.10　Minimum required stop band attenuation

Suppose the analog low-pass filter in the A/D converter system as shown in Figure 1.18 has a limited stop band rejection. Suppose, furthermore, that the analog bandwidth of the input amplifier, the sample-and-hold amplifier, or the comparator in the A/D converter is limited to f_{comp}. In analog-to-digital converter systems which use, for example, a successive approximation conversion algorithm (see Chapter 7) the comparator bandwidth must be at least N times larger than the sampling frequency. In such a system during the sampling time N trials are made to make the digital-to-analog converter output equal to the analog input signal. Moreover, to obtain an accurate settling of the comparator subtracter system even a larger bandwidth in the comparator is needed. At which point the aliasing occurs depends on the architecture of the analog-to-digital converter and the point in the system where the sampling is performed. Due to the sampling of the input signal all input frequencies that are in the aliasing signal bands are folded back into the baseband of the system. The number of aliasing signal bands that are folded back into the baseband are limited by the extra filtering introduced by the input signal amplifier, the sample-and-hold amplifier, or the bandwidth of the comparator in the A/D converter. In Figure 1.24 the attenuated signal at the output of the amplifier-comparator stage is shown as a function of frequency. In this configuration the sampling is performed by the comparator stage as is the case in, for example, a flash-type analog-to-digital converter (See Chapter 4).

Note that the bandwidth f_{comp} of the input amplifier or comparator in this example is about $3f_s$. When white noise or frequency components attenuated

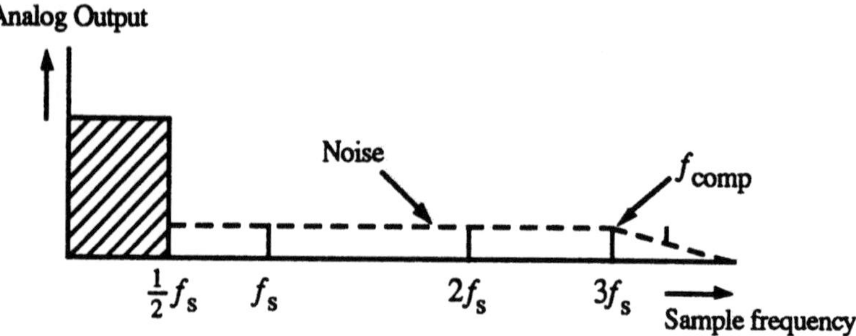

Figure 1.24 : Output signal of the amplifier-comparator stage in an A/D converter system

by the input anti-aliasing filter in the frequency range between $f = \frac{f_s}{2}$ and $f \gg 3f_s$ are input into the system, then the number of bands that are folding back noise or unwanted signal components into the baseband of the system is equal to:

$$N_{fold} = \frac{f_{comp} - \frac{f_s}{2}}{\frac{f_s}{2}}. \tag{1.52}$$

The fold-back noise and the unwanted signal components add to the quantization error and thus reduce the signal-to-noise ratio of the system. The "noise" in the baseband now increases to

$$\sqrt{N_{fold} + 1} = \sqrt{2 \times \frac{f_{comp}}{f_s}}. \tag{1.53}$$

If we want to have a condition in which the total fold-back noise in the base-band is equal to the quantization "noise" in the system when a signal is applied in the baseband, then the stop-band rejection of the low-pass filter must be increased by a factor equal to: $\sqrt{N_{fold}}$.

This increase in stop-band rejection ($A_{foldback}$) is equal to:

$$A_{foldback} = \sqrt{N_{fold}} = 10 \log N_{fold}\, dB. \tag{1.54}$$

This means that in a system with n bits the minimum stop-band rejection of the input filter must be:

$$A_{stopmin} = n \times 6.02 + 1.76 + 10 \log N_{fold}. \qquad (1.55)$$

Formula 1.55 gives a minimum anti-aliasing filter stop-band attenuation requirement under the condition that the total amount of fold-back noise is equal to the quantization noise.

Inserting formula 1.53 into formula 1.55 results in an expression for the minimum required stop-band rejection $A_{stopmin}$:

$$A_{stopmin} = n \times 6.02 + 1.76 + 10 \log(2 \times \frac{f_{comp}}{f_s}) \, dB. \qquad (1.56)$$

When a low-pass filter with a minimum stop-band rejection equal to formula 1.56 is used, then the dynamic range (S/N ratio) of the system is reduced by 3 dB.

If a smaller reduction is required, then the stop-band attenuation must be increased.

As an example, the minimum stop-band rejection of an input filter for an audio A/D converter will be calculated. This calculation involves the worst-case condition of signals outside the signal band. Suppose the bandwidth of the comparator is equal to 8.8 MHz, then with a sampling frequency of 44 kHz we obtain for the minimum stop-band rejection using formula 1.56 $A_{stopband} = 98 + 26 = 124$ dB. In practical situations, however, the noise and signals outside the baseband usually do not have the maximum amplitude equal to the signal amplitude. Furthermore, the basic A/D converter itself must have a signal-to-noise ratio measured over the signal band close to the theoretical value of 98.1 dB. The noise level of the analog filter must be well below the quantization noise of the system to avoid a reduction of dynamic range.

1.11 Conclusion

In this chapter the dynamic range of A/D and D/A converters is defined, and requirements for input and output filters are introduced. Different methods to implement anti-alias or reconstruction filters are shown. If a linear phase characteristic is required in a system, a combined analog-digital filtering solution gives the best results.

A general relation defining the sampling time uncertainty in relation to the resolution of the converter and the maximum input signal frequency is derived. This relation determines the stability of the sampling clock required over a short period of time.

A simple sampling clock time uncertainty calculation model is introduced, giving an understanding of the pulse-to-pulse time jitter of high stability clock systems.

An analysis of the amplitude reduction as a function of sample time for D/A converters is given. Oversampling or a reduction of the output sample time decreases this reduction to within acceptable values.

Chapter 2

Specifications of converters

2.1 Introduction

To obtain insight into the design criteria for converters it is important to arrive at a unanimous definition of specifications. These specifications must include the application of converters in conversion systems (see references [14,15]). Dynamic specifications of converters are needed to obtain insight in the applicability of a certain converter in a digital signal processing system: for example, digital audio or digital video. In a conversion system the complete conversion from analog into digital or digital into analog information is performed. Such systems include input or output amplification and anti-alias filtering.

Unanimous specifications for DC performance are well known in literature. Specifications for converters in signal-processing systems are more difficult to standardize. One of the reasons for this limitation in specification can be that in the early days of conversion the application of converters was in the area of digital voltmeters and control systems. These systems need the high dc performance at low signal speeds.

Digital audio in comparison with voltmeters, for example, applies for high-performance dynamic system specifications. The performance of converters in digital audio or digital video systems is at the limit of the possibilities of today's technology. In this chapter the most important specifications will be defined and discussed in more detail.

2.2   Digital data coding

Information in digital form appears at the input of a D/A converter or at the output of an A/D converter. This code can appear in parallel form or serially on a single line.

In the digital world several logic systems are used. Today's CMOS logic is overtaking the widely used Transistor Transistor Logic (TTL) in which a "1," ONE or "true" represents a minimum level of +2.4 V and "0," ZERO or "false" corresponds to a maximum level of +0.4 V. CMOS logic swings are determined by the applied supply voltage, which for today's logic blocks is 5 V but will in the future reduce to 3.3 V. In high-speed systems Emitter Coupled Logic (ECL) is used with levels of - 0.9 V for a logic "1" and - 1.8 V for a logic "0." The small logic swing of 0.9 V allows for high-speed applications. The advantage of ECL circuits is the possibility of a full differential system operation. This differential operation prevents large current spikes from flowing in the supply lines during switching. Thus a low interference between the analog signal path and the digital signal path can be obtained.

In systems with a serial digital data stream, a conversion from a serial stream into a parallel word must be performed. A shift register that performs serial to parallel conversion is added for this purpose. Furthermore, data latches are added which at the end of the serial-to-parallel conversion get the data word latched on the command of the latch clock. These data latches directly drive the current or voltage switches for example in a D/A converter. Analog data appear at the output of the converter after new data are latched in. The data latches together with the bit switches are optimized to obtain a minimal output glitch during switching of the converter. This is very important in a D/A converter system. Such a system can be operated without a so-called deglitcher when the glitch energy is small compared to the LSB energy. A "glitch" is a large output signal change (larger than the amplitude of an LSB step) appearing during a code transition of a converter. In Figure 2.1 an example of a serial-to-parallel conversion in a D/A converter is shown. The advantage of this circuit configuration is that the data latches together with the bit switches can be optimized for the best switching performance. Usually this means the smallest glitch error. D/A converters with parallel digital inputs often do not have data latches on board. Applying these converters requires an accurate board layout to match the delays between the output signals of the data latches and the switches. Users are not always

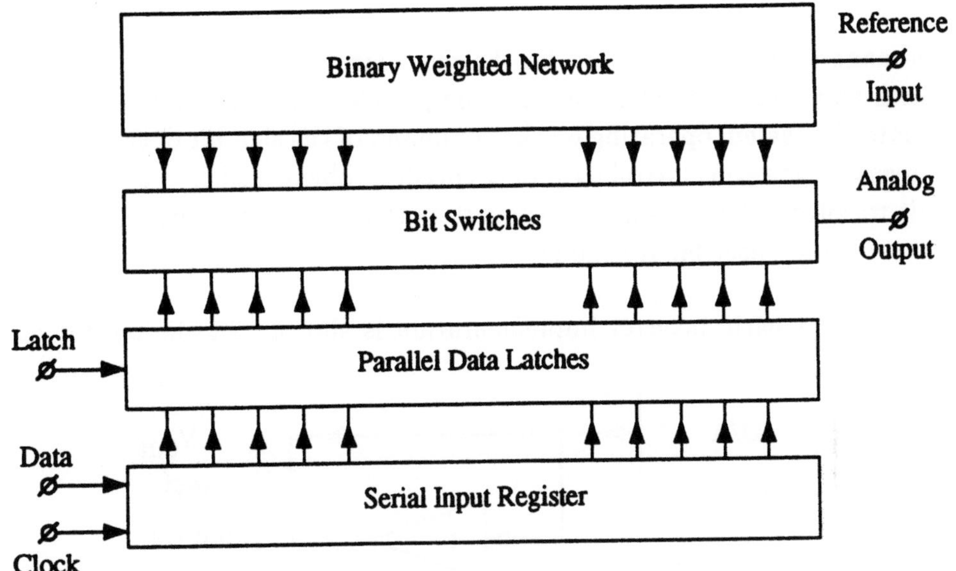

Figure 2.1 : Serial-to-parallel conversion in a D/A converter

aware of this problem and therefore end up with a low overall performance of the converter system. Large "glitches" may appear at the output of the converter because of the non-equal switching time of bit switches during code transitions.

2.3   Digital coding schemes

As a digital code the natural binary code (base 2) is used. Without additional measures it is difficult to obtain bipolar signals. Therefore, an offset binary code is introduced. This code is "offset" by turning on the MSB bit for an analog value of zero. In most cases an additional current is added to subtract the MSB bit current from the input or output current value to obtain analog zero. The principal drawback of the offset binary code is that a major carry transition occurs if a small bipolar signal is generated or converted around zero. Such a transition requires the highest linearity around zero and immediately gives rise to glitch problems. These problems are worst for D/A converters. In A/D converters the noise that is part of the offset current is added to the input signal and may result in a reduction of the dynamic range of the system. Furthermore, this offset must be stable with temperature, and variations in supply voltage should not change the zero setting of the A/D converter.

A sign-magnitude code seems to be the most straightforward approach to the glitch and noise problem. The sign-magnitude coding introduces an inverter into the system operating on the command of the sign bit of the signal. This inverter, however, introduces an accuracy problem which is even more difficult to solve in high-resolution converters than the dc "offsetting" of the converter in the offset binary mode.

In Figure 2.2 a basic inverter implementation is shown. At the output of

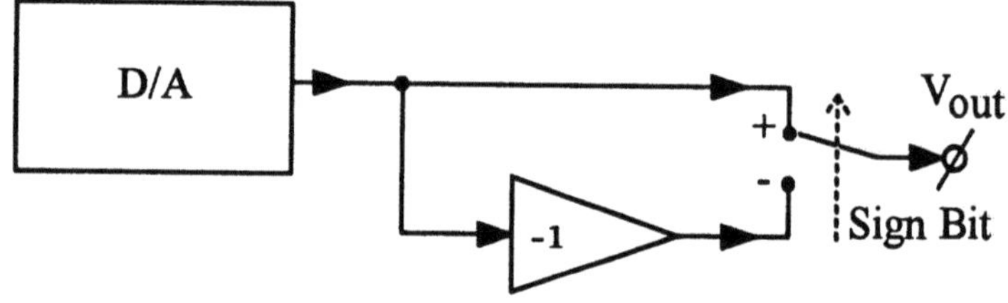

Figure 2.2 : Basic sign-magnitude inverter implementation

the digital-to-analog converter an analog inverting operational amplifier and a switch (S) controlled by the sign-bit is added to the system. During the positive output signal generation the output signal of the digital-to-analog converter is directly applied to the output terminal. During the negative signal generation the output signal of the digital-to-analog converter coming from the analog inverter is applied to the output terminal. The accuracy with which the signal is inverted determines the total accuracy of the system. In practical implementation an accuracy between 0.1% and 0.01% can be obtained. Other codes can also be used. When computational operations are needed the twos complement code is very useful. In Table 2.1 the different codes that can be used in converters are shown. From the table it can be seen that with the ones complement and the sign-magnitude code two code possibilities for zero are found. A computation has to be performed to avoid this double zero coding.

The preferred code implementation concerning noise and glitches is the sign-magnitude code. However, this code suffers from linearity problems around zero. As a result the offset binary code is the most commonly used code implementation in converter applications.

Number	Sign + Magnitude	Twos Complement	Offset Binary	Ones Complement
+7	0 1 1 1	0 1 1 1	1 1 1 1	0 1 1 1
+6	0 1 1 0	0 1 1 0	1 1 1 0	0 1 1 0
+5	0 1 0 1	0 1 0 1	1 1 0 1	0 1 0 1
+4	0 1 0 0	0 1 0 0	1 1 0 0	0 1 0 0
+3	0 0 1 1	0 0 1 1	1 0 1 1	0 0 1 1
+2	0 0 1 0	0 0 1 0	1 0 1 0	0 0 1 0
+1	0 0 0 1	0 0 0 1	1 0 0 1	0 0 0 1
+0	0 0 0 0	0 0 0 0	1 0 0 0	0 0 0 0
−0	1 0 0 0	(0 0 0 0)	(1 0 0 0)	1 1 1 1
−1	1 0 0 1	1 1 1 1	0 1 1 1	1 1 1 0
−2	1 0 1 0	1 1 1 0	0 1 1 0	1 1 0 1
−3	1 0 1 1	1 1 0 1	0 1 0 1	1 1 0 0
−4	1 1 0 0	1 1 0 0	0 1 0 0	1 0 1 1
−5	1 1 0 1	1 0 1 1	0 0 1 1	1 0 1 0
−6	1 1 1 0	1 0 1 0	0 0 1 0	1 0 0 1
−7	1 1 1 1	1 0 0 1	0 0 0 1	1 0 0 0
−8		1 0 0 0	0 0 0 0	

Table 2.1 : Different digital coding schemes

2.4 DC specifications

2.4.1 Absolute accuracy

Accuracy of converters should not be confused with linearity and resolution. It includes the errors of quantization, nonlinearities, short-term drift, offset, and noise.

The *absolute accuracy* of a converter is the actual full-scale input or output (analog-to-digital or digital-to-analog converter) signal (voltage, current, or charge) referred to the absolute standard of the National Bureau of Standards. This absolute accuracy is mostly related to the reference source used in the converter. Sometimes this reference source consists of a special temperature-compensated zener diode. In integrated circuits this zener diode is replaced by an integrable source which in modern systems is based on the band-gap voltage of silicon. This reference source should have a low-noise with respect to the resolution of the converter. Temperature coefficients in the ideal case should be so small that the accuracy of the reference over the specified temperature range stays within the resolution of the converter ($\frac{1}{2}$ LSB over the full temperature range).

2.4.2   Relative accuracy

The *relative accuracy* is the deviation of the output signal or output code of a converter from a straight line drawn through zero and full scale. Output signals or output codes must be corrected from a possible zero offset. This relative accuracy is called: *Integral Non Linearity* (INL) or sometimes *linearity*. Throughout this book the Integral Non Linearity will be used. In Figure 2.3 a graphical example is shown of the integral nonlinearity definition for a digital-to-analog converter. In the figure the boundaries for which the nonlinearity deviates not more than $\pm\,\frac{1}{2}$ LSB of a straight line through zero and full scale are shown.

The $\pm\,\frac{1}{2}$ LSB integral nonlinearity definition implies a *monotonic* behavior of this converter.

Monotonicity of a converter means that the output of, for example, a D/A converter never decreases with an increasing digital input code. A minimum increase of *zero* is allowed for a 1 LSB increase in input signal in a D/A converter. In Figure 2.3 the transfer curve of a monotonic converter is shown.

In an analog-to-digital converter *monotonicity* means that *no missing codes* can occur. It must be noted at this point that converters can be designed which are *guaranteed monotonic* but do not have the half LSB linearity specification. These converters are based on non-binary weighting of the bit currents. As an example of such a converter a tapped resistor with switches performs a monotonic digital-to-analog conversion without requiring an integral nonlinearity specification.

Examples of converters which are monotonic by design are presented in references [21,22,32].

Note that the *monotonicity* specification does NOT automatically include that a converter has a $\pm\,\frac{1}{2}$ LSB integral nonlinearity error!

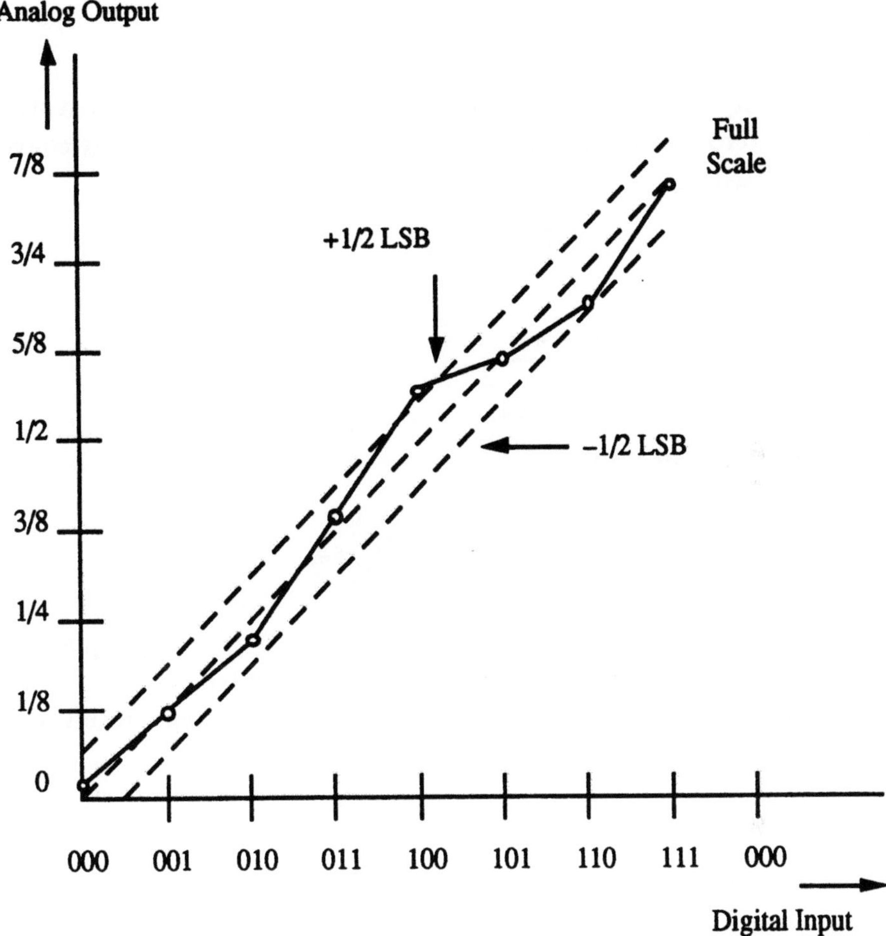

Figure 2.3 : Definition of the Integral Non-Linearity of a D/A converter

2.4.3   Nonlinearity calculation

A simple calculation can be performed to show that the nonlinearity specification of $\pm \frac{1}{2}$ LSB is sufficient to prove *monotonicity* of a binary-weighted converter. During this calculation all signals or codes are corrected for zero offset in advance simplifying the analysis.

Suppose we have an n-bit binary-weighted converter, then with ε_m corresponding to the error of the mth bit, the non-ideal weighting of this bit can be written as:

$$b_m = 2^m + \varepsilon_m.$$

$$(2.1)$$

Note that a unity LSB bit weight is used in this equation. This unity bit weight does not influence the nonlinearity calculations.

The nonlinearity will now be calculated as the total deviation from a straight line between zero and full-scale. The full-scale value B of an n-bit converter can be expressed as:

$$B = \sum_{m=0}^{n-1} (2^m + \varepsilon_m). \tag{2.2}$$

With

$$\sum_{m=0}^{n-1} 2^m = 2^n - 1, \tag{2.3}$$

this can be simplified into:

$$B = 2^n - 1 + \sum_{m=0}^{n-1} \varepsilon_m. \tag{2.4}$$

Note that the total number of quantization steps is given by equation 2.3. The "ideal" step size S of the converter can be calculated using the full-scale value of the converter and the number of quantization steps, resulting in:

$$S = \frac{B}{2^n - 1} = 1 + \frac{\sum_{m=0}^{n-1} \varepsilon_m}{2^n - 1}. \tag{2.5}$$

The linearity error Δ_k of the kth bit compared with a value obtained from the ideal straight line through zero and full scale becomes:

$$\Delta_k = 2^k.(1 + \frac{\sum_{m=o}^{n-1} \varepsilon_m}{2^n - 1}) - (2^k + \varepsilon_k). \tag{2.6}$$

In general errors will have positive and negative signs. Adding all errors together must result in a total error of zero, because the sum of all non-ideal bit weights is used as the full scale value. The following calculation proves this statement. Summing Δ_k results in:

$$\sum_{k=0}^{n-1} \Delta_k = \sum_{k=0}^{n-1} \frac{2^k}{2^n - 1} \cdot \sum_{m=0}^{n-1} \varepsilon_m - \sum_{k=0}^{n-1} \varepsilon_k. \tag{2.7}$$

Inserting again:

$$\sum_{k=o}^{n-1} 2^k = 2^n - 1, \tag{2.8}$$

this equation can be simplified into:

$$\sum_{k=o}^{n-1} \Delta_k = \sum_{m=0}^{n-1} \varepsilon_m - \sum_{k=0}^{n-1} \varepsilon_k = 0. \tag{2.9}$$

As a result of this calculation the absolute values of the sum of all positive errors Δ_k must be equal to the sum of all negative errors Δ_k. The integral non linearity of a converter (INL) is now specified as the total error of the positive or negative errors or:

$$INL = \sum_{k=0}^{n-1} Positive\Delta_k = -\sum_{k=0}^{n-1} Negative\Delta_k \leq \frac{1}{2} LSB. \tag{2.10}$$

To prove *monotonicity* of a binary-weighted converter, suppose that as a first step only the MSB bit value is too small. This means that the sum of the remaining bit errors must be equal to the MSB bit error but with the opposite sign. The nonlinearity error Δ_{n-1} of the MSB bit can be expressed as (2.6):

$$\Delta_{n-1} = 2^{n-1}.(1 + \frac{\sum_{m=0}^{n-1} \varepsilon_m}{2^n - 1}) - (2^{n-1} + \varepsilon_{n-1}), \tag{2.11}$$

or:

$$\Delta_{n-1} = (\frac{2^{n-1}}{2^n - 1} \cdot \sum_{m=0}^{n-1} \varepsilon_m) - \varepsilon_{n-1}. \tag{2.12}$$

As defined in equation 2.10 the MSB-bit error given by equation 2.12 is equal to the integral non linearity (INL) of the converter. Thus:

$$INL = \Delta_{n-1}. \tag{2.13}$$

In Figure 2.4, the minimum condition for *monotonicity* is shown. The dashed slanted lines indicate the $\pm \frac{1}{2}$ LSB boundaries determined by the integral nonlinearity specification. As is shown, this converter has a $\pm \frac{1}{2}$ LSB nonlinearity specification.

When the MSB-bit is smaller than the sum of the other bits, the converter becomes $non - monotonic$ as is indicated by the dashed curve in the figure. The major carry transition from all bits except MSB switched on (011) compared to only MSB switched on (100) is shown. Furthermore, it is assumed that the MSB bit value is too small compared to the rest of all bits.

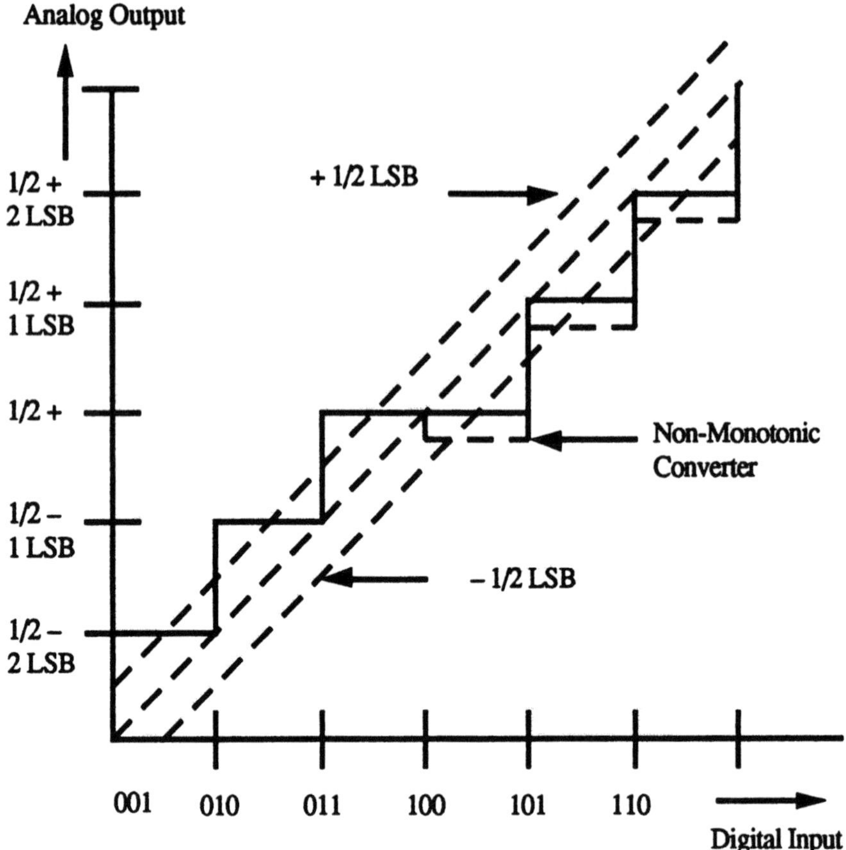

Figure 2.4 : MSB major carry transition

The linearity error given by equation 2.10 is a measure for *monotonicity* of
the converter. When in a D/A converter the digital input code is increased by
a value of 1 LSB, the minimum requirement for *monotonicity* is no increase
in analog output value as shown in Figure 2.4. This means that twice the
error given by equation 2.12 must be less or equal to 1 LSB or:

$$\Delta_{n-1} = \frac{2^{n-1}}{2^n - 1} \cdot \sum_{m=0}^{n-1} \varepsilon_m - \varepsilon_{n-1} \leq \frac{1}{2}LSB, \tag{2.14}$$

or:

$$INL \leq \frac{1}{2}LSB. \tag{2.15}$$

In an analog-to-digital converter an INL definition including the quantization levels equal to 2.15 is valid.

In general it can be said that the errors of the individual bit weights must be added according to the definition given in equation 2.10. To guarantee *monotonicity* of the converter, the sum of the errors must never exceed $\frac{1}{2}$ LSB value. An identical calculation can be performed for other bit weights in the converter. In this way the generalized *monotonicity* definition can be proven. Performing such calculations, it will be found that the error at the MSB transition shows the most stringent linearity requirement.

In Figure 2.5 the errors of the individual bit weights of a binary weighted converter are shown. The errors for full-scale and zero are calibrated to be zero. In this way a straight line between zero and full-scale is obtained. Starting from the left side of the figure, the errors start from the MSB-error and go down to the LSB-error. Finally the total of the positive and the negative errors are plotted at the most right side. This converter can be specified as having a $\pm \frac{1}{2}$ LSB integral nonlinearity error. In a converter

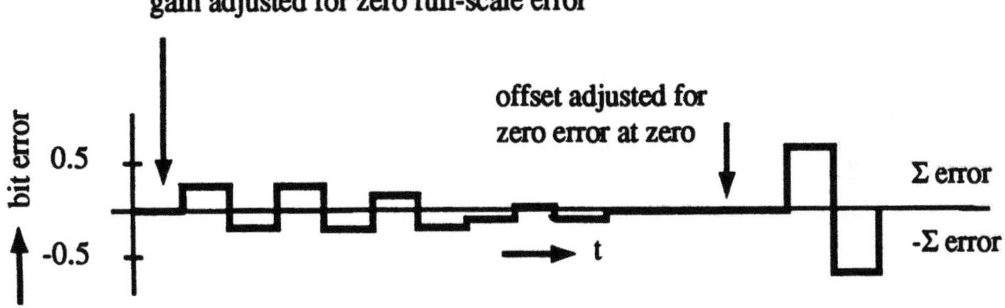

Figure 2.5 : Bit-weight error of a binary-weighted converter

it is always possible to generate a code that coincides with this worst-case error. Furthermore, a converter might be *monotonic* having a linearity error of more than $\frac{1}{2}$ LSB. In that case an increase in analog output larger than 2 LSBs is possible. A "missing" code value is introduced in this converter.

Conclusion: A converter is always *monotonic* when the integral nonlinearity specification (INL) is less than or equal to $\pm \frac{1}{2}$ LSB.

However, when a converter is specified to be always *monotonic*, then this specification does not automatically imply an integral nonlinearity error of less than or equal to $\pm \frac{1}{2}$ LSB.

2.4.4 Differential nonlinearity

Differential nonlinearity (DNL) error describes the difference between two adjacent analog signal values compared to the step size (LSB weight) of a converter generated by transitions between adjacent pairs of digital code numbers over the full range of the converter.

The differential nonlinearity is zero if every transition to its neighbors equals 1 LSB. In a *monotonic binary − weighted* converter an increase of the digital code value by 1 LSB can result in an increase in analog signal between 0 and 2 LSBs. The maximum differential nonlinearity in this case is $\pm$ 1 LSB. Writing down the differential nonlinearity for a digital-to-analog converter in a formula gives:

$$DNL = S_{out}(C_{m+1}) - S_{out}(C_m) - 1LSB. \tag{2.16}$$

C_{m+1} and C_m are two adjacent digital input codes. $S_{out}(C_n)$ is the output signal of the converter with an input code C_n.
The value of 1 LSB corresponds to the "ideal" step from the previous section. In case of an analog-to-digital converter the DNL can be written as:

$$DNL = A_{input}(Q_{m+1}) - A_{input}(Q_m) - 1LSB. \tag{2.17}$$

Q_{m+1} and Q_m are two adjacent quantization levels. $A_{input}(Q_n)$ is the analog input voltage corresponding to the quantization level Q_n. In Figure 2.6 the transfer curve of a 4-bit A/D converter is shown. The drawn line shows the ideal transfer characteristic, while a dashed line indicates the measured transfer curve of a practical converter. Integral nonlinearity (INL), differential nonlinearity (DNL), and full-scale range (FSR) are shown partly as a function of the LSB error between drawn line and actual measured dashed lines. Furthermore, an example of a missing code is given in the picture. In a visualized way the nonlinearity errors are shown to improve understanding.

2.4.5 Offset

Input amplifiers, output amplifiers, and comparators in practical circuits inherently have a built-in offset voltage sand offset current. This offset is caused by the finite matching of components. The offset results in a non-zero input- or output voltage, current or digital code although a zero signal is applied to the converter.

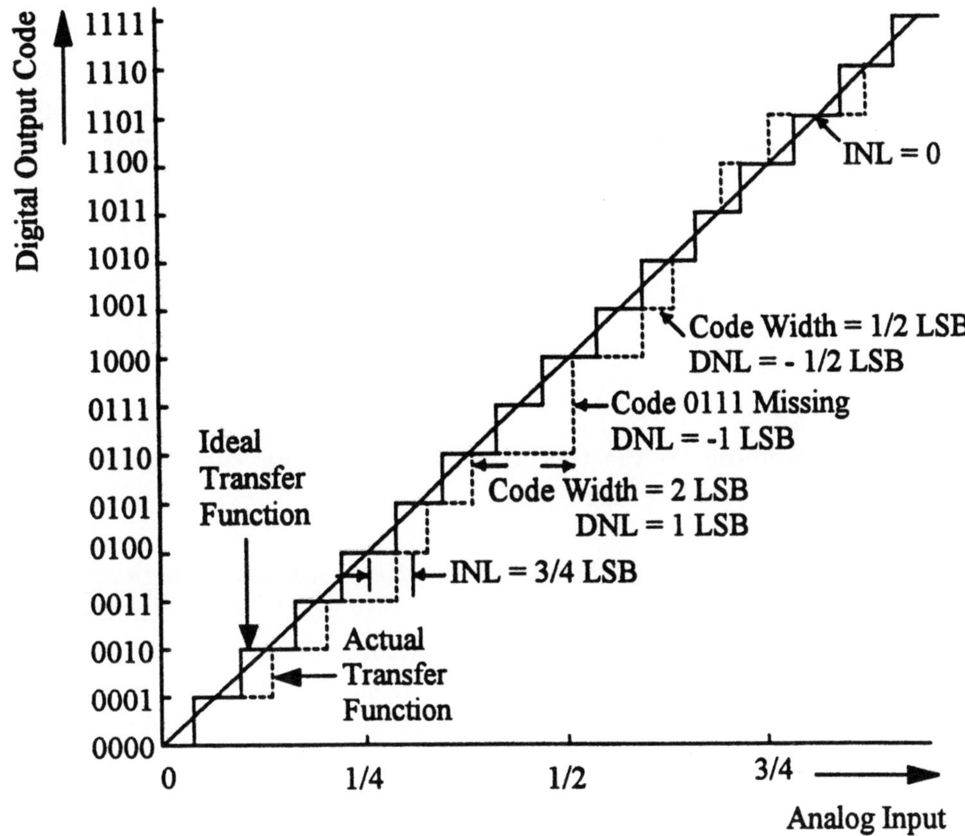

Figure 2.6 : Transfer curve of a 4-bit A/D converter

Offset is very important in dc systems. Temperature dependence of the offset must be small. Trimming or auto zero procedures can remove the offset in a system. Furthermore, care must be taken during the layout of the circuit to avoid thermal coupling and thermal gradients over an integrated circuit. If such a coupling exists, then the output code or output signal show slow changing components depending on the applied input signal or input code. Such signal components are unwanted.

2.4.6 Temperature dependence

Monotonicity and linearity must be maintained over a large temperature range to keep distortion and signal-to-noise ratio within the specified range. Mostly this requires a very good thermal tracking of components. In an n-bit system with a relative accuracy of $\pm \frac{1}{4}$ LSB this linearity may change maximally by $\pm \frac{1}{4}$ LSB over the full temperature range to maintain monotonicity of the system. A $\pm \frac{1}{2}$ LSB linearity over the full temperature range is obtained in this way, while in binary weighted systems this linearity specification automatically includes *monotonicity* of the converter over the full temperature range.

If we have a variation in temperature of ΔT, then the difference in thermal tracking for components must not exceed a value of

$$\frac{2^{-n}}{4\Delta T} \quad \text{per} \quad \text{degree} \quad \text{C}. \tag{2.18}$$

The error introduced by the thermal mismatch in this case amounts to $\frac{1}{4}$ LSB over a temperature range of ΔT.

Because the most significant bits in a converter must have a high accuracy, the thermal tracking of these bits must be very good. The estimate given in 2.18 is only needed for most significant bit matching. In a 16-bit system subjected to a temperature change of 105^0 C this means that the temperature tracking of components should not exceed 0.04 ppm per degree C. In this analysis it is supposed that the linear (first-order) temperature coefficient is much larger than the second-order term. Sometimes the first order temperature coefficient is reduced by a clever design. At that moment the second order coefficient takes over.
All specifications are performed at a specified temperature (usually 25°C) and over the full temperature range: $-20°$C to $+85°$C for consumer and $-55°$C to $+125°$C for military applications are commonly used temperature ranges.

2.4.7　Supply voltage

Specifications of converters are given at nominal supply voltages. Variations in supply voltages usually may not exceed a $\pm 5\%$ tolerance. To include these minimum and maximum supply voltage boundaries, a specification about the supply voltage dependence of a parameter is given, allowing the user to determine the error budget.

Additional specifications about power dissipation, input bias currents, reference impedance, clock input, output loading, and fan-in and fan-out of the logic system are needed.

2.5   Dynamic specifications

2.5.1   Signal-to-Noise Ratio

The most important dynamic specification of a converter is the signal-to-noise ratio. This signal-to-noise ratio depends on the resolution of the converter and automatically includes specifications of linearity, distortion, sampling time uncertainty, glitches, noise, and settling time. Over half the sampling frequency, this signal-to-noise ratio must be specified and should ideally follow the theoretical formula: $S/N_{max} = 6.02 \text{ n} + 1.76 \text{ dB}$.

This S/N ratio is calculated for a sine wave input with a maximum amplitude, and the ratio between the frequency of the sine wave, and the sampling frequency should be irrational. In case input signals with a smaller amplitude are applied, then the S/N ratio decreases in accordance to the input signal decrease.

Specifications must be given as a function of the signal frequency with various amplitudes and as a function of amplitude with a constant signal frequency. An even more stringent dynamic performance definition is obtained if the total harmonic distortion (THD) of the converter is added to the quantization error. Even in this case the maximum signal-to-noise ratio of a high-performance converter must be close to the above given formula. In Figure 2.7 the signal-to-noise ratio of a converter as a function of frequency from full scale to -30 dB in steps of -10 dB is shown. The sampling frequency of the 16-bit converter is 176.4 kHz and the measuring bandwidth is 20 kHz. Note that at lower output amplitudes the signal-to-noise ratio in the frequency range from 10 kHz to 20 kHz is at maximum. Distortion products of signals above 10 kHz are rejected by the output reconstruction filter.

In Figure 2.8 the signal-to-noise ratio as a function of amplitude is shown. The measuring frequency is 1 kHz, while in this situation a sampling frequency of 44.1 kHz is used. At full-scale output a larger deviation from the theoretical 16-bits line in observed. Measuring bandwidth is 20 kHz.

2.5.2   Spurious Free Dynamic Range

When converters are used with large oversampling ratios or the spectral purity of the converter is important, an additional specification determining

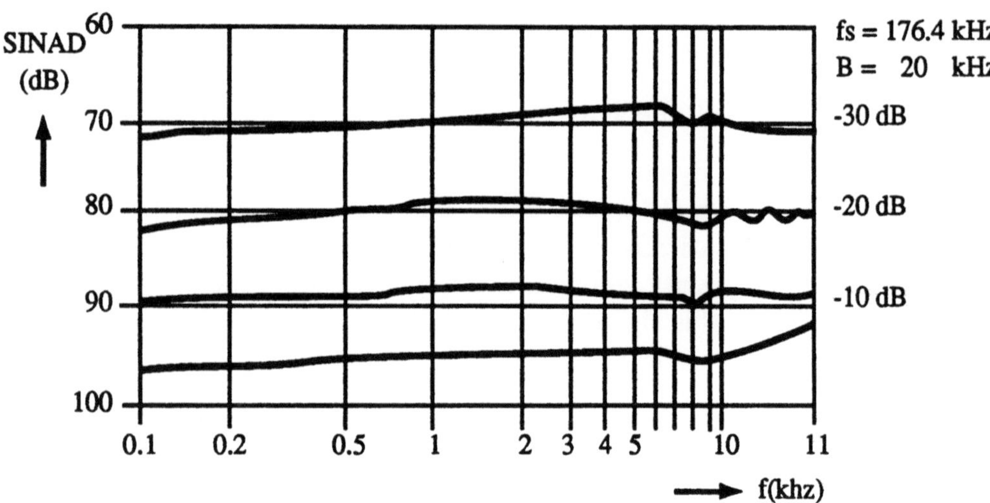

Figure 2.7 : Signal-to-(noise plus distortion) ratio as a function of frequency with various amplitude values

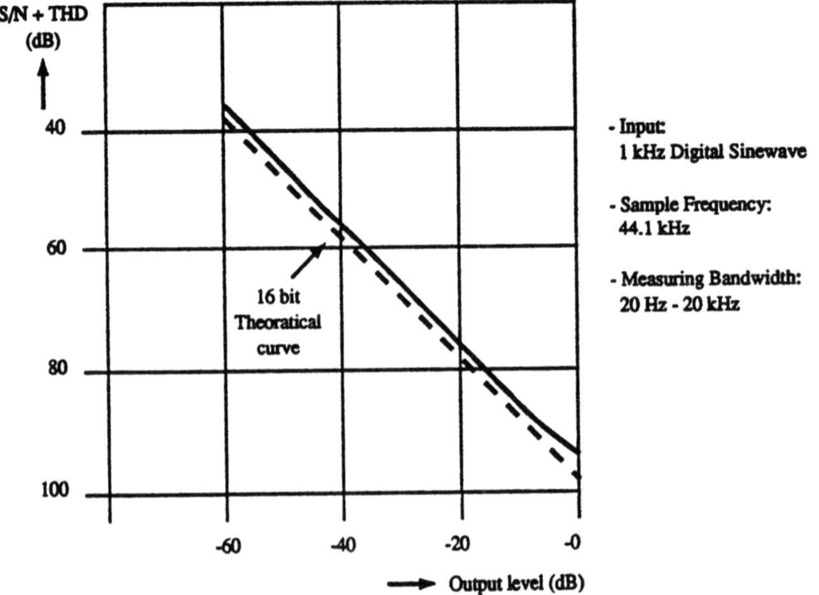

Figure 2.8 : Signal-to-(noise plus distortion) ratio as a function of amplitude

the ratio between the maximum signal component and the largest distortion component can be obtained. This ratio is called Spurious Free Dynamic Range (SFDR). In Figure 2.9 an amplitude spectrum of a converter is shown. From the figure it is seen that the SFDR ratio is obtained at the

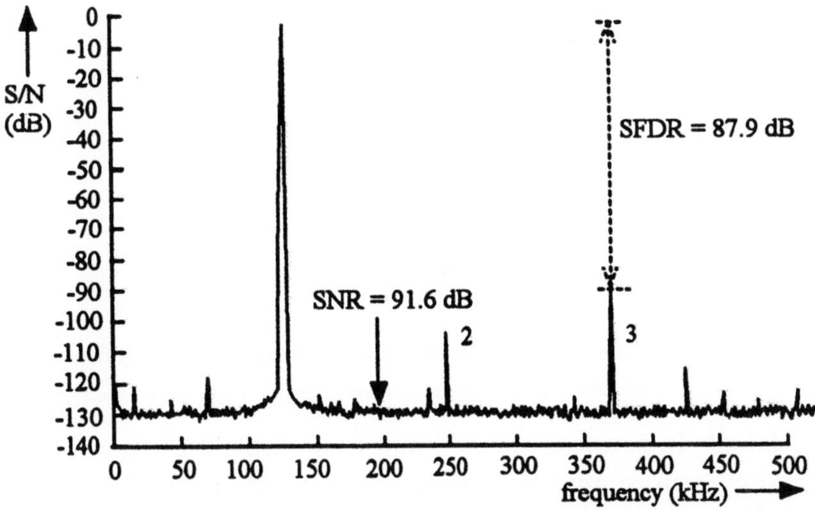

Figure 2.9 : Definition of Spurious free Dynamic Range

moment a maximum signal is applied to the converter. The ratio between the amplitude of the signal and the largest distortion component is defined as maximum distortion-free dynamic range of the converter.

Note that in Figure 2.9 the measuring bandwidth of the spectrum analyzer is much smaller than the total bandwidth (in this case 250 kHz) over which the signal-to-noise ratio is determined. Using such a small bandwidth results in an accurate determination of the different distortion products of a converter.

Converters with a good integral linearity usually give an SFDR that is larger than the signal-to-noise ratio of the system. To prove this statement, suppose a converter with n_d distortion components is analyzed. Furthermore, it is adopted that the n_d distortion components have equal amplitude and are a factor a_d attenuated with respect to the total noise of the system. Then we obtain for the total noise plus distortion (q_{nd}) in the system:

$$q_{nd} = q_n \sqrt{1 + \frac{n_d}{a_d^2}}. \tag{2.19}$$

The signal-to-noise ratio in this system will be reduced by the distortion of the converter. In the following table the influence of the distortion on the total performance of the converter is shown with $a_d = 3.162$ (10 dB) and n_d

varies from 1 to 4. In practical converters mostly second and third harmonic

n_d	q_{nd}/q_n
1	1.05
2	1.10
3	1.14
4	1.18

Table 2.2 : Signal-to-noise reduction ratio as a function of distortion products

distortion is found. In that case $n_d = 2$ and from Table 2.2 a loss of 0.8 dB in signal-to-noise ratio compared to an ideal converter gives a SFDR that is 10 dB larger than the signal-to-noise ratio.

Equation 2.19 shows that a large SFDR requires a converter with a signal-to-noise ratio as close as possible to the theoretical attainable value.

2.5.3　Dynamic range versus converter linearity

At the moment the signal-to-noise plus distortion (SINAD) of a converter is measured as a function of the converter construction, the results shown in Figure 2.10 are obtained. The dynamic range of an ideal 16-bit converter is a straight line. At the moment a converter uses an algorithm to obtain monotonicity without guaranteeing the integral nonlinearity specification of $\frac{1}{2}$ LSB then the curve called segmented D/A converter is obtained. The distortion in the system reduces the maximum dynamic range. At the moment a non-monotonic design is measured, then the third curve called basic D/A converter is obtained. Because of the non-monotonicity of the converter the distortion starts at about -40 dB from the maximum output level. This distortion reduces the dynamic range with about 20 dB when non-monotonicity is observed. The amount of dynamic range reduction by entering the non-monotonic relation between input and output signal of the converter depends on the matching of the components. Practical matching properties are about 0.1 % resulting in a monotonic range of about 60 dB. This result is obtained in Figure 2.10.

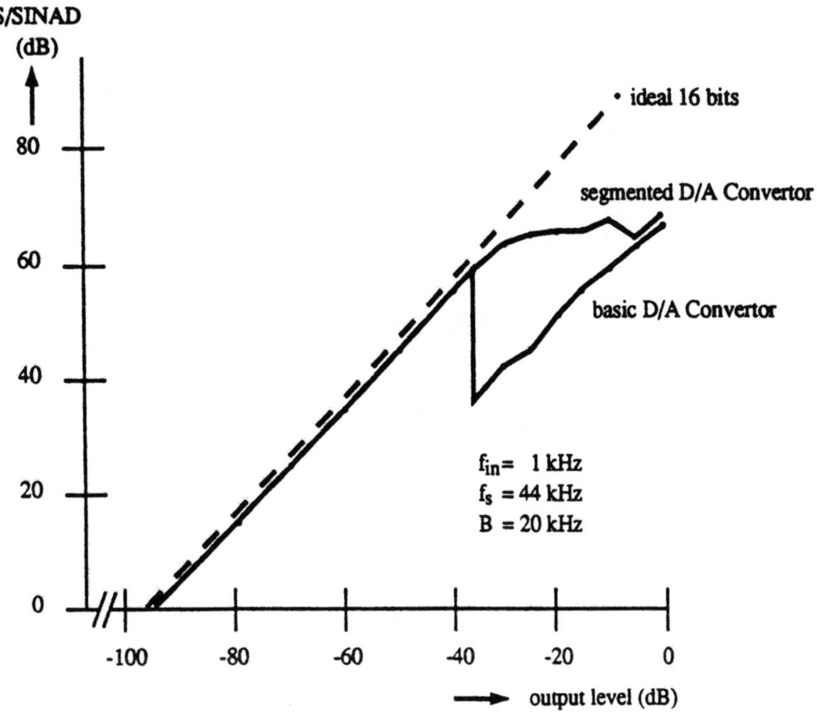

Figure 2.10 : SINAD as a function of converter construction

2.5.4 Glitches

Glitches are usually important for the performance specification of D/A converters. A glitch is generated when during a major carry code transition the new code signal value appears before or after the signal value of the former code disappears. The largest glitch is mostly generated by a major carry transition around the MSB level. In Figure 2.11 a glitch at the major carry transition is shown. The glitch obtained at the output of the D/A converter is generated by a slow switching on of the MSB bit value. As a result the output signal jumps from the value given by 01111 through 00000 to 10000. A sharp glitch is generated at the output which results in distortion giving a decrease is signal-to-noise ratio. To perform worst case calculations on the influence of timing on the glitch size a styled square glitch is used. In Figure 2.12 this styled glitch is shown. Suppose the full-scale (peak-to-peak) amplitude of the converter equals $2A$, then the MSB value is close to A. Furthermore, the difference in switching time is defined as t_{dif}.

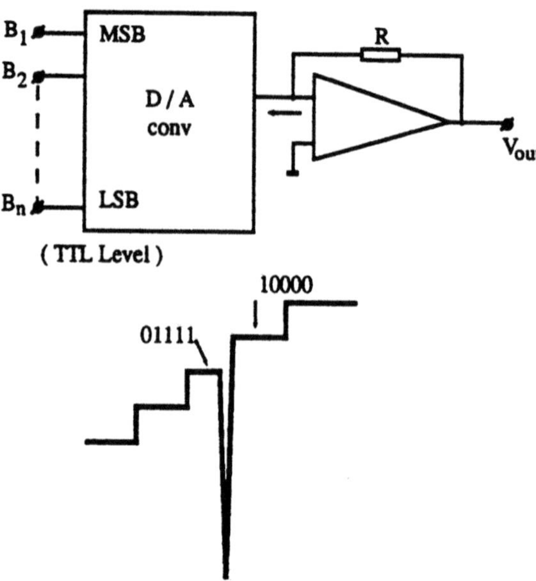

Figure 2.11 : Major carry glitch

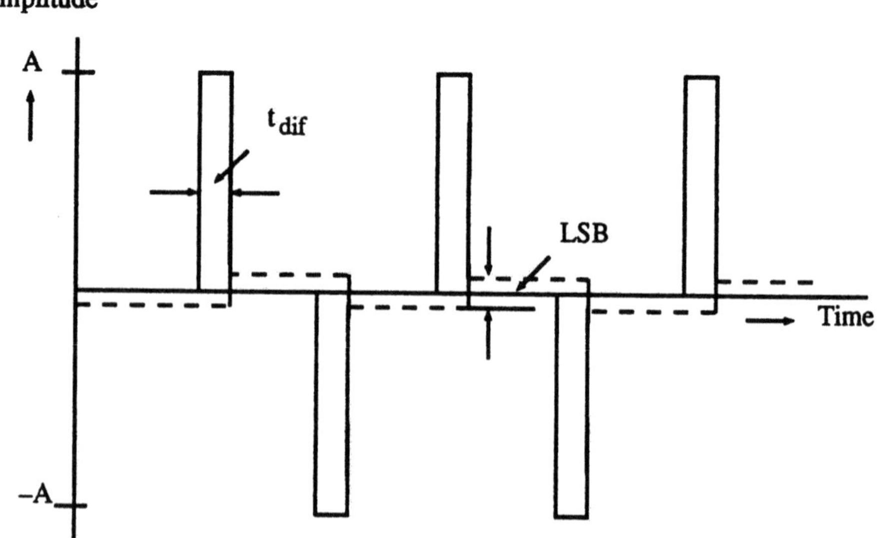

Figure 2.12 : Styled glitch of a converter

The glitch-area error is then equal to:

$$E_{glitch} = At_{dif} \quad V\,sec. \tag{2.20}$$

As an example, the glitch error of a 16-bit D/A converter will be calculated compared to the LSB bit value.

Suppose the switching time difference $t_{dif} = 1$ nsec and the full-scale value of the converter is 1 V, then the glitch error becomes: $E_{glitch} = 500$ pVsec.

The LSB area now equals:

$$E_{LSB} = 2^{-n+1} A t_s \quad Vsec. \tag{2.21}$$

With a sample time t_s of 20 μ sec for the converter and $n = 16$ we obtain:

$$E_{LSB} = 300 \text{ pVsec.}$$

The glitch error in this example is about one and a half times larger than the LSB area. To qualify this converter as "good," the glitch error must be at least reduced by a factor three.

From equations 2.20 and 2.21 the time difference t_{dif} can be expressed as:

$$t_{dif} = g 2^{-n+1} t_s \tag{2.22}$$

with:

$$g = \frac{E_{glitch}}{E_{LSB}}. \tag{2.23}$$

A solution to reduce the glitch error is the addition of a so-called "deglitcher" circuit at the output of a D/A converter. However, such a deglitcher circuit uses an analog storage element to hold the output signal during the code transition. Such a deglitcher circuit with a distortion level below the LSB level is difficult to design. In practical solutions it is simpler to optimize data latch and bit-switch for minimum glitch energy. Mostly in the latter case a better performance is obtained. Glitches exhibit in the output signal of a D/A converter as distortion, thus reducing the signal-to-noise ratio compared to the maximum theoretical value. When oversampling of a converter is used to obtain a larger dynamic range (S/N ratio), the distortion products introduced by the glitches counteract the increase in dynamic range, including distortion.

An example of a measured MSB-glitch of a 14-bit D/A converter is shown in Figure 2.13. The output signal is measured across a 25 ohm load resistor. Full-scale value under this measuring condition is 50 mV. The signal is directly applied to a 1 Ghz bandwidth oscilloscope without using an operational amplifier to convert the D/A current signal into a voltage. The

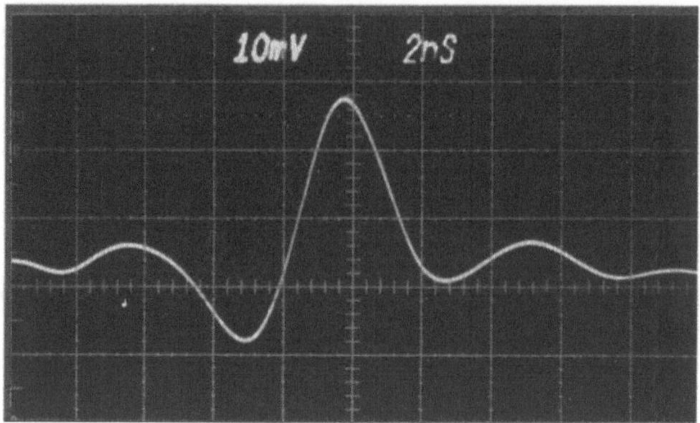

Figure 2.13 : Measured MSB-glitch error

glitch energy becomes even more important in systems that use a large oversampling ratio together with a noise-shaping technique. In such systems an *averaging* of signals occurs. This means that the number of MSB transitions that generate a glitch are increased to more than the minimum number required in generating, for example, a sine wave. At that moment the amount of glitch energy adds to the output signal. As a result the dynamic range and the distortion of the system do not decrease as much as expected by the theoretical analysis of the system.

2.5.5 Noise

Thermal noise (white noise) of bit-currents, amplifiers, resistors, and so on, adds to the quantization noise. This thermal noise exhibits itself as a deviation from the theoretical maximum signal-to-noise ratio that an ideal converter can have. When the thermal noise specification is given for a converter, this figure is not a measure for the (dynamic) signal-to-noise ratio of the system. However, the noise of the individual bit currents must be much lower than the quantization "noise" of the system.

A simple calculation demonstrates the decrease in signal-to-noise ratio of a system compared to the signal-to-noise ratio of the basic converter in that system. Define the noise power of the quantizer as $N_{quantizer}$ and the thermal noise power of the system as $N_{thermal}$. When there exists no correlation between the noise sources, then the total noise in the system (N_{system}) will be:

$$N_{system} = \sqrt{N_{quantizer}^2 + N_{thermal}^2}. \tag{2.24}$$

In rewriting equation 2.24 to the quantizer noise, the result becomes:

$$N_{system} = N_{quantizer}\sqrt{1 + \frac{N_{thermal}^2}{N_{quantizer}^2}}. \tag{2.25}$$

The ratio $\frac{N_{thermal}^2}{N_{quantizer}^2}$ can be expressed as a ratio between the signal-to-noise ratios of the individual parts of the system. Then equation 2.25 changes into:

$$N_{system} = N_{quantizer}\sqrt{1 + \frac{(S/N_{quantizer})^2}{(S/N_{thermal})^2}}. \tag{2.26}$$

Suppose further that the ratio between the signal-to-noise ratios of the thermal part and the quantizer part is equal to k or:

$$k = \frac{S/N_{quantizer}}{S/N_{thermal}}. \tag{2.27}$$

Inserting equation 2.27 into equation 2.26 and normalizing this equation into the signal-to-noise ratio of the total system, we obtain:

$$S/N_{system} = S/N_{quantizer} \times \frac{1}{\sqrt{1 + k^2}}. \tag{2.28}$$

This calculation is only valid if there exists no correlation between the quantization noise (error) and the thermal (white) noise.

With $k^2 \ll 1$ this can be rewritten as:

$$\frac{1}{\sqrt{1 + k^2}} \simeq (1 - \frac{1}{2}k^2). \tag{2.29}$$

Thus:

$$S/N_{system} = S/N_{quantizer}(1 - \frac{1}{2}k^2). \tag{2.30}$$

Inserting for $k = \frac{1}{3}$, then:

$$S/N_{system} = 0.95 S/N_{quantizer}. \tag{2.31}$$

This means that in a 16-bit D/A converter system with a theoretical signal-to-noise ratio of 98.1 dB, the thermal signal-to-noise ratio must be at least 108 dB ($k = \frac{1}{3} \approx -9.5$ dB) to get an overall signal-to-noise ratio loss in a system of not more than 0.5 dB.

In flash-type analog-to-digital converters a large comparator bandwidth is required to quantize an input signal with low distortion. In Chapter 5 the ratio between maximum analog input frequency and comparator bandwidth will be determined. As a result of the large comparator bandwidth it is difficult to keep the (thermal) noise in the system low enough not to deteriorate the dynamic range of the system. In A/D converters using a successive approximation conversion method, an identical phenomenon concerning noise is found.

The noise generated in the input circuits causes a "dithering" of the comparator at the transition levels between the successive quantization levels. We also have to add the aliasing noise. In Figure 2.14 a Gaussian distribution curve of noise is assumed. Here σ is the rms value of the noise. The effect of noise on the signal-to-noise ratio of an A/D converter is estimated using two dc biasing conditions for the converter.

The first estimate on the influence of noise on the maximum signal-to-noise ratio of an A/D converter is performed when this converter is operated at an input dc voltage equal to decision level A_j. In Figure 2.15 this situation is shown. In Figure 2.15 A_{j-n} represents the decision levels of a converter while Q_{j-n} are the corresponding quantization codes. Note that at a code level Q_{j-n} an input signal variation equal to $\pm\frac{1}{2}$ LSB does not change the output code of the converter. The condition given in Figure 2.15 is such that a comparator is always tripped resulting in a 1 LSB peak-to-peak output signal generation.

The next change in output code occurs when the levels A_{j+1} and A_{j-1} are tripped. This results in a peak-to-peak output code of 3 LSB. The following tripping levels result in a 5 LSB peak-to-peak output signal generation.

From Figure 2.15 it is found that the 2σ noise level is equal to 1 LSB or $e_{noise} = \sigma = \frac{1}{2}$ LSB. Looking at the distribution curve we can say that during 95 percent of the time the peak-to-peak quantization error is equal to 1 LSB. During the next 4.6 percent of the time the peak-to-peak error increases to

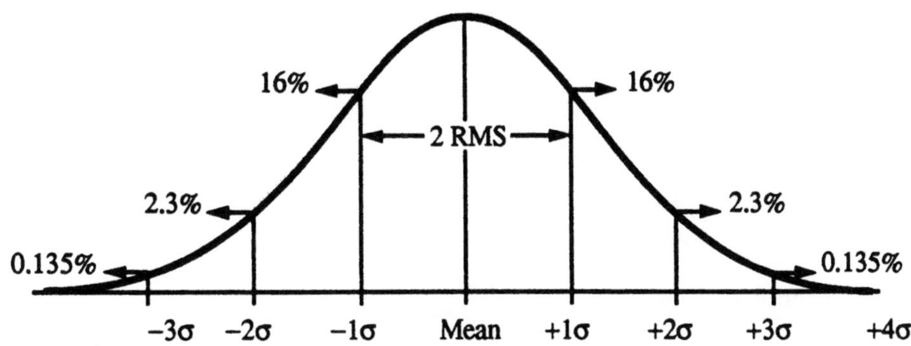

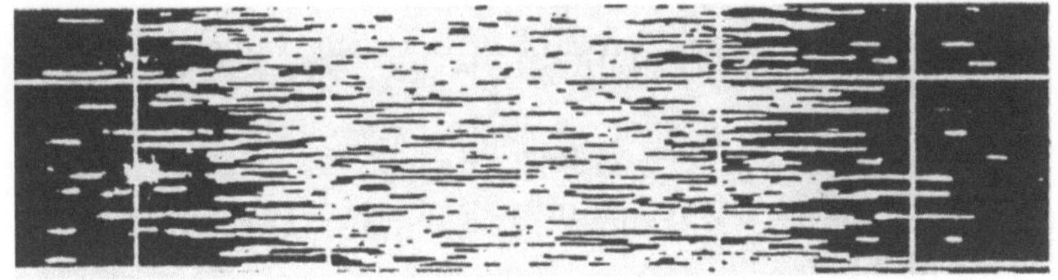

Figure 2.14 : Gaussian distribution curve of noise

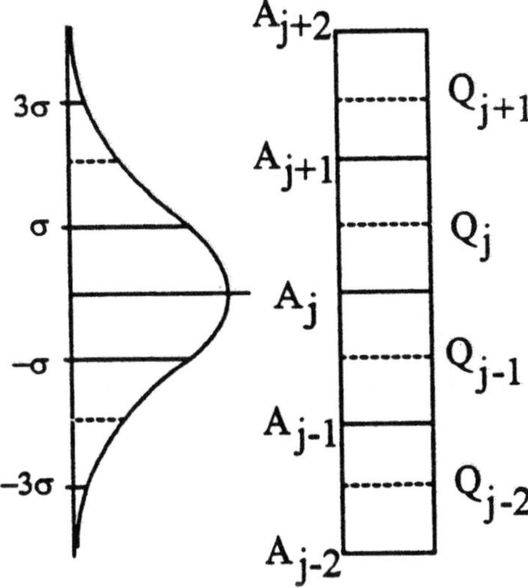

Figure 2.15 : A/D input noise dc biased at decision level A_j

3 LSBs. The next 0.006 percent of the time the peak-to-peak error increases to 5 LSBs. The input noise trips the different quantization levels with steps equal to 1 LSB.

The output noise of the converter can now be estimated with respect to the noise level of a quantization step which is equal to 1 LSB. Adding the quantization noise powers, we obtain:

$$P_{total} = .954 \times q^2_{qnsLSB} + .046 \times q^2_{qns3LSB} + .00006 \times q^2_{qns5LSB}. \qquad (2.32)$$

In this equation q^2_{qnsLSB} represents the quantization noise power of the LSB bit. The quantization error of the 3LSB level is 3 times larger (peak-to-peak value) than the quantization error of the LSB bit. Furthermore the quantization error of the 5LSB level is 5 times larger (peak-to-peak value) than the LSB bit. If the next quantization level is tripped by noise, then the peak-to-peak value increases with 2 LSB. Inserting values into formula 2.32 gives:

$$P_{total} = q^2_{qnsLSB} \times 1.37. \qquad (2.33)$$

This total power can be converted into a voltage across the same load resistance. This results in:

$$e_{total} = q_{qnsLSB} \times 1.17. \qquad (2.34)$$

When no input signal is applied to the analog-to-digital converter, then an output noise equal to 1.16 times the noise of an LSB bit is generated.

At the moment the dc biasing of the converter is changed with a value of $\frac{1}{2}$ LSB, a dead zone is found equal to $\pm \frac{1}{2}$ LSB.

In Figure 2.16 this situation is shown. The first term in formula 2.32 disappears. No output code appears because of the dead zone in the quantizer. An output code with a peak-to-peak value of 2 LSB appears at the output when the decision levels A_j and A_{j+1} are tripped.

As a result formula 2.32 now changes into:

$$P_{total} = .317 \times q^2_{qns2LSB} + .0027 \times q^2_{qns4LSB}. \qquad (2.35)$$

Inserting values for the two quantization errors as defined above results in:

$$P_{total} = q^2_{qnsLSB} \times 1.31; \qquad (2.36)$$

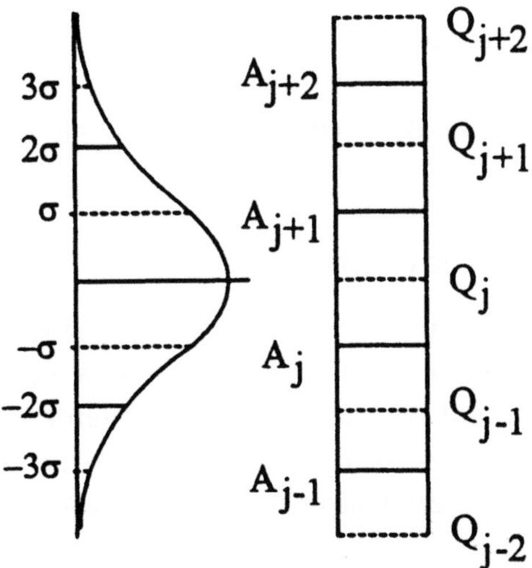

Figure 2.16 : A/D noise level dc biased $\frac{1}{2}$ LSB above quantization level

or, converted into a voltage, this value becomes:

$$e_{total} = q_{qnsLSB} \times 1.14. \tag{2.37}$$

So this analysis shows that the output noise nearly changes with changing input dc signal. At the moment the noise level of the input amplifier and comparator of an analog-to-digital converter is reduced with respect to the example given above, then the noise at the output of the converter in terms of LSB values shows larger variations when the input dc signal changes with respect to the quantization levels. This can be seen from the following example. At the moment the output noise of the input condition given by equation 2.34 tends to approach the quantization error q_{qnsLSB}, the noise given by equation 2.37 tends to approach zero because of the dead-band introduced at quantization level Q_j. In this manner a very simple test of an A/D converter can be performed by simply changing the input bias condition and analyzing the output signal. A small input referred noise of a converter is found when during the bias condition variations zero output codes follow 1 LSB output codes. The input noise level is well below $\frac{1}{2}$ LSB.

In this calculation it is supposed that the noise applied at the input of the A/D converter has a bandwidth that does not exceed half the sampling bandwidth. If the noise bandwidth is much larger than half the sampling frequency, the noise increases by a factor $\sqrt{2\frac{f_b}{f_s}}$. Here f_b is the bandwidth of the input amplifier-comparator stage. The results of formulas 2.34 and 2.36

must be multiplied by this factor.

Designers of high-resolution A/D converters must be aware of this noise phenomenon.

2.5.6　Minimum reference step size

In parallel type analog-to-digital converters a number of reference voltages for each comparator level are mostly generated by a resistor network. At the moment the input signal is smaller than the reference voltage the comparator gives a "0," while with a larger input voltage the comparator gives a "1." Although this sounds very simple, every comparator generates its own non-correlated thermal noise. These noise sources can result in inaccurate decisions, although compared to the reference voltage the input signal can be smaller for a certain comparator, because of noise a wrong decision may occur. The minimum step size needed in the reference network can be calculated referred to the noise generated in every comparator. For simplicity the $\frac{1}{f}$ noise component is neglected to simplify the calculations. In bipolar technologies this simplification is mostly allowed.

Suppose the noise shows a Gaussian distribution with probability density:

$$P(x) = \frac{1}{\sqrt{2\pi}\sigma} e^{-\frac{(x-m)^2}{2\sigma^2}}. \tag{2.38}$$

In equation 2.38 σ is the standard deviation and can be put equal to the rms value of the signal (noise). m is the mean value and is zero in the case of noise.

Integration of equation 2.38 gives the probability that the amplitude of a signal is within a predetermined range.
At the moment large values for σ are considered, an approximation can be used for the integral of the probability density. The one-sided (one tail) result of this approximation is given in 2.39.

$$Q(k) = \frac{1}{\sqrt{2\pi}k} e^{-\frac{k^2}{2}}. \tag{2.39}$$

This approximation is only valid for k≫ 1.

Furthermore, k gives the number of times a signal is larger than σ.
The possibility that the noise amplitude is larger than kσ is determined by

2Q(k), or

$$2Q(k) = \sqrt{\frac{2}{\pi}} \frac{1}{k} e^{-\frac{k^2}{2}}. \tag{2.40}$$

In Table 2.3 the probability as a function of k is given. From the Table it

k	2Q(k)
3	$2.9\ 10^{-3}$
4	$6.7\ 10^{-5}$
5	$5.9\ 10^{-7}$
6	$2.0\ 10^{-9}$
7	$2.6\ 10^{-12}$
8	$1.3\ 10^{-15}$

Table 2.3 : Amplitude probability as a function of k rms value

can be seen that the minimum reference voltage step size in a parallel type of converter must be at least between 6 to 7 times the rms noise voltage of a comparator. In that case the probability of a decision error due to noise tripping of a quantization level is below 10^{-9} with $V_{refstep} = 6\ e_{noise}$. Here e_{noise} is the input referred noise voltage of a comparator plus additional noise of the resistive reference divider.

2.5.7 Bit Error Rate (BER)

In analog-to-digital converters many decisions during a conversion process are performed. Especially in high-speed parallel-type converters this phenomenon is found [12]. At the moment a wrong decision has been taken, the internal code is converted into a wrong output code. Sometimes the internal code can be a meta stable condition of a comparator. A meta stable code of a comparator is an output code level that does not confirm the logical levels for "1" or "0." In such a case the conversion process from the internal code into the output code results in erroneous output codes. To obtain information about the error process the Bit Error Rate (BER) is defined. This figure defines the number of decision errors a converter makes. A high-quality analog-to-digital converter, for example, has BER numbers between 10^{-10} and 10^{-15}. In the design process of comparators this BER factor must be taken into account. In Chapter 5 an analysis of this phenomenon is presented.

2.5.8 Maximum sampling rate

The maximum sampling rate of a converter is a difficult to define number. In some cases the reduction in dynamic range (S/N ratio) can be used to define the maximum sampling rate. A definition can be:

The maximum sampling frequency of the converter for which the dynamic range measured over the Nyquist bandwidth is reduced with 3 dB or 0.5 bit.

This definition gives a rough indication about the maximum sampling frequency.

In systems which will use storage or digital signal processing it is better to relate the maximum sampling frequency with the bit error rate. In this case the definition of maximum sampling frequency can be related to the number of errors per second (EPS). We obtain:

$$EPS = BER.f_{sample} \tag{2.41}$$

In Chapter 5 is shown that the BER is related to the sampling frequency as:

$$BER = e^{-\frac{C_b}{f_{sample}}} \tag{2.42}$$

with C_b is a constant determined by the system. Equation 2.41 changes into:

$$EPS = f_{sample}.e^{-\frac{C_b}{f_{sample}}}. \tag{2.43}$$

Solving equation 2.43 for f_{sample}, the maximum sampling frequency can be found from:

$$f_{sample\ max} = \frac{EPS}{BER}. \tag{2.44}$$

Assume that for one error per minute a BER of 10^{-10} is obtained as a solution from equation 2.43, then the maximum sampling rate of a converter equals: $f_{sample\ max} = 160$ MHz.

Using this definition a more accurate number for the maximum sampling rate is obtained.

2.5.9 Digital signal feed-through

Digital signals in most of the converters are TTL- or CMOS compatible. The disadvantage of the TTL levels in, for example, high-resolution D/A converters is the feed-through of the TTL logic levels into the analog output (current) signal. This feed-through reduces the dynamic range of the converter and introduces harmonic distortion. These distortion phenomena can be explained by supposing that at the output of the D/A converter a sine wave is generated. In a system with offset binary coding the MSB bit changes with the sign of the sine wave. Feed-through in this case adds to the fundamental frequency, which can result in amplitude changes. The MSB -1 bit, however, changes twice as fast as the output sine wave. Feed-through in this case adds signals to the output, which results in harmonic distortion. To avoid this problem a serial coding of the input signal must be used. This serial coding reduces the number of input pins with TTL levels. Furthermore, the frequency spectrum of the input digital signal is far above the signal bandwidth of the converter, and therefore this spectrum will be removed by the reconstruction low-pass output filter.

2.5.10 Distortion

In a sampled system the signal band of interest is not only present as a baseband signal but is also reproduced around multiples of the sampling frequency f_s, $2f_s$, $3f_s$, and so on. If such a signal is applied to a linear amplifier, then, due to the finite linearity of such a system, inter modulation products can be generated. Apart from the harmonic distortion components introduced by the finite linearity of a converter, two dominating inter modulation products for an input signal with a frequency f_{in} can easily be found by mixing the lower band frequency $f_s - f_{in}$ with the upper band frequency $f_s + f_{in}$, resulting in a harmonic distortion $2f_{in}$. When the second harmonic $2(f_s - f_{in})$ of the lower band mixes with the upper band $f_s + f_{in}$, a non-harmonic product $| f_s - 3f_{in} |$ is obtained. As long as the frequency $f_s - 3f_{in} \geq f_b$ only the harmonic distortion $2f_{in}$ is found. Here f_b is the system bandwidth. The minimum sampling frequency at which only harmonic distortion products are found can be calculated. With $f_{in} = f_b$ we obtain $f_s = 4f_b$. In a digital audio system with $f_b = 20$ kHz the lowest sampling frequency to obtain only harmonic products must be at least $f_s = 80$ kHz. However, a lower sampling frequency ($f_s = 44$ kHz) is normally used, so harmonic and non-harmonic products must be encountered. In Figure 2.17 a graphic approach to the distortion problem in a D/A converter is shown. From Figure 2.17(b) it can be seen that signals close to half the sampling frequency have a large amplitude "mirrored" signal component ($f_s - f_{in}$). These signals require extreme linearity of the post-amplifier system to avoid

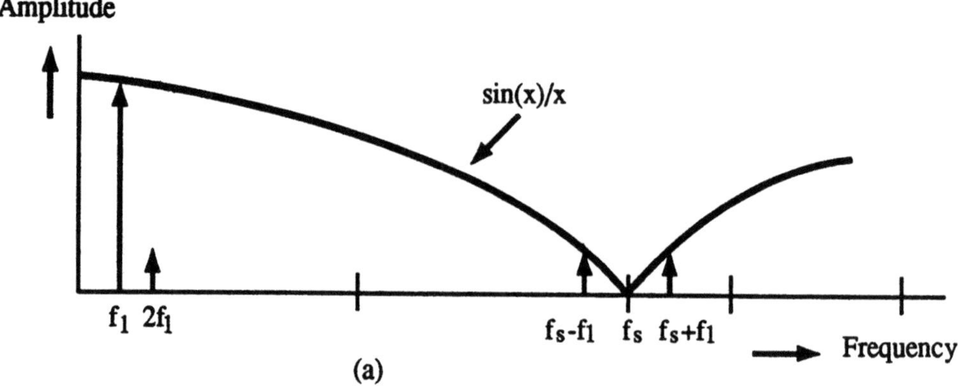

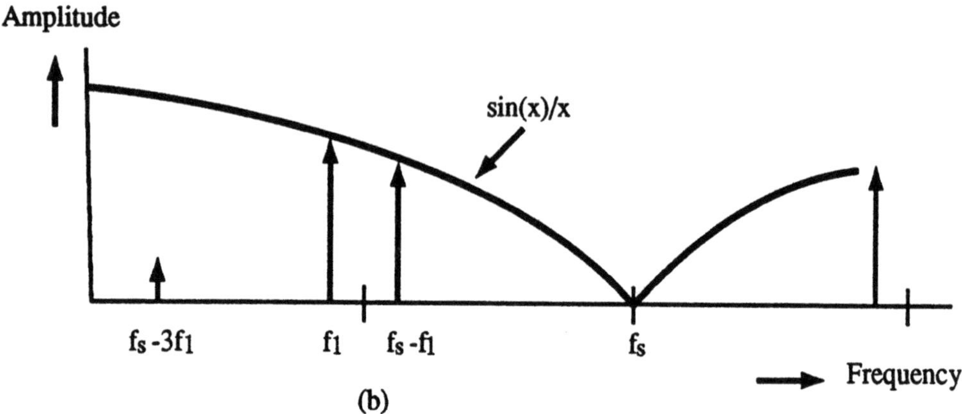

Figure 2.17 : D/A converter distortion model

mixing, resulting in non-harmonic distortion products appearing in the base-band. When, for example, the "hold" time of the converter is reduced, then the $\frac{\sin x}{x}$ amplitude distortion is reduced. At that moment the amplitude of the high-frequency signal components increases, the linearity requirements of the post amplifier system must be even more increased. Note that an identical result is obtained for a sample-and-hold amplifier.

2.5.11 Power supply rejection ratio

The power supply rejection ratio (PSRR) of a system as a function of frequency gives important data about the sensitivity of the system for noise induced on the power rails from, for example, the logic circuitry. Especially at high frequencies a good power supply rejection ratio is important to make the circuit immune from spikes introduced by fast logic circuits

(CMOS logic).

2.5.12   Settling time

The settling time of a system is defined as the time needed from the start of a transition until the time the output reaches the new value within the specified accuracy. The settling time specification of the full step of a digital-to-analog converter is important for applications of such a converter in a successive approximation analog-to-digital converter configuration.

2.5.13   Acquisition time

The acquisition time of a system is the time difference between the moment a command is given and the moment the system responds to this command. At the moment the system responds to the input signal, the error between the input signal and the output signal of the system must be within a specified number usually given in the data sheet. This time is important in, for example, sample-and-hold amplifiers. In Figure 2.18 the definition of the acquisition time is shown applied to a sample-and-hold amplifier. The acquisition time is defined in this system as the time difference from giving the "track" command until the output signal tracks the input signal. The moment of "tracking" is the time for which the output signal "track" the input signal with a well-specified accuracy. The acquisition time is also a measure of the maximum applicable sampling frequency of a sample-and-hold amplifier. After the acquisition time is elapsed, the system (sample-and-hold amplifier, for example) can be switched into the hold mode.

2.5.14   Aperture time

The aperture time of a sample-and-hold amplifier is specified as the time difference between the "hold" command and the moment the real sample is taken. In a sample-and-hold amplifier and flash-type A/D converters this specification is very important. Differences in aperture time, usually called *aperture time uncertainty*, determine one of the major errors in sampled systems (see, for example, Chapter 1, "Time Jitter").

In sample-and-hold amplifiers the aperture time determines the minimum time required to elapse before the "start conversion" command can be given. Usually an extra time called *hold mode settling time* is needed to be able to

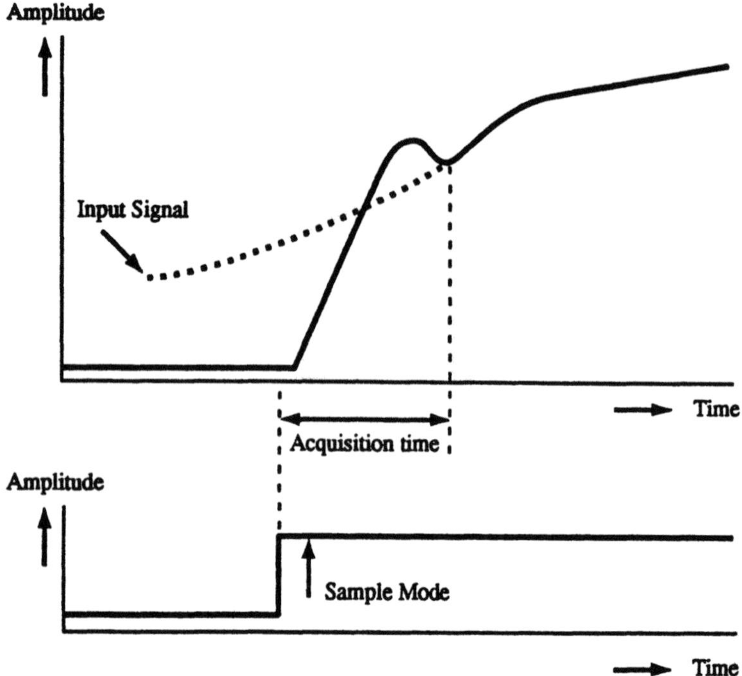

Figure 2.18 : Definition of the acquisition time of a S/H amplifier

specify the exact moment at which the hold output signal is within the specified accuracy. In a flash-type A/D converter the aperture time determines the difference between the sample command and the actual time the analog input signal is sampled and converted into a digital signal. In Figure 2.19 the definition of the aperture time is shown. A variation of the time difference between the "hold" command and the "start conversion" command in an A/D conversion system can result in fractional changes of the signal-to-noise ratio. Depending on the type of A/D converter, the start conversion command can be given before the final settling of the hold amplifier. As a result, a fractional change in maximum accuracy of the sample-and-hold amplifier is found. This accuracy variation gives rise to small changes in the maximum signal-to-noise ratio of a system. An optimum signal-to-noise ratio is mostly obtained when the time difference between the sample command and the start conversion command is more than the minimum time required according to the specification sheet. The hold mode settling time must be added.

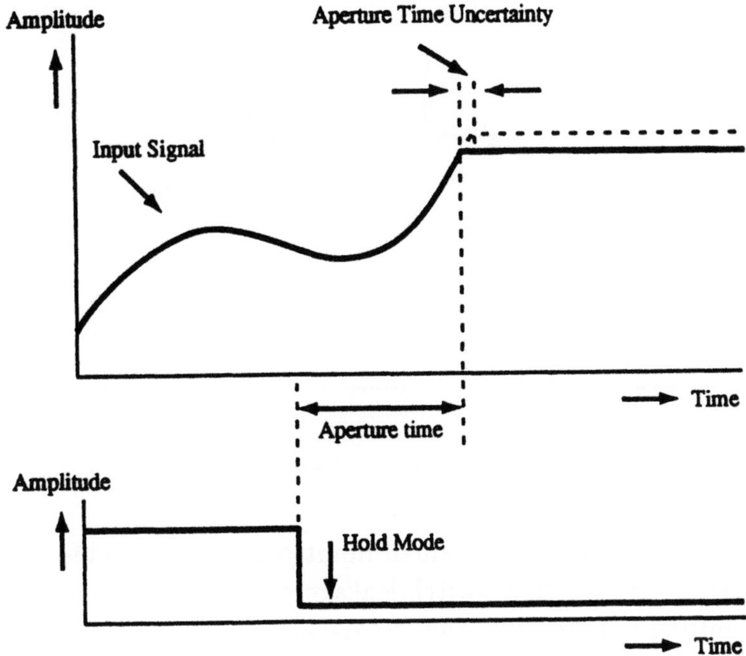

Figure 2.19 : Definition of aperture time

2.5.15 Sample-to-hold step

The sample-to-hold step is a change of the analog output signal in a sample-and-hold amplifier at the moment the circuit changes state from the sample to the hold mode. Due to the charge feed-through in the switch an extra amount of charge which is not a measure of the analog input signal is added to the hold signal. This *hold* step introduces an error. In Figure 2.20 the sample-to-hold step of a sample-and-hold amplifier is shown. When the *hold* step is independent of the signal level, no nonlinear distortion is introduced and the hold step can be seen as an extra dc offset. Sometimes a control signal can be applied to the sample-and-hold amplifier to zero the hold step. Basically, designs can be made which minimize the sample-to-hold step and make this step independent of the input signal level. (See chapter 8 for a detailed description of sample-and-hold amplifiers.)

2.5.16 Droop rate

At the moment a sample-and-hold amplifier is in the hold mode, the output signal is stored on the hold capacitor. The voltage across the hold capacitor

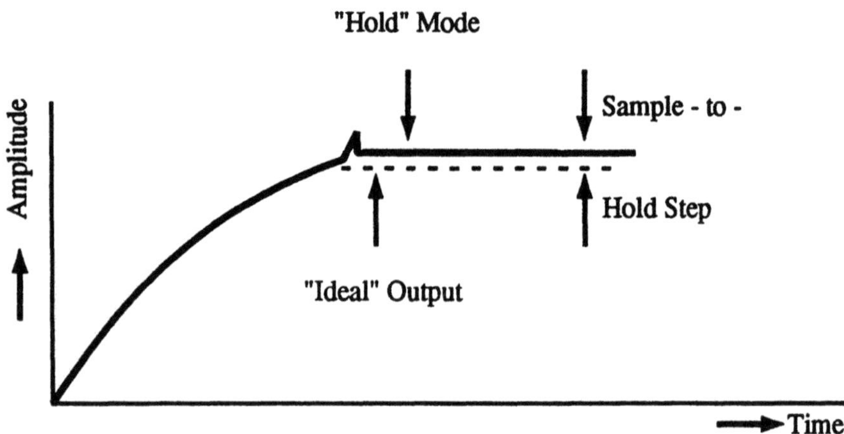

Figure 2.20 : Sample-to-hold step

is sensed with an amplifier with a small input bias current. This input bias current, together with possible switch leakage currents, discharges the hold capacitor, resulting in a droop of the voltage across this capacitor. The rate with which the capacitor is discharged must be small. Then the total droop during the conversion time of a converter can be kept small, usually under 1 LSB. In this way variations in the conversion time do not change the accuracy of the system. Mostly the droop rate is specified in mV/μsec. In some cases when the droop rate cannot be made small enough with respect to a 1 LSB value, then a fixed conversion time is used at which the final decision is taken. The droop can be measured and added to the output result of the analog-to-digital converter. In this way a reasonable accuracy of the total system is obtained. An estimation between the total leakage current $I_{leakage}$, the conversion time of an analog-to-digital converter $T_{conversion}$ and the droop rate of the system is found from:

$$V_{droop} = \frac{I_{leakage}}{C_{hold}} T_{conversion}. \tag{2.45}$$

When the droop rate is too high, a larger hold capacitor can be used to keep the droop rate within a 1 LSB specification.

2.5.17   Signal feed-through during hold mode

When the sample-and-hold amplifier is in the hold mode, then the input signal must be disconnected from the stored signal on the hold capacitor. In a practical system the sampling switch shows a finite impedance at the moment it is in the "off" mode. Especially at high signal frequencies the capacitive coupling over the sample switch results in a finite feed-through of the input signal onto the signal of the hold capacitor. The architecture of a sample-and-hold amplifier must be configured in such a way that a maximum attenuation between input signal and signal across the hold capacitor is obtained. The attenuation must be larger than the dynamic range of the analog-to-digital converter that is used for the conversion of the analog input signal into a digital output value. As an example, a feed-through attenuation between 70 to 80 dB is required for high-speed 8- to 10-bit analog-to-digital converter systems.

2.5.18   Noise in sample-and-hold amplifiers

In sample-and-hold amplifiers the noise introduced will be tracked as accurately as possible. As a result the output signal at the moment the system is switched from track into hold mode contains the momentary noise value present at the sampling moment. Usually it is not possible to introduce a noise filtering operation in such a system because at that moment the high frequency signal components are not accurately sampled. It is impossible to avoid the sampling of noise. Therefore, it is necessary to design circuits in such a way that the peak value of the noise generated over the applicable signal bandwidth of the system is below an LSB value of the succeeding analog-to-digital converter.

2.5.19   Overview of sample-and-hold specifications

To obtain a quick overview about the various specifications of a sample-and-hold amplifier these different parameters are shown in Figure 2.21. The top of the figure shows the input signal as it is applied to the sample-and-hold system. The middle part shows the control signal as a function of time as it is applied to this system. In the bottom part of figure 2.21 the output signal as a function of time is shown. Note that during large signal transients the analog output slews until it reaches the final value. Furthermore, during the hold mode the feed-through of the analog output signal is shown. The slow decrease of the output signal as a function of time shows the droop. The ringing shown in the figure when the system is switched from sample to hold mode increases the aperture time. In some designs the output voltage of the sample-and-hold can show an overshoot at the moment the slew operation in the amplifiers is finished and the signal level comes within the linear

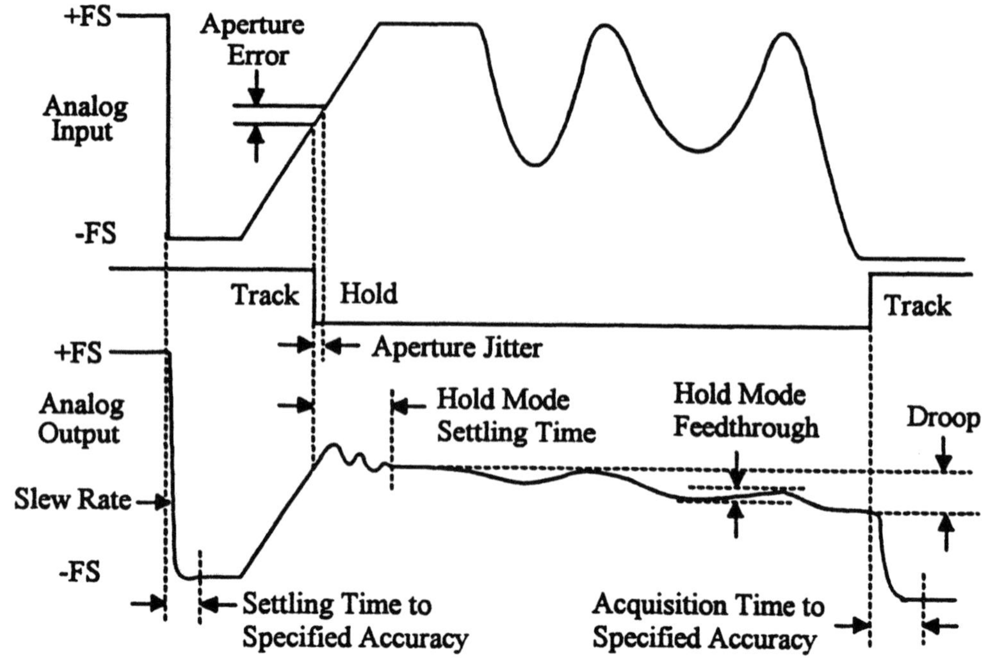

Figure 2.21 : Overview of sample-and-hold parameters

operation range. This overshoot can be caused by a limited overall system stability introduced by a maximization of the bandwidth. Note furthermore that the input settling is a different specification than the aperture time. This settling time is responsible for signal transients during track mode. In the same way hold mode settling corresponds to the time needed for the system to settle within a specified accuracy during the final "off" switching into hold mode.

2.5.20   Analog system bandwidth

In high-speed A/D converter systems the analog bandwidth with a full-scale input signal must be defined. In ideal converter systems the maximum analog bandwidth is equal to half the sampling bandwidth. In practice, however, there are various reasons why this theoretical value is not obtained. In Chapter 4 a number of reasons will be given and explained. Therefore, it is necessary to give a good specification for the analog signal bandwidth of a converter.

A very precise specification of this analog bandwidth is found by specifying the maximum analog frequency for which the signal-to-noise ratio of the system decreases by 3 dB or $\frac{1}{2}$ LSB with respect to the theoretical value of the system defined over a bandwidth equal to half the sampling frequency [53].

By definition the bandwidth obtained in this way is called Effective Resolution Bandwidth (ERB).
The number of bits obtained in this way are called Effective Number Of Bits

During the determination of the effective resolution bandwidth of a system a fixed sampling frequency is used. This sampling frequency is usually more than two times the analog effective bandwidth.

In Figure 2.22 the results of resolution bandwidth measurements are shown. The 3 dB decrease in signal-to-noise ratio of the system is used to define the (full-scale) analog signal bandwidth of the system. The effective resolution

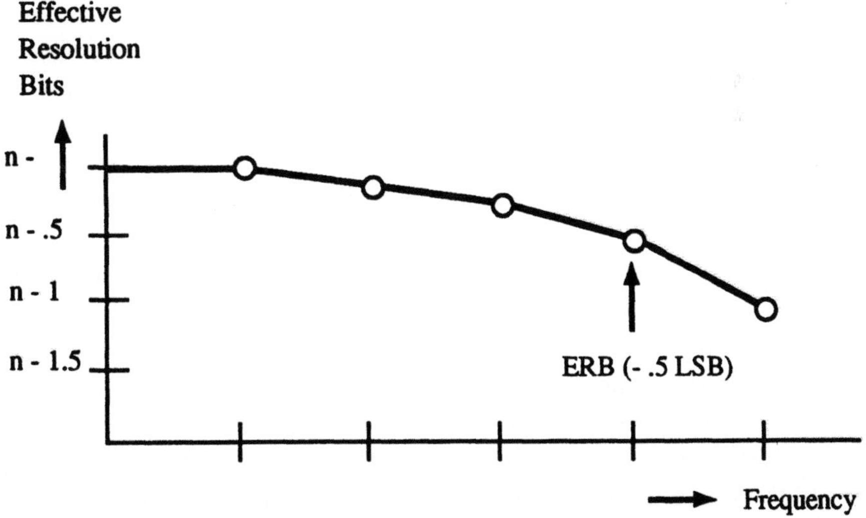

Figure 2.22 : Effective resolution bandwidth of a converter

bandwidth must be measured at full scale to include time jitter, noise, and linear and nonlinear distortion products.

2.5.21 Differential gain and differential phase

In the beginning phase of digital video systems, the analog specification of differential gain and differential phase using a stepped well-defined series of color sub-carrier bursts between zero and full scale of a converter has been used to specify the linearity of the converter at high frequencies. A typical specification is: 1% gain variation and 0.5 degree phase variation at color sub-carrier frequency.

2.6 Conclusion

In this chapter the basic requirements for A/D and D/A converters are defined. These definitions lead to basic circuit design constraints which must be fulfilled to obtain high-performance converters. In a binary-weighted converter the $\pm \frac{1}{2}$ LSB linearity specification is the only necessary specification to guarantee *monotonicity*.

In high-resolution binary-weighted converters the linearity requirement implies a very high accuracy of the most significant bit weights to obtain a monotonic converter. Signal-to-noise and signal-to-(noise plus distortion) specifications give a quick and reliable qualification of a converter. It includes noise, glitch error, sampling time uncertainty, and distortion. These dynamic specifications are very important in digital-signal-processing systems such as, for example, digital audio and digital video systems.

The glitch energy is an important measure for the application of a D/A converter in an oversampled and noise-shaped system. Due to the increase in sampling frequency the number of *zero crossings* in an offset binary-coded converter increases, especially when a noise-shaping operation is included in the filter section. As a result, the distortion increases, and the gain and dynamic range is not in accordance with the theoretically expected values. A low glitch design is needed in this case. In Chapter 10 the noise-shaping techniques will be explained.

The Bit-Error-Rate specification is an important specification for high-frequency analog-to-digital converters. It determines the accuracy and repeatability of high-speed comparators. Apart from the Effective Resolution and the Effective Resolution Bandwidth of a converter, the BER specification determines the errors made by the converter during a longer period of time. Applications in oscilloscopes are requiring a low BER to allow error free measurements over a long period of time (1 hour, 1 day).

Chapter 3

Testing of D/A and A/D converters

3.1 Introduction

To verify the different specifications of converters and converter systems it is important to set up test facilities and structures and to agree unanimously on the test procedures. In general, static performance tests can be performed by using digital voltmeters which can be a part of an Automatic Test Equipment set (ATE). Dynamic tests, especially in the case of high-dynamic-range tests, require special equipment. Furthermore, these dynamic tests are generally more difficult to standardize and consume more test time (see [36,37,16, 17]). Up until now dynamic test results have been listed very briefly in specification sheets. In this chapter we will determine test configurations and test procedures in order to arrive at a unanimous qualification of converters.

3.2 DC testing of D/A converters

DC specifications are obtained by applying a digital signal source to the input of the D/A converter. This digital source can be a personal computer (PC) or a specially developed device. In Figure 3.1 a test set-up for DC measurements is shown. At the output an IEEE bus programmable digital voltmeter (part of ATE) is connected. The measured data can be applied to a (personal) computer to process these data. How the measured data are processed depends on the configuration of the test system. Zero offset, full-scale accuracy, integral and differential nonlinearity can be measured.

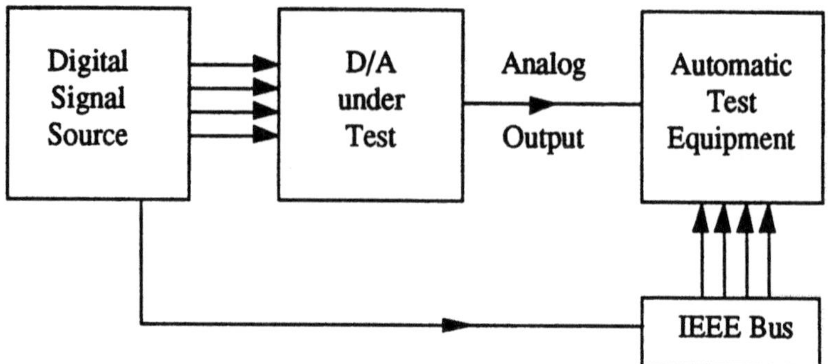

Figure 3.1 : DC measurement test set-up

It is clear that the accuracy of the digital voltmeter must be much higher than the accuracy of the converter under test. In a binary-coded converter the individual bit weights are measured. By using equations 2.6 and 2.10 shown in Chapter 3 it is possible to calculate the measured nonlinearity of the converter. The nonlinearity of a converter is defined as the deviations from a straight line drawn through zero and full scale. In Figure 3.2 these deviations from a straight line through zero and full scale for the bit weights in a binary-weighted converter are shown. Zero error (offset) is trimmed to zero to obtain a straight line between zero and full scale according to the linearity specification. The errors of the individual bit weights are shown, starting with the MSB error at the left side of the figure to the LSB error at the right side of the figure. The total positive and negative error is shown at the most right side of the figure as $\sum error$. This converter has a total nonlinearity error of $\pm \frac{1}{2}$ LSB. In a *monotonic by design* type of converter construction

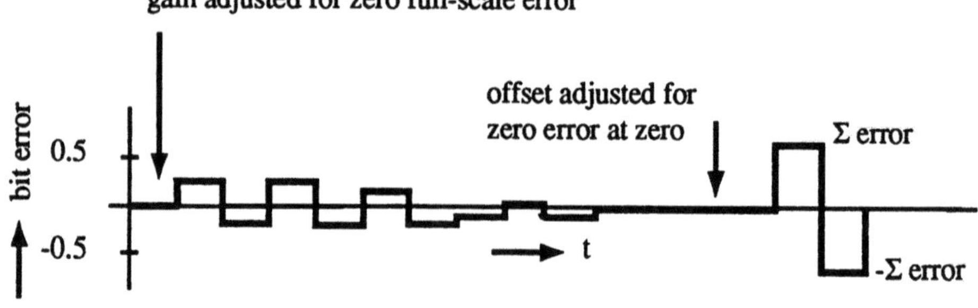

Figure 3.2 : Bit-weight error of a binary-weighted converter

every code must be measured and deviations from the straight line through zero and full scale determined. Testing these *monotonic by design* types is more difficult and therefore requires much more test time. Furthermore, the converter implementation determines how many data points are needed to verify the linearity specification. A resistor string having as many taps as the number of digital codes that can be applied shows a good example of a *monotonic by design* type of D/A converter. From this example it is easily understood that many converter codes must be tested to guarantee a linearity specification. Sometimes it is possible to reduce the amount of test samples to be taken for the linearity specification when the converter implementation is accurately known. Use can then be made of a certain repetitive character which is mostly found in converters. Mostly linearity is determined by the linearity of a number of the most significant bits. The remaining information interpolates between values of the most significant bits resulting in a repetitive character.

3.2.1 Temperature relations

To obtain information about the temperature dependence of DC specifications, the offset, full-scale accuracy, integral and differential nonlinearity tests must be performed at different temperatures. Temperature dependence can then be calculated and specified.

3.2.2 Supply voltage dependence

Information about supply voltage dependence is obtained by performing offset, full-scale, and linearity measurements at the minimum, nominal, and maximum supply voltage specifications of the converter. Mostly information about the supply voltage sensitivity as a function of frequency is added to these specifications. At frequencies around the clock frequency and multiples of this clock frequency the supply voltage rejection ratio must be high to suppress noise generated by the digital circuitry that surrounds the converter (internal and/or external logic circuits). In series with the supply voltage a small signal is added to perform the high-frequency power supply rejection ratio (PSRR) measurements. When these measurements are performed at different temperatures, a full specification is obtained.

3.2.3 Bit weight noise

By applying input codes that switch on the individual bit weights of a binary weighted D/A converter, the noise superimposed on these bit weights can be measured. The value of the total noise of all the bit weights is sometimes specified as the maximum noise of the system. However, it is important to know the bit weight noise values, but this does not define the dynamic performance (signal-to-noise) of the system. Mostly the noise of the bit weights is well below the total quantization noise value determined by the number of bits in the system. When in some cases bit weights show a high noise value, then these bit weights contribute too much to the total system noise. A reduction in the final signal-to-noise ratio of the system is found.

3.3　Dynamic testing of D/A converters

By using a digital programmable sine wave source at the input of a D/A converter, dynamic tests can be performed. A digital sine wave generator applies a practically ideal sine wave to the D/A converter. When an A/D converter with an analog sine wave generator is used at the input of the D/A converter, then the performance of the A/D converter system influences the accuracy of the measurements. In Figure 3.3 a test set-up is shown. A low-

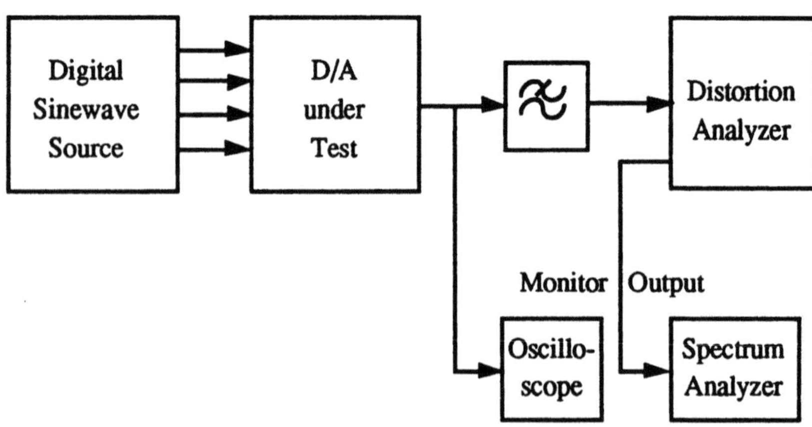

Figure 3.3 : Dynamic test set-up

pass filter applies the baseband to the analog distortion analyzer. Usually such a distortion analyzer is equipped with a monitor output, which contains all the signal information except the fundamental signal. This monitor output signal can be analyzed using a spectrum analyzer. Information about distortion and spectrum of the error signal is obtained in this way. The test set-up is extended with an oscilloscope to analyze the high-speed analog output of the converter. The glitch error can be measured by adjusting the input signal at a *major carry* transition in a binary-weighted converter. Only a 1-bit output signal is generated in this case (e.g. 1000... to 0111... code transition). With a distortion analyzer the signal-to-noise plus distortion ratio (SINAD) is measured over a bandwidth equal to the low-pass filter bandwidth. At the monitor output of the distortion analyzer a spectrum analyzer is connected to obtain information about the spectrum of the quantization error and the distortion of the output signal of the converter.

3.3.1 Dynamic integral nonlinearity test

By measuring the signal-to-noise ratio with a full-scale sine wave at the input and over the full input signal bandwidth, information about the dynamic linearity is obtained. In a very well-designed converter the signal-to-noise ratio must be close to the theoretical value given by:

$$S/N = n \times 6.02 + 1.76 \text{ dB}. \tag{3.1}$$

This measurement includes sampling time uncertainty, glitches, and integral nonlinearity. Additional information about the linearity of the converter is obtained when at a constant frequency the signal-to-noise measurement is performed as a function of the signal amplitude. Usually close to the maximum signal amplitude small deviations from the ideal (theoretical) value are found. When close to maximum signal level large deviations are found, then these deviations may be introduced by a large distortion. Therefore, it is important to analyze the quantization error signal as a function of frequency with a spectrum analyzer. If no large distortion components are found, then the clock stability must be determined. Using a spectrum analyzer, the noise level of the clock signal source can be measured. The best information about the noise level of an oscillator is found at $f_m \approx 0$ or $f_m \approx 2f_{osc}$. Determining the signal-to-noise ratio of the clock signal under these conditions must result in a high value per root Hertz. Comparing the results obtained with a crystal oscillator must give small deviations in case a high resolution or high-speed system is measured. Today high-speed oscilloscopes with special software exist to measure the pulse-to-pulse stability of a clock generator. The results of these measurements must comply with converter resolution and maximum analog input signal frequency.

3.3.2 Spurious free dynamic range

The spurious free dynamic range (SFDR) of a D/A converter is measured using a spectrum analyzer. The bandwidth of the analyzer is much smaller than the signal bandwidth of the converter system. When a full-scale signal is applied, the amplitude of the largest distortion component is measured. The ratio between the maximum signal and the largest distortion component determines the SFDR. Mostly the SFDR is signal frequency and amplitude dependent.

3.3.3   Differential nonlinearity

Dynamic differential nonlinearity is measured by applying a very low frequency signal of, for example, 0.01 Hz superimposed with a 2- to 3- LSBs-sized signal with a higher frequency, for example, 400 Hz. The amplitude of the 0.01 Hz signal is close to the full-scale value of the converter. In this way all the levels in the converter pass by. The variations of the 400 Hz output signal are a measure for the differential nonlinearity of the converter. This measurement can be performed in a short time. A maximum variation of 2 LSBs peak-to-peak is allowed for a binary-weighted converter with an integral nonlinearity of $\pm \frac{1}{2}$ LSB.

3.3.4   Glitches

The glitch energy can be measured using an oscilloscope. At the input of a binary weighted converter a code giving a major carry transition must be applied. For example, a code transition between 100... and 011... can be used to measure the MSB glitch. Usually the largest glitches occur at these major carry transitions. In Figure 3.4 the result of a major carry glitch measurement is shown. The output glitch of a current type D/A converter is measured across a 25 ohm load resistor. Full-scale value is 50 mV equivalent to 2 mA. By measuring the total energy and comparing this glitch energy with the LSB energy, it is possible to determine the influence of glitches on the total transfer function of a D/A converter. Especially in oversampled offset-binary-coded D/A converter systems, a low glitch energy is very important. In these applications a larger resolution than the basic converter is obtained by using oversampling. When the glitch error cannot be ignored with respect to the LSB size of the emulated higher resolution converter, this oversampling is mostly paid off with a larger distortion. In offset-binary-coded converter systems this problem is encountered when small output signals are reproduced around zero. The addition of a deglitcher circuit can improve the performance of the system. However, such a deglitcher circuit is a critical stage in the D/A conversion process, too.

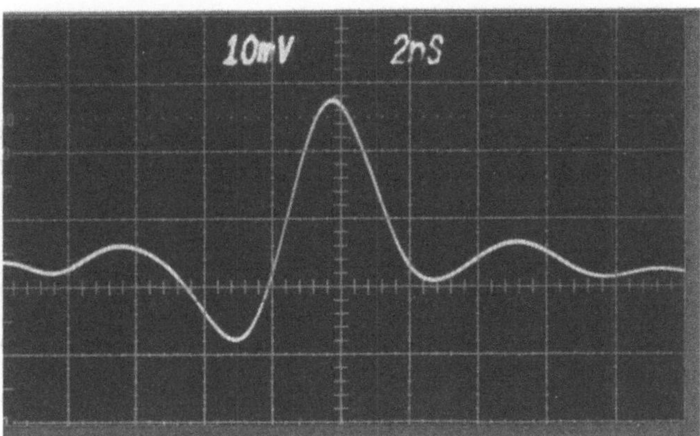

Figure 3.4 : Major carry glitch measurement result

3.3.5 Distortion measurement

The distortion of a converter system can be measured by applying an appropriate full-scale input test signal. In non-oversampled D/A converters the largest distortion mostly occurs at the high-frequency end of the frequency band. This distortion results from the finite slew rate of output amplifiers. In these amplifiers, the input signal mixes with the "mirrored" signal obtained through the sampling process, resulting in non harmonic distortion products. By analyzing the monitor signal of a distortion analyzer with a spectrum analyzer, it is easy to see if this phenomenon occurs. In oversampled converters distortion must be measured at the low-frequency band edge. Nonlinear distortion occurs here because of the final accuracy of the converter. This final accuracy results in harmonics of the signal frequency as long as these harmonics are within the passband of the system. The oversampling operation transforms the frequency spectra to a much higher sampling frequency. Therefore, it is less probable that non-harmonic mixing products appear in the baseband of the system. Harmonics of the input signal are rejected by the reconstruction low-pass filter for signals above 10 kHz in an audio converter. A better performance is obtained for frequencies between 10 kHz and 20 kHz in such an oversampled audio converter.

3.3.6 Settling time measurement

A dynamic measurement of the settling time of a converter is possible by measuring the signal-to-noise ratio over a fixed bandwidth and with a fixed measuring frequency. By varying the sampling frequency, the signal-to-noise ratio should increase linearly with the square root out of the sampling frequency. At the point when there is no increase in signal-to-noise ratio, the settling time of the converter has been reached. To obtain an accurate measurement of this condition, a small glitch error is required. Adding a deglitcher, however, can overcome this glitch problem.

Direct measurement of settling time is also possible by applying a special test circuit. In Figure 3.5 this test set-up is shown. The output signal of the

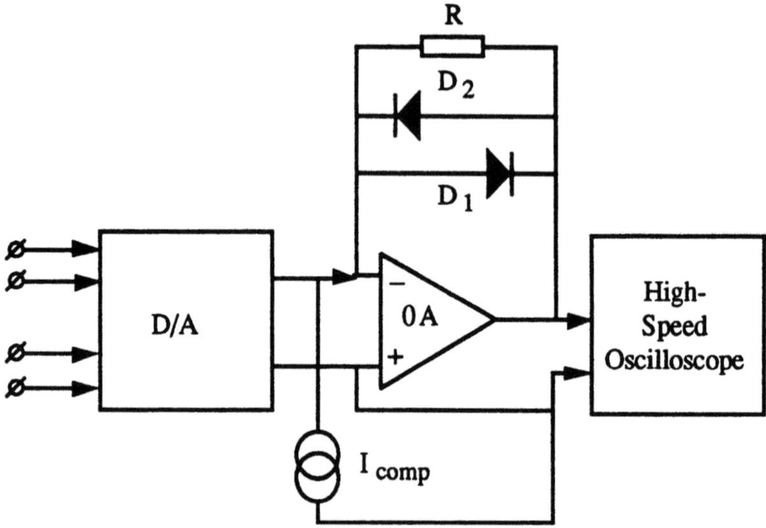

Figure 3.5 : Direct settling time measurement set-up

D/A converter is applied to a high-speed Schottky diode (D_1, D_2) clamped operational amplifier (OA), which performs a current-to-voltage conversion in this current D/A converter example. Furthermore, a current I_{comp} is subtracted from the output current of the D/A converter. This compensation current determines the current value at which the direct settling time measurement must be performed. Schottky diode clamping is used to prevent overloading of the oscilloscope. The measurement is performed by switching the digital D/A converter input value from zero or full scale to the desired current setting. With the oscilloscope the boundary within which the output signal must settle is determined. In this way a direct measurement is obtained. However, the settling time of the operational amplifier and the Schottky-clamped circuit must be much smaller than the values to be measured. Therefore, a careful layout of the circuit board and a good choice of devices are needed.

3.4 DC testing of A/D converters

In Figure 3.6 a DC test set-up is shown. If DC tests are performed, the sample-and-hold function does not need to precede the converter. An accuracy-limiting circuit can be avoided in this way. The set-up consists of a pro-

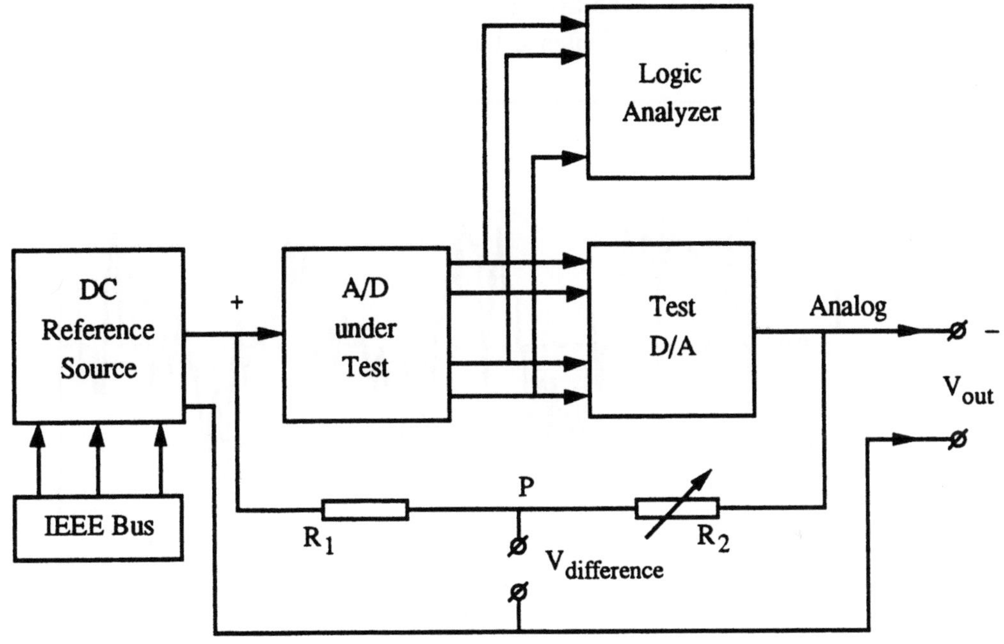

Figure 3.6 : DC test set-up for A/D converters

grammable high-accuracy DC voltage source, a logic analyzer, and a controller. The output of the A/D converter can be analyzed using a logic analyzer. If a D/A converter with a much higher accuracy than the A/D converter is connected at the output of the A/D, it is advantageous to connect the output data of the A/D to the D/A converter in such a way that an opposite polarity of the output signal compared to the input signal is obtained. By connecting two resistors R_1 and R_2 in a bridge circuit between input and output of the system, the linearity can be measured by monitoring the signal marked $V_{difference}$ at test point P between the resistors R_1 and R_2 as a function of the input voltage. Scale errors can be calibrated by changing the value of one of the two resistors in such a way that only differences appear at the node. By programming the analog input source, offset and full-scale errors can be monitored with the logic analyzer. Differential and integral nonlinearity measurements are performed by slowly varying the analog input signal from zero to full scale. By connecting an X,Y plotter at test point P, $V_{difference}$ as a function of the input signal can be plotted. A hard copy can be made to determine differential and integral nonlinearity of the converter. The difference between every quantization level shows the differential nonlinearity. The deviation of $V_{difference}$ from zero gives the integral nonlinearity. All steps are verified with this measurement. In Figure 3.7 a measurement result of such a test is shown. These measurements can

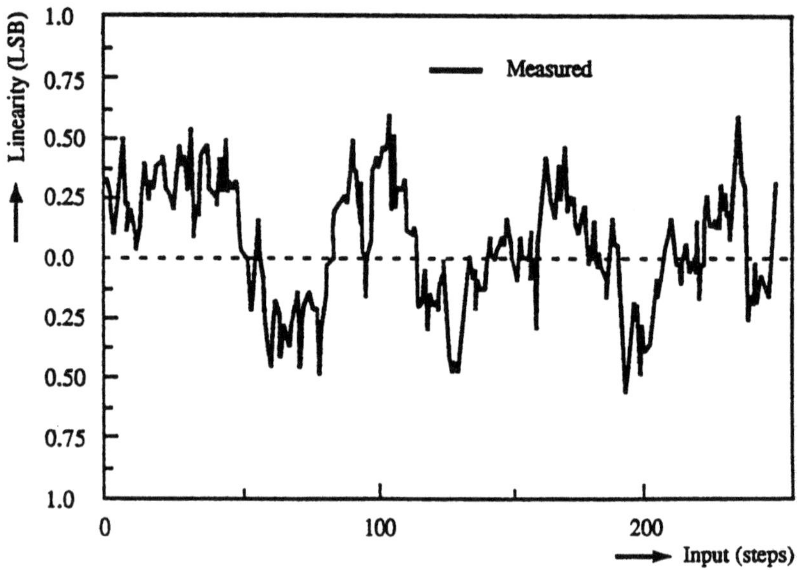

Figure 3.7 : Integral Non-Linearity measurement result

be performed over a large temperature range and with varying supply voltages. In this way the temperature variations and supply voltage dependence of offset, full-scale value and linearity can be determined.

3.5   Dynamic testing of A/D converters

In Figure 3.8 a test set-up for dynamic performance tests of A/D converters is shown. When dynamic tests are performed, a sample-and-hold amplifier must in most cases be added to the input circuitry of the A/D converter to sample the analog input signal. This sample-and-hold amplifier must have a much better performance than the A/D converter to be tested. The sample-and-hold amplifier samples the input signal and keeps this signal constant during the time the A/D conversion takes place. At the input of the test

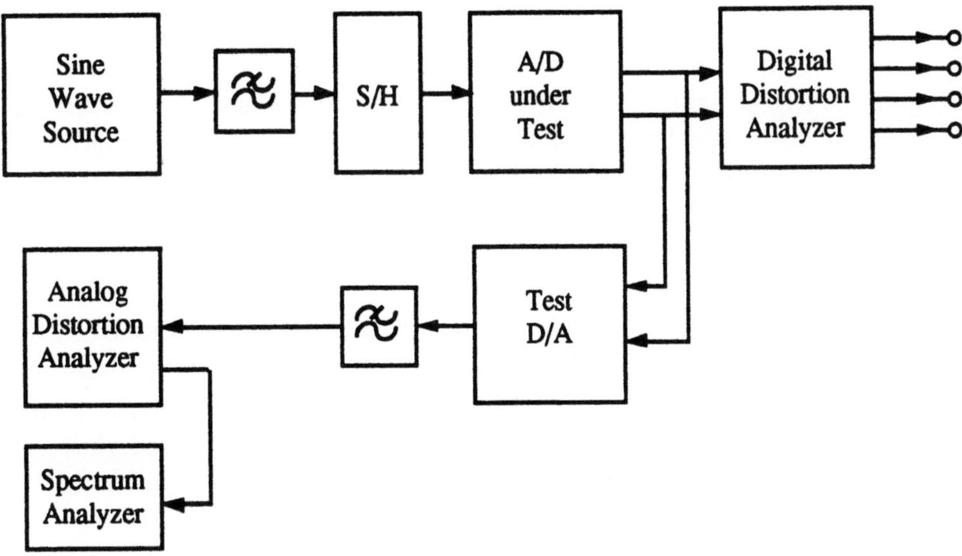

Figure 3.8 : Dynamic test set-up for A/D converters

set-up a low-distortion sine wave source is applied. Moreover, a low-pass input filter is used to filter out possible high-frequency interference and noise signals that may be present on the output signal of the sine wave generator. Digital data can be analyzed using a specially built digital distortion analyzer. If such an instrument is not available, then a high-accuracy D/A converter can be used to convert the digital data into analog signals again. This analog signal is filtered by the reconstruction filter and then analyzed using a distortion analyzer. In most cases distortion analyzers have a monitor output showing the error signal after the fundamental has been removed. This monitor signal can be analyzed with a spectrum analyzer to obtain information about harmonics and glitches. On the other hand, the converted digital signals can be analyzed by using a fast Fourier transform (FFT). In such a case enough samples must be taken to obtain accurate information about the signal to be analyzed. Furthermore, the limited number of samples that can be analyzed calls for a completely random choice of input signal frequency with respect to the sampling frequency. To minimize the problems with "leakage" in the fast Fourier transform, a window function must be applied. It is important to let the ratio between sampling frequency and input frequency be an irrational number. In that case the quantization errors are randomized so a correct measurement and interpretation of the results is possible. Signal-to-noise measurements as a function of frequency and amplitude must be performed to characterize the A/D converter.

The spurious free dynamic range of the converter is obtained by determining at maximum input signal the amplitude of the largest distortion component. The SFDR is the ratio between the signal amplitude and the largest distortion component. In case more components with nearly equal amplitude are present, the SFDR does not change because of the increase in distortion power. By measuring at a high sample frequency the signal-to-noise ratio of a system, the *effective resolution bandwidth* can be determined. This bandwidth is the maximum analog input frequency for which the resolution (signal-to-noise ratio) of the system decreases with $\frac{1}{2}$ LSB compared to the low-frequency result. During this measurement it must be noted that the low-frequency signal-to-noise ratio does not deviate more than a fraction of an LSB from the theoretical value. At the same time the Effective Number of Bits (ENOB) as a function of input frequency is determined.

3.5.1   Conversion speed

The conversion speed of an A/D converter can be measured by varying the conversion time of the converter and keeping the sampling frequency and input signal conditions constant. Particularly at high input signal frequencies, the measurement is sensitive to conversion speed variations because of the small amount of samples per period of input signal. Additional quantization errors introduced by the A/D converter reduce the signal-to-noise ratio, while mixing of the input signal with the sampling frequency can also result in non-harmonic distortion. In Figure 3.9 the result of such a conversion time measurement is shown. When a long conversion time is used, the optimum signal-to-noise ratio of such a system is found.

A definition of the minimum conversion time of an A/D converter can be specified as: the minimum conversion time for which the signal-to-noise ratio is reduced by 1 dB compared to the optimum long conversion time value.

A 1 dB signal-to-noise reduction is not seen as a disturbing loss. This specification gives a clear definition of the minimum conversion time of a converter.

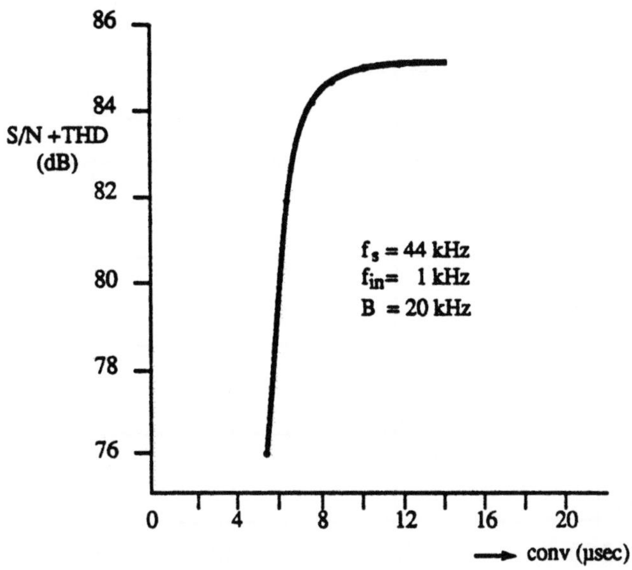

Figure 3.9 : Conversion time measurement result

3.6 Bit Error Rate

The bit error rate (BER) of an analog-to-digital converter can be tested
by applying an appropriate DC input offset voltage together with an ac
input signal of arbitrary frequency. To clarify this measurement the BER
of the MSB with the MSB-1 will be explained. The input signal of the
converter is shifted to the MSB transition level. The peak amplitude of the
ac signal must be smaller than a quarter of the full scale of the converter.
At the output of the converter the code transition between the MSB and
MSB-1 are always opposite:, for example, 01... and 10..., because of the
adjustment of the input DC bias signal. With an Exclusive OR function
the MSB and MSB-1 are compared. At the moment an equal code, for
example, 00... or 11..., is found a signal appears at the output of the EXOR.
This output signal is counted using a counter, and the number of times
such a code is found is a measure for the BER. These measurements are
performed with increasing sampling frequency. Normally the errors per time
unit increase with increasing sampling frequency. The frequency of the input
signal is of minor importance, because the measurements concern the number
of decisions that have to be taken during the analog-to-digital conversion by
the MSB and MSB-1 comparators. In Figure 3.10 the result of a BER

measurement as a function of sample frequency is shown. It is clear that,

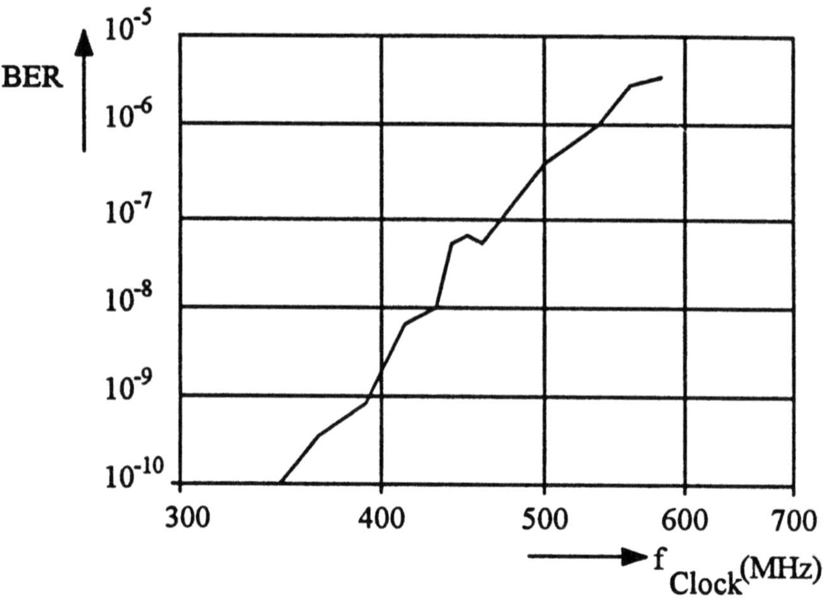

Figure 3.10 : BER measurement result

depending on different DC levels, the same operation can be performed to obtain information about the MSB-1 and MSB-2 error rate. In this case a DC level of one-fourth or three-fourths of full scale must be applied to obtain the transition between 010.. and 001.. with an appropriate smaller ac input signal. Again the EXOR determines coincidences of bits 0100. or 0011. which may not occur during this measurement. In high-quality A/D converters, BER values between 10^{-10} and 10^{-15} are required. Applications of these converters are in digital oscilloscopes.

3.7 Testing very high-speed A/D converters

Testing very high-speed A/D converters is a different subject. A digital distortion analyzer is difficult to build because of the high sampling rates of these converters. As a result, an A/D-D/A converter loop is the preferred test environment. The test configuration and the test boards that contain the different system elements must be optimized for the best dynamic performance. The influence of small changes in the test board (e.g., wiring) can immediately be analyzed with the spectrum analyzer and the oscilloscope. However, D/A converters that show the required linearity and low glitch error at these speeds are difficult to obtain. Both A/D and D/A converters usually operate at the limit of a technology used to build the devices. Sub-sampling of output signals of the A/D converter is therefore applied. At the lower sub-sampling rates D/A converters with a much better linearity and glitch error specification are available. In this manner the question about errors introduced by the D/A converter can be postponed. In Figure 3.11 a test set-up using a sub-sampling technique is shown. From Figure 3.11 it can

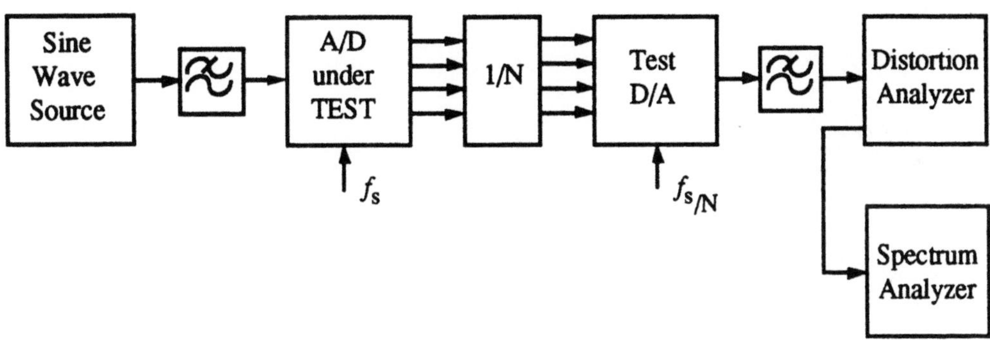

Figure 3.11 : Very high-speed A/D converter test set-up using sub-sampling

be seen that the output samples of the A/D converter are reduced by a factor N before they are applied to the D/A converter. The maximum amount of reduction in samples N depends on the speed of the D/A converter used in the test set-up. Moreover, during measurements the sub-sampling rate N can be optimized to get the best performance of the D/A converter over a large sampling frequency range.

Sub-sampling of signals is allowed but the signals, quantization errors, and distortion are folded back into the baseband. It is already known that a reduction in sampling rate by a factor N increases the quantization noise folded back into the baseband by a factor equal to $\sqrt{N}$. At the same time the measurement bandwidth is reduced by a factor N. As a result of this operation the signal-to-noise ratio of the system does not change. In Figure 3.12 the sub-sampling operation is shown, along with a four times sub-sampling factor. Sub-sampling with a factor two results in a "shifting" of frequencies around the sub-sampling frequency $2f_s$ into the baseband from zero to $\frac{1}{2} f_s$. An identical operation takes place on the subsamples signal when again a factor two reduction in sampling frequency is performed. Frequencies around f_s are "shifted" into the baseband again. As a result of this operation the quan-

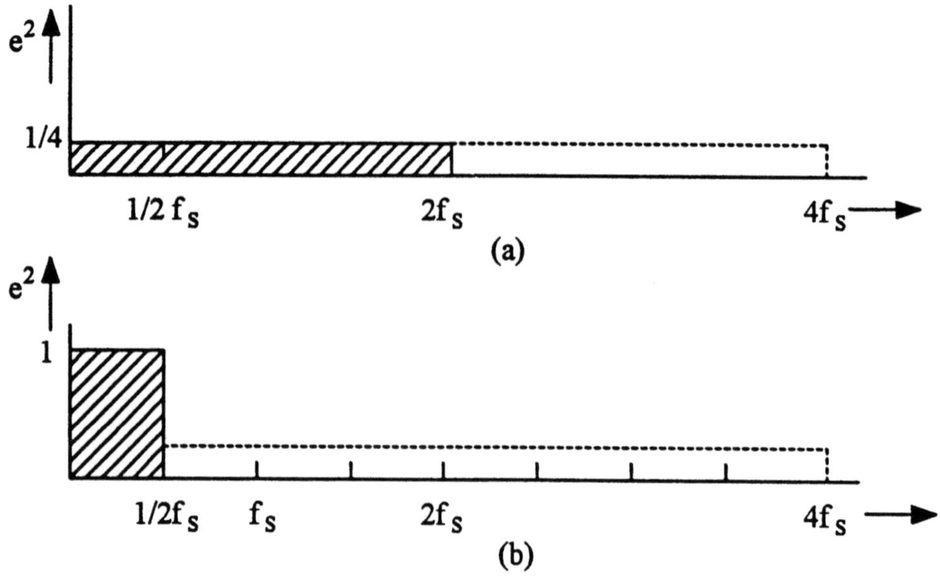

Figure 3.12 : Sub-sampling in converter systems

tization noise density is increased by a factor four as long as no filtering operation is performed before the sub-sampling takes place. Higher sub-sampling rates than two can be successfully applied to an A/D-D/A system to obtain accurate information about distortion and signal-to-noise ratio. The D/A converter in such a system can be repeatedly used over the same sampling frequency by changing the sub-sampling factor N depending on the sampling frequency range of the A/D converter. The *Effective Resolution Bandwidth* (ERB) and the *Effective Number Of Bit* (ENOB) of the system can be determined with this test method.

From the measurements on, for example, high-speed A/D converters that do not use a sample-and-hold amplifier at the input, a dominating third-order signal distortion is found. This third-order distortion is a measure for the maximum bandwidth of the input circuitry of the converter under test. In Chapter 5 a theoretical analysis of this phenomenon will be given. The increase in third-order distortion determines the ERB when signal-to-noise plus distortion (SINAD) is measured as a function of frequency. In Figure 3.13 the result of a distortion measurement as a function of input frequency is shown. This result is measured on an A/D converter that does not need a sample-and-hold amplifier.

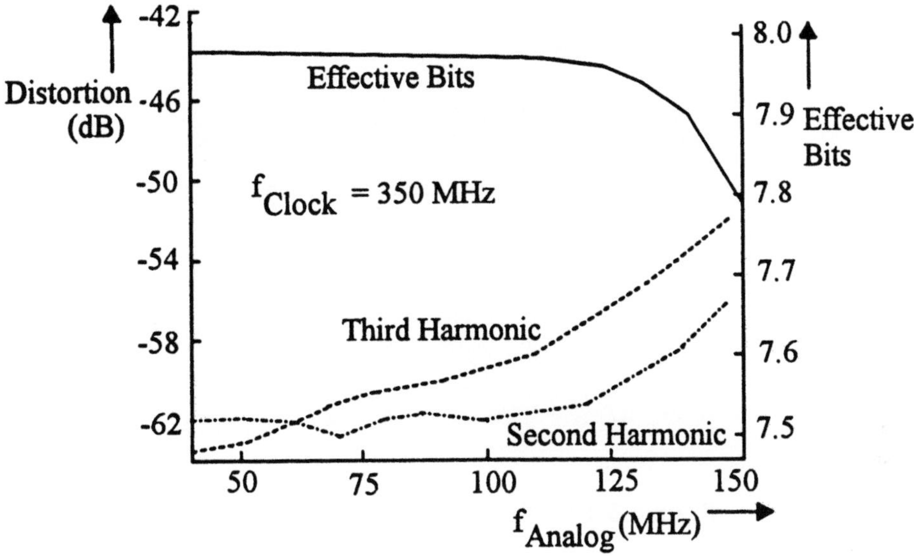

Figure 3.13 : High-speed A/D converter distortion measurement result

With a high-speed logic analyzer it is possible to measure a large amount of samples. These samples can be stored in the internal memory of the analyzer and then transferred to a computer system to calculate the spectral response. Care must be taken that no correlation exists between the input signal frequency and the sampling frequency in the converter. In that way the quantization errors are randomized and a correct result is obtained. Care must be taken to avoid "leakage" in the fast Fourier transform by applying an appropriate window function.

3.8 Beat frequency test configuration

A quick method to obtain information about the high-frequency performance of a converter can be obtained when the input frequency (e.g., 100.00 MHz) and the sample frequency (e.g., 100.10 MHz) differ only by a small amount. A low-frequency (100 kHz) beat frequency is obtained. In Figure 3.14 an example of a beat frequency signal is shown. This beat frequency signal can be analyzed visually to get information about missing codes or possible repeating signal parts. Timing errors that may occur in the converter appear as distortion of the beat frequency. Furthermore, step size and quantization levels can be determined accurately. The distortion of the beat frequency is a measure for the linearity and timing errors of the converter. Missing code

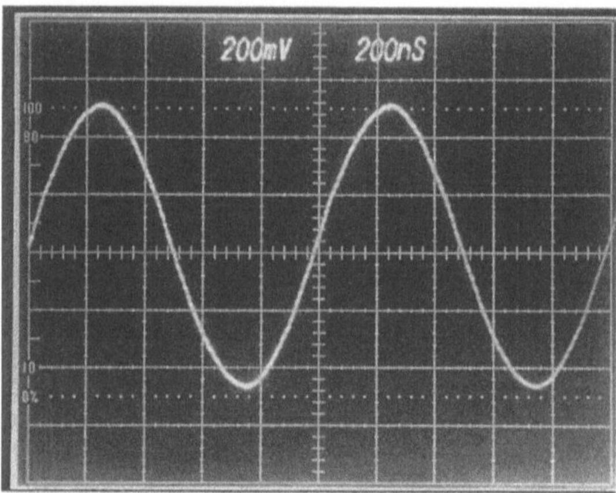

Figure 3.14 : Beat frequency output signal

performance of the system is also shown at these frequencies. Note that in this test condition the sample frequency cannot be much higher than the maximum analog input frequency. This frequency is about half the minimum sampling frequency of the converter. Digital analysis of the beat frequency is preferred because of the high accuracy possible with this measurement method. A D/A converter with good specifications over the total analog input signal bandwidth must be used. Glitches must be small compared to the LSB value to avoid measurement errors.

3.9　Code density DNL and INL measurement

The dynamic differential (DNL) and integral nonlinearity (INL) can be tested using a code density measurement. At the input of an analog-to-digital converter a triangular waveform is applied. The amplitude of this triangular waveform is slightly larger than the full-scale value of the converter. With a logic analyzer the output code is stored. These codes are analyzed for the number of times a code value appears at the output of the converter. To obtain an accurate measurement at least an average of codes per level between 8 and 16 are needed. This results in a storage capacity for the logic analyzer of about 4000 codes, when an 8-bit analog-to-digital converter is tested. With a triangular input signal, every code in the converter should have an equal density. When a code is missing, then no output occurrence of this code is found during this measurement. In case the differential nonlinearity shows a large step, the number of code occurrences at that specific point in the converter shows an increase. In Figure 3.15 the result of such a code density measurement is shown. No accuracy problems

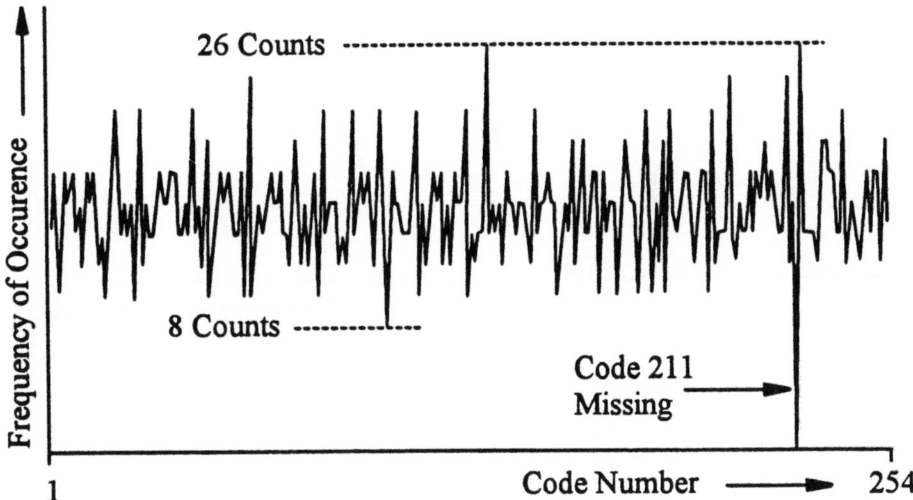

Figure 3.15 : Code density (DNL) measurement result with a triangular input signal

exist in the reproduction of the output signal. The result of the code density measurement gives a direct information about the differential nonlinearity (DNL) of the converter. Integrating the measurement result shows the integral nonlinearity (INL). After integration of the measurement result of Figure 3.15 the result shown in Figure 3.16 is obtained. At the moment a sine wave is used as an input source, the density result changes. Because the sine wave shows a flat top and bottom part, the amount of code transitions during these times increases, resulting in a signal level dependent result. In Figure 3.17 the result of a sine wave density (DNL) measurement is shown. The curved result can be declared using the following simplified analysis. Suppose the input sine wave is equal to:

$$V_{in} = V_{max} \sin \omega t, \tag{3.2}$$

then the time difference (Δt) between two quantization levels is a measure for the number of times a code is generated.

Using V_{fs} to define the full scale value of the converter, in an n-bit converter

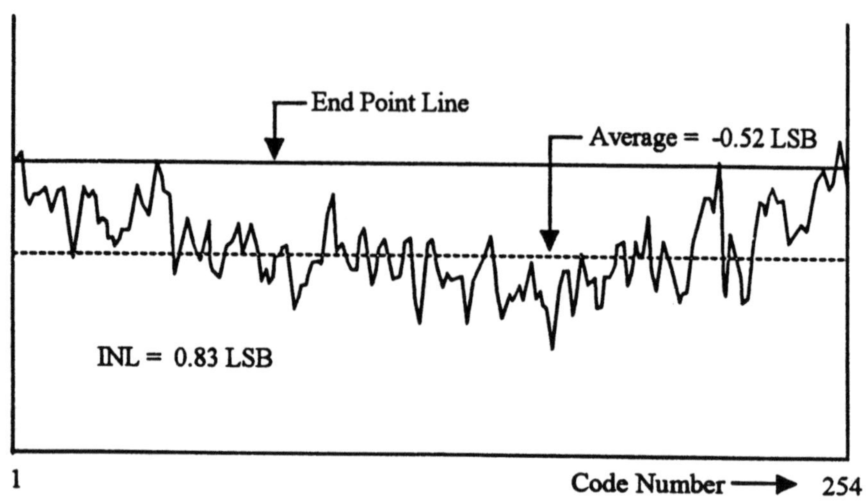

Figure 3.16 : Integral nonlinearity density measurement result

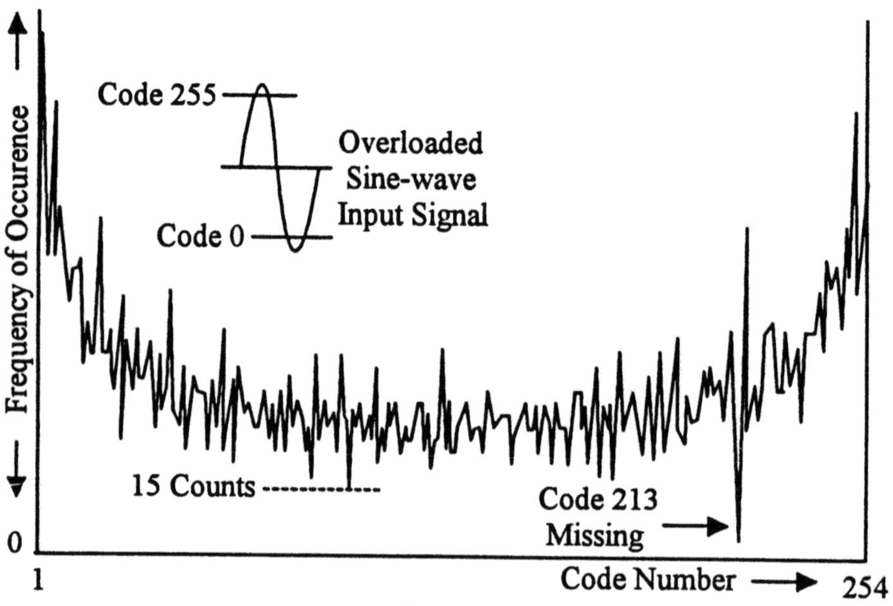

Figure 3.17 : Code density (DNL) measurement result with a sine wave input

the step size (ST) equals:

$$ST = \frac{V_{fs}}{2^n - 1}. \tag{3.3}$$

The time difference Δt can be determined from:

$$ST = V_{max} \sin \omega(t + \Delta t) - V_{max} \sin \omega t = \frac{V_{fs}}{2^n - 1}. \tag{3.4}$$

When a rather large number of bits is used in the converter, then $\Delta t \ll 1$ and equation 3.4 can be approximated by:

$$\omega \Delta t \approx \frac{V_{fs}}{V_{max}} \frac{1}{(2^n - 1)} \frac{1}{\cos \omega t}. \tag{3.5}$$

The equation will be validated over the range:

$$-\frac{V_{fs}}{V_{max}} \frac{\pi}{2} \leq \omega t \leq \frac{V_{fs}}{V_{max}} \frac{\pi}{2}. \tag{3.6}$$

With $\Delta \phi \approx \omega \Delta t$ and $\phi = \frac{V_{fs}}{V_{max}} \pi$ equation 3.5 can be rewritten into:

$$\frac{\Delta \phi}{\phi} = \frac{1}{\pi (2^n - 1)} \frac{1}{\cos \omega t}. \tag{3.7}$$

Suppose that over the total range of ϕ determined by 3.6 a total of Nπ measurement pulses are used, then per quantization level the number of pulses N_q which can be counted is:

$$N_q = \frac{N}{2^n - 1} \frac{1}{\cos \omega t} \tag{3.8}$$

with ωt stepping over the range defined by 3.6 with discrete steps equal to:

$$\frac{V_{fs}}{V_{max}} \frac{\pi}{2^n - 1}. \tag{3.9}$$

From equation 3.8 the curved relation found during the measurement is explained.

Normally the amplitude of the sine wave is made larger than the full-scale value of the analog-to-digital converter. This small over-drive of the input full scale results in a smaller curving of the output data during the code density test. The differential nonlinearity measurement data depend on the time difference a sine wave needs to trip two adjacent levels in the analog-to-digital converter. Integration of the differential nonlinearity measurement result gives the specification for the integral nonlinearity.

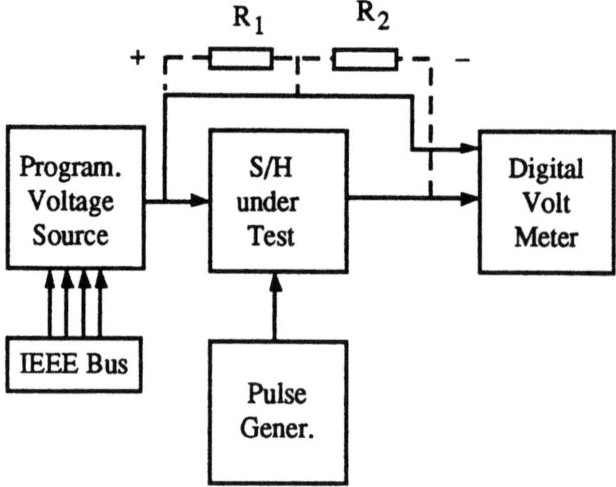

Figure 3.18 : Test set-up for measuring S/H amplifiers

3.10   Testing of sample-and-hold amplifiers

In Figure 3.18 a test set-up for measuring the performance of a sample-and-hold amplifier is shown. The set-up consists of a programmable voltage source, a sample pulse generator, and a (differential) digital voltmeter. Depending on the architecture of the sample-and-hold amplifier, a floating digital voltmeter or a bridge type measurement configuration will have to be used.

A floating digital voltmeter is used to measure the difference between input and output signal of the sample-and-hold amplifier if the circuit performs a non-inverting transfer operation.

When an inverting sample-and-hold amplifier is tested, then the configuration shown in dashed lines can be used. In this way a more accurate testing of the linearity is possible by connecting the digital voltmeter at the interconnection of the resistors R_1 and R_2 (test point P) and the common signal reference.

3.10.1   Testing DC characteristics

During the *track* mode DC offset and nonlinearity can be measured. If an inverting type of sample-and-hold is tested, then the resistor bridge may be used to measure nonlinearity. In a non-inverting system a floating measurement instrument must be used to measure the difference between input and output signal as a function of the input signal. Information about nonlinearity is obtained in this way. The "track-to-hold step" of a sample-and-hold amplifier is measured by monitoring the difference between the DC input signal during the *track* mode and the signal during the *hold* mode. Most frequently the track-to-hold step is independent of the input signal when an inverting type of sample-and-hold amplifier is used. In the non-inverting type, the switch drive voltage can change with varying input signal which results in a signal dependent feed-through giving distortion. The temperature dependence of all these parameters is obtained by performing the above defined measurements at different temperatures.

3.10.2 Dynamic measurements

In Figure 3.19 a dynamic performance test set-up is shown. The measurement set-up consists of a programmable sine wave source, a pulse generator with an extra delayed output signal terminal, an oscilloscope, and a distortion analyzer to measure the output signal of the sample-and-hold amplifier under test. A test A/D and D/A loop is added to the system to perform additional tests.

Without special measures it is difficult to analyze the sample-and-hold amplifier on itself. Usually a "glitch" is generated when the system switches from *track* to *hold* mode. Therefore, it is not possible to analyze the sampled *analog* signal at the output by simply filtering off the harmonics introduced by the sampling process by using a low-pass filter and analyzing the resulting signal with a distortion analyzer and a spectrum analyzer. When the sample-and-hold amplifier is designed as a *deglitcher* at the output of a D/A converter, then the above-mentioned method can be used to obtain information about the dynamic performance of the system.

The acquisition time of a sample-and-hold amplifier is the time needed to switch from the hold mode into a full-accuracy following (tracking) of the analog input signal. The acquisition time of the system is measured by varying, at a fixed sampling frequency, the track-and-hold time. A small acquisition time is a measure of the performance of a sample-and-hold amplifier. It determines the maximum sample frequency of the sample-and-hold amplifier. The signal-to-noise ratio of the system is measured as a function of the acquisition time. In Figure 3.20 a test result is shown.

The minimum acquisition time is determined at a point where the signal-to-noise of the system is reduced by 1 dB with respect to a measurement value obtained with a large acquisition time.

The aperture time of a sample-and-hold amplifier is the time difference between the start of the *hold* command and the moment the output signal of the *hold amplifier* is settled within the specified accuracy of the amplifier. The aperture or *track to hold* time is measured by varying the delay time T_d between the hold command, applied to the sample-and-hold amplifier, and the start conversion command, applied to the A/D converter. During

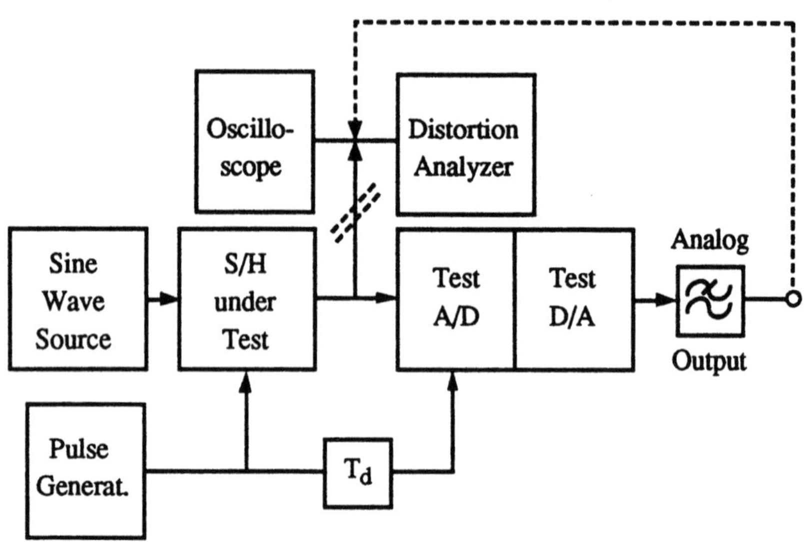

Figure 3.19 : Dynamic test set-up for sample-and-hold amplifiers

the *track mode* the distortion of the sample-and-hold amplifier is measured. This measurement includes the distortion of the *hold amplifier*, too. A very important parameter is the *aperture uncertainty* time of a sample-and-hold amplifier. This aperture uncertainty determines the timing accuracy with which analog samples are taken. Especially at high analog input frequencies, this parameter determines the performance of a system, that is signal-to-noise ratio. This *aperture uncertainty* of a sample-and-hold amplifier is difficult to measure. With a large acquisition time the signal-to-noise ratio of the total system is analyzed, and results close to the theoretical value must be obtained. If large differences occur, then the clock jitter has to be reduced and the clock generation circuits improved.

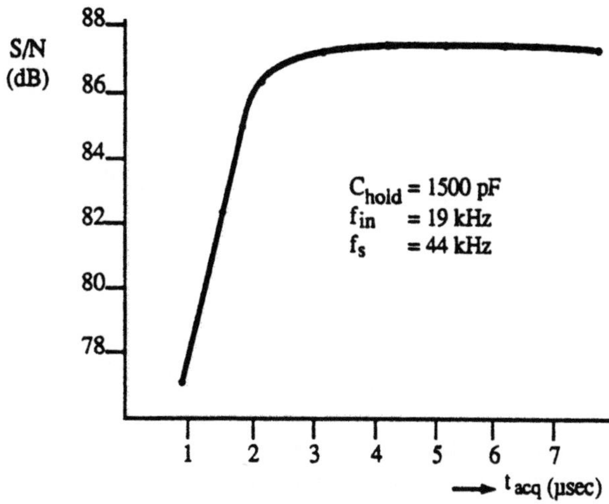

Figure 3.20 : Acquisition time measurement result

The small signal bandwidth of the system must be measured to obtain information about the noise bandwidth. Noise tests are performed by analyzing the output noise of the system during the track mode as a function of the DC input signal. The frequency spectrum of the output noise must be analyzed up to very high frequencies to obtain information about the fold-back noise which is finally added to the sampled output signal. The measuring bandwidth is equal to the -3 dB small signal bandwidth of the sample-and-hold amplifier.

Slew rate is part of the acquisition time specification and may be specified as an additional parameter.

Input signal feed-through is measured at the moment the sample-and-hold amplifier is in the hold mode. Parasitic capacitance across the switch limits the signal feed-through in the system at high frequencies. Over the specified signal bandwidth the feed-through must be larger than the dynamic range of the system to avoid quantization errors. The droop rate of the system is measured during hold mode. The change in output signal as a function of the hold time determines the droop rate. In Figure 3.21 an overview of the measured parameters needed to characterize a sample-and-hold amplifier are shown. Note that some of the parameters are easy to specify, but a measurement setup and a verification of the parameter sometimes needs an

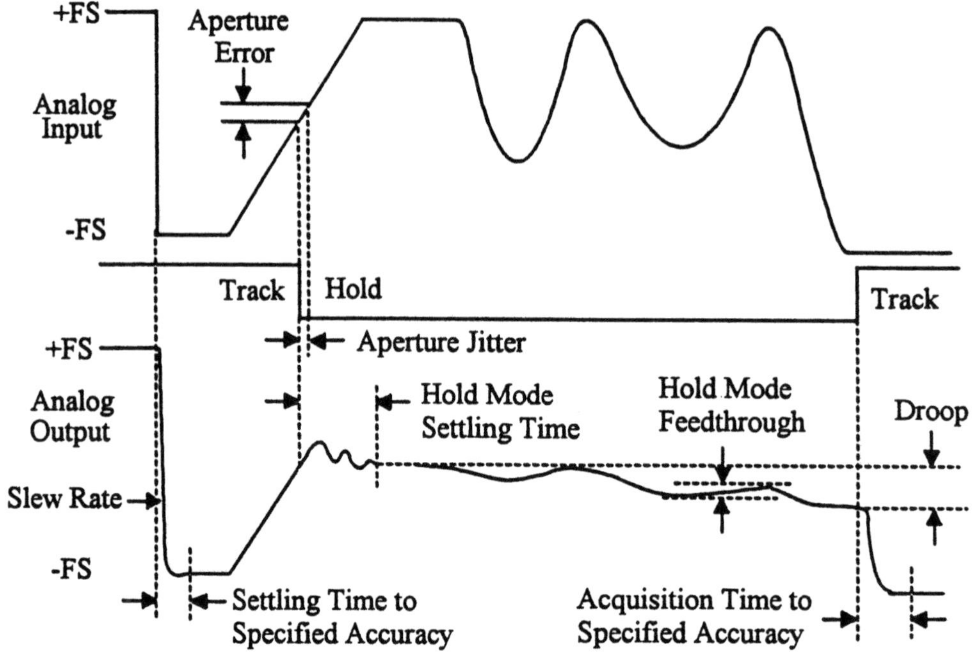

Figure 3.21 : Overview of parameters to measure sample-and-hold amplifier

indirect test method.

3.11   Cascading sample-and-hold amplifiers

In Figure 3.22 a cascade of two sample-and-hold amplifiers clocked with different clock signals is shown. Accurate information about the sampled output (hold mode) signal of a sample-and-hold amplifier can be obtained when two of these units are cascaded and operated with different clock signals. This hold information is important for applications with analog-to-digital converters which are only active during the hold mode of the sample-and-hold amplifier. The basic idea behind the cascaded operation is the use of beat frequencies obtained by the different clock signals of the two sample-and-hold amplifiers. At the moment an input frequency close to f_{s1}/n is applied with n = 2, 3, 4, and so on, a beat frequency occurs when the second sample-and-hold amplifier is clocked with a sampling frequency equal to f_{s1}/n. Note that the beat frequency is chosen at the low-end of the frequency spectrum of the system. The sub-sampling performed by the second sample-and-hold amplifier results in small changes in steps between succeed-

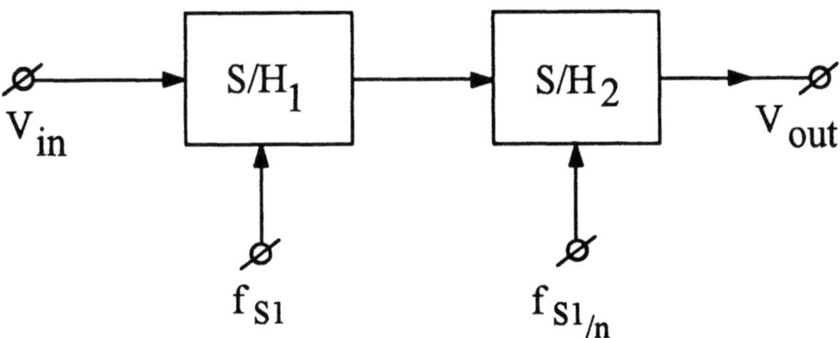

Figure 3.22 : Cascade of two sample-and-hold amplifiers

ing samples. The acquisition time accuracy of the second sample-and-hold amplifier is therefore increased while aperture time errors do not contribute to the final sampling accuracy. As a result, a much larger capacitor to store the analog information can be used in the second unit. Furthermore, when the glitches of the second system are minimized, the total waveform can be analyzed, and accurate information about the hold mode performance of the first sample-and-hold amplifier at the high frequency band edge is obtained without using an analog-to-digital converter.

3.12 Conclusion

In this chapter the measurement set-ups and some measurement results of the most important specifications of converters are shown. Many of the DC parameters are fairly easy to measure in contrast with the more difficult to perform dynamic tests. Some results, however, depend on a conglomerate of parameters and are therefore measured as a total. In high-speed A/D converters sub-sampling is used to obtain a high measurement accuracy without using D/A converters operating at the speed limit of a technology. A quick test can be done by using the beat frequency test between the sampling and input signals. This beat frequency signal can be analyzed using a distortion and a spectrum analyzer to obtain information about the distortion and signal-to-noise ratio at these high frequencies. Aperture-time and acquisition-time definitions for sample-and-hold amplifiers are introduced again, together with specific measurement set-ups to perform these tests. A cascade of two sample-and-hold amplifiers operating at different sampling frequencies resulting in a beat frequency at the output of the second unit results in an accurate and independent test method for sample-and-hold amplifiers.

Chapter 4

High-speed A/D converters

4.1 Introduction

The best-known architecture for a high-speed analog-to-digital converter is the flash converter structure. In this structure an array of comparators compares the input voltage with a set of increasing reference voltages. The comparator outputs represent the input signal in a digital thermometer code which can be easily converted into a Gray or binary output code. The flash architecture shows a good speed performance and can easily be implemented in an integrated circuit as a repetition of simple comparator blocks and a ROM decoder structure. However, this architecture requires 2^N-1 comparators to achieve an N-bit resolution. The parallel structure makes it difficult to obtain a high-resolution while maintaining at the same time a large bandwidth, a low power consumption, and a small die area. An alternative to the full-flash architecture is the multi-step A/D conversion principle. In high-speed converters the two-step architecture is the most popular because of the ease of implementation. However, a two-step architecture must be preceded by a sample-and-hold amplifier which performs the sampling of the analog input signal. In the two-step architecture a coarse and fine quantization takes place. These succeeding conversion steps need time. The sample-and-hold operation on the signal keeps the sampled signal constant. During this "hold" time the conversion takes place, making it virtually "timeless." After the coarse quantization is performed, the digital signal is applied to a D/A converter to reconstruct the analog signal. This reconstructed signal is subtracted from the analog input signal which is held by the sample-and-hold amplifier. After subtraction has taken place the residue signal can be amplified and is then applied to the fine quantizer which performs the conversion into a digital value. The coarse plus fine output code with, in many cases, an error correction operation results in the final digital output word. A good balance between circuit complexity, power consumption, and die size is obtained in this type of converter. The final dynamic performance, however, depends on the quality and dynamic performance of the sample-and-hold amplifier.

In a bipolar technology a continuous-time mode of operation of circuits is found. Base currents of bipolar devices do not allow the use of hold circuits without introducing a signal droop. Furthermore, switches are difficult to implement in this technology.
In MOS technology circuits can be operated in a continuous-time mode or in a discrete-time mode. Most architectures in MOS use a discrete-time mode of operation. In such a solution the sample-and-hold operation is combined with the system function. Mostly the discrete-time operation includes an automatic offset cancellation technique in the comparator stages. In a full-flash system, 2^N-1 small sample-and-hold amplifier-comparators are therefore used to perform the conversion function. Because of the small value of the hold capacitor, offsets induced by switching transients and channel charges of the switching devices limit the resolution of the total system.

In MOS using a discrete-time circuit solution a two-step system implementation can be easily obtained by using a set of coarse sampled data comparator-amplifiers stages and a set of identical fine stages. The coarse conversion is performed by comparing the input signal with the coarse tap voltages on the input ladder. After the coarse value is determined, a block of fine reference voltages is switched on. These reference voltages are in-between the two already determined coarse ladder taps. Again a comparison is performed and the final output code is obtained. In this system the ladder is used for coarse and fine quantization without needing an extra D/A subtracter circuit to drive the fine converter stage. To overcome some problems of the sample-and-hold amplifier, design alternatives have been worked out that have the advantage of the digital sampling used in the full-flash converter and the die size of the two-step system but do not require a sample-and-hold amplifier. This architecture is called a *folding architecture*, which is capable of achieving a large analog bandwidth and high resolutions without incurring the power and area penalties associated with the flash architectures (see references [50,51,52,53,5]).

By using folding techniques, an A/D converter can be designed in which each comparator detects the zero crossings of the input signal through a number of quantization levels, thus reducing the number of comparators required for a given resolution. The number of comparators is reduced by the number of times that the input signal is folded by the folding stages. However, each reference level requires a folding stage generating the folding signal. This folding signal is a combination of output signals from folding stages. By combining a number of output signals from folding stages, a repetitive folded signal is applied to the decision-making comparators. The folding factor, which determines the number of signals to be combined reduces at the same time the number of comparators. This reduction in comparators, however, is offset by the number of folding stages that are needed to obtain the resolution of the converter. The number of folding stages can be further reduced by interpolating between outputs of folding stages to generate additional folding signals without the need for more folding stages. The interpolation stage in this way reduces the number of folding stages by the interpolation factor. The folding architecture results in a compact, low-power system with small signal and clock delays over the interconnection lines. Therefore, a larger analog input bandwidth can be obtained than would be possible with a standard flash converter. However, the folding operation increases the internal signal frequencies with the folding factor.

Different system implementations will be described using bipolar, CMOS or biMOS technologies.

Up until now no accurate model has existed that describes the frequency limitations of flash-type converter structures. A high-level model has been developed that describes the nonlinear behavior of amplifier stages. This behavior results in a signal-dependent delay giving third-order distortion. Keeping in mind the clock timing accuracies needed in accordance with the converter speed, the third-order distortion term is thought to impose a basic speed limitation on flash converters. This model is used to optimize the analog bandwidth of the converter with respect to the technology used.

Depending on the way the circuits are used, a direct implementation from a bipolar technology into a MOS technology is possible if a discrete-time MOS solution is not used. In time-continuous MOS designs the same imperfections and limitations as in bipolar systems are found. Auto-zero systems or offset compensation systems might be easier to design in MOS and show a better performance when these circuits are not operated at the maximum clock rate of the system but at a much lower sampling rate. An auto zero system must therefore be designed to operate independently of the sampling of the converter.

4.2 Design problems in high-speed converters

There are two main problems that impair the dynamic performance of high-speed A/D converters: *timing* inaccuracies and *distortion*. [44,5].

4.2.1 Timing errors

In most A/D converters there are four main sources of timing errors:

1. Sampling clock jitter.

2. Limited rise and/or fall time of the sampling clock.

3. Skew of the clock and input signal at different places on the chip.

4. Signal-dependent delay.

The sampling clock jitter can originate both inside and outside the A/D converter. The outside sampling clock must be designed to have a very small (short-term) jitter. Internally, a small rise or fall time of the sampling clock avoids additional jitter caused by (white) noise of the clock amplifier circuits. Furthermore, crosstalk from other circuits must be minimized to avoid modulation of the sampling clock. The skew between the clock and the analog signal introduces timing errors between the same signal at different places on the die. The clock signal at the top comparator stage, for example, may be slightly out of phase with the clock signal at the middle comparator. This difference in time causes a quantization error that results in nonlinear distortion. As an example, in about 12 ps a signal can travel only 3.6 mm over an interconnection line at the speed of light. On a die the transmission speed is lower because of the high dielectric constant of the oxide layers and the finite conduction of the epitaxial layer or the substrate material. In practice a speed between half and one-third of the speed of light is obtained on the die. This means that 1 ps equals a distance of 100 μm to 200 μm between elements in an integrated circuit. Therefore, the sampling clock lines and the signal lines in the converter must be laid out very carefully. If the paths of clock and signal lines include different processing circuits, the delays of these processing elements have to match within a fraction of the required timing accuracy. It implies, furthermore, that a small total die size is best for minimum timing performance errors.

Finally, many circuits introduce a signal-dependent delay. For example, each amplitude-limiting circuit followed by a bandwidth-limiting circuit introduces a delay that is slope-dependent. These circuits are invariably found in input and comparator stages of high-speed A/D converters. A behavior model is developed that can be used for optimization of the analog bandwidth of A/D converters. A relation between the maximum analog input frequency and the small signal bandwidth of the input amplifier-comparator stage is determined with this model. The signal-dependent delay of a signal exhibits itself as a third-order distortion of the quantized signal.

4.2.2 Distortion

The distortion of a quantized signal can be caused by four main reasons:

1. Sampling comparators aperture time.

2. Distortion in the linear part of the input amplifier.

3. Changes in the reference voltage values and comparator offsets.

4. Delays of analog signal and clock signal.

A large comparator aperture time may be caused by the architecture of the comparator or by a large rise or fall time of the sampling clock. Such a large aperture time results in high-frequency sampling errors and causes an averaging effect in the time domain. This phenomenon exhibits itself as a third-order distortion. An enhanced small signal bandwidth of the comparator stages reduces this effect.

Non-linear distortion in the input amplifier introduces harmonics and mixing products of the input signal. These harmonics may give aliasing products in the baseband due to mixing of these components with the sample frequency.

Most high-speed A/D converters require a large number of reference voltages that are normally generated by a resistive divider and a reference source. Errors in these reference voltages introduce a nonlinear distortion equivalent to a nonlinear distortion in the input amplifier. Comparator offset voltages must be small with respect to the reference voltage steps to obtain a non-linearity which is only dependent on the accuracy of the resistive divider. An additional problem, especially in converters with a large bandwidth, is the kickback from the comparator stages (clock feed-through) on the reference voltages. During sampling of the input signal, the reference voltage temporarily deviates from the nominal value, resulting in additional quantization errors. Timing and distortion problems are common to all high-speed A/D converter architectures. However, the following sections show that these problems can be minimized more easily in folding architectures than in full-flash or two-step converters.

4.3   Internal converter coding schemes

In converters different internal coding schemes are used before the final (binary) code is generated at the output of the system.

4.3.1   Thermometer code

In full-flash systems an internal code called *thermometer code* is mostly used. Every time a reference level has been tripped by the input signal an output "1" is added to the output code of the comparators. In Table 4.1 the output binary code and the (internal) thermometer code are shown for a value up to 8. From the table it is easy to see why this code is called *thermometer code*. With an increasing binary number the number of "1"s is increasing. By using a simple gate function and an ROM the binary output code can be implemented in a simple manner. Looking at the table, for the binary code 0010, for example, we only need to encode in an *and* gate the values of $b.\bar{c}$. When the output value is true, then a true is applied to the ROM function and the binary output code 0010 is obtained. This operation is performed for every binary output signal using other combinations of the thermometer code.

4.3.2   Gray encoder

At the moment a special analog encoding is used at the input of the comparators then the *Gray* shows to be an efficient code for implementation. In Table 4.2 the binary-Gray code implementation is shown. The specific characteristic of the Gray code is that from one code to the next code only

| BINARY CODE | THERMOMETER CODE |
d c b a	h g f e d c b a
0 0 0 0	0 0 0 0 0 0 0 0
0 0 0 1	0 0 0 0 0 0 0 1
0 0 1 0	0 0 0 0 0 0 1 1
0 0 1 1	0 0 0 0 0 1 1 1
0 1 0 0	0 0 0 0 1 1 1 1
0 1 0 1	0 0 0 1 1 1 1 1
0 1 1 0	0 0 1 1 1 1 1 1
0 1 1 1	0 1 1 1 1 1 1 1
1 0 0 0	1 1 1 1 1 1 1 1

Table 4.1 : Binary-thermometer code implementation

INPUT LEVELS	BINARY CODE d c b a	GRAY CODE d c b a
0	0 0 0 0	0 0 0 0
1	0 0 0 1	0 0 0 1
2	0 0 1 0	0 0 1 1
3	0 0 1 1	0 0 1 0
4	0 1 0 0	0 1 1 0
5	0 1 0 1	0 1 1 1
6	0 1 1 0	0 1 0 1
7	0 1 1 1	0 1 0 0
8	1 0 0 0	1 1 0 0
9	1 0 0 1	1 1 0 1
10	1 0 1 0	1 1 1 1
11	1 0 1 1	1 1 1 0
12	1 1 0 0	1 0 1 0
13	1 1 0 1	1 0 1 1
14	1 1 1 0	1 0 0 1
15	1 1 1 1	1 0 0 0

Table 4.2 : Binary-Gray code implementation

one bit changes. This is shown in Table 4.2. As a result, the value for a in the Gray code is obtained by combining the levels 1, 3, 5, 7, 9, 11, 13, and 15. For b in the Gray code a combination of levels 2, 6, 10, and 14 is used. In the same way for c a Gray code combination of levels 4 and 12 is used.

When in a system a Gray code is generated, then a binary output code is obtained by using an exclusive OR function between a and b. The output of the exclusive OR function is put in an AND function with the $\bar{c}$ of the Gray code to obtain the binary a output code. An identical operation is needed for binary b using Gray b, c, and d. Binary c is obtained from d Gray code directly.

A circuit implementation of a Gray code analog-to-digital converter will be shown in one of the following sections.

4.3.3   Circular code

At the moment signals are repeating in a converter implementation, then a circular code is mostly generated. In table 4.3 the binary code and a corresponding circular code are shown. The Table shows that with an increasing

BINARY CODE	CIRCULAR CODE
c b a	d c b a
0 0 0	0 0 0 0
0 0 1	0 0 0 1
0 1 0	0 0 1 1
0 1 1	0 1 1 1
1 0 0	1 1 1 1
1 0 1	1 1 1 0
1 1 0	1 1 0 0
1 1 1	1 0 0 0

Table 4.3 : Binary-circular code implementation

binary number in the circular code the number of "1"s is increasing until the row is completely filled with "1"s at code 1111. Then the "1"s are moving out until finally the code 1000 appears at the end of the scale. When the code is increased with one step, the starting code 0000 appears again. Because of this phenomenon this code is called circular code. The conversion of the circular code into a binary one is performed by using an exclusive OR function $(a.\bar{b} + \bar{a}.b)$ and a ROM encoder. The output of the exclusive OR results in a true state at two code transitions. The MSB determined by c in the binary code is equal to d in the circular code. The remaining three bit codes are determined by a, b,and c. The result of this operation is that with only four comparators a three bit binary output code can be determined.

In one of the following sections attention will be paid to circuit implementations for circular codes. Such a code appears in "folding" analog-to-digital converters, too.

4.4 Full-flash converters

In an N-bit flash A/D converter, $2^N - 1$ reference voltages and comparator stages are used to convert the analog input signal into a thermometer-like digital output code (see Figure 4.1). This code is converted into a binary output code using an ROM structure. In today's technologies, 8-bit converters having a reasonable die size and consuming moderate power are available. Increasing the resolution to 10 bits increases the die size and power dissipation roughly four times. In practice, however, there is a limit to the power dissipation that can be handled in IC packages. Therefore, the power per comparator stage must be drastically reduced to keep the overall power dissipation at the same level as the 8-bit unit. As a result, the bandwidth of the comparators has to be reduced, resulting in a much lower effective analog bandwidth for the converter. The bandwidth of a system is mostly related to biasing current, which in turn results in power dissipation. Because of the increase in size, it is more difficult to distribute clock and input signal lines without introducing delay-induced errors exceeding $\pm \frac{1}{2}$ LSB, and to match the properties of all these comparators within the same specification. The input capacitance of the system increases linearly with the number of comparators, making it impractical to incorporate an input signal buffer on the chip. Even external buffers are difficult to design and need a large power-driving capability at high frequencies. The large number of comparators results in a heavy loading of the clock driving circuits. Small rise and fall times of the clock signals are difficult to obtain and therefore external clock drivers are often required.

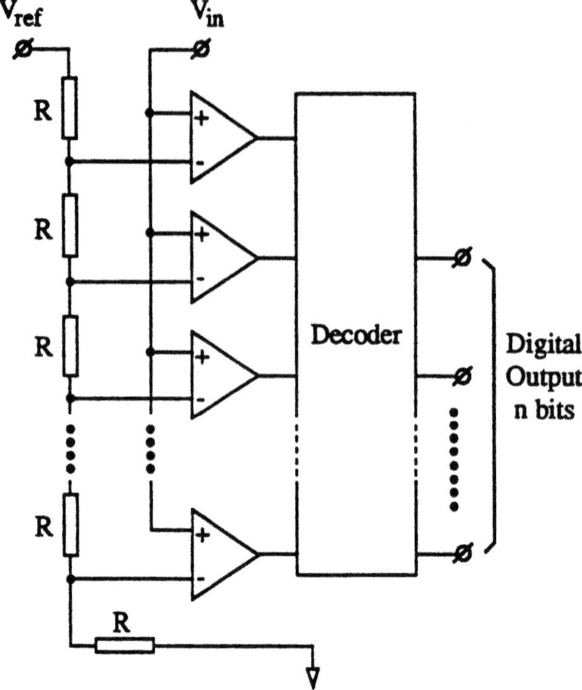

Figure 4.1 : Full-flash A/D converter structure

4.4.1   Bipolar comparator

The basic circuit diagram of a clocked comparator is shown in Figure 4.2. The circuit consists of a differential amplifier T_1, T_2 which compares the analog input signal with the reference voltage generated across a tapped resistor. The input signal is amplified and appears across the collector resistors R_1, R_2. The cross coupled differential pair consisting of T_3 and T_4 regenerates the input difference signal into a logic output signal. The differential pair T_5, T_6 switches the comparator from the amplification mode into the regeneration (=comparison) mode. When the clock signal is low, then the current J_1 flows through T_5 and the input amplifier is active. At the moment a decision is required, then the clock signal is made high and transistor T_6 conducts the current J_1. The cross coupled differential pair T_3, T_4 becomes active. The positive feedback in this system gives a large gain resulting in a high comparator sensitivity. Optimizing the system for a large bandwidth allows accurate comparison at high clock frequencies. Furthermore the bit error rate of this comparator can be small (see chapter 5 section 5).

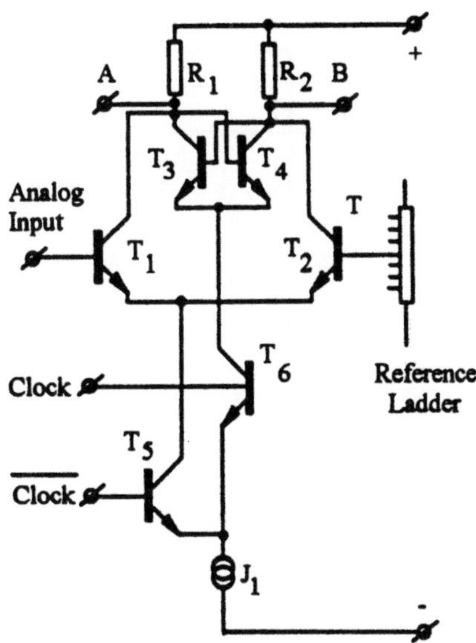

Figure 4.2 : Basic bipolar clocked comparator

4.4.2 Bipolar full-flash converters

In Figure 4.3 an example of a master-slave comparator that can be used in a full-flash converter is shown. This circuit is a clocked master-slave flip-flop [38,39,42,43,45,47,48]. At the input of the comparator stage an emitter follower is added to increase the input impedance, thus reducing the loading of the circuit on the reference ladder and the input source. The operation of the circuit is as follows. When the clock signal is high, current flows through the input amplifier of the master flip-flop, and during the same time the slave flip-flop is in the latch mode. The input signal is amplified by the input differential amplifier. At the moment the clock changes state, the input amplifier is disconnected and the difference in signal that is present across the collector resistors of the master flip-flop is amplified by the positively feedbacked master flip-flop. If small signal differences are present, then these differences are amplified until the flip-flop is in a stable condition, giving a ONE or ZERO depending on the input signal difference. This signal is amplified by the input amplifier stage of the slave flip-flop and the decision is transferred to the output. By clocking the flip-flop, an analog signal comparison with a large gain and a large bandwidth is obtained. Cascading two sensitive clocked comparators gives a reduction in the bit error rate resulting in a high performance clocked comparator.

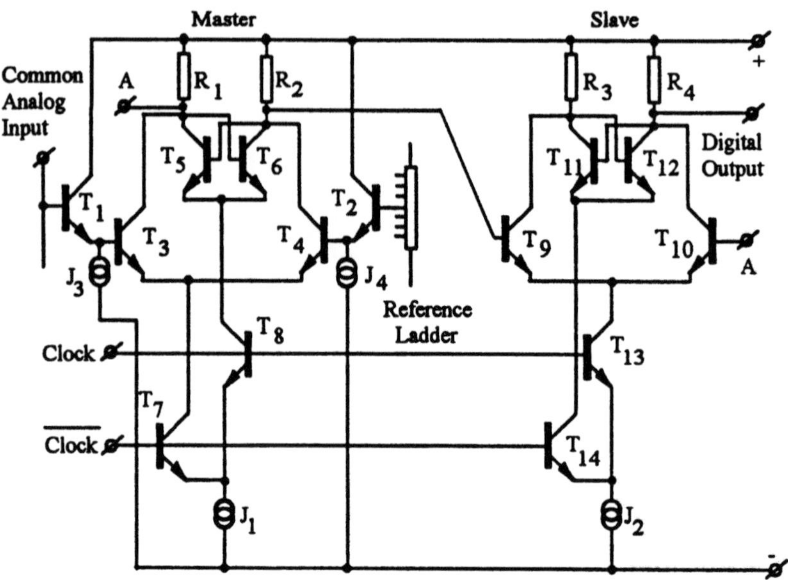

Figure 4.3 : Master-slave comparator circuit

Offset of the input amplifier is an important parameter. In Figure 4.4 the offset of a differential pair as a function of the collector current is shown. A parameter is the emitter size of the transistors. From the figure it can be seen that with increasing collector current the offset linearly increases. This increase is caused by the emitter resistor difference of the differential pair. An example of a three-stage comparator cell is shown in Figure 4.5. In the first stage the input signal is continuously amplified. This signal is applied to a clocked comparator consisting of a master-slave flip-flop as shown in Figure 4.5. Note that in all amplifier and flip-flop stages cascode stages consisting of differential common-base stages are used to minimize the influence of the collector-base capacitance of the amplifying devices on the overall circuit performance. Between the master and the slave flip-flop an extra coupling from the above and below neighbor of this comparator is obtained. Therefore, the signals "to $n-1$" and "to $n+1$" are obtained via emitter followers. These emitter followers together with the output emitter follower of the second stage form a *wired OR* function. Depending on the signals from the neighboring comparators, a correction of the decision signal is possible. In case the above and below neighbor would generate positively a ONE signal and the master comparator would give a ZERO decision, then the wired OR function transfers a ONE signal to the slave flip-flop which

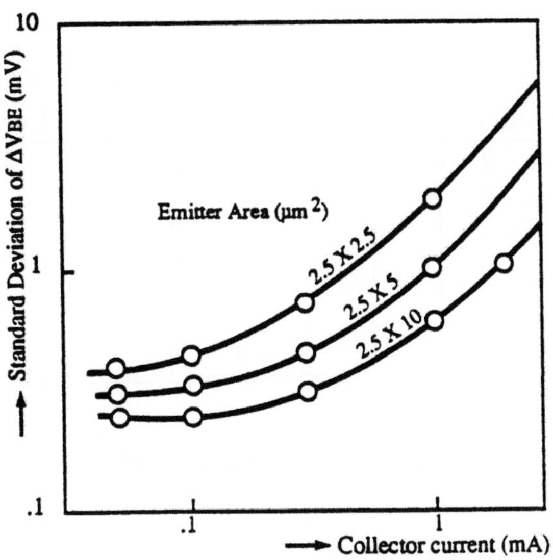

Figure 4.4 : Offset of a differential pair as a function of collector current

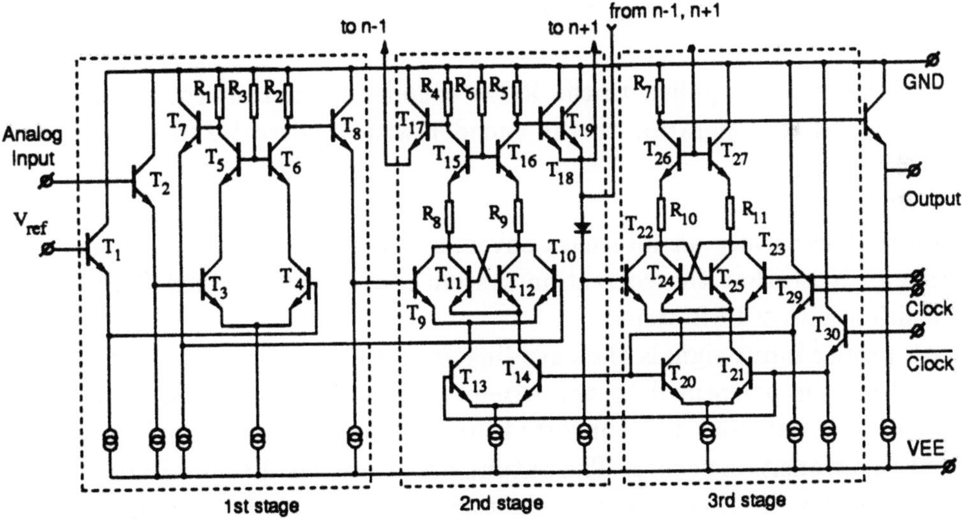

Figure 4.5 : Three stage comparator cell with error correction

is in the third stage. Because of the inverter operation in the third stage this ONE signal is converted into a ZERO at the output of the comparator system. No encoding takes place. In this way a correction of the decision signal is possible, resulting in an improvement in performance.

Code Word	Latch Number	Input Correct	Output	Input Error	Output Corrected
f	n - 2	0	0	0	0
e	n - 1	1	1	1	0
d	n	1	0	0	0
c	n + 1	1	0	1	1
b	n + 2	1	0	1	0
a	n + 3	1	0	1	0

Table 4.4 : Encoding Table

The logical function applied by the encoding system equals:

$$n_{out} = (n - 1).(n).(\overline{n + 1}). \tag{4.1}$$

The signal n_{out} is applied to the ROM encoder for output code generation. In Table 4.4 the operation of the encoder with a correct and an erroneous code is shown.

Note that vertically an internal converter codeword (f e d c b a) has been written. Note that the correction system moves the encoded ZERO-ONE transition with 2 LSB steps.
With proper input signals the encoding from the thermometer into a *single* ZERO level is performed by the wired-OR function, too. The ZERO-ONE transition is detected and converted into a HIGH output signal at the moment this transition occurs. The output signal is applied to the ROM encoder that performs the encoding from the thermometer code into the mostly used binary output code. In Figure 4.6 an example of such an encoder is shown. Again this ROM is a wired-OR function. When the signal, for example, from "comparator out n" is a ONE, then all other signals are ZERO. At that moment the bases of the ROM transistors are made high and a code conversion appears at the output of the encoder. This encoder is very fast and simple to design. In the example shown, the 400 μA current is shared over all the emitter followers which apply the high level to the ROM encoder. In this way a reduction in dissipation is obtained while maintaining a high circuit speed.

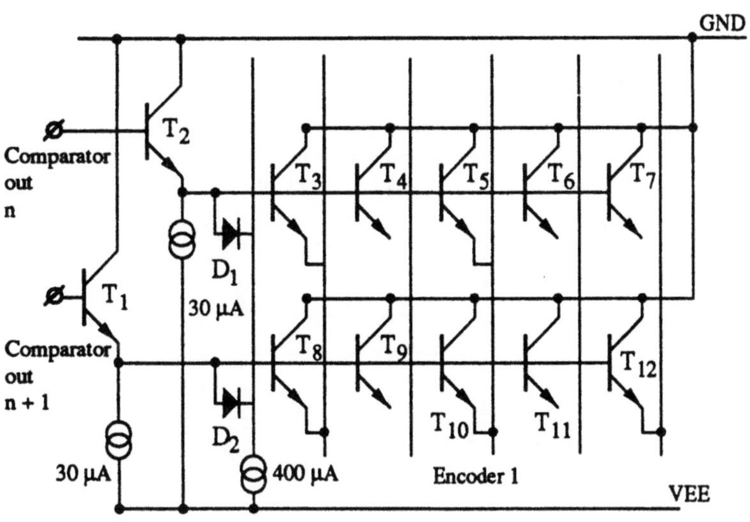

Figure 4.6 : ROM encoder circuit

4.4.3 Improved bipolar comparator

To reduce the bit error rate (BER) of the comparator shown in Figure 4.3 a small modification is needed. The improved circuit is shown in Figure 4.7 [44]. In the circuit diagram it is shown that the collector load resistors in the master flip-flop are split-up. During the "amplification" phase the input signal is amplified, using the input differential amplifier and the load resistors R_1 and R_2. A small signal settling with a large signal bandwidth is obtained. At the moment the circuit is in the "decision" phase, the load resistors R_3 and R_4 are added to the input load resistors R_1 and R_2 to increase the gain of the flip-flop. The split-up in collector resistors reduces the collector time constant for the amplifier and the flip-flop mode. Practically equal resistors show a good optimum for the collector resistor split-up, resulting in about a 30% increase in speed. As a result of this operation, the accuracy and the number of meta-stable conditions of the comparator is significantly reduced. See Chapter 5 for a detailed analysis of the meta stability of comparators.

4.4.4 MOS full-flash converters

In MOS technology a different comparator architecture is used. An example of a basic MOS comparator stage is shown in Figure 4.8 [49,41]. The circuit consists of a CMOS inverter stage using transistors NM_1 and PM_1, a capacitor C to store the information and three switches S_1, S_2, and S_3,

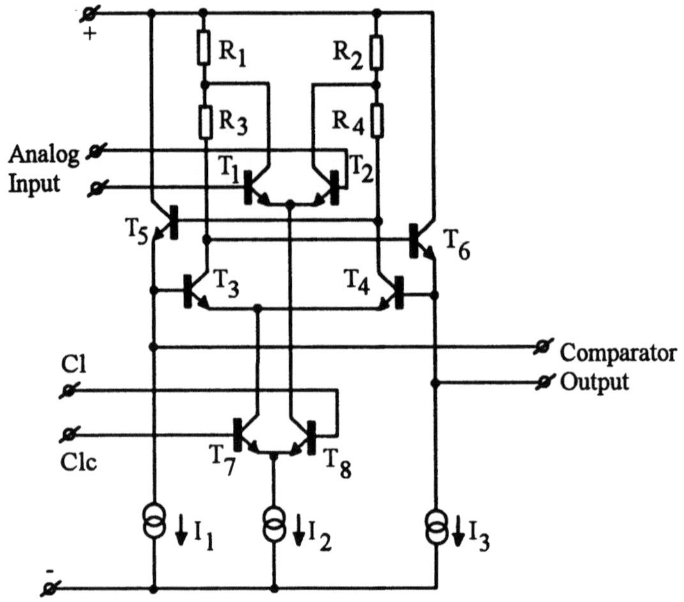

Figure 4.7 : Improved bipolar comparator

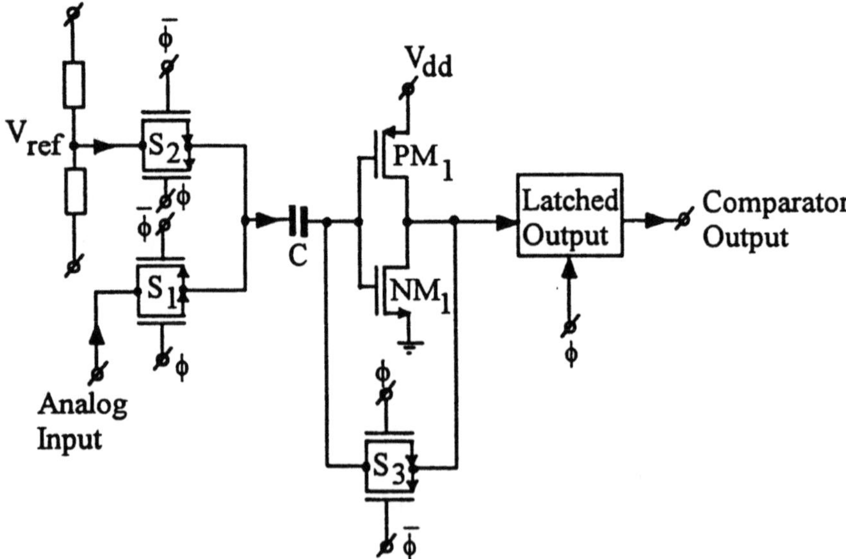

Figure 4.8 : Basic MOS comparator circuit

consisting of transmission gates controlled by the clock signals ϕ and $\bar{\phi}$. A transmission gate is built up using a parallel connection of a PMOS and an NMOS device. Such a switch is less dependent on the input and comparator signal levels within the supply voltage range.

A comparison starts by closing switches S_1 and S_3. The switch S_3 shorts the drains with the gates of the CMOS inverter. As a result a low impedance determined by the transconductance of the stages is found at the gate terminals of the inverter. This means that the capacitor C is grounded with a low differential impedance. The dc biasing of the CMOS inverter is chosen in such away that with S_3 shorted about half the supply voltage is generated across the drain-source of the individual devices. In practice this means that the W/L ratio of the MOS transistors is different, depending on the β_{square} of the individual devices.

Via S_1 the input signal is connected to the opposite site of the capacitor C. The input signal is sampled, and the signal sample is stored on the capacitor at the moment the switches S_1 and S_3 are opened. With S_3 open, the full gain of the CMOS inverter becomes active. By closing switch S_2, the reference voltage, which is generated across a resistive divider network, is applied to the capacitor. When there is a difference between the input voltage and the reference voltage, this difference is applied to the gates of the high-gain CMOS inverter. The inverter amplifies the signal, and the output signal of this inverter is applied to a latch circuit. In the latch the final decision is taken, and the output signal is amplified into a standard logic level that is used in the succeeding logic operation. Because the input signal plus the offset of the inverter stage are sampled on the capacitor C and this totally sampled signal is compared with the reference signal using the same circuit, no influence of offset is found in the final decision. The only errors occur because of the non-ideality of the switches. These switches introduce additional charge on the capacitor C which operates as a "hold" capacitor, resulting in clock frequency dependent offset voltages. Furthermore, this circuit uses a single-ended signal amplification and is therefore sensitive to supply and substrate noise.

Another problem with the comparator circuit from Figure 4.8 is the rather low gain and the storage of switch channel charge and switch feed-through signals on the capacitor C when the switches are opened. To avoid part of this problem, transmission gate switches consisting of a PMOS and a NMOS transistor in parallel are used. The counter-phase clock signals which control the switches compensate partly for the capacitive feed-through of the practical devices (gate-source capacitance, gate-drain capacitance and channel charge). Furthermore, two of the same stages are cascaded to increase the gain of the total comparator. In Figure 4.9 a circuit diagram of a two-stage system is shown. The circuit consists of two capacitively coupled CMOS in-

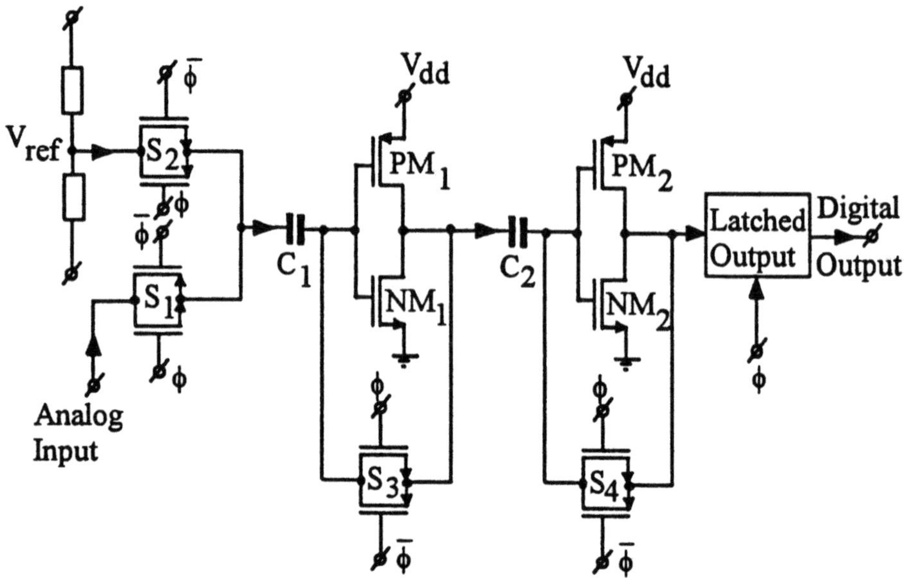

Figure 4.9 : Two-stage CMOS comparator circuit

verter stages, which are bypassed by switches S_3 and S_4. By closing switches S_1, S_3 and S_4 the input signal is sampled on capacitor C_1, while additional offsets and errors can be sampled on capacitor C_2. When switches S_1 and S_3 are opened at the same time, a part of the channel charge and charge feed-through of the switch S_3 can be sampled on capacitor C_2 as long as switch S_4 remains closed. A compensation for the non-ideality of switch S_3 is obtained. Then switch S_4 is opened and switch S_2 is closed to perform the comparison operation. The larger gain in this system allows for a more accurate comparison with less meta-stable conditions.

This comparator system is again followed by a latch that gives a well-defined logical output signal to the (thermometer-to-binary) encoder circuit.
The advantage of these systems is the per comparator built-in sample-and-hold function. This feature can be used advantageously in simplifications of full-flash converter stages. The maximum sampling frequency of this system is determined by the comparator speed.

4.4.5 Interleaved comparator full flash converter

A doubling of the sampling speed is obtained by multiplexing the input comparator part. In Figure 4.10 a simplified circuit diagram of the comparator system is shown. The circuit consists of two comparator circuits as shown in

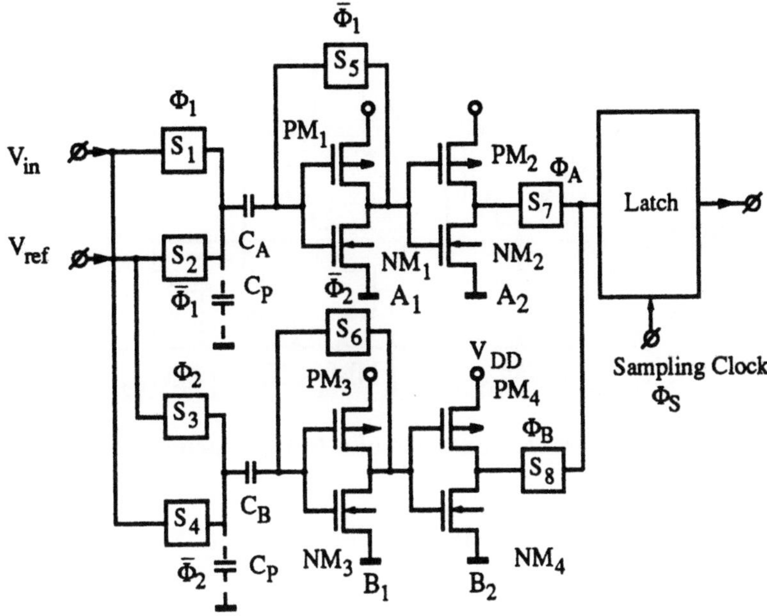

Figure 4.10 : Interleaved comparator A/D converter

Figure 4.8 in parallel. Two extra CMOS inverter stages are added to increase the comparator gain. The output signal of the comparators is applied to the output latch circuit via multiplexer S_7, S_8. The operation of the circuit is as follows. At the moment the input comparator part marked A_1 and A_2 is in the sampling mode by closing switches S_1 and S_5, the comparator part marked B_1 and B_2 is in the decision phase. Switches S_4 and S_8 are closed and the decision information is applied to the output latch. At the next clock phase the function of comparator A_i and B_i is interchanged and the decision information from A_2 is applied to the latch via switch S_7. Using the dual input system, effectively a doubling of the sampling clock frequency is possible without loosing converter performance. However, sampling occurs on the rising *and* falling edge of the clock requiring an accurate square wave clock with small jitter on both edges. In Figure 4.11 an example of a latch circuit in NMOS depletion-enhancement mode technology is shown. The basic latch is a MOS equivalent of the bipolar comparator in Current Mode

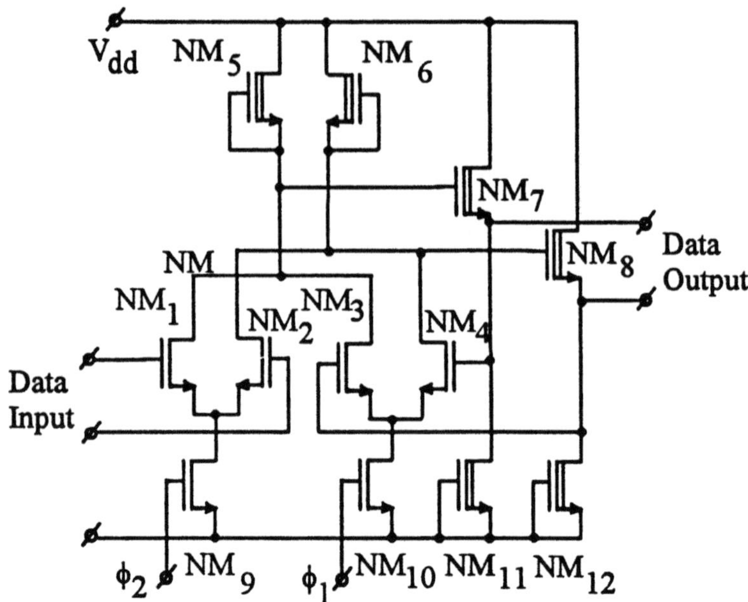

Figure 4.11 : Example of a NMOS latch circuit

Logic (CML) from Figure 4.2. Transistors NM_1 and NM_2 perform the data input. The latch consists of NM_3 and NM_4 with the buffer stages NM_7 and NM_8 and the drain loads NM_5 and NM_6. A high clock signal ϕ_2 biases the input data amplifier allowing data to be read in the system. The data are latched when the clock signal ϕ_1 is made high. The MOS equivalent of the bipolar system is designed to operate with a minimum digital signal swing. As a result, a maximum speed is obtained for this system with a minimum in substrate bounce.

4.4.6 Differential auto-zero comparator

To overcome the problems of power supply noise sensitivity of MOS comparators, a fully differential auto-zeroed comparator circuit has been designed [57]. This circuit is shown in Figure 4.12. The circuit consists of the input differential pair NM_1, NM_2 with the PMOS devices PM_1 and PM_2 as active loads. Switches NM_5 to NM_{10} perform the switching between amplifying and comparison mode. Furthermore, buffer stages NM_3 and NM_4 are added to reduce the loading of the storage capacitors C_1 and C_2 on the input amplifier stage NM_1 and NM_2. A continuous biasing of the system is performed.

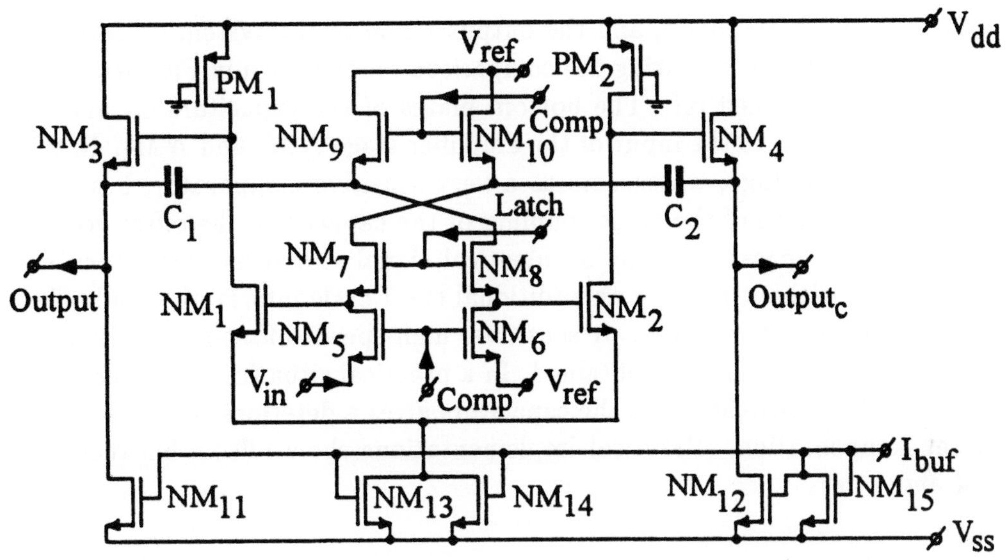

Figure 4.12 : Differential auto-zero comparator circuit

To make the understanding easier of the operation of the circuit a simplified diagram showing separately the amplifying and the latching mode are given in Figure 4.13(a, b). In analyzing Figure 4.13(a) the input signal plus the

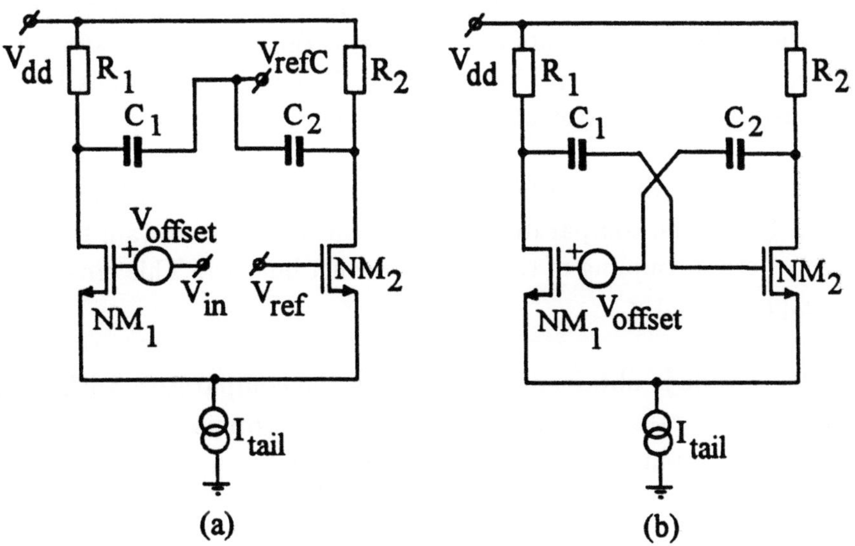

Figure 4.13 : Amplification (a) and latching mode (b) of the comparator

offset signal are amplified, and the output signal of the system is stored on capacitors C_1 and C_2. During the comparison mode shown in Figure 4.13(b) the input is switched off. The bottom plates of the capacitors C_1 and C_2 are cross-coupled to the input of the amplifier stage NM_1 and NM_2. In this way a flip-flop stage, which acts as a very sensitive comparator, has been obtained. Because of the cross-coupling of the capacitors, the offset voltage V_{offset} is compensated for by an identical signal stored on the capacitors. As a result, without needing an additional clock cycle information and offset are separated and removed. A sensitive, high-speed, offset-free differential comparator stage has been obtained. In a practical situation, however, feed-through of the switches onto the capacitors gives a deterioration of the ideal offset compensation. Practical implementations show offsets between 100 μV and 1 mV.

4.5   Gray code full flash converters

To reduce the number of comparators an analog encoding of the comparator input signals is used [54,51]. A low device count implementation is obtained with a Gray coding of the signal. The advantage of the Gray code is that with a linear increasing input signal only one comparator may change state. In Table 4.5 the Gray code and the offset binary code are repeated for ease of understanding the total circuit implementation. Note from the table that in the Gray encoding there exists a mirroring of the code and that with every one bit increase only one bit changes state. The most significant bit can be encoded using a comparator at reference level 8.

The MSB-1 code is obtained by combining the comparator with input reference level 4 with the comparator with input level 12. To obtain the right information a cross-coupling of comparators must be used. This cross-coupling will be explained later.

The MSB - 2 code is obtained by combining the comparators with reference level 2, the comparator with reference level 6, the comparator with reference level 10, and the comparator with reference level 14. Again a cross-coupling of the stages is required to obtain the right encoding.

Finally, comparators with input levels of 1, 3, 5, 7, 9, 11, 13, and 15 are cross-coupled to obtain the LSB code information.

DECIMAL VALUE	BINARY CODE	GRAY CODE
0	0 0 0 0	0 0 0 0
1	0 0 0 1	0 0 0 1
2	0 0 1 0	0 0 1 1
3	0 0 1 1	0 0 1 0
4	0 1 0 0	0 1 1 0
5	0 1 0 1	0 1 1 1
6	0 1 1 0	0 1 0 1
7	0 1 1 1	0 1 0 0
8	1 0 0 0	1 1 0 0
9	1 0 0 1	1 1 0 1
10	1 0 1 0	1 1 1 1
11	1 0 1 1	1 1 1 0
12	1 1 0 0	1 0 1 0
13	1 1 0 1	1 0 1 1
14	1 1 1 0	1 0 0 1
15	1 1 1 1	1 0 0 0

Table 4.5 : Gray code table

Note that an even number of encoding stages is needed except for the MSB. In Figure 4.14 an example of the encoding stages for the MSB, MSB-1, and MSB-2 codes are shown. The analog encoding of the comparator signals is performed by differential pairs. From Figure 4.14 it can be understood that every time a differential amplifier changes state, the output comparator which is connected between the output load resistors (e.g., R_L) must sense this difference voltage. The cross-coupling between the stages is needed to perform the 0 to 1 and 1 to 0 transition every time a level is detected. Because of the even amount of comparators, a middle level is not detectable. Therefore, an extra current with a value I_0 must be connected to the output load resistor to obtain an accurate code transition detection. The output code then changes state as shown in Figure 4.15. Note that only the MSB requires a single comparator stage giving an odd number of stages. Because of the increasing number of encoding stages that are loading the output load resistors R_L the bandwidth of the system is reduced, therefore decreasing the

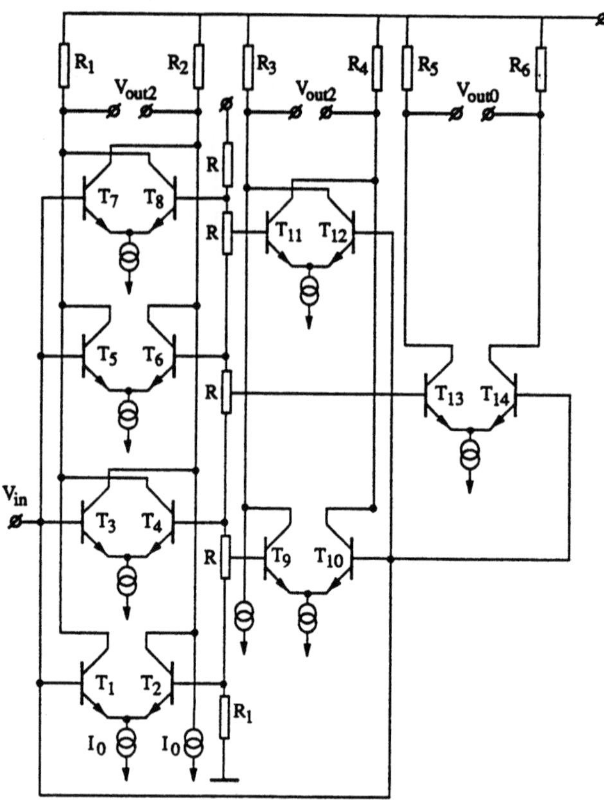

Figure 4.14 : Analog Gray encoding for MSB, MSB-1, and MSB-2

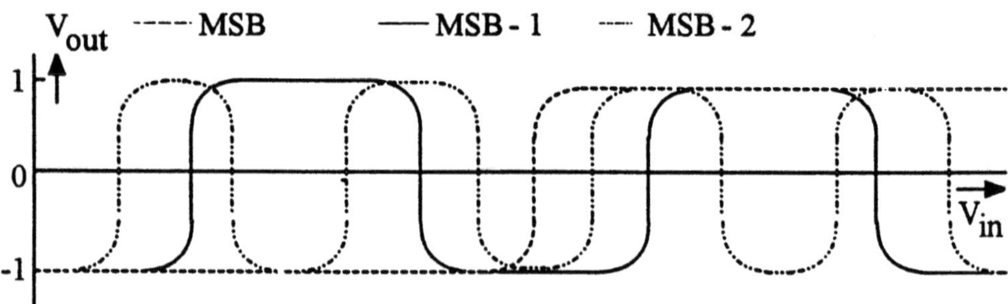

Figure 4.15 : Output signal of the analog encoder

accuracy with which a high-frequency signal can be encoded. To overcome
this problem a two-step encoding using two clocked comparators for the LSB

must be used. The final LSB is then obtained by a logical combination of the two output codes.

The advantage of this system is the small amount of clocked comparators needed for the conversion, resulting in a small die size and a low power consumption. The disadvantage of the system is the difference in comparator delay of the different encoding stages, resulting in coding errors and distortion. With a sample-and-hold function in front of the system this problem can be overcome.

4.6 Circular code flash converters

To obtain equal comparator loading, an implementation of a circular code can be used [5]. Table 4.6 shows again the circular code implementation. To make the understanding easier of the encoding, a decimal code is added. Using Table 4.6 the MSB of the circular code is equal to the MSB of the

DECIMAL VALUE	BINARY CODE	CIRCULAR CODE
0	0 0 0	0 0 0 0
1	0 0 1	0 0 0 1
2	0 1 0	0 0 1 1
3	0 1 1	0 1 1 1
4	1 0 0	1 1 1 1
5	1 0 1	1 1 1 0
6	1 1 0	1 1 0 0
7	1 1 1	1 0 0 0

Table 4.6 : Circular code table

binary code. All other circular codes shown in this simple example have only two transitions. These transitions are implemented by using two cross-coupled differential amplifiers as shown in Figure 4.16. The MSB-1 is obtained by cross-coupling differential amplifiers at decimal level 3 with the amplifier at level 6. MSB-2 is obtained by cross-coupling amplifier at level 2 with the amplifier at level 5. In case of the LSB code, the amplifier at decimal level 1 is cross-coupled with the amplifier at level 4. At the output of the cross-coupled amplifiers a comparator is added to obtain at the clock transi-

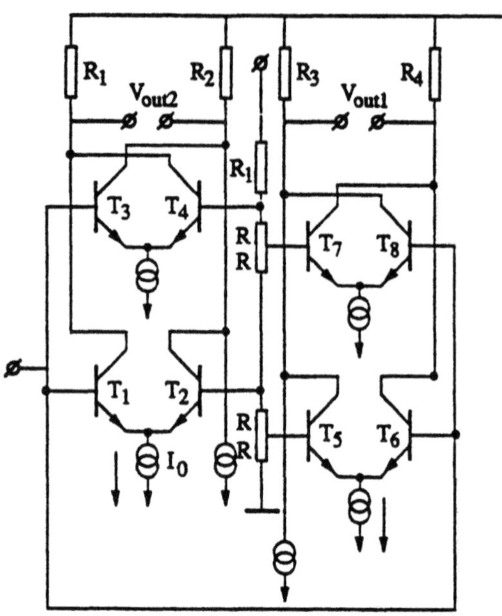

Figure 4.16 : Circular code circuit implementation

tions a digital output code. With an EXOR function the code transitions are
determined and converted into a binary code using a ROM structure. The
advantage of the circular code implementation is a reduction in the amount
of required comparators of at least a factor two. The loading of the "folding"
(cross-coupled amplifier stages) is identical for all output "bits" except the
MSB bit which only needs one amplifier-comparator stage. Up until a much
higher input frequency a more equal delay in the system is obtained. Similar circuits will be used in "folding" analog-to-digital converters described
in one of the following sections.

4.7 Two-step flash converters

To avoid some of the problems encountered with a full-flash converter the
two-step architecture was developed. This two-step method uses a coarse
and fine quantization to increase the resolution of the converter (see Figure
4.17). Examples of converters using the two-step architecture are given in
these references: [40,55,56,58,59,81,84]. Consider, for example, an 8-bit system that uses a 4-bit coarse quantization. After the coarse quantization has
been performed, the 4-bit digital data are converted into an analog value

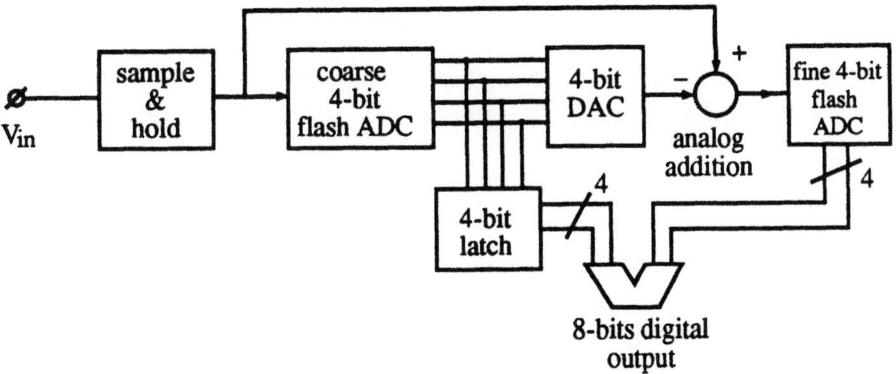

Figure 4.17 : Two-step A/D converter structure

again using a 4-bit D/A converter. This analog value is subtracted from the input signal and the difference is applied to a 4-bit fine converter which generates the fine code.

In this system an ideal coarse-fine signal level matching is expected. In practical applications, however, timing and accuracy limitations can result in conversion problems resulting in "missing" codes. To avoid such a problem, the full-scale range of the fine converter is increased with respect to the LSB step size of the coarse system. In this way a compensation for errors between coarse and fine conversion is obtained.

In this system only 40 comparators are needed to achieve 8-bit resolution. The (4-bit) D/A converter in these applications, however, needs to have an 8-bit accuracy and linearity to obtain the full 8-bit overall linearity. Furthermore, a sample-and-hold amplifier is needed to compensate for the time delay in the coarse quantization/reconstruction step. In this way the dynamic performance of the A/D converter is mostly determined by the performance of the sample-and-hold amplifier. In pipe-lined architectures a second sample-and-hold amplifier is used to store the analog remainder of the input signal for fine quantization. In the meantime a new input sample is coarse-quantized. This architecture results in a higher throughput rate for the total system. Higher resolutions can easily be obtained without a drastic increase in hardware and power dissipation. The applicability of this converter type depends on the availability of high-performance sample-and-hold amplifiers.

4.7.1   Bipolar two-step A/D converters

In Figure 4.18 an example of a two-step bipolar converter circuit is shown.
The sample-and-hold amplifier is followed by two 6-bit quantizer circuits

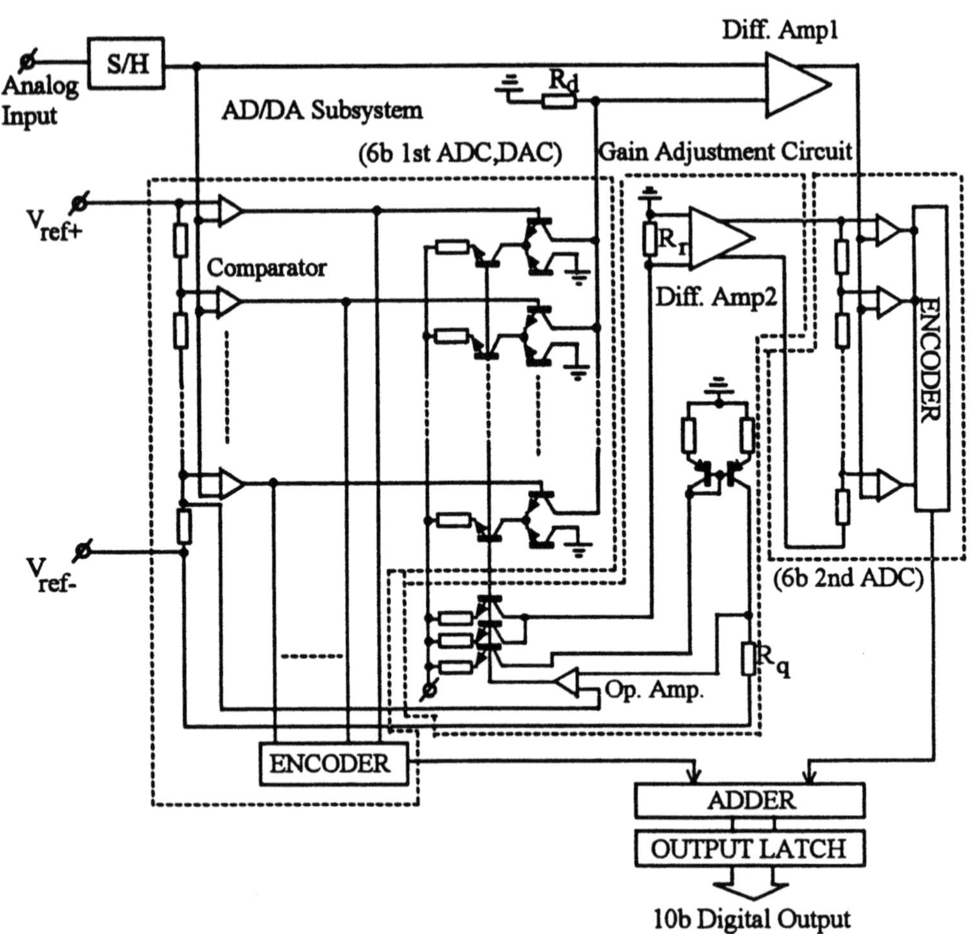

Figure 4.18 : Two-step bipolar A/D converter circuit

from which the first quantizer uses a 6-bit D/A converter to reconstruct the
analog quantized signal. The 6-bit quantizers are full-flash type A/D converters. At the output of the first A/D converter the thermometer code is
used to drive the segmented D/A converter. In this converter equal well-
matched currents are used to obtain the overall accuracy. Monotonicity can
be guaranteed because of the thermometer-code-driven D/A converter. The
INL of the D/A converter must be less than 0.05% to obtain $\frac{1}{2}$ LSB overall

linearity.

The output current of the D/A converter is applied to resistor R_d to reconstruct the quantized analog signal. This signal is subtracted from the input signal and amplified. This amplified signal is applied to the second 6-bit flash converter. The outputs of both 6-bit flash converters are decoded, and the overlap between the first and the second step is corrected. At the output a 10-bit binary code is obtained after adding the coarse and fine signals.

The reference voltage for the first flash converter is applied directly across the reference division resistors as V_{R+} and V_{R-}. The first coarse voltage reference step starting from V_{R-} is used to fix the D/A current value. A current mirror inverts the D/A current and generates a voltage equal to the coarse step across the resistor R_g. The operational amplifier controls the base voltage of the D/A current sources to fulfill this condition. As a result of this operation the D/A output signal is correlated with the full-scale value of the analog-to-digital converter. An extra D/A current is used to generate across R_r the reference voltage for the second flash converter. Differential amplifier(2) applies this reference voltage across the second string of reference resistors. In this way variations due to process parameters are canceled. A good matching of components in the D/A converter and the subtraction amplifiers are the most important design parameters.

4.7.2 MOS two-step A/D converters

In Figure 4.19 a MOS implementation of a pipe-lined A/D converter is shown. The system is built-up using a first 4-bit coarse quantizer with built-in D/A converter function and a second 5-bit fine quantizer. The first sample-and-hold amplifier samples the analog input signal. This signal is converted into a digital value using the first 4-bit flash A/D converter. The information of the first quantizer is stored and the D/A converter segment is switched on. The D/A converter consists of a string of resistors having a double function. The first function is the generation of the reference voltages for the 4-bit coarse flash converter. When a coarse quantization has been performed, then the output data are stored, and the thermometer encoding in the flash converter allows an easy control of the reconstruction D/A converter. This is the second function of the resistor string. From the figure it can be seen that in the D/A converter a double amount of taps is used. The subtracter circuit has a gain of two for signals applied at the non-inverting

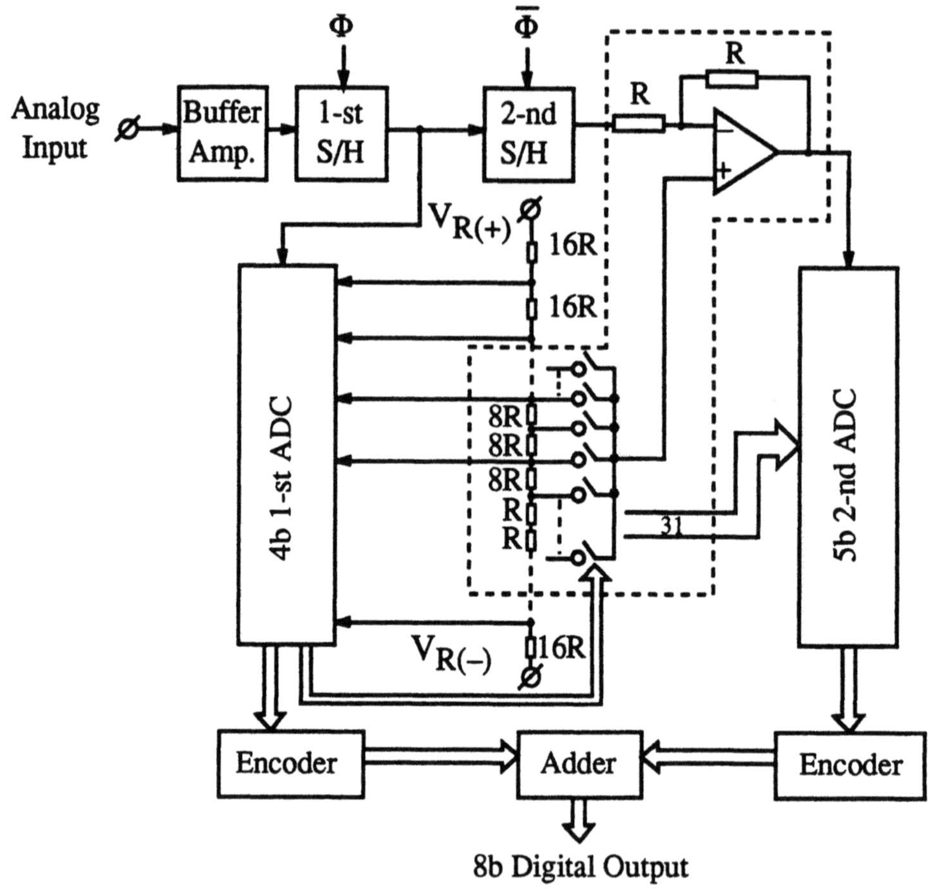

Figure 4.19 : Pipe-lined MOS A/D converter

terminal. The D/A voltage is applied at this side and therefore the step sizes of the D/A must be reduced to half the step sizes of the 4-bit coarse quantizer. The extra tap voltages are generated over the resistors $8r$. After completion of the coarse quantization, the input signal is transferred into the second sample-and-hold amplifier. Then the D/A output signal is subtracted from the analog input signal and the remainder is applied to the 5-bit fine quantizer. Next, fine quantization takes place and the data are stored in latches. A 5-bit quantization is used to allow corrections for overflow and offset errors from the second sample-and-hold amplifiers and gain errors in the subtraction circuit. Finally, the two digital signals of the coarse and the fine quantizer are added to obtain a corrected 8-bit output signal.

In Figure 4.20 a more detailed circuit implementation of the sample-and-hold amplifiers and the subtracter circuit is shown. The sample-and-hold

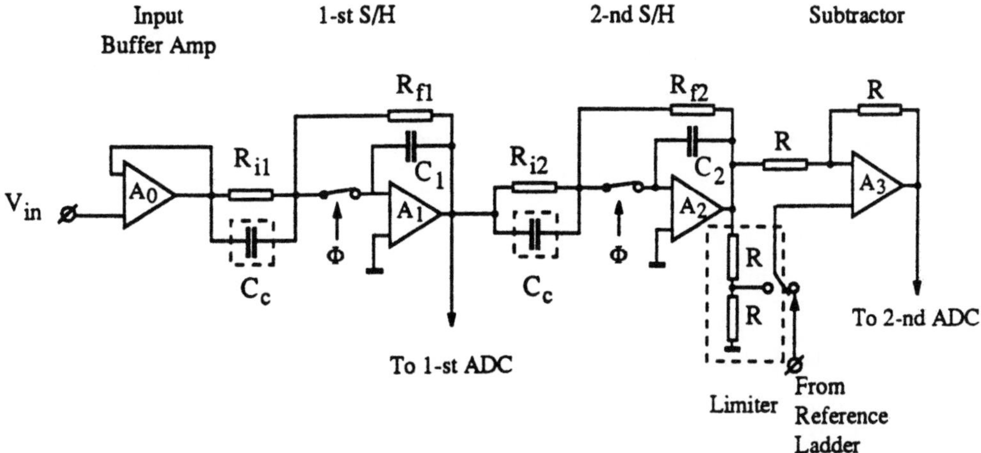

Figure 4.20 : Detailed circuit diagram of the S/H and Subtracter circuit

amplifiers are, as discussed above, of the inverting type. The connection of the hold capacitor in parallel with the feedback resistor R_F introduces a frequency limitation that is not allowed. Therefore, across the input resistor R_I a compensation capacitor C_C is connected. With equal time constants the signal bandwidth of the sample-and-hold amplifiers is only dependent on the bandwidth of the amplifiers. However, with a limited accuracy in compensation a not exact pole-zero cancellation is introduced. This pole-zero error results is slow settling components in the output signal.

4.7.3 Two-step A/D with sample-and-hold comparator

The ability of the comparator circuit from Figure 4.8 to sample the input signal accurately can be used advantageously. At the input an extra switch is needed to allow in succession a coarse and a fine comparison of the input signal sample with the reference voltages (see Figure 4.21). In this system a conversion requires at least three clock cycles (see Figure 4.22. During the first clock period (AZ = auto zero) the input signal is sampled and stored on the input capacitors. In the second clock cycle a coarse comparison (CC) is performed with, for an 8-bit system, 16 coarse quantization levels. The coarse quantization determines the two levels between which a sign change in data is obtained. This sign change indicates that between these coarse quantization levels the fine quantization must take place. During the third

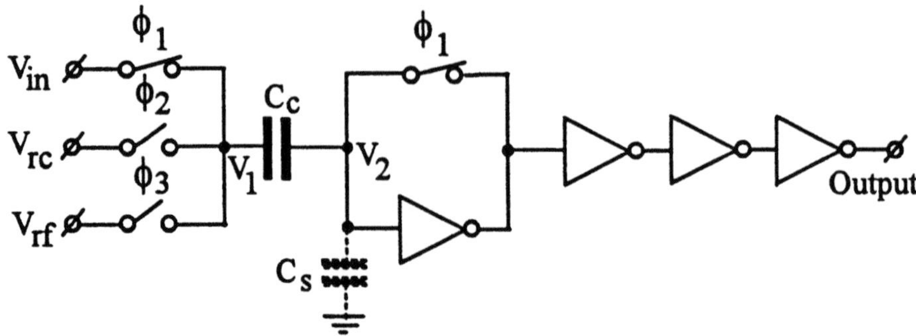

Figure 4.21 : Three input MOS comparator stage

clock cycle (FC) the fine reference levels between the previously determined coarse levels are applied to the same comparators and the fine conversion is performed. With 15 comparators an 8-bit converter can be designed. Such

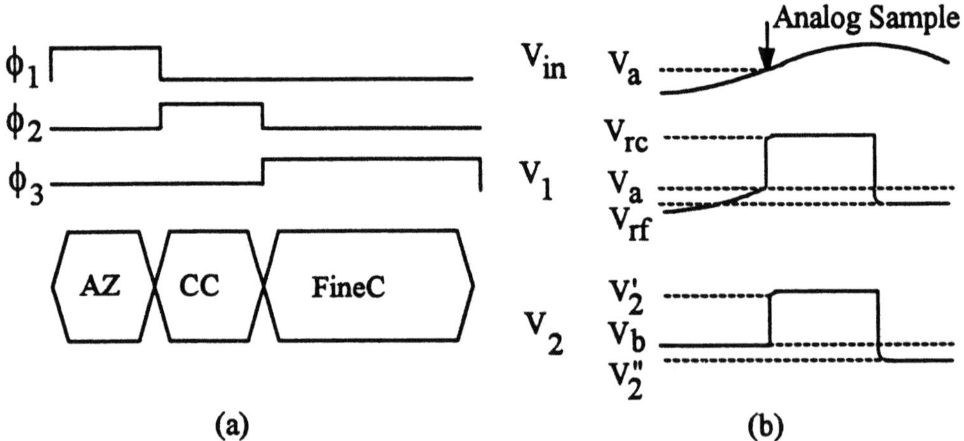

Figure 4.22 : Conversion process as a function of time

a converter, however, does not have any correction for decision errors made during the coarse conversion. When more comparators are used, an over-ranging during the fine conversion cycle can be implemented, resulting in a possible correction of the output codes. Different methods exist to generate the coarse and fine quantization reference signals.

In Figure 4.23 an example of a practical circuit using a resistor string is shown. The resistor string consists of $2^n - 1$ resistors. In an 8-bit approach 255 resistors are used. Splitting up the system into two nearly equal parts a 4-bit quantizer is needed. The coarse switches in the comparators from

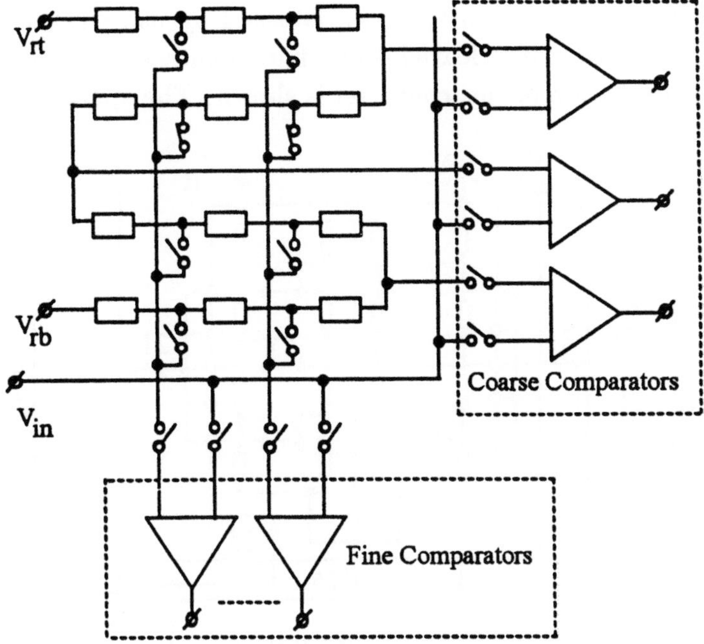

Figure 4.23 : Reference voltage generation system

Figure 4.21 are connected to the 15 coarse quantizing levels while the fine switches are addressed by a block selection system obtaining information from the coarse quantizer. In the 16 blocks 15 switches are used to generate the fine reference voltages between two coarse quantizer levels. The fine switch from Figure 4.21 consists of 16 switches in a practical application. One switch out of 16 is addressed by the block selection signal from the coarse quantizer. The construction of the reference generating system as discussed above insures monotonicity of the converter. A minimum amount of comparators is used, and as long as the switch feed-through and channel charge induced by the switches are well below the LSB level, then monotonicity can be guaranteed without the need of over-ranging comparators in the fine quantizer. A disadvantage of this system is the need for three clock cycles to obtain one complete conversion.

4.7.4 Interleaved comparator two-step A/D converter

To increase the conversion speed of an A/D converter with comparators running at maximum speed, the number of coarse and fine quantizer blocks must be increased. An interesting optimum between the number of components and the maximum conversion speed is obtained by using one 4-bits coarse quantizer and two time interleaved 4-bits fine quantizers. In Figure 4.24 the system implementation of this converter is shown. The operation of the con-

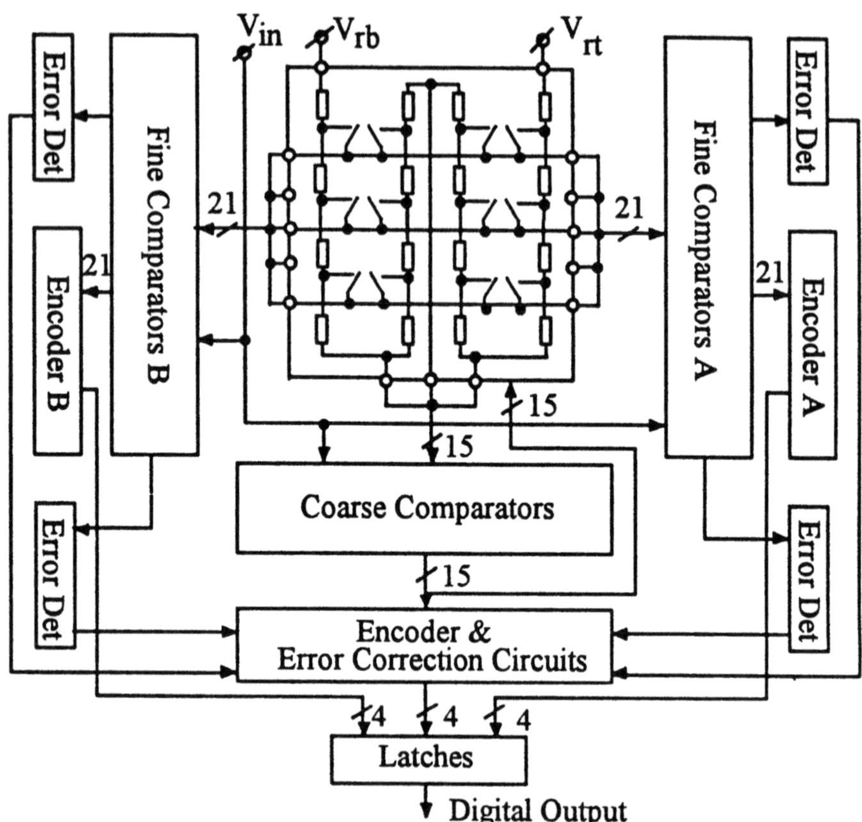

Figure 4.24 : Interleaved comparator two-step A/D converter system

verter with the time interleaving of the fine comparators is as follows. During the sampling and coarse quantization of the input signal the coarse quantizer and the fine quantizer (A) are connected to the input terminal. After the input signal has been sampled, the coarse quantization is performed. Then the information from the coarse quantizer is used to generate the fine reference levels to perform the fine quantization. Fine quantization is performed during a period equal to the time it takes to sample the input signal and perform a coarse quantization. The interleaving now occurs by taking the second input sample using the coarse quantizer and the second fine quantizer (B). At the moment the fine comparison of the first input sample is finished, the second fine converter (B) can start processing the second sample. During this processing the third input sample is taken using the coarse quantizer and the fine quantizer (A). This procedure repeats and the conversion speed can be increased to about two times the speed of the system from Figure 4.22. During the switching of the comparators an extra averaging period is used to increase speed and obtain a more accurate decision. In Figure 4.25 the operation of the fine quantizer is shown. At the moment the sampling

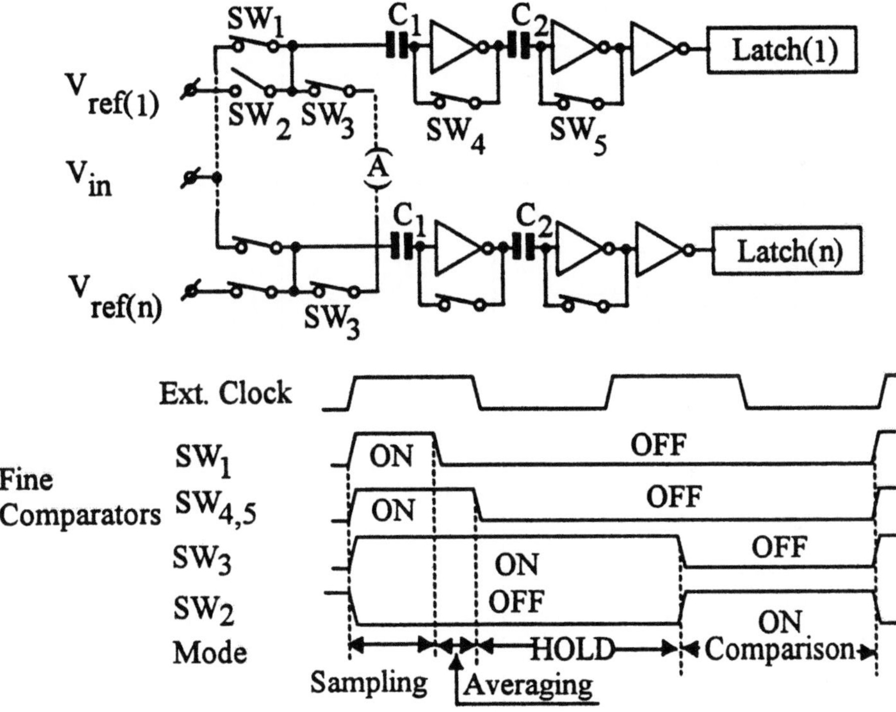

Figure 4.25 : Fine quantizer circuit configuration and timing diagram

switch SW_1 is switched off, then the comparators still remain during a short period of time into the short circuited sampling mode. The switches SW_3 remain closed which results in an averaging of the channel charge and feed-through charge of the sampling switches SW_1. These charges are sampled on the capacitors C_1. Then switches SW_4 and SW_5 are switched off. A short time later switches SW_3 are opened and switches SW_2 are closed to perform the comparison. By this circuit implementation a more accurate comparison of signals is possible. The averaging operation divides capacitive crosstalk and channel charge equally over all the sampling comparators resulting.

4.7.5   Two-step recycling A/D converter

By incorporating a binary-weighted capacitor D/A converter into a sample-and-hold amplifier system, a two-step recycling A/D converter has been designed. In Figure 4.26 the basic system is shown [86]. From Figure 4.26 the

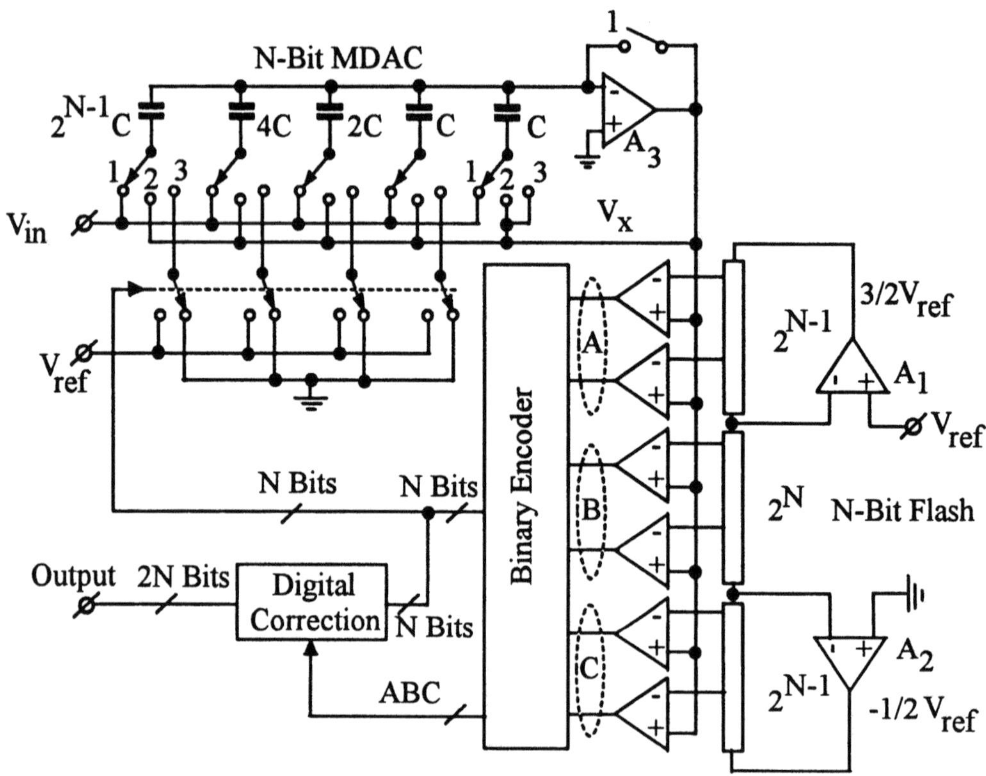

Figure 4.26 : Two-step recycling A/D converter system

multiplying binary-weighted capacitor D/A converter can be seen. This D/A converter with the switched buffer amplifier A_3 operates as a sample-and-hold amplifier, a residue amplifier, and D/A amplifier. The output signal of the sample-and-hold amplifier is applied to the quantizer, consisting of a parallel A/D converter with overflow. To simplify the understanding of the operation of the system, the three different phases of the quantization are shown in Figure 4.27 In the top figure the input signal sampling is shown. The input signal is sampled on the binary-weighted capacitor D/A converter array.

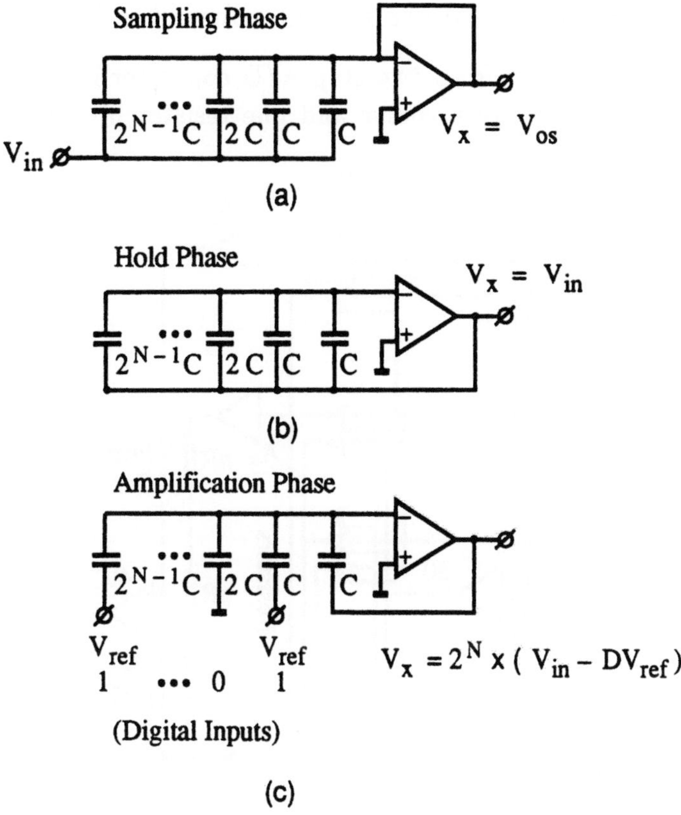

Figure 4.27 : Three operation phases of the two-step A/D converter

The middle figure shows the hold phase of the system. The capacitor D/A converter array is switched between the input and the output of the buffer amplifier. During the hold mode the coarse quantization is performed using the parallel quantizer. The coarse data are stored, and the D/A operation is performed which subtracts the coarse quantized signal from the analog input sample. In the bottom part of the figure this operation is shown. The residue signal is amplified by connecting a capacitor C between the input and the output of the buffer amplifier. An amplification with 2^N of the residue signal is obtained. Then the residue is quantized again to obtain the fine output code. The total output code is obtained by combining the fine and the coarse data.

Accuracy and linearity of this system are basically only determined by the accuracy of the binary-weighted capacitor array.

4.7.6 BiCMOS two-step A/D converter implementation

A BiCMOS solution for a two-step A/D converter is shown in Figure 4.28. The circuit consists of a coarse and fine resistor divider structure. During

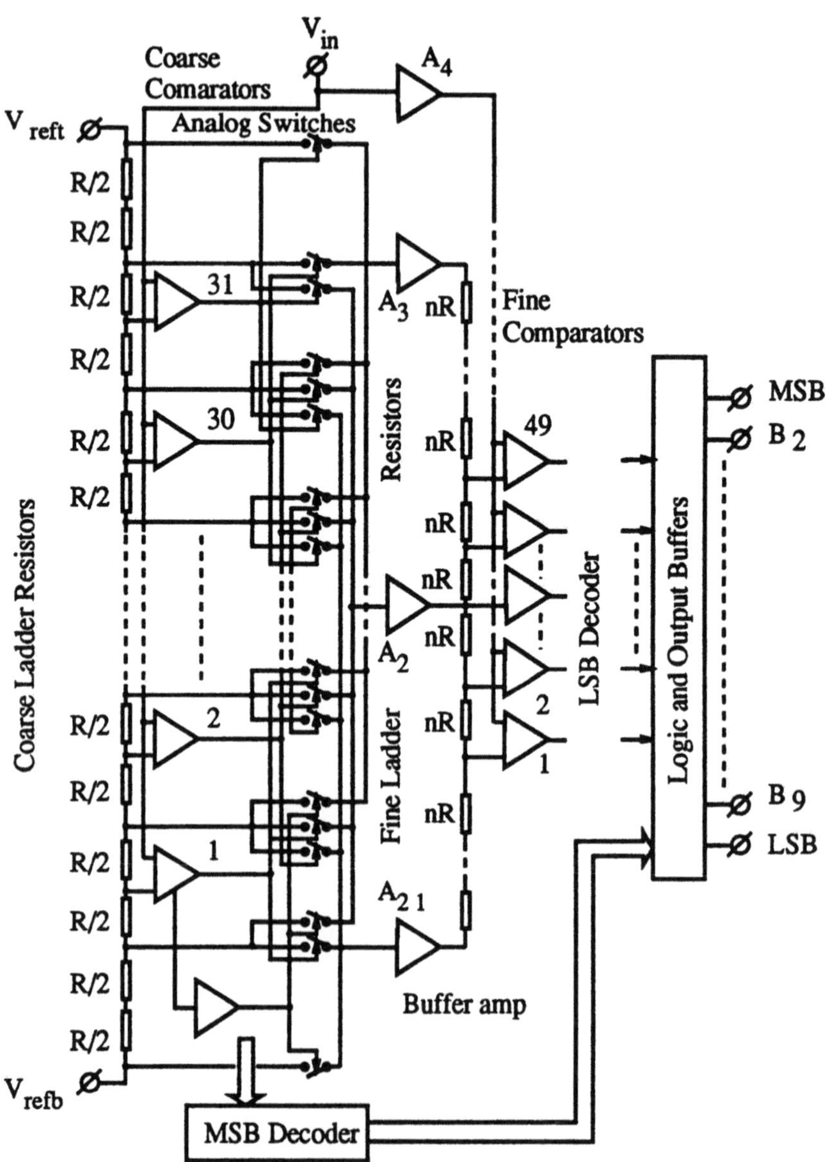

Figure 4.28 : BiCMOS two-step A/D converter

the coarse conversion the input signal is compared with the coarse reference voltage taps. When a decision is made, the fine reference voltage divider is connected across two coarse taps which have a voltage level just above and just below the input signal. In the coarse quantizer a resolution of 5-bit is used. In the fine ladder 49 comparators are used to have a correction possibility in case a small error is made during the coarse conversion. The fine reference voltage is applied via buffer amplifiers. The coarse code determines the ladder taps to which these buffer amplifiers must be connected. MOS switches are used for tap voltage switching while in the comparator stages bipolar transistors are used because of their low offset voltage. Note that the tap voltages to which the fine buffer amplifiers are connected are exactly in-between of the coarse reference levels. In this way an overlap between coarse and fine quantization equal to $\pm \frac{1}{2}$ coarse level is obtained. The additional levels are obtained by doubling the amount of coarse reference level generating resistors (two resistors of value $\frac{R}{2}$ instead of R are used). In the output logic circuit the over-range or under-range corrections are made to obtain a 10-bit output code.

4.8 Multi-step A/D converter

A basic system implementation of a multi-step analog-to-digital converter is shown in Figure 4.29 [82,83,34]. The system uses a cascade of parallel analog-

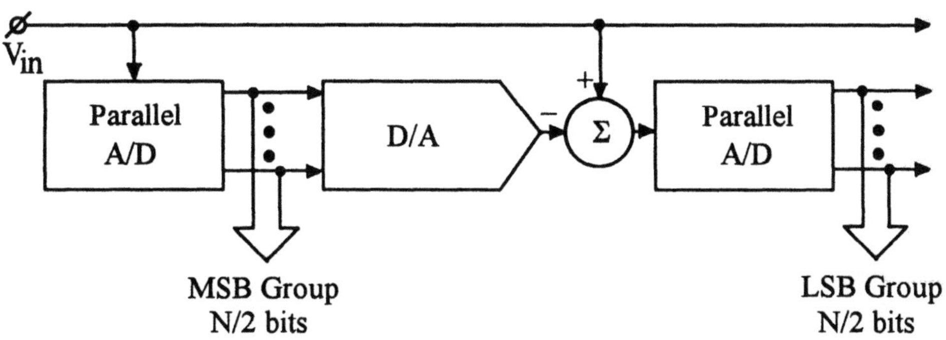

Figure 4.29 : Basic multi-step converter implementation

to-digital converters coupled with digital-to-analog converters to reconstruct the quantized analog signal and subtracters. The minimum resolution of the parallel A/D and coupled D/A can be 1 bit. Using the minimum 1-bit quantizer resolution, a multiple of N operations is needed to obtain a N-bit resolution. Mostly when the 1-bit resolution is used, sample-and-hold amplifiers are added to store the intermediate analog quantization error. A pipe-lined operation of the coder is obtained which results in a conversion speed corresponding to the time required for one basic quantization operation. With multi-bit conversion units the number of steps for a total conversion can be reduced. The advantage of this system is the high conversion speed and the small die area required. A disadvantage is the complexity of the system and the implementation difficulty of a fast high-accuracy subtracter circuit and a pipe-line sample-and-hold amplifier circuit.

4.8.1 Bipolar five-step A/D converter system

In Figure 4.30 an example of a multi-step A/D converter is shown [82]. The

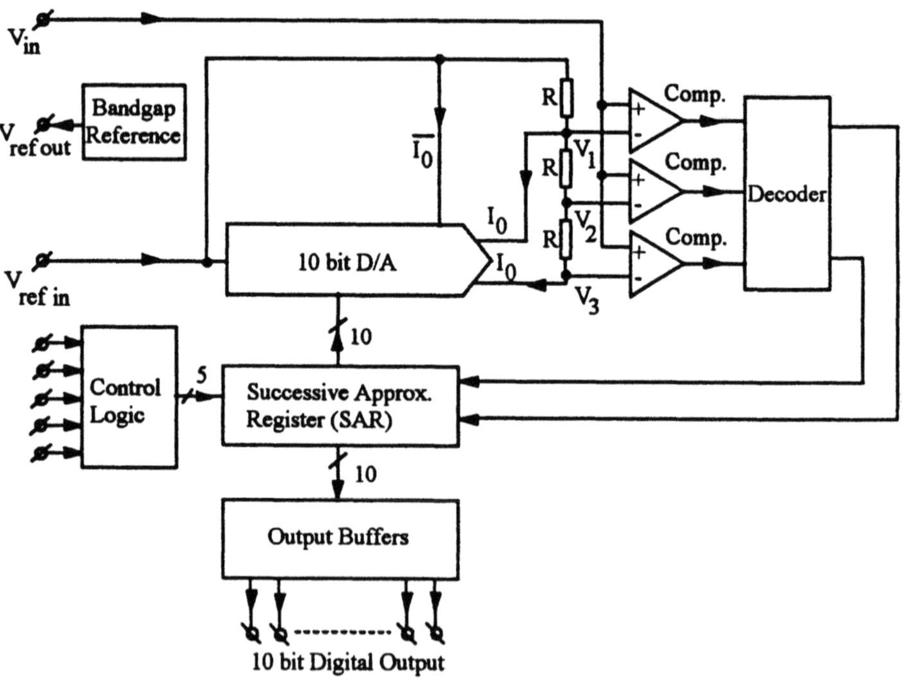

Figure 4.30 : Multi-step A/D converter

circuit consists of a three comparator quantizer stage with decoder, a successive approximation register which can handle more than one bit at a time, and a 10-bit D/A current converter. The output current of the D/A converter can be switched through the three reference resistors R at point V_3 indicated as current I_{01} or through the subtraction resistor R at the node V_1

indicated as current I_{02}. At the start of the conversion the input signal V_{in} is applied at the high input impedance nodes of the comparators. The current I_{02} is made zero and the output current of the D/A converter is switched as I_{01} through the three resistors R generating the three reference voltages. After a decision has been made, the D/A output current is switched as I_{02} through the resistor R to subtract a quantized voltage from the input signal V_{in}. The reference voltage level applied to the three comparators is now reduced by decreasing the D/A output current I_{01} so the next quantization step can be performed. This operation is furthermore repeated until the signal at the node V_1 is as close as possible to the input signal V_{in} and all the bit currents of the D/A converter are tried out during the conversion process. The digital output of the successive approximation register is applied to the output buffers to store the output code of the converter.

In Figure 4.31 a graphical approach of the conversion process during the five steps is shown. The table below the figure shows the values of the currents I_{01} and I_{02} for the five successive steps of the conversion process. The

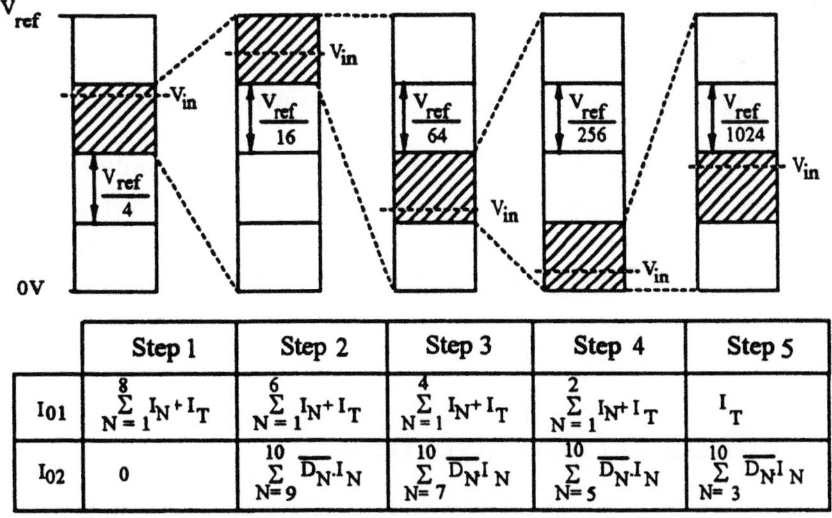

	Step 1	Step 2	Step 3	Step 4	Step 5
I_{01}	$\sum\limits_{N=1}^{8} I_N + I_T$	$\sum\limits_{N=1}^{6} I_N + I_T$	$\sum\limits_{N=1}^{4} I_N + I_T$	$\sum\limits_{N=1}^{2} I_N + I_T$	I_T
I_{02}	0	$\sum\limits_{N=9}^{10} \overline{D_N} \cdot I_N$	$\sum\limits_{N=7}^{10} \overline{D_N} I_N$	$\sum\limits_{N=5}^{10} \overline{D_N} \cdot I_N$	$\sum\limits_{N=3}^{10} \overline{D_N} I_N$

Figure 4.31 : Graphical analysis of the five conversion steps

linearity of the system is determined by the linearity and accuracy of the 10-bit D/A converter used in the system. Furthermore, the reduction of the reference voltages of the 3-level quantizer requires a low offset voltage of the input differential pairs used in this quantizer. As a result, the maximum resolution of the system depends on the matching of transistors in the process and the maximum input voltage used. In this system the offset of the 3-level quantizers is the dominant limiting factor for low-level applications. In Figure 4.32 the basic construction of the 10-bit D/A converter used in this system is shown. Note that the bit switches use three point switches. It is

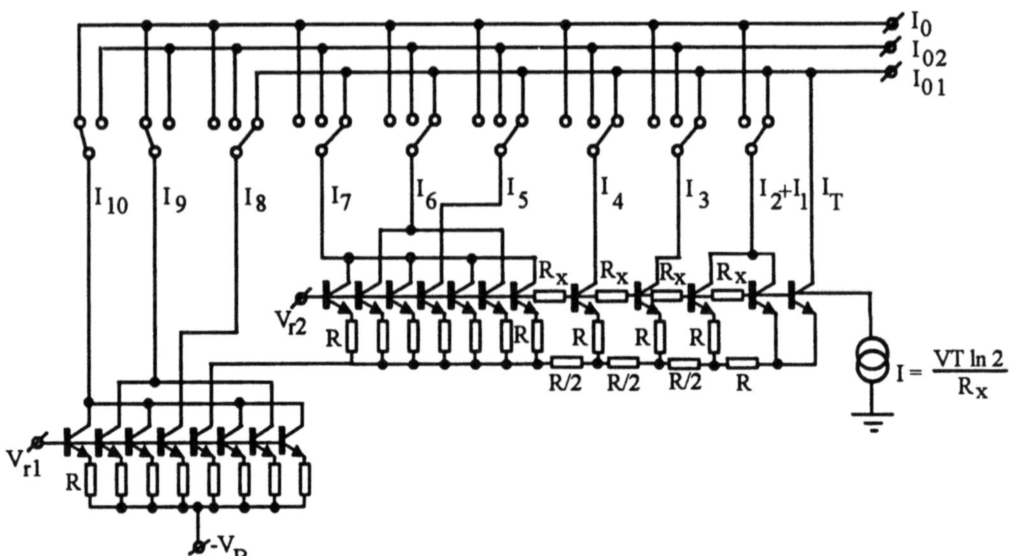

Figure **4.32** : 10-bit D/A converter system configuration

clear that by increasing the number of comparators the number of approximation cycles can be reduced, which results in a smaller conversion time. At the input of the converter, however, a sample-and-hold amplifier is needed to sample the analog input signal and hold this signal during the time the conversion takes place.

4.9   Folding A/D converters

In a folding analog-to-digital converter the advantages of the digital sampling of signals used in a full-flash system are combined with the component savings architecture of the two-step system. No sample-and-hold is required in this system [50,52,53]. The architecture uses analog preprocessing to transform the input signal into a repetitive output signal to be applied to the fine converter (see Figure4.33). In this system the most significant bits are determined by the coarse quantizer, which determines the number of times a signal is folded. The fine bits are determined by the fine quantizer which

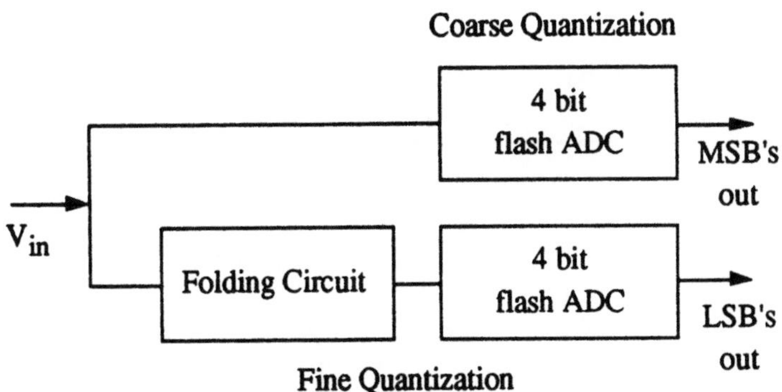

Figure 4.33 : Folding A/D converter architecture

converts the preprocessed "folded" signal into the fine code. In this way it is possible to obtain an 8-bit resolution with only 30 comparators (4-bit coarse plus 4-bit fine). Sampling of the analog signal at the same clock edge does not require the need for a sample-and-hold amplifier. The low component count results in a small die area, while more power can be expended into the system to extend the bandwidth of the comparator and folding stages resulting in a higher sampling speed and a larger analog input bandwidth. On the other hand a reduction in power can be obtained when sampling rate and analog input bandwidth are fixed. One drawback, however, is the higher repetition rate of the folded input signals that can result in rounding-off the tips of the folded signal. This rounding problem can result in a loss of resolution at the high-frequency end of the input spectrum when amplitude quantization is used in the conversion process. The moment a sample-and-hold amplifier is added to the system, the rounding problem is eliminated and the only speed limitation is the settling time of the system.

4.9.1 Current-folding A/D converter system

A simple example of a circuit implementation of a 4-level (2-bit) folding circuit is shown in Figure 4.34. The most important parts are the reference current sources I, the common-base transistor stages T_1, T_2, T_3, and T_4, and the diodes D_1, D_2, and D_3. The input signal current I_{in} is compared with the four reference currents I. The collectors of the odd-numbered transistors of the common-base stages T_1 and T_3 are connected and so are the even numbered transistors T_2 and T_4. If the input current I_{in} is equal to zero, then through all common-base transistors a current I will flow. This results in a current $2I$ flowing through each of the resistors R_1 and R_2. With equal-

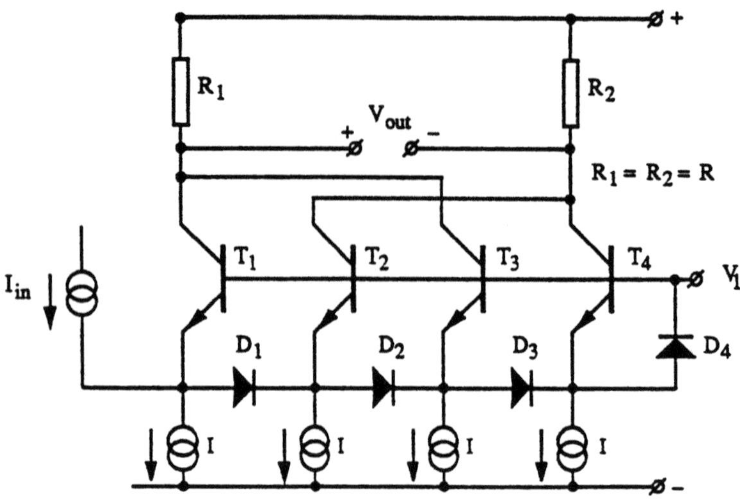

Figure 4.34 : Current-folding 2-bit A/D converter structure

valued resistors, the differential output voltage V_{out} will be zero. If an input signal current I_{in} is applied, for example, $1\frac{1}{2}I$, then this current is subtracted from the reference current flowing through T_1. The difference in current, being $\frac{1}{2}I$, will forward-bias the diode D_1 and will be subtracted from the reference current I flowing through T_2. As a result, the current through R_1 is reduced to I and the current through R_2 is reduced to $1\frac{1}{2}I$. In Figure 4.35 the input signal changing from 0 to $4I$ and the corresponding output signal of the folding stage as a function of time are shown, respectively. The result of the operation is an output signal with a frequency that is a multiple of the input frequency. In this particular case with a full-scale input signal the output frequency is four times higher than the input frequency. Moreover, the output amplitude is reduced from $4I$ into I. The output signal of the circuit can be applied to a following quantizer which may consist of a flash A/D converter. The digital output signals of the coarse quantizer are determined by the conduction or non-conduction of the coupling diodes D_1, D_2, or D_3. The voltage excursion at the input of the circuit is equal to the voltage across the number of conducting diodes. In the example given in Figure 4.34 the maximum voltage excursion will be between zero and $3V_D$ plus the voltage across the overload clamp diode D_4. When for an increasing resolution of this system the number of reference sources is increased, the voltage excursion at the input increases and may exceed the maximum reverse bias of the base-emitter voltage of the transistors. Reverse biasing of high-speed bipolar transistors with voltages above the reverse zener voltage will deteriorate

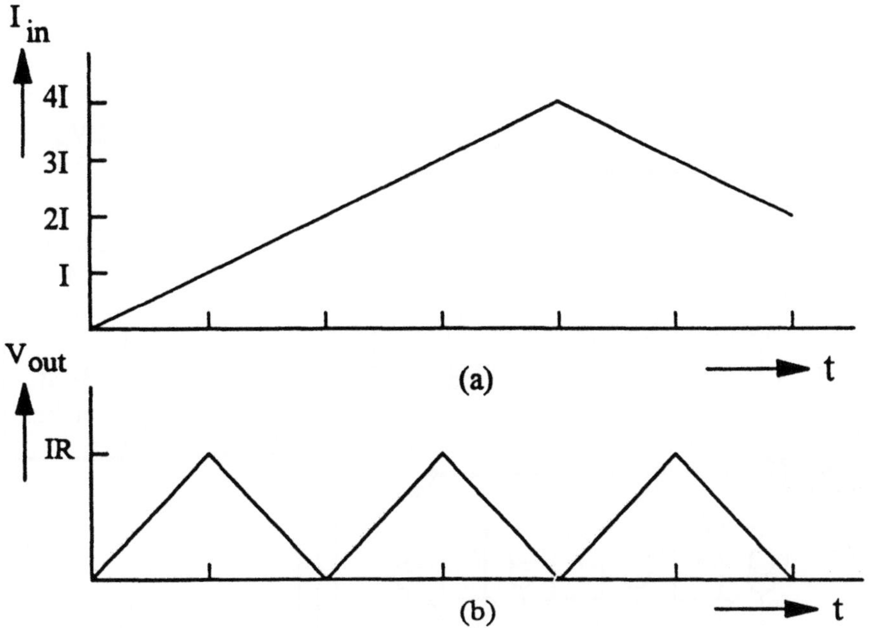

Figure 4.35 : Input and output signal as a function of time

these devices. To overcome this problem, more quantizer circuits can be connected in parallel to interpolate additional levels within one diode voltage. The diode switching quantizes the input signal with a very high accuracy. The overall conversion accuracy is thus determined by the accuracy of the reference current sources I. The current gains of transistors T_1, T_2, T_3, and T_4 only influence the accuracy of the "folded" fine output signal. With a practical value for the current gain β of 100 and a matching between the current gains of about 10% a maximum error in the fine output signal of about .1% is obtained giving about 10-bits (fine) accuracy.

4.9.2 Parallel connection of quantizers

The parallel connection of two coarse quantizers of the type shown in Figure 4.34 that interpolate within one diode voltage is shown in Figure 4.36. Diode D_7 and the voltage sources V_1 and V_2 are added for this purpose. The voltage difference between V_1 and V_2 is equal to $\frac{1}{2}V_D$. So the voltage V_1 must be adjusted to: $V_2 = V_1 + \frac{1}{2}V_D$. The parallel connection of the second coarse quantizer also eases the interconnection of the odd or even numbered collectors as shown in Figure 4.34. As can be seen from Figure 4.36, collectors T_1, T_2, T_3, and T_4 are connected to resistor R_1. The same operation is

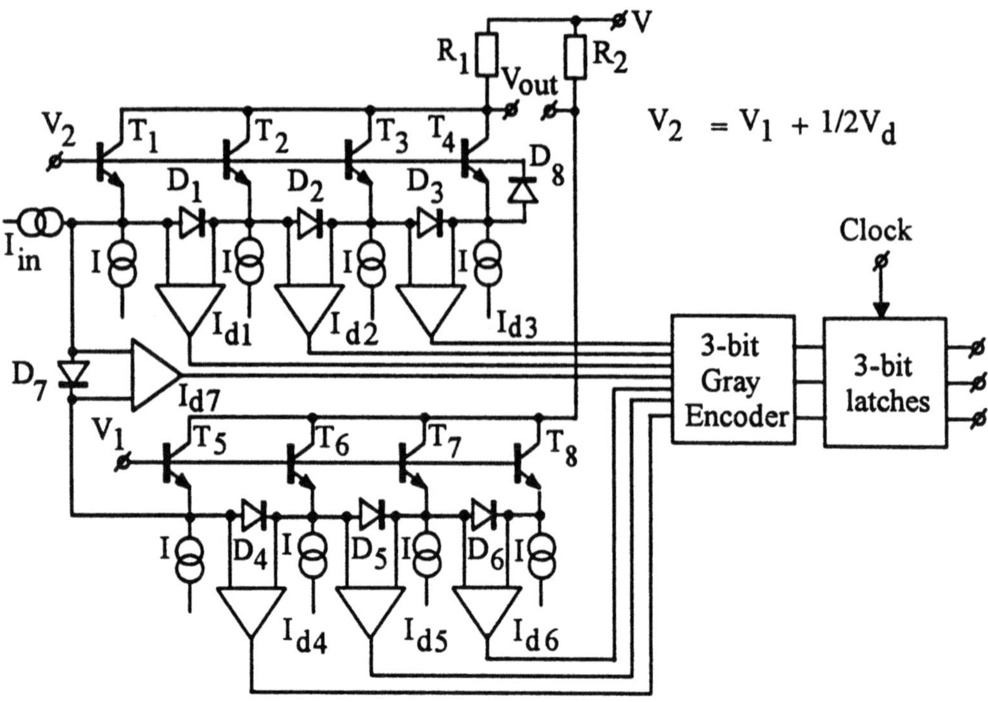

Figure 4.36 : 3-bit coarse quantizer configuration with digital signal encoding

performed with the collectors of T_5, T_6, T_7, and T_8 which are connected to resistor R_2. The differential output voltage between resistors R_1 and R_2 shows again a triangular form in case of an input current linearly increasing from 0 to $8I$. This can be explained by supposing that the input current I_{in} increases from 0 to $8I$. At first the current I flowing through T_1 is subtracted. Then diode D_7 starts conducting and the current through T_5 is subtracted from the input current. The voltage difference between V_1 and V_2 which was equal to half a diode voltage performs the switching in conduction of diodes from one coarse quantizer to the other and vice versa. Diodes D_1, D_4, D_2, D_5, D_3, D_6, and D_8 will become active alternately. In this way the eight quantization levels are tripped by the input signal, and the maximum input voltage span is equal to four diode voltage levels. The digital signals are detected by the level detectors I_{d1} to I_{d7} which are activated by the forward bias voltages of the conducting diodes D_1-D_7. The output signals of the level detectors are fed to an encoder circuit which converts the seven input signals into a 3-bit Gray output code. This output code is latched to give a digital output signal.

4.9.3 Fine quantizer circuit

The coupling of the coarse quantizer with a 4-bit fine quantizer is shown in Fig. 4.37. In the fine quantizer system a flash-type A/D converter structure

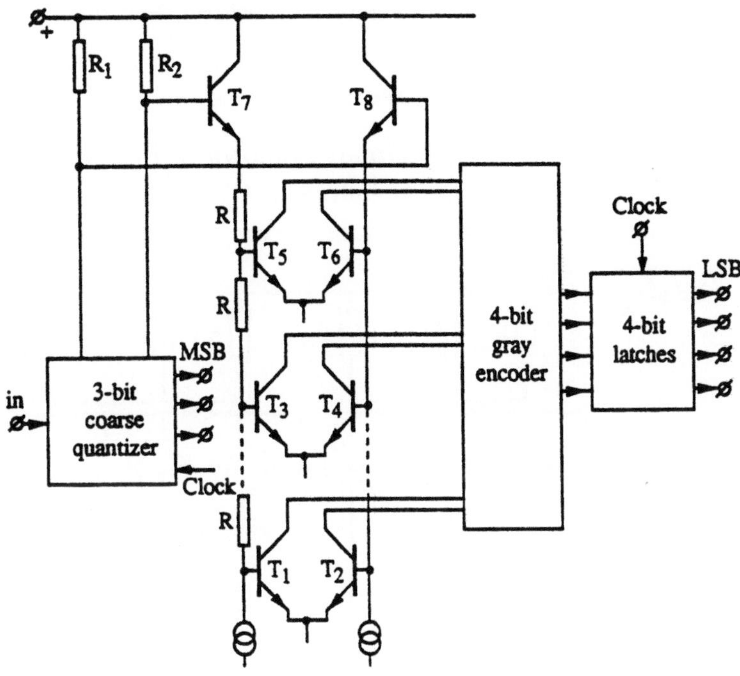

Figure 4.37 : 4-bit fine quantizer system

is used. The fine encoder is driven via emitter followers to avoid loading of the coarse system by the large amount of differential amplifiers. The differential amplifier stages are used as comparator stages. The output of these stages is encoded and converted into a 4-bit binary code. This binary code is latched in four latches to give the digital output code of the circuit. The small delay in the coarse quantizer, which is determined solely by the switching on and off of the diodes and the common-base transistors, allows digital sampling of the input signal by latching comparators. Furthermore, this circuit configuration processes the analog input signal independently of the sampling frequency. There is, therefore, no coupling between the maximum analog input frequency and the maximum sampling frequency that can be used. Usually a much higher sampling frequency can be used without affecting the accuracy of the system. Maximum sampling frequency is determined by the bit error rate of the comparators. Supposing that in a system only 1 error per second is allowed, then the maximum sampling frequency is $f_{max} = \frac{1}{BER}$. As is known, oversampling increases the resolution of a system without needing additional comparators or other circuitry. Note also that the differential fine comparator stages reject the common mode signal part that is present on the voltage across R_1 and R_2.

4.9.4　Fine encoder-latch circuit

A detail of the fine encoder-latch circuit is shown in Fig. 4.38. To minimize the amount of comparators, the encoding of the analog signal into a Gray code is performed first. From Fig. 4.38 it can be seen that the differential

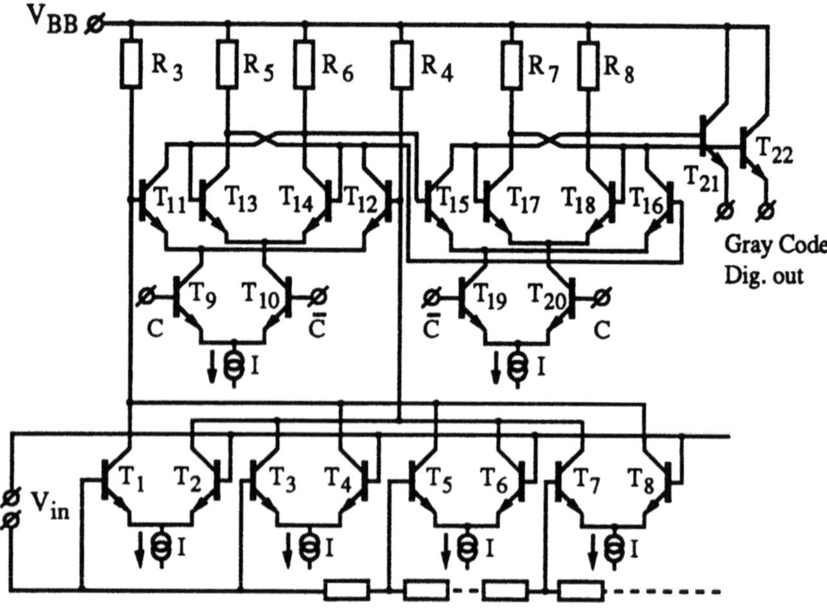

Figure 4.38 : Detail of fine encoder-latch circuit

comparator- amplifier stages are cross-coupled and connected to the input of a master-slave flip-flop. This flip-flop acts as a sensitive clocked comparator. When four stages are cross-coupled, then information about the zero crossing of one differential stage is obtained as long as the other three stages are in the over-drive position. In this way the number of comparator stages can be reduced by a factor four. Note, however, that due to the fact that the other three differential stages are over-driven, the bias currents of these amplifiers still flow through the collector resistors. As a result, a much better matching of these currents I is needed, while an additional requirement is made for the matching of the collector resistors to obtain an accurate level comparison. The master flip-flop is latched when the sampling clock signal becomes low. When the clock signals becomes high, the information of the master flip-flop is transferred to the slave flip-flop. At the output a Gray code is obtained which can be easily converted into a binary output code.

4.9.5 Complete A/D converter

Using the current folding architecture, a 7-bit A/D converter IC has been designed [51]. At low frequencies this system performs fairly well. The simplicity of this system results in a small die size of 2.4 x 2.5 mm. A drawback of this system at high frequencies is the rounding of the folding signal. This rounding results in a loss of fine codes. Also the delay between the coarse and the fine quantizer decisions results in code errors at the moment when the most significant bits are changing. These code errors result in missing codes that are not allowed. In the next part of this chapter improved systems will be discussed that do not suffer from these drawbacks.

4.10 Double folding system

To avoid missing codes resulting from rounding of folding signals at high frequencies a double folding system has been developed [52]. A block diagram of the double folding system is shown in Fig. 4.39. The system consists, for example, of a 3-bit coarse quantizer that determines the coarse bit and a folding system which generates two analog output signals with a 90-degree phase shift. The analog output signals from the folding circuit are applied to two 4-bit fine quantizers to determine the fine output code. The basic idea for using two fine quantizers with 90-degree phase shifted input signals is that at the moment the signal applied to, for example, fine quantizer 1 runs out of its linear range, fine quantizer 2 will come into its linear range and vice versa. Furthermore, the connection of the folding signals to two 4-bit fine quantizers increases the resolution to 5 bits. In this way a total resolution of 8 bits for this converter structure is obtained. In Fig. 4.40 the output waveforms of the folding circuit are shown. The figure shows that when output 1 is in the linear transfer range, output 2 is in the non-linear range. When these waveforms are compared with the single folding system, the coarse bit should be determined at the tips of the first folding signal. In this double folding system these "tips" coincide exactly with the zero crossing of output signal 2. The output signals of the coarse and

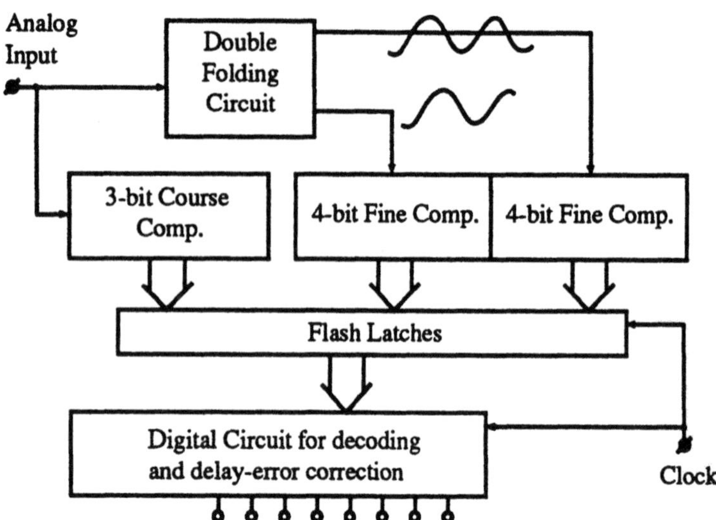

Figure 4.39 : Block diagram of double folding system

fine quantizers are latched and applied to an error correction circuit. In this circuit, delays due to the different transit times in the fine and coarse parts of the converter are eliminated. The delay-error correction is based on the idea that the coarse quantizer determines which bit transitions can be erroneous. At that moment a correction is applied which depends on the code of the fine quantizer. The fine quantizer information is accurate to the least significant bit and therefore the exact coarse code transition can be determined.

4.10.1   Practical double folding circuit

A simplified circuit diagram of the double folding system is shown in Fig. 4.41. The "phase" shift between the two analog output signals of the folding circuit is obtained by connecting one folding processor to a shifted reference voltage with respect to the other folding processor. In practice this shifting is obtained by connecting the input differential amplifiers of the first analog processor to resistor taps that are in-between the resistor taps to which the second analog processor is connected. When an input ramp signal is applied, then due to the time difference it takes to reach the reference levels for the first and the second analog processor block, the exact signal shift is obtained. The collectors of the differential amplifiers in the first or the second processor are cross-coupled to obtain a repetitive output signal. This output signal

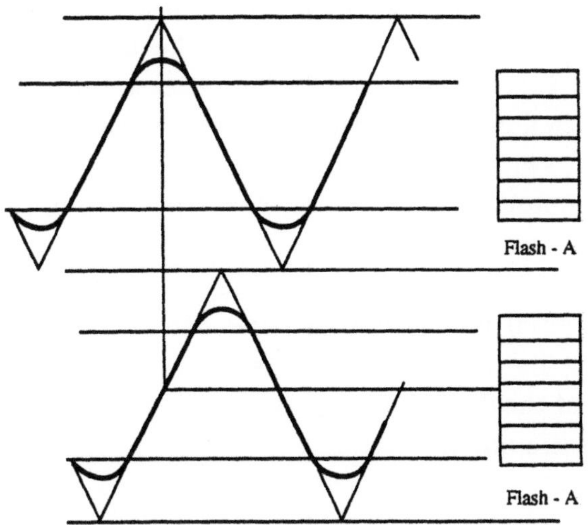

Figure 4.40 : Output waveforms of the double folding circuit

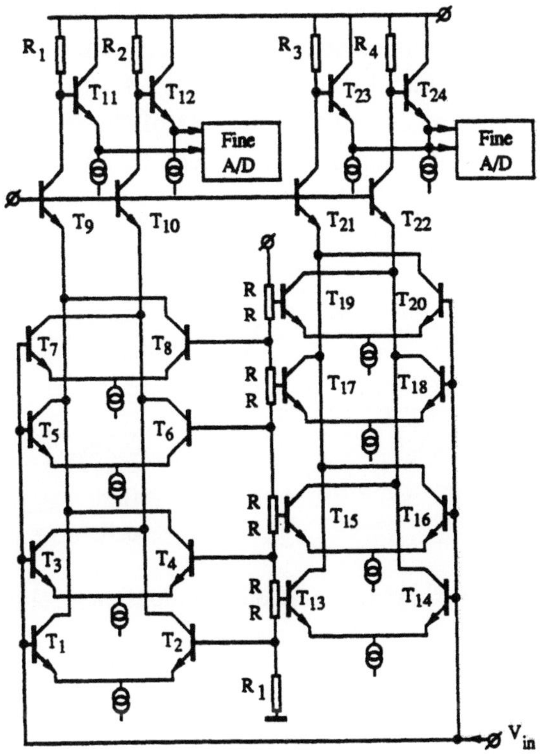

Figure 4.41 : Practical double folding circuit arrangement

depends only on the switching of one differential amplifier pair. All other pairs are over-driven so no interaction on the output signal is obtained. One drawback of this connection of differential amplifier pairs is the large bias currents which flow through the collector resistors. Signal information is obtained from one switching differential pair, while the remaining over-driven pairs add a "common mode like" offset to the output signal. These bias currents can cause extra offsets in case of a poor matching of the tail currents of the differential amplifiers or the matching of the resistors. Therefore, the number of differential amplifier pairs that can be combined in this way is limited. The two fine quantizer systems are fully differential. To avoid an extreme loading of the double folding system by the two fine quantizer stages emitter followers are added.

4.10.2   Fine quantizer stages

The rounding of the folding signals that are applied to the two fine quantizer systems requires only comparator stages over the linear operation range of the folding stage. The total number of comparators does not need to increase more than the resolution required in the fine converter. The following explanation will clarify this statement. In Fig. 4.42 the quantization of the nonlinear output signal at one of the outputs of the folding circuit is shown. Suppose an ideal triangular signal is generated by an ideal folding circuit. In

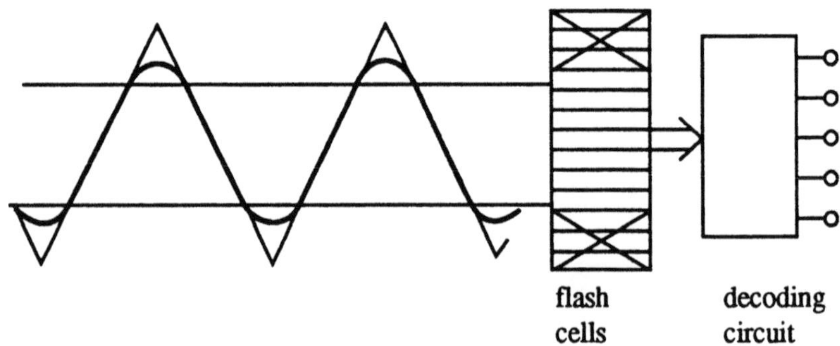

Figure 4.42 : Fine quantization of a nonlinear folding signal

this case $2n$ fine comparator cells are needed for quantization of the signal. Now apply the nonlinear folding signal to the same fine comparator system. Due to the rounding of the signal a reduction in the useful linear range of the system is obtained. In practice this means that the number of effective comparator cells is reduced. At the point when the two non-linear folding signals are each other's complement, for each fine quantizer system the number of effective comparators is reduced by a factor two. This results in n effective comparators cells per branch. The $n/2$ units at the top and bottom shown in the figure can be deleted because they do not add significant information to the system. As a result, in the two fine quantizer systems a total of $2n$ comparators is needed to quantize the two folding signals. No difference in the number of comparators between an ideal single folding system and the double folding system is found.

4.10.3 Delay-error correction

In Fig. 4.43 the two fine quantizer input signals are shown. In the top

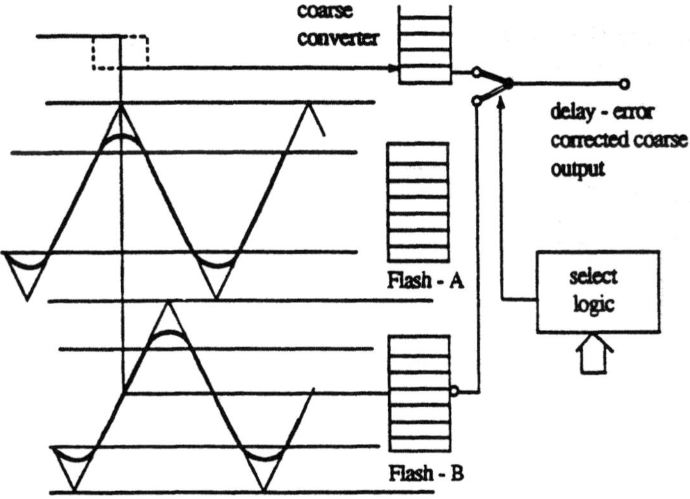

Figure 4.43 : Two fine quantizer input signals

part of Fig. 4.43 the switching of the coarse bits is shown. The dashed boxes show the maximum switching time uncertainty that can be corrected by the delay-time correction system. The exact switching of the coarse bits must coincide with the switching of the middle comparator of the second fine comparator array. Differences in switching moments occur because the folding signal and the coarse quantizing signal pass through different amplifier stages. These delays can introduce erroneous output codes. A system whose basic operation is shown in Fig. 4.43 overcomes this problem. In principle, the two fine quantizers contain all the important information needed for the delay correction. The only additional information required is the coarse region in which the fine quantization occurs. This information is determined by the coarse quantizer, but the exact moment of transition of the coarse bit is not transferred to the output as digital data. A logic selection circuit now determines the weak code combinations. Those MSBs that have to be corrected are in turn determined. A correction is obtained by inserting the zero crossing information of the second fine quantizer as shown in Fig. 4.43 instead of the coarse code information. An exact and corrected output code is now obtained. In the figure the second fine quantizer is denoted as "flash-B." A time correction equal to plus or minus half the coarse region can be obtained.

4.10.4   Complete double folding A/D converter

On a 3x4.2 mm^2 chip area an 8-bit A/D converter has been implemented using a standard bipolar process having double metal interconnections. f_t = 350 MHz and the minimum emitter size is 10x10 μm^2. The performance obtained is shown in Table 4.7. The chip contains an input amplifier and

Resolution	8 bits
Maximum analog input frequency	6 MHz
Input voltage range	1 V_{pp}
Maximum clock frequency	20 MHz
Digital output signals	ECL compatible
Supply voltage	5.2 V
Supply current	100 mA
Chip size	3 $\times$ 4.2 mm^2
Differential phase	0.5 degree
Differential gain	1.5 %
No sample-and-hold amplifier required	

Table 4.7 : Converter data

digital output buffers.

4.11   Folding and interpolation systems

One drawback of the double folding converter system is the accurate matching between the coarse and fine quantizers that is needed to avoid missing codes. The two fine quantizers also have a matching requirement at high frequencies. This problem is introduced by the rounding of the folding signals at high frequencies. At that moment the rounded signal undergoes an amplitude-limiting action because of the limited practical bandwidth of a system.

The problems caused by the rounding of the triangular waveform at higher frequencies can be avoided by not quantizing these waveforms in amplitude. In that case several folding signals with suitable offsets are required to ensure that there is always at least one folding signal that is close to its zero crossing. Each folding signal has to be generated by its own folding circuit. This results in the same number of folding amplifiers as the number of levels that must be compared. Should an 8-bit circuit be designed, 255 folding amplifiers are needed. No difference between a full-flash system and this folding system is found concerning input capacitance of the circuit. An important difference of this architecture with respect to the previously given folding systems is that only the zero crossings of the folding signals are used to determine the value of the least significant bits. This has the advantage that the results no longer depend on the shape of the folding signals. There is, therefore, no need to create an exactly linear triangular folding signal, even at low frequencies. The folding signals actually used in this converter do not resemble triangular signals at all except for a small, almost linear, region around the zero crossings. Also, the use of zero crossings eliminates the need for accurate handling of the input signals through the analog input folding system. Furthermore, the input range of the fine-flash converters does not need to match accurately the range of the folding signals. The folding principle is used in this system to obtain a multiple of zero crossings that can be detected by a single comparator per folding signal. In this way a reduction in hardware is obtained.

The coarse quantization system can be described using a repeating circular code implementation as shown in Table 4.8. The repeated use of a circular code implementation results in a reduction of hardware in the coarse quantizer. The only indication needed to obtain a full output code is the number of repetitions. An extra encoding is needed for this purpose as will be shown in the detailed description of the folding encoder circuit.

In Table 4.8 a circular code generation is obtained by four systems (a, b, c, and d) each having a one coarse step reference level shift. A four times rep-

BINARY CODE	REPEATING CIRCULAR CODE
a b c d	a b c d
0 0 0 0	0 0 0 0
0 0 0 1	0 0 0 1
0 0 1 0	0 0 1 1
0 0 1 1	0 1 1 1
0 1 0 0	1 1 1 1
0 1 0 1	1 1 1 0
0 1 1 0	1 1 0 0
0 1 1 1	1 0 0 0
1 0 0 0	0 0 0 0
1 0 0 1	0 0 0 1
1 0 1 0	0 0 1 1
1 0 1 1	0 1 1 1
1 1 0 0	1 1 1 1
1 1 0 1	1 1 1 0
1 1 1 0	1 1 0 0
1 1 1 1	1 0 0 0

Table 4.8 : Repeating circular code implementation

etition is obtained for every encoding system. A designer can make a choice concerning the amount of systems to be used in parallel or about the number of repetitions of the circular code. A repetition number between four and eight gives a good optimum between the bandwidth in the system and the amount of hardware to be used. Furthermore, the number of stages which practically can be used in parallel is between four and eight. An example of an eight by eight function will be described in more detail in the upcoming sections of this chapter.

As already mentioned, the number of folding amplifiers is equal to the number of reference levels to which the signal must be compared and thus the advantages of the folding architecture seem to be limited. However, this problem can be solved by creating only a small number of folding signals and deriving the other folding signals by resistive interpolation between the outputs of two adjacent signals. This requires a limited linearity specification for the folding amplifiers. A linear interpolation between the two adjacent signals with an accuracy less than $\frac{1}{2}$ LSB of the interpolated system is required. In this case, the 32 folding signals necessary for the five least significant bits are derived from a four times interpolation between eight output signals of the folding encoder. An extension of this interpolation system to 10 or 12 bits by increasing the number of times the input signal is folded, the number of folding circuits, and/or the number of interpolated folding signals versus folding circuits, is relatively straightforward. This makes this architecture very suitable for high-resolution converters that require a large analog bandwidth. In practice, a maximum of eight times interpolation can be used to increase the resolution. The repetition rate comparators need to detect zero crossings increases, but the time accuracy to perform a zero comparison remains the same as in a full-flash converter.

4.11.1 Voltage-folding A/D converter system

The explanation of the folding and interpolation system will be performed using simplified circuit blocks. After the basic explanation of the operation of the system the description of a practical implemented system will be given. A very good optimum between performance, power consumption, and die size is obtained.

4.11.2 Most Significant Bit generation

As a start the generation of the MSB bits will be described. In Fig. 4.44 the basic system of a differential amplifier with the transfer curve from input to a single ended output is shown. With the simple differential pair only one signal level decision is obtained. In the folding system a repetition of decision signals is required. To perform such an operation, a coupled differential pair as shown in Fig. 4.45 is used. When the input signal increases from a level smaller than V_{lo} to a level above V_{hi}, then the output signal changes from low to high and returns to low as shown in the figure. A repetition of decision signals is obtained.

The MSB signal generation in the system is obtained by using two coupled differential pairs with reference 0, $\frac{1}{2}$, and 1. In Fig. 4.46 the use of the two coupled differential pairs is shown. Note from Fig. 4.46 that the lower coupled differential pair generates the inverted MSB signal while the upper

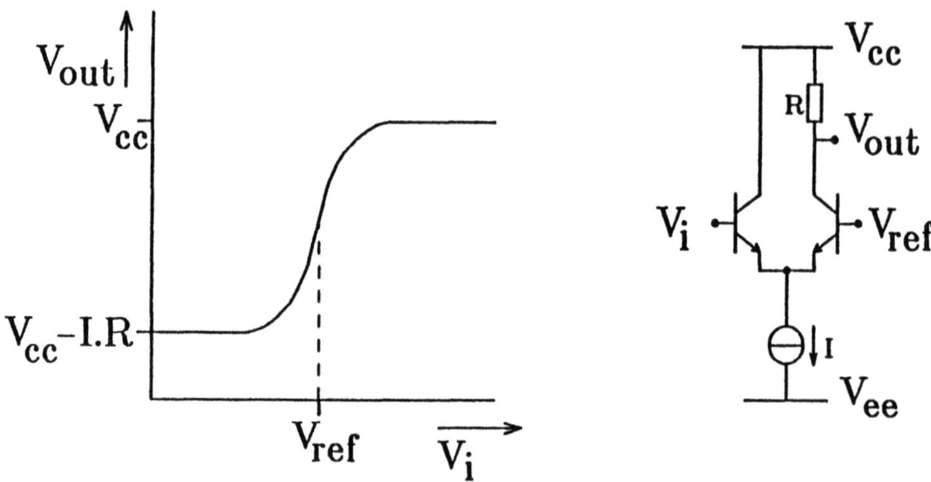

Figure 4.44 : Single differential pair

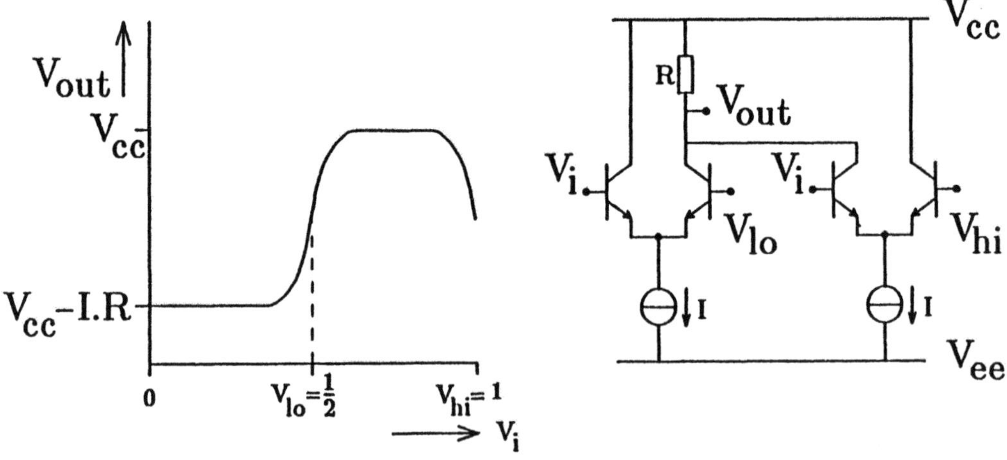

Figure 4.45 : Coupled differential pair

coupled differential pair generates the MSB signal. The difference between the two output signals exactly gives the transition moment of the MSB. Fully differential decision signals are obtained using this construction. By increasing the number of coupled differential pairs to four, the number of most significant bits that can be generated increases to two. In Fig. 4.47 this basic operation is shown. From the figure it can be seen that the MSB bit is generated using the summation (wired OR operation) of signal C + D, while MSB-1 is obtained by combining signals B + D. At the same time

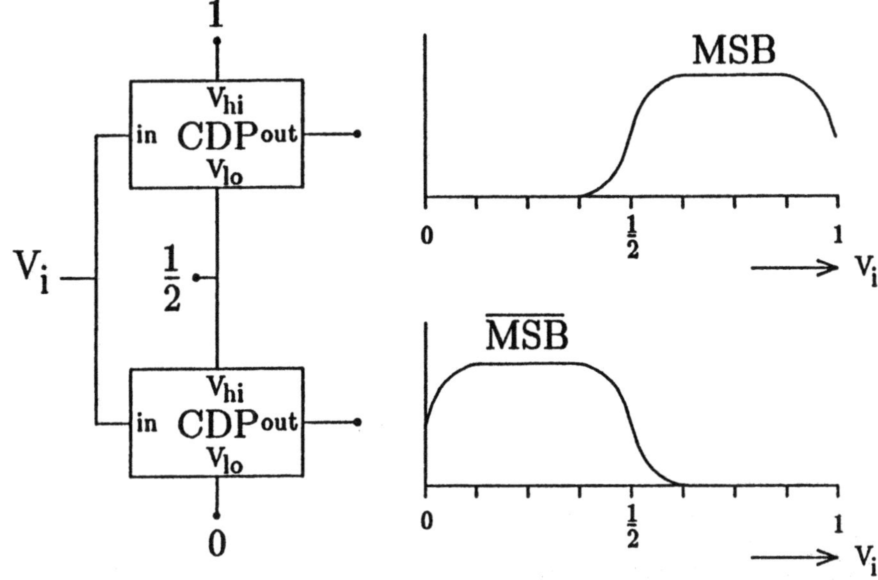

Figure 4.46 : MSB generation

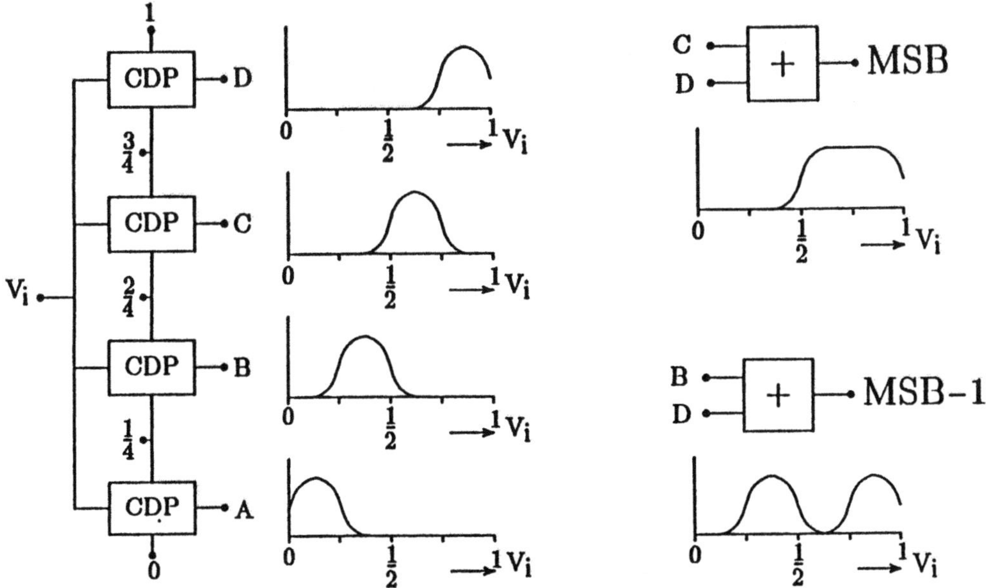

Figure 4.47 : MSB and MSB-1 bit generation

it is also possible to generate the complimentary signals of MSB and MSB-1 so a full differential comparison of zero crossings is possible. The MSB-2 bit generation is obtained by increasing the number of coupled differential amplifiers to eight as shown in Fig. 4.48. In Fig. 4.48 only four of the eight

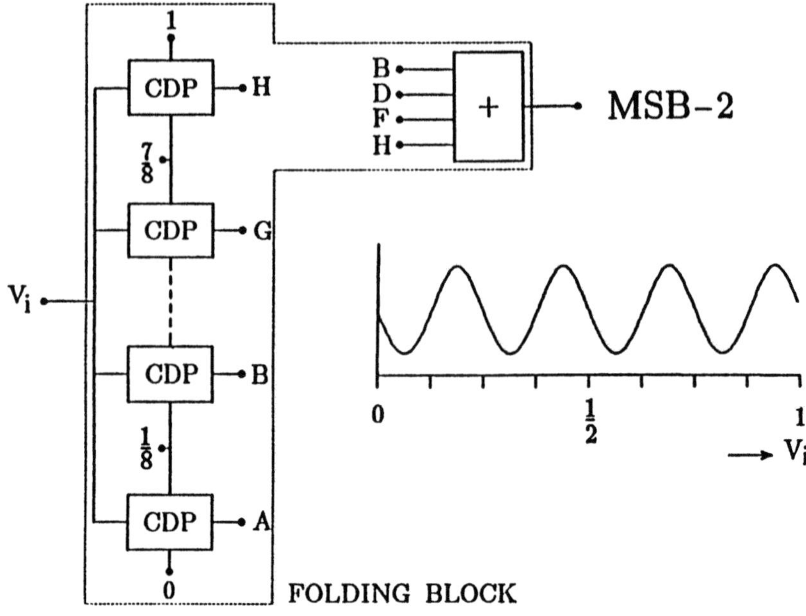

Figure 4.48 : MSB-2 bit generation example

coupled differential pairs are shown to simplify the system. Combining signals B+D+F+H gives the information for the MSB-2 zero transitions. Note that the repetition of the MSB-2 signal increases with a factor eight when an input full scale sine wave is applied. As a result a very high internal signal frequency is obtained. In practice the maximum number of folding operations is limited to about eight to avoid very high internal signal frequencies. During the following system descriptions a block consisting of eight coupled differential amplifier pairs will be called *Folding Block*. To increase the resolution of the system more folding blocks will be connected in parallel as shown in Fig. 4.49 The parallel connection of two folding blocks using an offset between the inputs of 0 and $\frac{1}{16}$ results in two series of in time shifted eight times folded output signals. With four blocks in parallel an increase in resolution of the converter with 2 bits is obtained. Basically the number of parallel connected blocks can be increased in such a way that all levels (2^n - 1) are covered. The disadvantage of this large amount of parallel blocks is the increase in hardware to the level of a full-flash system. All the disadvantages of such a system are then found.

To overcome this problem, an interpolation between the in time shifted out-

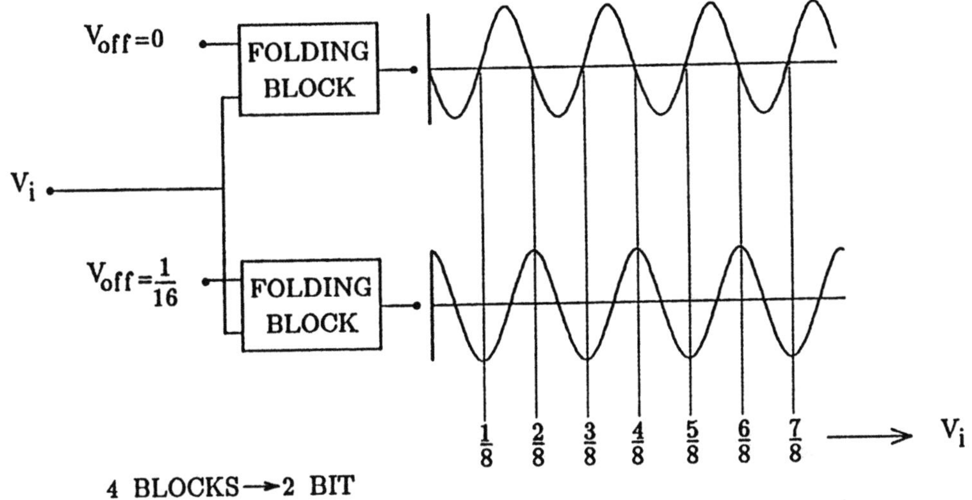

Figure 4.49 : Parallel connection of folding blocks

puts of two folding blocks will be used. In the example given in Fig. 4.50 an

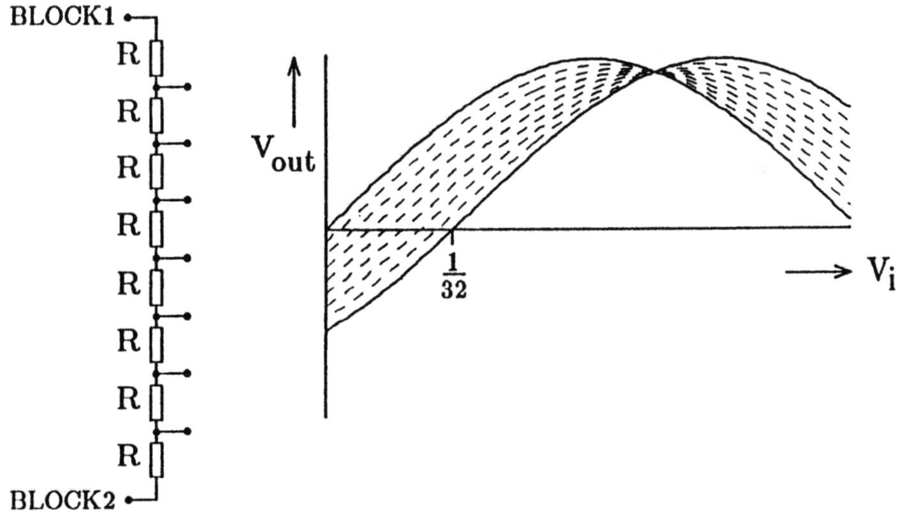

Figure 4.50 : Interpolation between time-shifted folding signals

eight times interpolation of signals is shown. Such an eight times interpolation increases the resolution of the converter with 3 bits without needing much hardware. Only a string of equal resistors is used. The error that is introduced due to a limited signal accuracy of the folding signals is below 0.1 LSB for an 8-bit converter. The result of a linearity calculation is shown in Fig. 4.51. The interpolated signals are applied to a comparator block

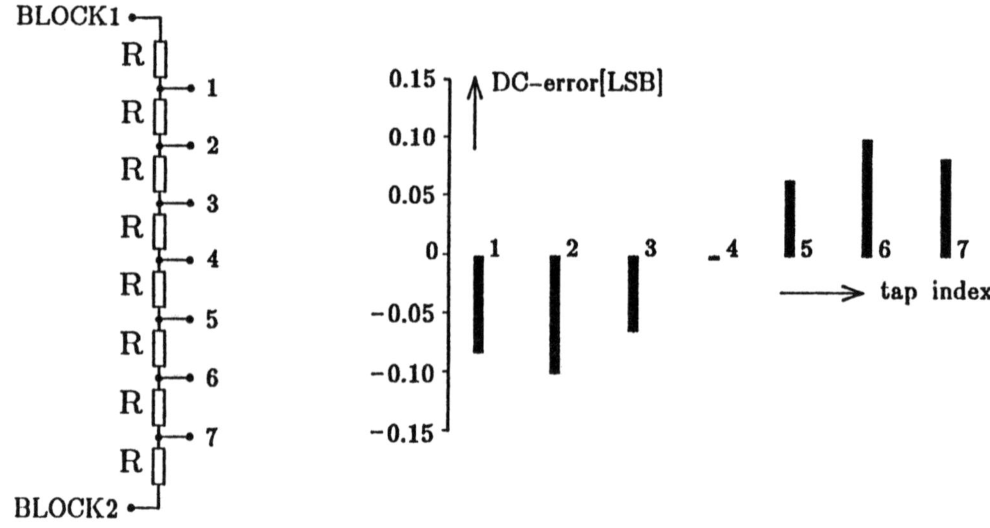

Figure 4.51 : Calculated 8 bit interpolation error

consisting of master-slave flip-flops. The outputs of the coarse levels and the fine levels are at the output converted into binary weighted output signals that form the output code of the converter. The outputs of the interpolated folding block signals result in a circular output code. This code can be uniquely distinguished using an EXOR gate. This operation is shown in Fig. 4.52. At the moment with increasing input signal only one single zero-one transition is obtained as a comparator decision, then a correct operation of the EXOR gate is obtained. When an extra 0 embedded in 1s is found, the the EXOR gate results in a wrong output code as shown in the right part of Fig. 4.52. In general it is not allowed to have codes as shown in the right part of the figure. To avoid decision errors, a correction system has been introduced. In Fig. 4.53 the basic configuration is shown. In this system an *analog averaging* of (digital) master comparator signals is performed. This analog averaging results in a code correction at the best possible code position compared to the digital code correction scheme from Table 4.4. Such an operation also results in valid decisions in case a master comparator would give a meta-stable output code. A meta-stable code gives an output code that does not correspond to the digital one or zero value but shows a value in between. To avoid code errors, information from the neighbor comparators is used to make a majority decision. Note that at the moment a comparator which will be corrected is at its decision point, then in an averager circuit the dominant one or zero will be determined. This operation is performed be-

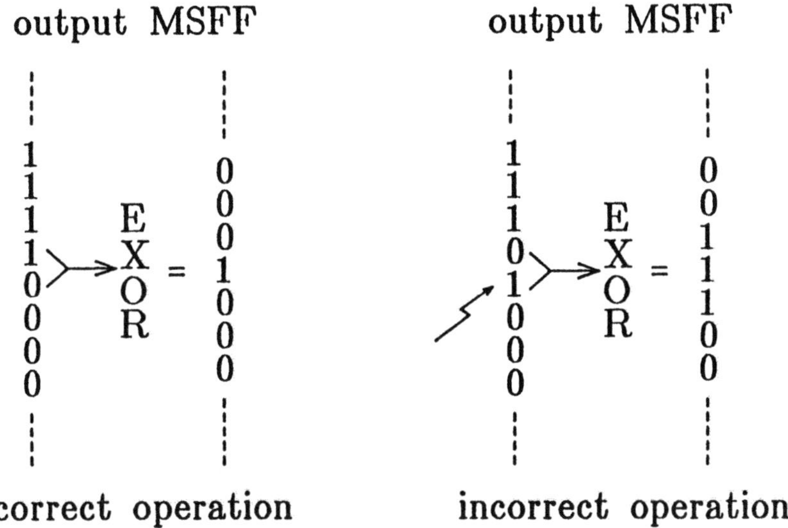

Figure 4.52 : Wanted and erroneous output codes of fine encoder

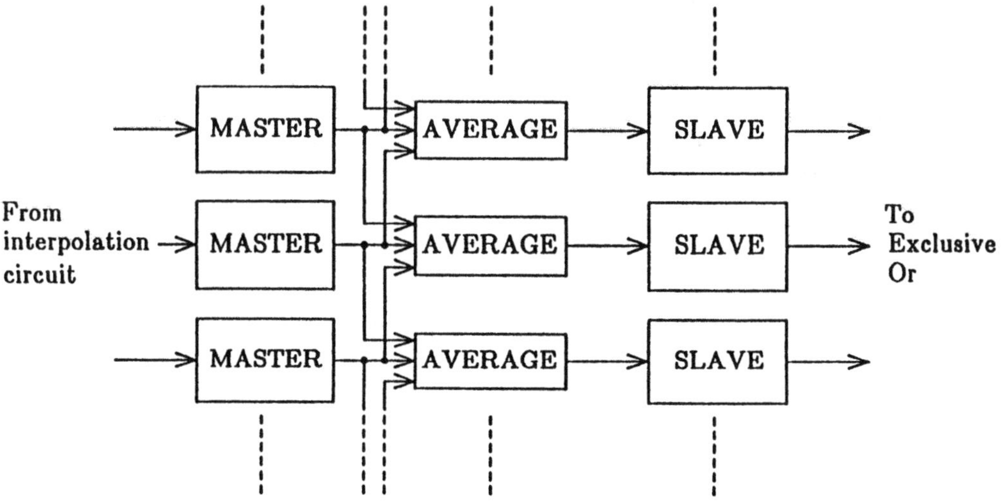

Figure 4.53 : Basic error correction scheme

tween the master and slave operation of the comparator stages. The outputs
of the slave are applied to the EXOR stages to obtain a unique decision. As
a result a code error over 1 bit can be corrected. In Fig. 4.54 the basic circuit
implementation of the error correction is shown. The information from the
neighboring comparator stages is added to the summation point to obtain
the democratic majority. As a result, the right decision is obtained. Note

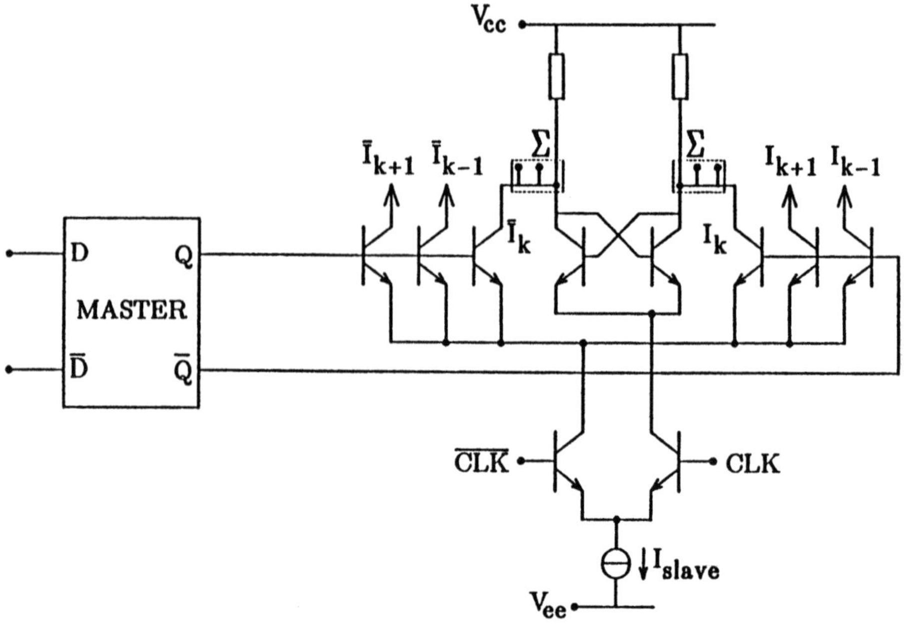

Figure **4.54** : Implementation of error correction

that the signal current of the slave input pair is split-up into three equal currents. Two of these currents are connected to the appropriate upper and lower neighbor for correction purposes. The complete system diagram of the folding and interpolation analog-to-digital converter is shown in Fig. 4.55. In the figure the analog preprocessing block is shown generating the two MSB signals. After interpolation, decision, and error correction, the binary-encoded five LSB's are obtained. Because of delays between the coarse and fine bit generation a timing correction for the MSB's is introduced using the so-called Bit Sync(hronization) operation. As a result, the full corrected 8 bits output signals are latched and can be applied to the output bus.

4.11.3   Practical folding and interpolation system

In Fig. 4.56 a block diagram of a complete folding and interpolation A/D converter system is shown [6]. The input signal is applied to the buffer amplifier that drives the folding stages and includes a level-shift. The input signal level is adjusted to run from 0 to $-V_{in}$ V. A reference current source generates across a resistive ladder the required reference voltages which are applied to the folding circuit block. In the folding block, multiple ampli-

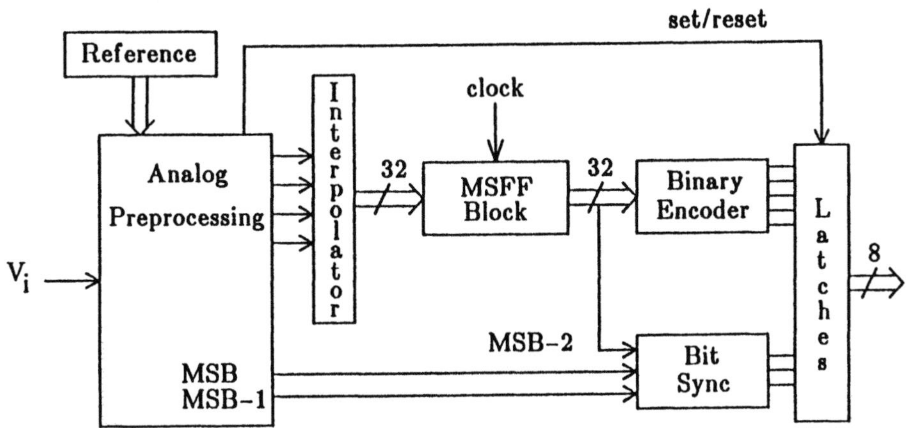

Figure 4.55 : Complete basic converter system

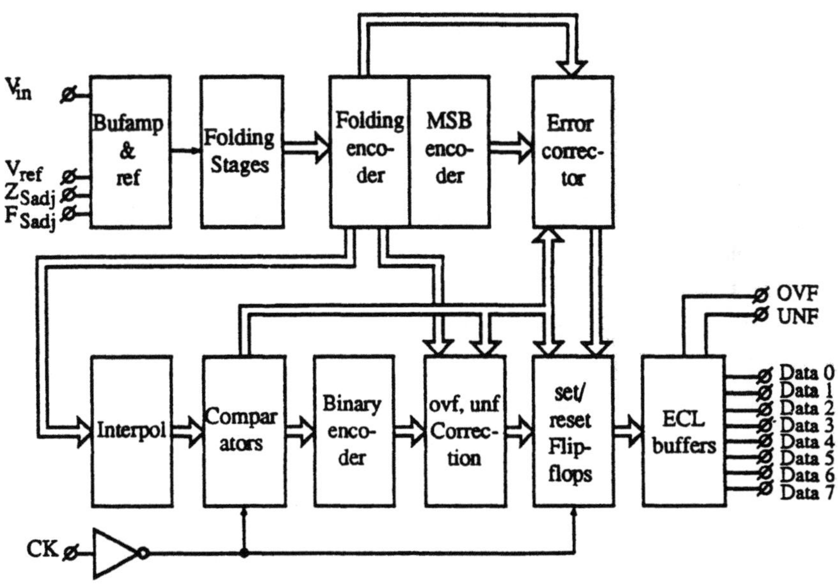

Figure 4.56 : Block diagram of folding and interpolation system

fier/comparator stages generate multiple zero crossings which are applied to
the folding encoder. This folding encoder is an analog ROM-like structure
that combines a number of zero crossings into a repetitive folding signal. At
the same time, signals are combined to obtain information about the most
significant bits. This operation is performed in the MSB encoder part of the
system.

The folding encoder output signals are applied to the interpolation circuit which generates, in this particular application, three additional folding signals for each output of the folding encoder. A four times interpolation is obtained in this way. The output signals of the interpolation circuit are applied to comparator stages which, on the command of the sampling clock, determine the position of their respective input signals compared to zero. A circular output code is obtained. This circular output code is converted into a binary form using a binary encoder ROM structure. The outputs of this ROM encoder are the least significant bits of the converter. Information from the circular fine code is used in the error correction block to correct for inconsistencies (e.g., caused by different delays between the coarse and fine zero crossing/encoding/comparison blocks) in the most significant bits before they are applied to the set/reset flip-flop blocks which drive the ECL output buffers. At the same time, from the folding encoder, overflow and underflow information is used to set the output signals to ZERO when the underflow bit is tripped, or to ONE when the overflow bit is activated. This information overrides the information available in the fine code when overflow or underflow is activated. Use of this system avoids large code transitions when the converter is over-driven. An accurate clipping at zero and full-scale is also obtained.

The clock driver supplies the sampling clock to the different points in the system and does not increase the timing uncertainty. The gain in the clock driver speeds up the clock and makes the system less dependent on noise.

4.11.4   Optimal folding circuit implementation

In an 8-bit folding architecture different choices between the number of input folding amplifiers, the number of times an interpolation is performed and the number of fine quantizers required are possible. Suppose in the folding block 64 reference levels are generated, then the system can use 8 coupled differential pairs with an 8 times folding or 4 coupled differential pairs with a 16 times folding. A 4 times interpolation is needed in this system approach to obtain 8 bit resolution. A 16 times folding will introduce a 16 times higher internal signal frequency and will find its most important application in low-power low-frequency converters. When only 32 reference levels are used, then the system can consist of 4 folding stages with an 8 times folding. In this system an 8 times interpolation is required for the total 8-bit resolution.

When 128 reference voltages are used then a folding scheme consisting of 8 coupled differential pairs with a 16 times folding can be used to increase the resolution of the system to 9 bits with a 4 times interpolation or to 10 bits with an 8 times interpolation.

In Fig. 4.57 a more detailed diagram of a 32 reference level folding system is shown. The system consists of four blocks each built-up from 8 folding amplifiers. The outputs of every block of folding amplifiers are combined in the folding encoder to obtain the repetitive folding signal. The most significant bits are obtained by combining the signals of the four folding encoder blocks in the coarse encoder. The encoder system is built-up using wired OR gates. The 8 times interpolation increases the total resolution to 8 bits. In

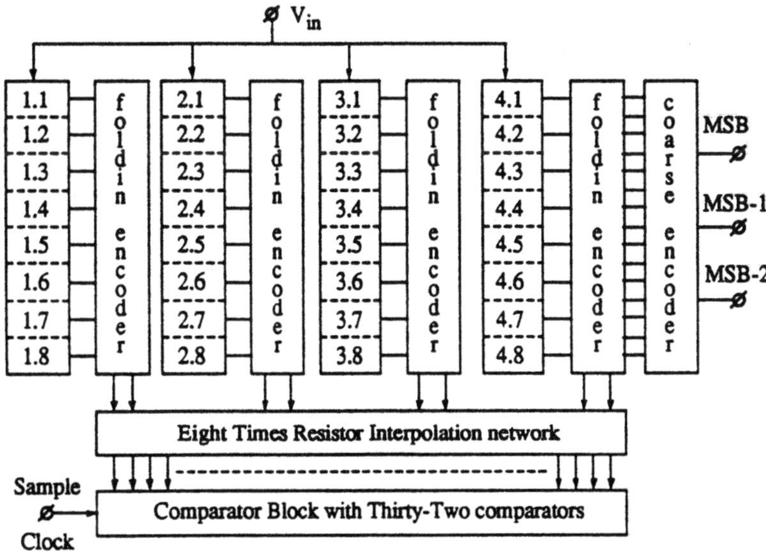

Figure 4.57 : 32-level folding converter system

Fig. 4.58 an example of three coupled folding amplifier stages is shown. The differential pairs T_1/T_2, T_3/T_4, and T_5/T_6 compare the input voltage with the reference signals V_{r1}, V_{r2}, and V_{r3}. Furthermore, the collector of T_2 is cross-coupled with the collector of T_3. The same yields for T_4 and T_5. Cascode stages T_7 to T_9 are added to minimize the parasitic collector-substrate capacitance and to maintain an equal collector-base voltage for all cascode stages in the folding amplifiers. The capacitances in the collectors of T_7 to T_9 remain more constant and vary identically over the different stages.

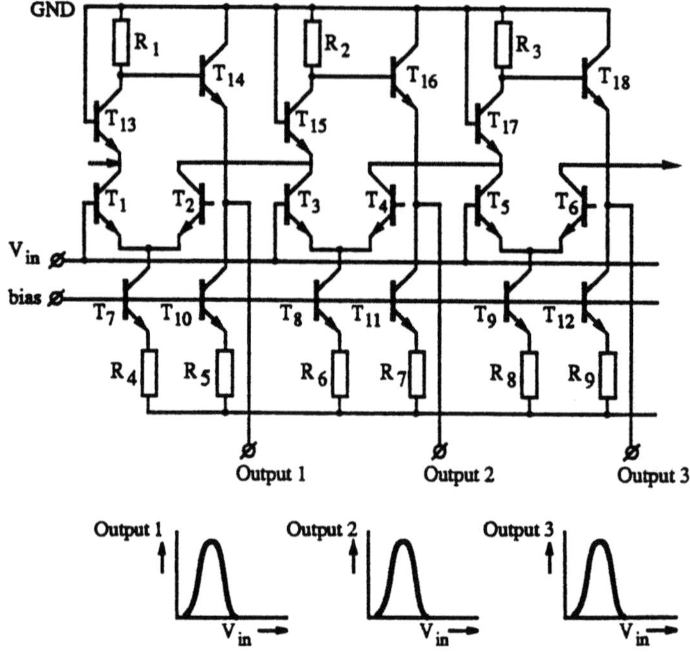

Figure 4.58 : Folding amplifier circuit

The basic function performed by the folding block is the conversion of an increasing (or decreasing) input signal into a number of bell-shaped output signals V_{out1}, V_{out2}, and V_{out3}, which have different levels for V_1, V_2, and V_3. Adjacent output signals are complementary near their zero crossings, which results in a full differential operation of the system. This can be explained as follows. Suppose the input signal V_{in} increases from a level below V_1, with V_1 being the smallest and V_3 being the largest value of the three reference voltages. The differential tail currents supplied by T_{12}, T_{15}, and T_{18} flow through transistors T_2, T_4, and T_6, respectively. As a result V_{out1} is high and V_{out2} and V_{out3} are low. When the input voltage reaches V_1, the current through T_2 and therefore through T_8 decreases, and the current through T_1 and T_7 increases, in such a way that V_{out1} and V_{out2} show a differential output voltage change equivalent to the complementary output voltage variation of a differential pair. The same situation is found for V_{out2} and V_{out3} when the input voltage equals V_{r2}. The bell-shaped output signals which are finally found at the output terminals differ in time by the length of time the input signal needs to change from V_{r1} to V_{r2} to V_{r3}.

The operation of the circuit is valid as long as no interaction exists between the coupled differential pairs due to large differences between the reference voltage sources V_1, V_2, and V_3. It can be shown that differences of 120 to 150 mV between the reference voltage levels give some interaction but do not influence the operation of the folding stages in a negative way.

Comparison of the input signal level with the reference level is still performed by a single differential pair, giving a possible system implementation with low offset voltages. From the folding amplifier stages, 64 output signal lines are supplied to the folding encoder circuit.

4.11.5 Folding encoder circuit

The folding encoder circuit combines the 32 signals from the folding amplifiers into four complementary output signals. Each folding signal consists of two complementary signals, and one such signal is created by concatenating the outputs of, for example, the odd stages in a row of folding stages. Other folding signals are generated by similar rows of folding stages, but with slightly different reference levels to create an offset between the folding signals. This is achieved by interleaving the connections of the V_r on the resistive divider: folding stage 1 for folding signal 1 is connected to tap 1, folding stage 1 for folding signal 2 is connected to tap 2, folding stage 1 for folding signal 3 is connected to tap 3, and so on. In case of eight folding signals, folding stage 2 of folding signal 1 is then connected to tap 9. In order to avoid confusion, folding stages are numbered according to their connections to the reference voltage ladder. Folding stage 1 and folding stage 17 therefore generate subsequent periods of the same folding signal.
The B_1, B_2 and B_3 lines in Fig. 4.59 show the combinatorial circuit that combines the outputs of the different folding stages into folding signals. The input and output waveforms associated with this circuit are shown in Fig. 4.60. Here the individual outputs of folding amplifiers i, $i+16$, and $i+32$ are shown in the top part. The lower part of the figure shows the combined output signal which is a concatenation of the bell-shaped signals.

Returning to the top part of Fig. 4.59, we see that the MSB and MSB-1 signals can also be generated by a proper combination of the folding amplifier outputs. Up until half the reference voltage the not-MSB signal is kept high and the MSB signal line is kept low by a combination of the bell-shaped folding amplifier signals. At half the reference voltage level, a single differen-

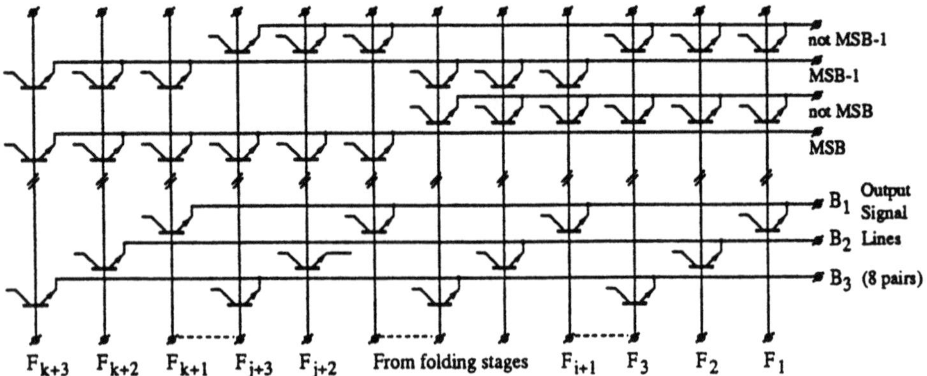

Figure 4.59 : Folding encoder circuit diagram

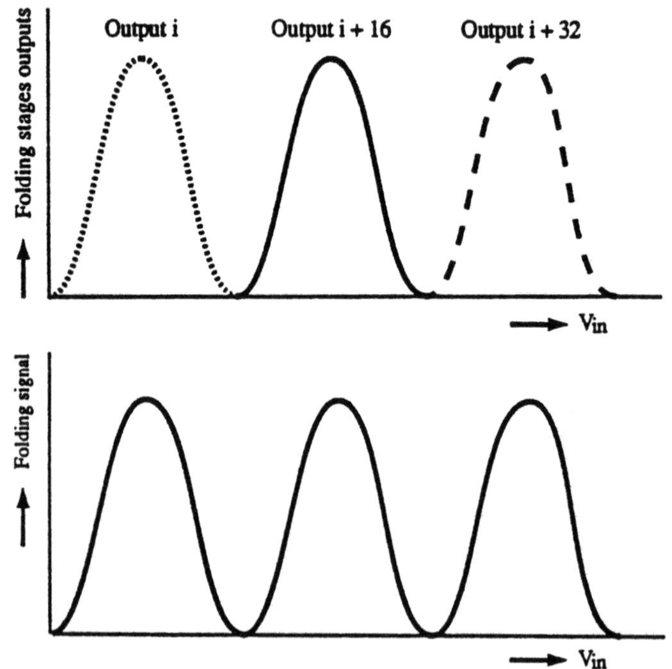

Figure 4.60 : Input and output signals of the folding encoder

tial amplifier pair switches and in this way determines the MSB change. No extra offsets or accuracy problems are found relative to the fine conversion signals that are generated in a later stage of the system.

A similar operation can be performed for the MSB-1 signals. Moreover, the MSB-2 signal is equal to the output signal in the lower part of the circuit which combines stages 0, 16, 32, 48 for example.

4.11.6 Interpolation circuit

Up to this point, 32 reference levels have been converted into a repetition
of zero crossings that could drive four differential comparator stages, and
thus generate two least significant bits. Together with the three MSBs this
corresponds to a 5-bit system. To increase the resolution to 8 bits without
adding folding stages, an eight times interpolation of zero crossings is introduced at the outputs of the folding encoder. In Fig. 4.61 a simple resistor
network that performs the interpolation of zero crossings is shown. For the

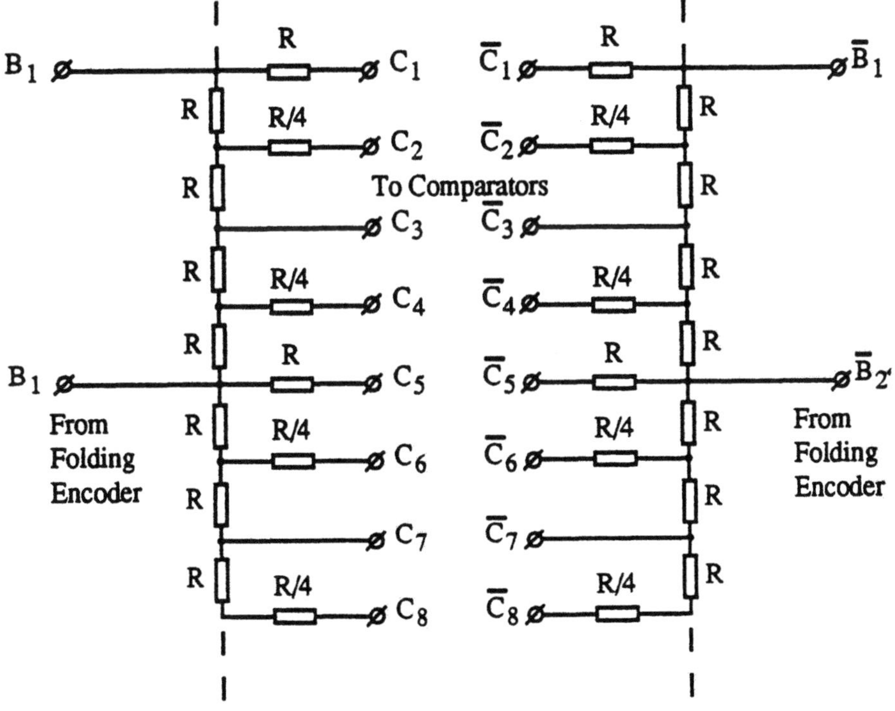

Figure 4.61 : Interpolation circuit diagram

32 folding signals created this way, 32 differential comparators are needed to
derive the five least significant bits.

Because the interpolation network is driven from emitter-follower stages, the
output impedance of the network that drives the comparators varies from
zero to R (R is the value of the interpolation resistors). The capacitive load
of the comparator input stages on the interpolation network results in a

variable signal delay which can easily be more than is allowed in this system (12 psec). Therefore, additional resistors of values R and $\frac{R}{4}$ are added in series with some outputs to adjust the output impedance of the interpolation network equal to R. All comparator stages now have an equal signal delay.

A cross-coupling between the beginning and the end of the interpolation network is needed to obtain interpolated signals "around the corner," which is necessary because of the repetitive properties of the folding signal (see Fig. 4.61).

When an 8 times interpolation is used, then the number and values of the compensation resistors increases. A simple analysis will give the exact required values.

4.11.7 Comparator architectures

Most flash converters use clocked flip-flops for the analog comparators. The basic circuit diagram of such a comparator stage is shown in Fig. 4.62. The operation of this circuit is as follows. When the clock signal is high,

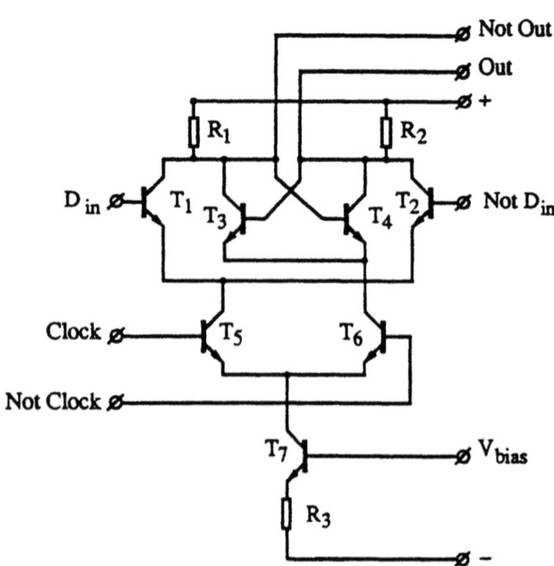

Figure 4.62 : Basic clocked flip-flop comparator circuit

transistors T_1 and T_2 are conducting and the input signal is amplified by this amplifier stage with R_1 and R_2 as load resistors. A decision is made when the not-clock signal is made high. This simple circuit shows an extremely good performance. However, every time data have to be entered into the flip-flop, base charge must be applied to T_1 and T_2. These transistors were switched off during the decision making and are switched on again to apply input information to the collectors of T_1 and T_2. The base charge results in large charging currents flowing through the interpolation network. These charging currents disrupt the exact zero-crossing information which results in a reduction in effective bits. To overcome this problem a different type of clocking scheme is used. The "high-clocking" is shown in Fig. 4.63. In

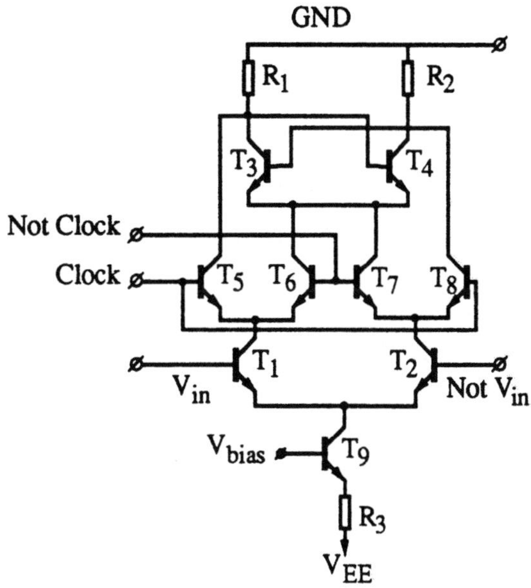

Figure 4.63 : Comparator with "high-clocking" operation

this circuit the input amplifier stage is continuously biased. Kickback effects from the sampling clock are reduced considerably, while the "cascoding" by the clock switches T_5, T_6, T_7, and T_8 of the input signal and the comparator output signal reduces the Miller effect of the collector-base capacitance of the amplifier transistors T_1 and T_2.

4.11.8   Circular-to-binary encoder

Because of the repetitive effect in the fine converter, a circular code, instead of a linear thermometer code, is obtained. In Table 4.9 examples of the thermometer and circular code are shown. When implementing an encoder,

	Thermometer	Circular
0	0 0 0 0 0 0 0	0 0 0 0
1	0 0 0 0 0 0 1	0 0 0 1
2	0 0 0 0 0 1 1	0 0 1 1
3	0 0 0 0 1 1 1	0 1 1 1
4	0 0 0 1 1 1 1	1 1 1 1
5	0 0 1 1 1 1 1	1 1 1 0
6	0 1 1 1 1 1 1	1 1 0 0
7	1 1 1 1 1 1 1	1 0 0 0

Table 4.9 : Thermometer- and circular-code representation

the transition between a group of ONEs and a group of ZERO's must be determined first. In the case of the circular code this function is performed by an EXCLUSIVE-OR function. The output of the EXCLUSIVE-OR function $A_0, A_1, A_3 \ldots A_7$ is applied to a ROM structure which converts the transition into a binary code (Fig. 4.64). When a coupling between the inputs and the encoded output is required a bipolar transistor is inserted. The differential output signals of the encoder ROM are applied to differential amplifier stages which restore the current mode logic level.

4.11.9   Bit synchronization between MSB, MSB-1, and LSB's

Although care has been taken to ensure equal delays for the MSBs and the LSBs in the converter, noise, and decision uncertainties at the zero crossings of the MSB comparators can result in erroneous output codes. To overcome this problem an error correction system is included. This correction is achieved by checking the outputs of a set of folding amplifiers that have the maximum output voltages for input signal values where such errors can occur, that is, near transitions of the MSBs. If such a folding amplifier output indicates that the signal is in the transition region, then the MSBs are replaced by an appropriate value based on the most significant bit of the LSB bits. Whether one of the MSBs is replaced by the most significant bit of the LSBs or by its complementary value depends on the transition region of the input signal. In Fig. 4.65 these dangerous regions are shown. A timing error correction corresponding to $\pm\frac{1}{16}$ of the full scale can be obtained with this method.

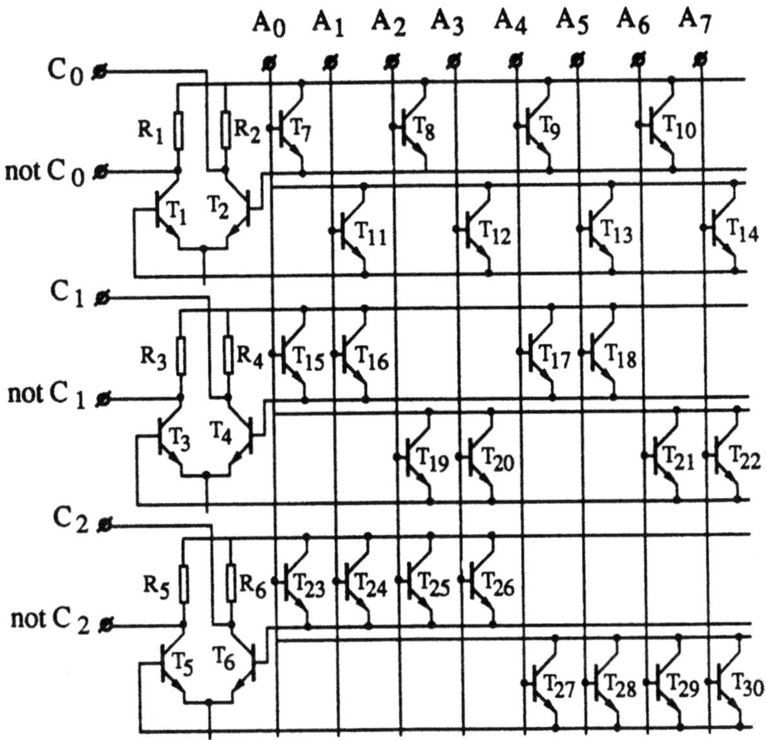

Figure 4.64 : Binary encoder ROM

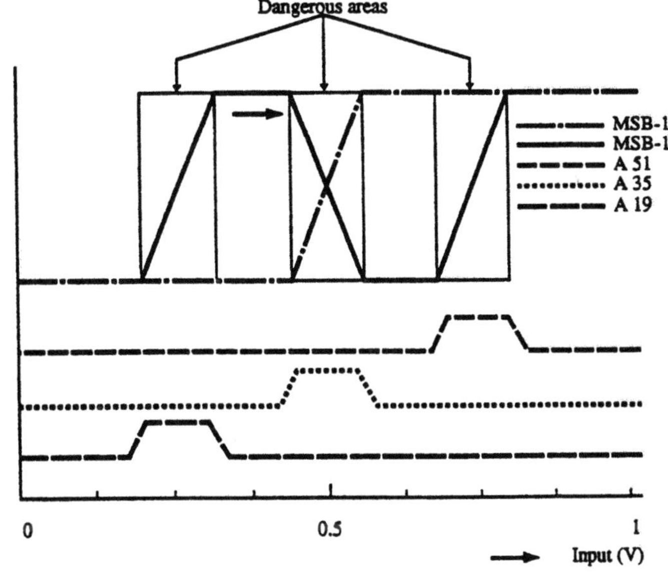

Figure 4.65 : Dangerous regions of MSB and MSB-1 signals

A similar problem occurs at the top and the bottom of the input range. Because of the circular properties of the folding signals, the MSB bits saturate at the end of the input range while the LSBs wrap around the corner. This may result in large code errors, which is not allowed. For example, with five LSBs and the three MSBs an input signal decreasing from a value corresponding to the output code "00000010" would go through the following output codes: "00000010", "00000001," "00011111," "00011110," and so on. These large signal glitches are not found in the input signal and therefore must be limited to a maximum or minimum code depending on "overflow" or "underflow" indication. A similar correction as described for the MSBs is performed to force the output code to ONEs in the case of overflow and ZERO's in the case of underflow.

4.11.10   A/D converter implementation

The A/D converter was implemented in an advanced oxide isolated bipolar process. Important device parameters are shown in Table 4.10. In the final

H_{fe}	170
Emitter size	$2 \times 6 \ \mu m^2$
f_t	9 GHz
C_{be}	22 fF
C_{bc}	43 fF
C_{cs}	68 fF

Table 4.10 : NPN device parameters

system design one of the design goals was a minimum number of external components to operate. The total die size of the A/D converter function needed only 3.2x3.8 mm^2. Although this is a small die compared to a full-flash converter, delays over wires of lengths comparable to the die size are still unacceptable. A 1 psec delay over a wire corresponds to a length of 100 μm, assuming that the transmission speed is $\frac{1}{3}$ the speed of light on the oxide/air interface. However, only delay differences in identical operations in the system influence the high-frequency performance. These delay differences can be kept small using proper layout structures. For example, tree-type wire structures as shown in Fig. 4.66 are used to ensure virtually identical path lengths for clock and signal distribution. A die photograph is

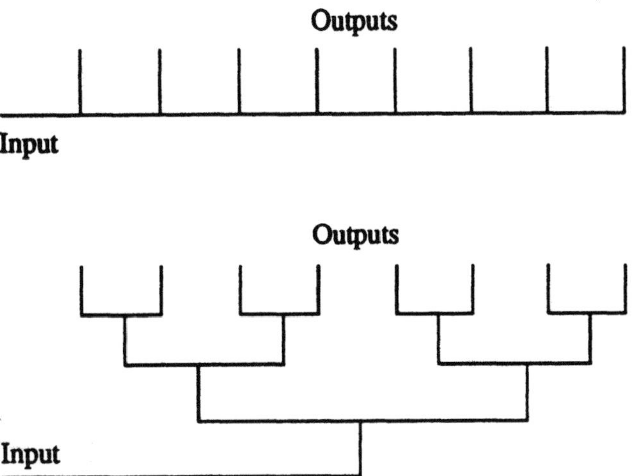

Figure 4.66 : Conventional wiring layout and tree-type wiring structure

shown in Fig. 4.67

4.11.11 Measurements

To verify the performance of the A/D converter, a test set-up using an A/D-D/A loop as shown in Fig. 4.68 was used. In this loop a D/A converter is used to reconstruct the analog value. The performance of this D/A converter must be much better than the performance of the A/D converter to be measured. If DC tests are carried out a 14-bit linear low-speed D/A of the type TDA 1540 can be used. The D/A converter can be hooked up in such a way that the input signal to the A/D converter being tested and the output signal of the D/A converter are in counter phase. Using a half resistor bridge between the input and output of this system, the nonlinearity can be easily measured and recorded using a plotter. No additional accurate elements are needed in this test procedure. The result of the linearity measurement is shown in Fig. 4.69. In dynamic tests a high-performance high-speed D/A converter is needed. Because of the availability of a slower speed D/A converter, sub-sampling is used to allow measurement results up to 60 MHz analog input signals. The measured data are converted to the overall bandwidth of the converter at the used measurement condition. A result of the effective bits as a function of frequency is shown in Fig. 4.70 At an analog input frequency of 40 MHz 7.5 effective bits are obtained. In-

Figure 4.67 : Die photograph

put signals with a low distortion (below 60 dB) and a sampling clock with an effective jitter below 12 psec are necessary. Further performance data are given in Table 4.11. The results of the measurements show that with a folding architecture a high-speed high-performance A/D converter can be designed. This converter needs a moderate die size and consumes less than 1 W of power.

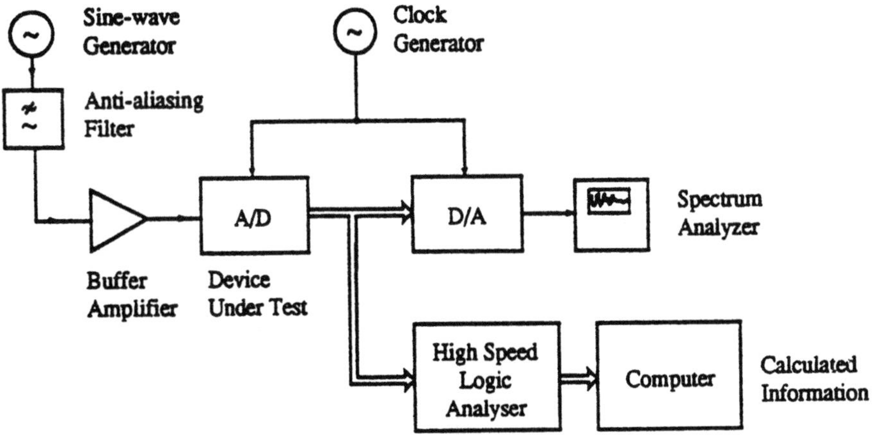

Figure 4.68 : A/D-D/A converter test set-up

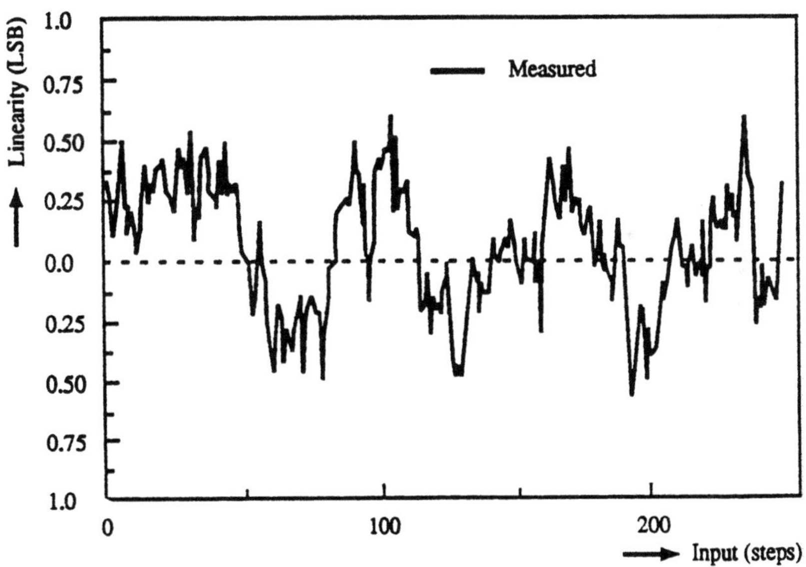

Figure 4.69 : Linearity measurement result

4.11.12 Conclusion

In this chapter high-speed full-flash, two-step,and folding A/D converters
have been discussed. In time-continuous systems, not much difference in
problems between a bipolar or an MOS circuit solution is seen. When, however, time-discrete solutions are used, then only special MOS systems exist
because of the special character of this technology.

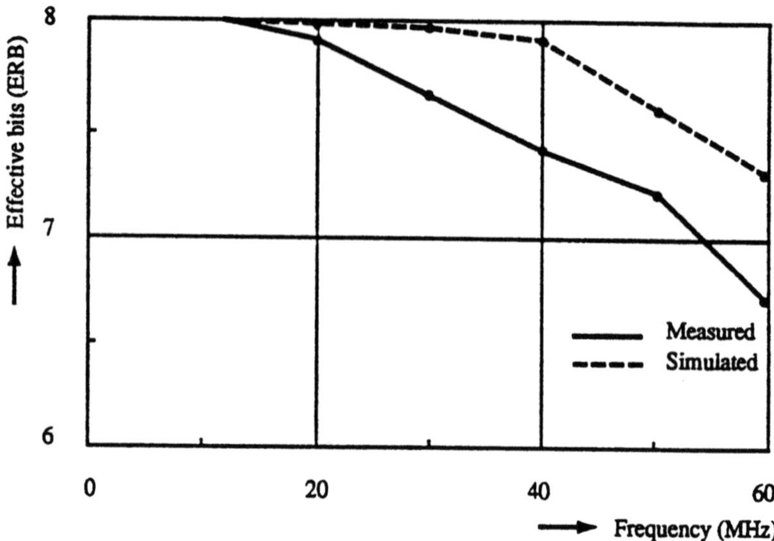

Figure 4.70 : Effective bits as a function of input frequency

Resolution	8 bits
Linearity	± 0.5 LSB
Conversion rate	100 MHz
Effective Resolution Bandwidth	40 MHz
Input signal voltage	0 to -1 V
Input Capacitance	5 pf
Supply voltage	-5.2 V
Power dissipation	800 mW
Output logic levels	ECL 10K compatible
Outputs	8 data bits, and
	overflow and underflow
Built-in:	clock driver
	input buffer
	voltage reference
Die size	3.2×3.8 mm^2
Package	24 pin

Table 4.11 : A/D converter performance data

A special type of two-step system that uses a continuous coarse-fine operation is discussed. The advantages of folding systems over full-flash types without a sample-and-hold amplifier at the input are clearly shown. With a lower power consumption and a smaller die size, the performance of the folding-type converter is better than the best-known full-flash type. Comparator/amplifier stages can run at higher current levels, increasing the analog bandwidth of the system. This larger bandwidth results in a smaller signal-dependent delay, which results in a lower third-order distortion. Design criteria for full-flash or folding-type converters are identical concerning the signal-dependent delay. In the next chapter a detailed analysis of the signal-dependent delay will be given.

Chapter 5

Limitations of comparators

5.1 Signal delay in limiting amplifiers

5.1.1 Introduction

In most A/D converters the input signal is applied to an amplitude-limiting circuit. This amplitude-limiting circuit can be a differential amplifier stage or the input stage of a comparator. The amplitude of the input signal is usually much larger than the linear range of the input amplifier stages. In this way a limitation of signals for the individual amplifier stages occurs. This amplitude limitation results in variations of the delay times of the zero crossings of the differential stages. This can be explained in the following way. When a sine wave is applied at the input of the A/D converter, it looks as if this sine wave is cut into pieces with a variable slope. This slope depends on the level at which the input signal is equal to a reference voltage level. The variable slope introduces a variable delay of the zero crossing of the output signal. These delays are accordingly signal-slope-dependent. As a result, a nonlinear distortion of the input signal occurs while it travels through the A/D converter system. The moments at which ideal amplifiers would show zero crossings are shifted in time. These time shifts result in errors in the output code of the A/D converter. This delay variation is caused by the frequency limitation which is always present in practical amplifier stages. This distortion occurs in particular in bipolar amplifier stages that have a limited linear range. MOS amplifier stages have a larger linear range because of the much higher threshold voltage of the individual devices with respect to a bipolar transistor. A simple calculation and design model will be presented in this chapter. For additional information see [60].

A/D converters which use a sample-and-hold amplifier at the input are less sensitive to this slope-dependent delay phenomenon. In reference [61] the improvements in performance of flash-type A/D converters proceeded by a high-speed sample-and-hold amplifier are shown. The time constant with which the sampling of the analog signals occurs can be made smaller than is the case with the large number of amplifier stages used in parallel or folding types of converters.

When master-slave flip-flops are used as clocked comparators, it is very important to know the sensitivity of such a comparator. Furthermore, the number of meta-stable conditions such a comparator can have determines the errors an analog-to-digital converter introduces as a function of time. A meta-stable condition is defined as the output signal a comparator obtains after a well-defined comparison time and cannot be used as a logical "1" or "0". The encoding logic under such a condition does not give a unique output code, which results in quantization errors. A simple model of the master-slave flip-flop is used to calculate these meta-stable conditions and the design criteria needed to overcome these problems. Basically this method can be applied for a bipolar as well as a MOS technology.

5.2   Definition of the delay problem

In a flash-type A/D converter usually the input signal amplitude is much larger than the linear range of the individual input amplifiers or the input stages of a clocked comparator. The input stages are over-driven by the input signal, and the zero crossings at the outputs of the amplifiers-comparators are delayed in time. In Figure 5.1 the input signal is shown with the reference voltages V_{refn} drawn at the moment the sine wave amplitude equals the input signal level. In the circle the nonlinear input characteristic of the input amplifier-comparator stage is shown. Here V_{lin} is the linear range of the amplifier stage. Depending on the level of the zero crossing, the shaped ramp function which can be used to model the system shows a signal level dependent slope with a maximum around the mid value of the sine wave and a minimum slope in the tops and valleys of the sine wave. This graphic approach to the over-drive of amplifier-comparator stages leads to a simple calculation model to analyze the behavior of these stages as a function of input frequency and signal level.

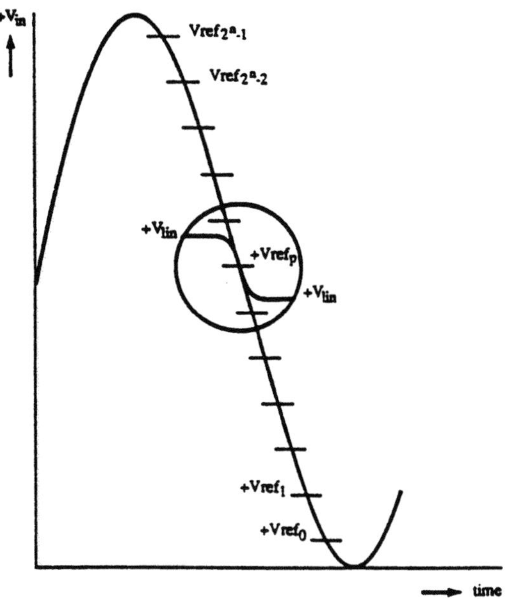

Figure 5.1 : Over-drive model of a flash converter

5.3 Delay calculation model

A simple model that represents the nonlinear behavior of an amplifier pair
with a frequency-limiting circuit is modeled in Figure 5.2. The model con-

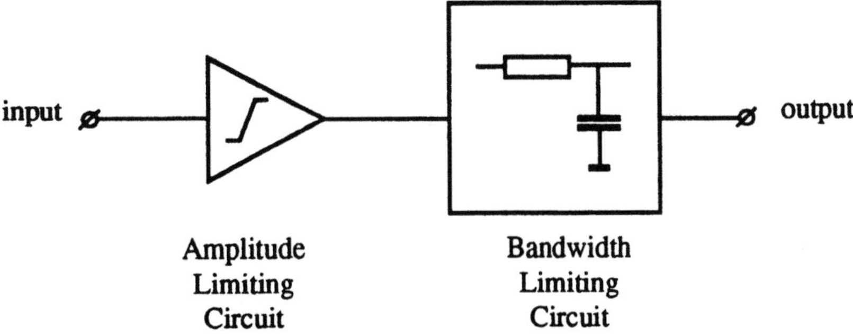

Figure 5.2 : Simple nonlinear model of a limiting amplifier stage

sists of an amplitude-limiting stage followed by a bandwidth-limiting circuit.
The amplitude-limiting circuit is modeled by a linear transfer curve that
saturates on both sides without any smooth transitions. Although practical
amplifier stages show a certain smoothing in saturation, this simple model
turns out to be good enough for understanding the problem and deriving design requirements for the amplifiers. The bandwidth limitation is modeled

by a simple *RC* network. Now assume that the input signal is increasing
with a constant slope within the linear region of the amplitude-limiting circuit. Then the first part of the output signal of the *RC* network equals the
response of an *RC* network to a ramp function (see Figure 5.3). After the

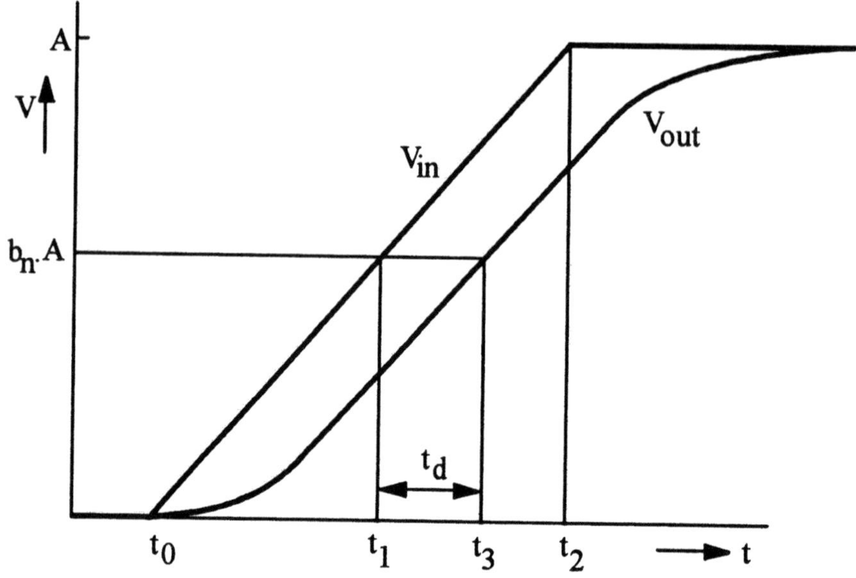

Figure 5.3 : Output signal of an amplitude-limiting circuit with a ramp input
signal

output signal of the amplitude-limiting circuit reaches its maximum value,
the output of the *RC* network equals the response to a step function. During
normal operation of an A/D converter, the input signal source, which can be
a sine wave, over-drives the differential comparator amplifier stages. When
the input signal passes a differential stage at the zero crossing of the signal, a
waveform like an input step is applied to the *RC* network. The delay in this
case equals the response of an *RC* network to a step input signal. This delay,
at half the output signal value, equals about 0.7 *RC*. When the input signal
reaches the tips of the sine wave, a signal that can be modeled by a ramp
signal is applied to the *RC* network. In this case the delay will become *RC*.
The result of this operation is that, depending on the slope of the input signal, the output zero crossings will have a variable delay, which changes from
0.7*RC* to *RC* when the signal changes from maximum to minimum slope.
The input slope variations occur because at the high end frequency range a
sine wave is always compared with the reference levels. The anti-alias filter,

which must precede a practical converter to avoid aliasing, performs this signal filtering. This variable delay will cause signal distortion, which is not allowed. A more exact analysis of this phenomenon will be presented in the next section.

5.4   Variable delay calculation

In a practical situation a step input response is never obtained. Therefore, a ramp with variable slope is used to model the limited input signal. Using Figure 5.3, the following notations for the variable signal delay are used:

V_{in} input signal
V_{fs} full-scale value of the A/D converter
G_A amplifier gain
$V_{lr} = \frac{A}{G_A}$ input amplifier linear range
f_{in} input signal frequency
$f_b = \frac{1}{2\pi RC}$ -3 dB bandwidth of the input amplifier
$\omega = 2\pi f_{in}$
S slope of the A/D converter input signal
S_{out} slope of the output signal of the amplitude-limiting circuit
A maximum output voltage
b_n relative output voltage level at which the signal delay is determined
t_0 start time of the ramp signal
t_1 time at which the ramp signal reaches the level $b_n A$
t_2 time at which the output signal of the limiting stage reaches the maximum value A
t_3 time at which the output ramp-shaped signal reaches the level $b_n A$
$t_d = t_3 - t_1$

Suppose an input signal equal to

$$V_{in} = \frac{V_{fs}}{2} \sin \omega t \qquad (5.1)$$

is applied to the input of the A/D converter. Because of the anti-alias filtering that is required at the input of an A/D converter, this sine wave is a good model for an input signal at the high end of the input frequency band. The slope of the signal depends on the level at which the reference level crossing occurs. After differentiating the input signal, we obtain for the slope of the signal

$$S = \frac{V_{fs}}{2}\omega \cos \omega t. \tag{5.2}$$

The input slope is increased by the gain factor G_A of the amplifier/limiter circuit. As a result, the output slope that is applied to the RC network becomes

$$S_{out} = G_A \frac{V_{fs}}{2}\omega \cos \omega t. \tag{5.3}$$

The output slope of the amplitude-limiting circuit can be compared with the slope of the ramp function of Figure 5.3. This means that:

$$G_A \frac{V_{fs}}{2}\omega \cos \omega t = \frac{A}{t_2 - t_0}. \tag{5.4}$$

The value of $\frac{A}{G_A}$ can be replaced by the linear input voltage range of the amplifier v_{lr}, so then equation 5.4 can be rearranged into:

$$t_2 - t_0 = \frac{2V_{lr}}{V_{fs}\omega \cos \omega t}. \tag{5.5}$$

With the use of Laplace transforms, the response of the ramp signal applied to the RC network can be calculated. We obtain using the notations of Figure 5.3:

$$V_{out}(t) = (\frac{t - t_0}{t_2 - t_0}A - \frac{RC}{t_2 - t_0}A(1 - e^{-\frac{t-t_0}{RC}}))U(t - t_0) \tag{5.6}$$

$$- (\frac{t - t_2}{t_2 - t_0}A - \frac{RC}{t_2 - t_0}A(1 - e^{-\frac{t-t_2}{RC}}))U(t - t_2). \tag{5.7}$$

In this equation $U(t - t_0)$ is the unit step response. The time delay t_d between the input and output ramp signal can be expressed in the circuit parameters when the output signal crosses the amplitude level $b_n A$.

After some rearranging and inserting $t = t_3$ we obtain:

$$b_n(t_2 - t_0) = (t_3 - t_0 - RC(1 - e^{-\frac{t_3-t_0}{RC}}))U(t_3 - t_0) \tag{5.8}$$

$$- (t_3 - t_2 - RC(1 - e^{-\frac{t_3-t_2}{RC}}))U(t_3 - t_2). \tag{5.9}$$

The time t_3 can be expressed as follows:

$$t_3 = t_0 + b_n(t_2 - t_0) + t_d. \tag{5.10}$$

Depending on the value of the time t_3, two cases can be distinguished:

1. $t_3 < t_2$, and

2. $t_3 \geq t_2$.

Inserting the boundary value $t_3 < t_2$ into equation 5.10 results in:

$$t_2 - t_0 > \frac{t_d}{1 - b_n}. \tag{5.11}$$

Combining equation 5.5 and equation 5.11 and rearranging results in:

$$\frac{t_d}{RC} < 2(1 - b_n)\frac{V_{lr}f_b}{V_{fs}f_{in}}. \tag{5.12}$$

Equation 5.12 determines the switching between case 1) and case 2). Suppose $t_3 < t_2$, then only equation 5.8 is valid to express the output voltage. Inserting equation 5.10 into equation 5.8 results in:

$$t_d = RC(1 - e^{-\frac{b_n(t_2 - t_0) + t_d}{RC}}). \tag{5.13}$$

When $\cos \omega t$ equals 0 or 1, then using equation 5.4 a value for $t_2 - t_0$ of ∞ or $\frac{2V_{lr}}{V_{fs}\omega}$ is obtained, respectively. Inserting the value of ∞ for $t_2 - t_0$ into equation 5.13 gives:

$$t_d = RC. \tag{5.14}$$

Now an estimate of the switching point between case 1) and case 2) can be made. Inserting equation 5.14 into equation 5.12 gives the estimate:

$$\frac{f_{in}}{f_b} < 2(1 - b_n)\frac{V_{lr}}{V_{fs}}. \tag{5.15}$$

When the slope of the input signal varies, the interesting parameter is the variation of the delay time with respect to the slope of the signal. Define:

$$t_d = RC - \delta t_d. \tag{5.16}$$

Proceed with case 1): Substitution of equation 5.16 into equation 5.13 results in the following relation between the input signal and the delay variation δt_d:

$$\delta t_d = RCe^{-\frac{2b_n V_{lr}}{V_{fs}RC\omega \cos \omega t} + \frac{\delta t_d}{RC} - 1}. \tag{5.17}$$

In Figure 5.4 the variation of the delay as a function of the input signal slope is shown. In the figure V_{in} represents the input sine wave which is offsetted

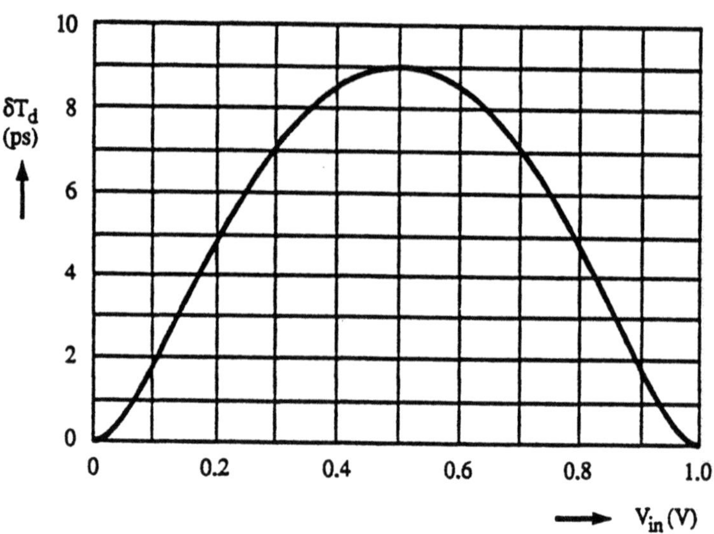

Figure 5.4 : Variation of the delay as a function of the input signal

by 0.5 V at which level the maximum slope is found. 0 and 1.0 V are the bottom and top parts of the sine wave, respectively. The following values for the different parameters are substituted: RC = 200 psec

$b_n = .5$

$V_{lr} = 140$ mV

$V_{fs} = 1$ V

$\omega = 2\pi 50 MHz.$

Inserting $\cos \omega t = 1$ into equation 5.17 results in a simplified expression for the maximum delay variation:

$$\delta t_d = RCe^{-2b_n \frac{V_{lr} f_b}{V_{fs} f_{in}} + \frac{\delta t_d}{RC} - 1}. \tag{5.18}$$

A further useful simplification is introduced by replacing $\frac{\delta t_d}{RC}$ by the normalized delay variation g. Then equation 5.18 becomes:

$$g = e^{-2b_n \frac{V_{lr} f_b}{V_{fs} f_{in}} + g - 1}. \tag{5.19}$$

From equation 5.19 the following can be concluded for a minimum delay variation:

1. The linear range V_{lr} of the input amplifier must be large.

2. A large amplifier bandwidth f_b is needed.

3. A small full scale range of the converter is needed.

The linear range V_{lr} of an amplifier is determined by the technology used. In a bipolar system without using emitter degeneration elements a linear range of about 120 to 180 mV is obtained. In MOS devices this linear range depends on the threshold voltage of the devices.

The bandwidth of a system is mostly determined by the dc bias condition of the amplifier stage. A large bias current results in a large power dissipation. Due to the complexity of the overall system, bias currents must be kept within certain limits, preventing the power dissipation of the total circuit from growing to excessive values. A small-size high-frequency technology is needed to increase the analog bandwidth of the amplifier system. The minimum full-scale range of the converter is determined by the resolution and the matching of the input devices of the input amplifier. In bipolar devices in general a much better matching is obtained than is possible in an MOS technology. On the other hand, the noise generated in the input amplifiers determines the minimum tap spacing in a converter. In practice a minimum full-scale value of 1 V can be used for an 8-bit converter in a bipolar technology.

To solve the implicit equation 5.19 a computer program is needed. However, it is possible to simplify equation 5.19 by putting $g \ll 1$:

$$g \approx e^{-2b_n \frac{V_{lr} f_b}{V_{fs} f_{in}} - 1}. \tag{5.20}$$

In Figure 5.5 the relation between g and the ratio $\frac{f_b}{f_{in}}$ is shown. Note that this figure is only valid for case 1).

Proceeding to case 2), the combined equation for the output voltage $V_{out}(t)$ to reach the crossing point $b_n A$ is obtained by adding equations 5.8 and 5.9. The result becomes:

$$(1 - b_n)(t_2 - t_0) = RC(e^{-\frac{t_3 - t_2}{RC}} - e^{-\frac{t_3 - t_0}{RC}}). \tag{5.21}$$

Inserting equation 5.10 for t_3, we obtain:

$$(1 - b_n)(t_2 - t_0) = RC(e^{\frac{t_2 - t_0}{RC}} - 1)e^{-\frac{b_n(t_2 - t_0) + t_d}{RC}}. \tag{5.22}$$

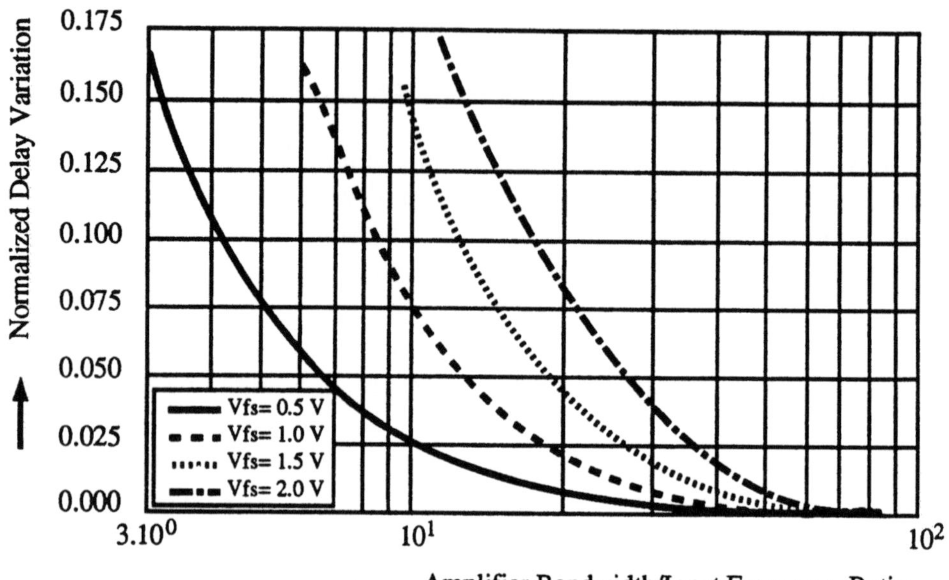

Figure 5.5 : Relation between the delay variation g and the ratio bandwidth/ input frequency

Rearranging the terms in equation 5.22 results in:

$$e^{-\frac{t_d}{RC}} = (1 - b_n)(\frac{t_2 - t_0}{RC}e^{b_n\frac{t_2-t_0}{RC}}(e^{\frac{t_2-t_0}{RC}} - 1)^{-1}. \tag{5.23}$$

The limit value of $\frac{t_d}{RC}$ when t_2 reaches t_0 equals:

$$\frac{t_d}{RC} = -\ln(1 - b_n). \tag{5.24}$$

Inserting a value of 0.5 for b_n gives $t_d = RC\ln(2)$. This value equals the delay of an RC time constant on a unity step response as expected.
A further simplification of formula 5.23 is obtained by inserting equation 5.16 and equation 5.5 for $\cos\omega t = 1$. After rearranging the terms we obtain for a maximum in the normalized delay variation:

$$g = 1 - \ln(e^{2\frac{Vlrf_b}{V_{fs}f_{in}}} - 1) + 2b_n\frac{Vlrf_b}{V_{fs}f_{in}} + \ln(2(1 - b_n)\frac{Vlrf_b}{V_{fs}f_{in}}). \tag{5.25}$$

In Figure 5.6 the normalized delay variation ($g = \frac{t_d}{RC}$) for the combination of case a) and b) are shown as a function of the ratio between the amplifier bandwidth and the input frequency. The full-scale value of the converter

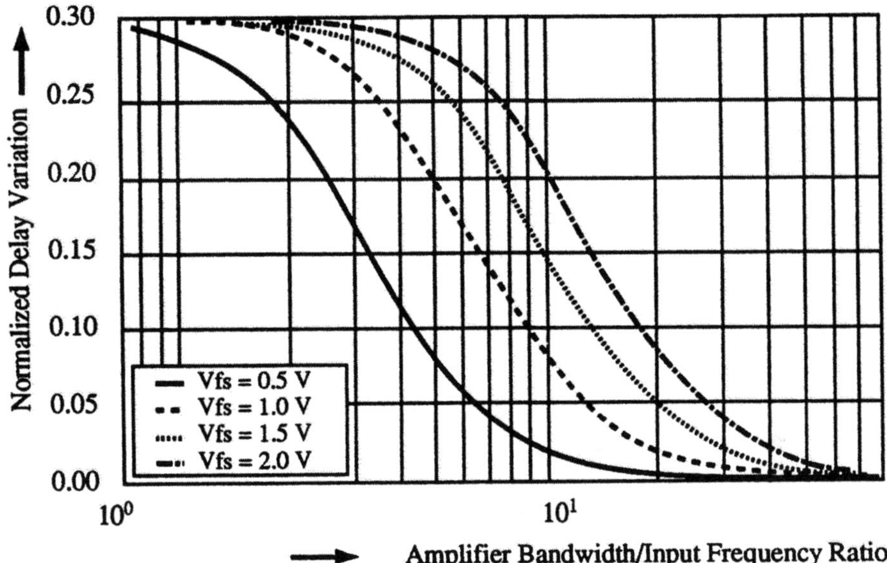

Figure 5.6 : Total normalized delay variation as a function of input amplifier bandwidth/input frequency

is inserted as a parameter. The linear range of the input amplifier $V_{lr} = 160$ mV. This model is verified by extensive simulations at transistor level of chords of input amplifiers performing an A/D input signal operation. These results can be used in an equation which will determine the third-order distortion in the converter system. This equation will be evaluated in the next section.

5.5 Distortion calculation

Suppose an input signal $V_{in} = \sin(\omega t + \psi)$ is applied at the input of the converter. In this formula ψ is an arbitrary phase shift introduced at the input to compensate for the various delays in the converter system. This phase shift is constant and independent of the slope of the input signal. At the output of the converter a digitized signal is found that obeys the following equation:

$$V_{out} = \sin(\omega(t + \delta t)). \tag{5.26}$$

From Fig 5.4 it can be seen that a good approximation for the delay difference δt_d as a function of the input frequency is:

$$\delta t \cong -\delta t_d \mid \cos \omega t \mid . \tag{5.27}$$

The absolute value for the cosine is inserted because regardless of the signal polarity the delay must have the same sign. In Figure 5.7 the input sine wave is shown that is applied to the signal dependent delay function of the amplifier. The delay function is normalized for zero variations at the tops of the sine wave. This results in a negative delay for the mid-points of the sine wave as shown in the figure. At the output of the system a sine wave appears with timing errors equal to the δt_d curve. The expectation of this

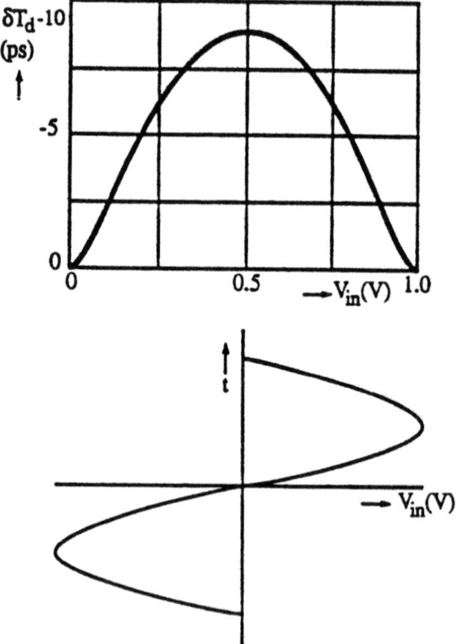

Figure 5.7 : Signal shifting of the output sine wave by the signal dependent delay

shifting of the time axis of the sine wave is a third-order distortion. When the input signal is subtracted from the time shifted output curve then the result shown in Figure 5.8 is obtained. The error signal contains still an important part of the input fundamental sine wave. When an amplitude and phase adjustment is applied during the subtraction of the input signal from the time shifted output signal then the result shown in Figure 5.9 is obtained. From this figure it is clear that a dominant third order distortion is obtained. This third-order distortion will be calculated now. Inserting

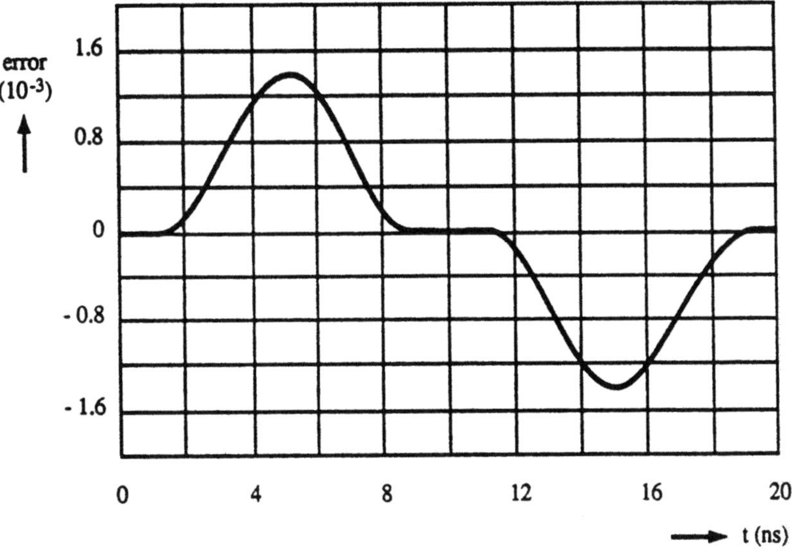

Figure 5.8 : Error signal after subtraction of the input signal

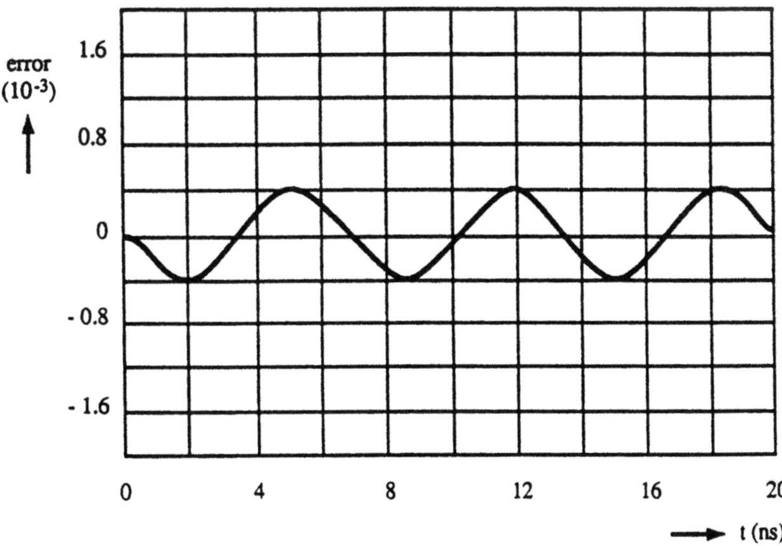

Figure 5.9 : Error signal after subtraction using an amplitude adjustment

equation 5.27 into equation 5.26 results in:

$$V_{out} = \sin \omega t \cos(\omega \delta t_d \mid \cos \omega t \mid) - \cos \omega t \sin(\omega \delta t_d \mid \cos \omega t \mid). \tag{5.28}$$

Equation 5.28 can be simplified using series expansion and assuming that δt_d is small with respect to t. When only the first-order terms in the series expansion are used, equation 5.28 becomes:

$$V_{out} = \sin \omega t - \cos \omega t \, \omega \delta t_d \, |\cos \omega t| . \tag{5.29}$$

A further simplification is obtained when $|\cos \omega t|$ is expressed in a Fourier series. We obtain:

$$|\cos \omega t| \simeq \frac{2}{\pi} + \frac{4}{3\pi} \cos 2\omega t. \tag{5.30}$$

Substitution of the series expansion from equation 5.30 into equation 5.29 results in:

$$V_{out} = \sin \omega t + \frac{2}{\pi} \omega \delta t_d \cos \omega t + \frac{4}{3\pi} \omega \delta t_d \cos \omega t \cos 2\omega t. \tag{5.31}$$

Reworking the product of $\cos \omega t$ and $\cos 2\omega t$ into a sum of cosine terms gives:

$$\frac{2}{3\pi} \omega \delta t_d \cos \omega t + \frac{2}{3\pi} \omega \delta t_d \cos 3\omega t. \tag{5.32}$$

Inserting this simplification into equation 5.31 results in:

$$V_{out} = \sin \omega t + \frac{8}{3\pi} \omega \delta t_d \cos \omega t + \frac{2}{3\pi} \omega \delta t_d \cos 3\omega t. \tag{5.33}$$

From equation 5.33 it is shown that the signal-dependent delay results in a third-order distortion of the input signal after the conversion takes place in an absolutely linear A/D converter.

The small cosine term can be included in the sine wave expression by substituting $\tan \eta = \frac{8}{3\pi} \omega \delta t_d$. This results in a small phase shift of the sine term; we obtain:

$$V_{out} = \sqrt{\tan^2 \eta + 1} \sin(\omega t + \eta) + \frac{2g f_{in}}{3\pi f_b} \cos 3\omega t. \tag{5.34}$$

The third-order term is a direct expression for the distortion in the A/D converter. In Figure 5.10 the distortion as a function of the ratio $\frac{f_b}{f_{in}}$ for the limited case 1) is shown. This curve gives a general insight into the effective bits of a converter as a function of the ratio bandwidth/input frequency. The distortion of the system is shown in Figure 5.11. Here again, V_{fs} is a parameter to get information about the distortion dependence of the full-scale range of the converter. Care must be taken in using the total distortion equation at the high-frequency end limit. It must be verified that the comparators or amplifiers at those high frequencies settle to the final signal value. Otherwise wrong conclusions are drawn.

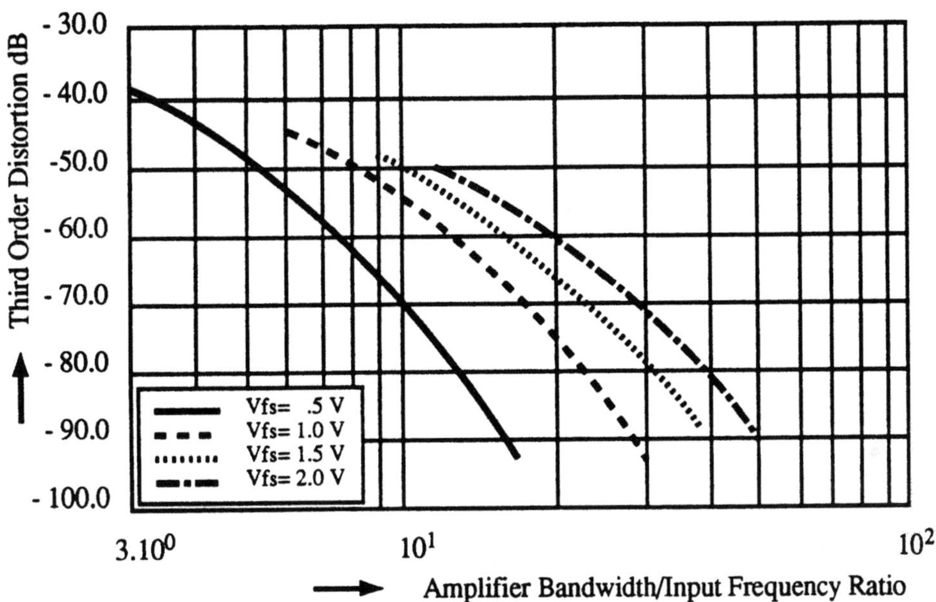

Figure 5.10 : Relation between distortion and bandwidth/input frequency ratio of an A/D converter

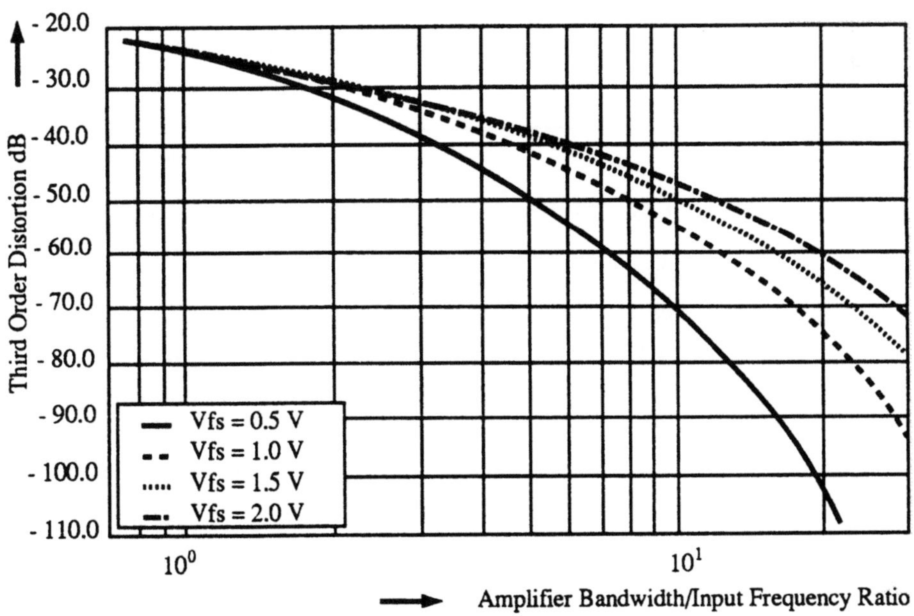

Figure 5.11 : Total distortion versus bandwidth/input frequency ratio of an A/D converter

5.6 Failure analysis of comparators

When flip-flops are used as comparators, at the beginning of a decision cycle the input signal can be so small that at the end of the decision cycle no digital "1" or "0" value is obtained. Such a condition is called a meta-stable state (MSS) [12]. In Figure 5.12 a generalized curve showing the two logical states V_1 and V_2 and the meta-stable state of a flip-flop are shown. During the design of comparators it is very useful to obtain a criterion that

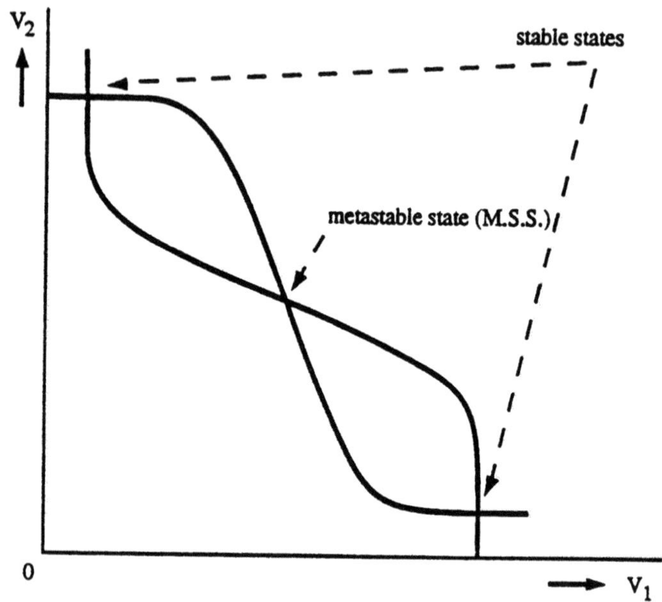

Figure 5.12 : States of a flip-flop used as a comparator

determines the number of times such a state occurs. An analysis will be given that determines the failure rate of a comparator with respect to the sample frequency and the unity gain bandwidth of the amplifiers used in the flip-flop. Noise in the system is supposed to have basically no influence on the number of meta-stable states of the comparators. Using the superposition principle the noise can be referred to the input of the circuit and adds to the input signal. In this way a disturbance of the comparison by noise is obtained but no influence on the number of meta-stable states is found.

5.6.1 First-order model of a flip-flop

In a first approximation a flip-flop can be modeled as two positively feed-backed first-order amplifiers. The small signal model is shown in Figure 5.13.

The amplifier stages are modeled as inverters with a gain $-A$ and a time

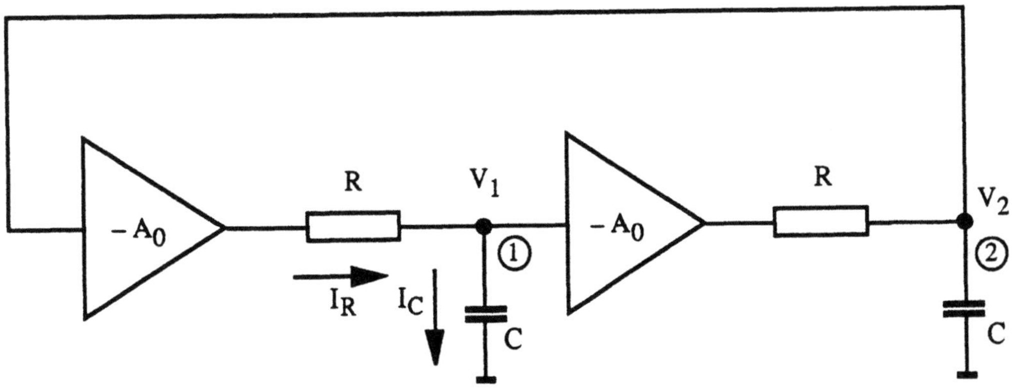

Figure 5.13 : Small signal model of a flip-flop

constant $\tau = RC$. Using feedback theory, the condition for "oscillation" can be used to determine the meta-stable condition. This results in:

$$1 - A^2 = 0. \tag{5.35}$$

In equation 5.35 A is expressed as:

$$A = \frac{A_0}{1 + s\tau}. \tag{5.36}$$

In solving equation 5.35 with respect to time, the following general relations are found:

$$V_1 = \lambda_1 e^{\frac{A_0-1}{\tau}t} + \lambda_2 e^{-\frac{A_0+1}{\tau}t} \tag{5.37}$$

and

$$V_2 = -\lambda_1 e^{\frac{A_0-1}{\tau}t} + \lambda_2 e^{-\frac{A_0+1}{\tau}t}. \tag{5.38}$$

V_1 and V_2 are the output voltages of amplifiers 1 and 2, respectively. In these equations λ_1 and λ_2 are integration constants. These constants can be determined by inserting the initial conditions at $t = 0$ into equations 5.37 and 5.38. These initial conditions are:

$$V_1 = V_{01} = \lambda_1 + \lambda_2 \tag{5.39}$$

and

$$V_2 = V_{02} = -\lambda_1 + \lambda_2. \tag{5.40}$$

In solving λ_1 and λ_2 from equations 5.39 and 5.40, the following result is obtained:

$$\lambda_1 = \frac{V_{01} - V_{02}}{2} \tag{5.41}$$

and

$$\lambda_2 = \frac{-V_{01} + V_{02}}{2}. \tag{5.42}$$

Furthermore, by analyzing equations 5.37 and 5.38 with increasing time t, only the terms with the positive exponent show an increase in signal. Putting

$$V_{01} - V_{02} = 2\delta V_0, \tag{5.43}$$

the solutions for the dominating output voltages can be obtained. The result is:

$$V_1 = \delta V_0 e^{\frac{A_0 - 1}{\tau} t} \tag{5.44}$$

and

$$V_2 = -\delta V_0 e^{\frac{A_0 - 1}{\tau} t}. \tag{5.45}$$

In general the input signal values δV_0 are uniformly distributed over the decision interval, the same is now found for the output signal values V_1 and V_2 after a time t_d ($=$ decision time) has been elapsed. Note that the differences between the final states of V_1 and V_2 are equal to the logical swing V_l of the system.

This can be related to the probability that a meta-stable state occurs after the sampling time t_d:

$$P(t > t_d) = e^{-\frac{A_0 - 1}{\tau} t_d}. \tag{5.46}$$

Equation 5.46 is valid when one sample is taken. If we take $f_s = \frac{1}{t_s}$ samples, then the number of meta-stable states per second becomes:

$$M_n = f_s e^{-\frac{A_0 - 1}{\tau} t_d}. \tag{5.47}$$

Equation 5.47 can be rewritten into general design parameters. Mostly $t_d = \frac{t_s}{2}$ when a symmetrical clocking scheme is used. During half the sampling time the flip-flop is in input position while during the second half sampling time the decision is taken. Furthermore, the time constant τ can be expressed into the unity gain bandwidth of the individual amplifiers. This results into:

$$\tau = \frac{1}{2\pi f_{3dB}} \tag{5.48}$$

and with $f_{ugb} = A_0.f_{3dB}$ equation 5.47 can be rewritten into:

$$M_n = f_s e^{-(1-\frac{1}{A_0})\frac{f_{ugb}}{f_s}\pi}.$$
(5.49)

When a moderate gain is used in the amplifier stages, then equation 5.49 gives a direct relation between the unity gain bandwidth (f_{ugb}) of the amplifiers and the sampling rate which can be applied to the system. At the moment the number of meta-stable states needs to be determined over longer time periods (1 minute, 1 hour, 1 day, or even 1 year), then equation 5.49 becomes:

$$M_n = T_p f_s e^{-(1-\frac{1}{A_0})\frac{f_{ugb}}{f_s}\pi}.$$
(5.50)

Here T_p is the time period over which the number of meta-stable states is determined. In equation 5.50 the relation between the unity gain bandwidth of a system and the sampling frequency is shown for different time intervals T_p. Figure 5.14 shows a very powerful relation in designing flip-flops with a high

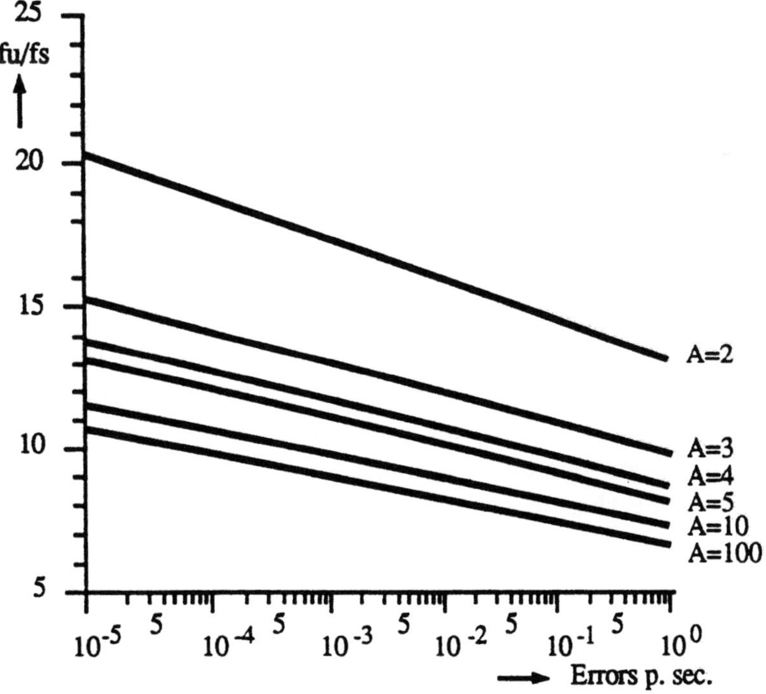

Figure 5.14 : Relation between unity gain bandwidth and sampling frequency ratio and the number of meta-stable errors per second

decision accuracy. A sampling frequency of 1 GHz is used in the calculations.

When the tap voltage of an A/D converter is not equal to the logic swing of the output signal of the flip flop, then an extra correction of equation 5.50 is required. Suppose that the tap voltage of the A/D converter is equal to V_{tap} and the gain of the input differential amplifier during sampling of the input signal is equal to A_{dif}, then equation 5.50 must be modified into:

$$M_n = T_p f_s \frac{V_l}{A_{dif} V_{tap}} e^{-(1-\frac{1}{A_0}) \cdot \frac{f_{ugb}}{f_s} \pi} \tag{5.51}$$

In a practical application the introduced correction factor between the logical swing V_l and the tap voltage of the converter V_{tap} gives only a small modification on the result of Figure 5.14 because of the very steep exponential relationship in the formula.

5.7   Input frequency decision moment variation

When clocked comparators are used to quantize varying input signals, the sampling moment changes with increasing input frequencies. Usually the sampling moment is shifted a small time, depending on the clock rise or fall time, if sampling occurs at the rise or the falling edge of the clock signal. This change in sampling moment finds its origin in the gain decrease at high input frequencies. When clocked flip-flops are used as a comparator then a decision takes place at the moment the total loop gain equals 1. At that moment no change in decision can be made anymore. In a bipolar system a larger tail current in the decision flip-flop is needed to compensate for the loss in gain at high frequencies. By using the slope of the sampling clock pulse, this larger tail current occurs at a later time moment. As a result a frequency dependent sampling moment is obtained which will result in third-order distortion components in the output signal of an analog-to-digital converter using these comparators.

5.8   Conclusion

In this chapter a generalized relation between amplifier bandwidth and maximum input frequency for an A/D converter is calculated. When no other distortion products due to ladder nonlinearity, and so on, occur, then the given expressions determine the effective resolution bandwidth of a converter. There is no difference in analysis between a full-flash converter or a folding converter. The relations given are verified using extensive simulations on

A/D converter circuits. These simulations are performed at transistor level. Furthermore, measurements on a flash-type A/D converter show the considerable increase in third-order signal distortion at high frequencies.

A simplified analysis about decision failures of flip-flops used as comparators has been carried out. This analysis results in design criteria for master-slave flip-flops as a function of the maximum analog input frequency and the maximum clock rate to be applied to the converter with the failure rate as a parameter.

Chapter 6

High-accuracy D/A converters

6.1 Introduction

High-resolution monolithic D/A converters are subject to growing interest due to the rapidly expanding market for digital signal processing systems. An example of such a market is digital audio. The large dynamic range of a digital audio system requires converters with resolutions of 16 to 20 bits. Monolithic converters with such a high linearity are difficult to design and require special circuit configurations. The most simple types of D/A converters are obtained with pulse-width modulation systems. These systems require fast logic circuits. In a pulse-width D/A converter structure an output low-pass filter reconstructs the analog signal and removes the modulation signal. Maximum speed of these types of converters is limited to the kHz range. The advantage of these systems is the small amount of accurate components that are needed in a practical implementation.

In integrating types of converters, an analog signal is reconstructed by integrating a current during a signal-dependent time. The output signal of the converter is applied to a sample-and-hold amplifier. The sample-and-hold amplifier reconstructs the stepped quantized waveform. With a low-pass filter the final analog output signal is reproduced again.

With matched components (resistors, capacitors, or transistors) it is possible to directly convert a digital number into an analog quantized signal.

However, the limited accuracy with which components can be matched maximizes the resolution to about 10 to 12 bits. In that case a converter still fulfills the linearity specification. Modifying the design can result in *monotonic by design* types of converters that do not have a $\frac{1}{2}$ LSB linearity specification but may have an excellent differential nonlinearity specification.

In this Chapter a special circuit configuration called "Dynamic Element Matching" will be described. This technique combines an accurate passive current division with a time division method to obtain the very high accuracy needed in monolithic converters without using trimming techniques. This system is, furthermore, insensitive to element aging and remains accurate over a large temperature range and with varying elements. In references [63,1,2] examples of circuits realized in practice are given.

Furthermore, a MOS solution in which currents are calibrated is described. This current calibration technique results in a very high matching accuracy without the need for an extra filtering operation to reduce a calibration ripple.

A third self-calibration scheme is shown, which uses an additional cycle to calibrate the complete converter. During calibration, however, it is not possible to use the converter effectively in the system.

6.2　Pulse-width modulation D/A converters

In Figure 6.1 an example of a pulse-width D/A converter is shown. The system consists of a digital buffer in which the input data are stored. These data are compared with a ramp signal that is generated by a counter operating with a clock frequency f. The digital comparator compares the buffer data with the ramp signal generated by the counter. When the counter signal is smaller than the buffer signal, then the output of the comparator is high. At the moment the counter signal is equal or larger than the buffer signal, then the comparator output is low. The output signal of the converter is a pulse-width modulated signal. This signal is filtered by a low-pass output filter and the analog signal is reconstructed again. A simple calculation gives for the output signal:

$$V_{out} = \frac{T_x}{T} \times V_{ref}. \qquad (6.1)$$

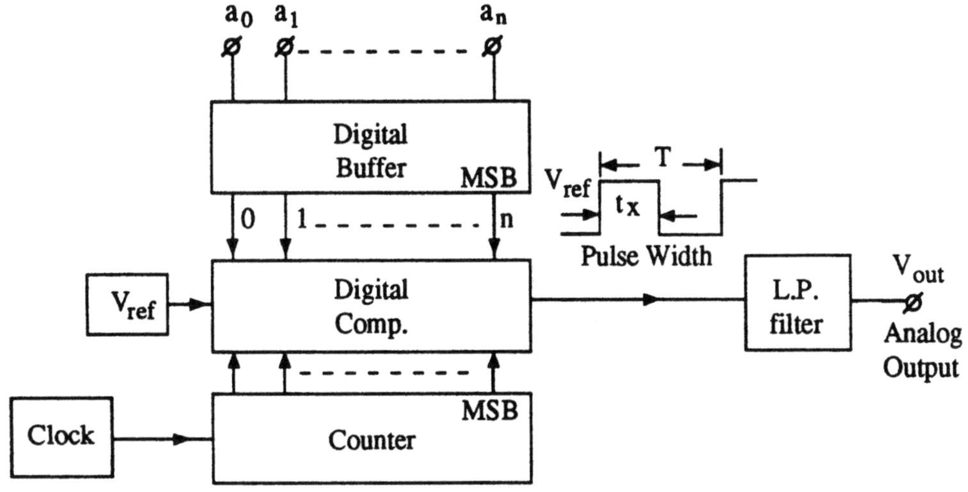

Figure 6.1 : Pulse-width modulation D/A converter

Here T_x is the time that the comparator output signal is high and T is the repetition time of the ramp signal. V_{ref} is the reference signal that is modulated by the comparator output signal. Moreover, it can be shown that the conversion time is equal to:

$$T_{conversion} = \frac{2^n}{f_{clock}}.$$ (6.2)

Here n is the number of bits of the counter. f_{clock} is the frequency of the counter clock.

A disadvantage of this system is the low repetition rate of the pulse-width modulated signal. To reconstruct the analog output signal a low-pass output filter with a large stop-band attenuation is required. The stop-band attenuation must be at least equal to the dynamic range of the converter.

In Figure 6.2 a system implementation is shown that has the ability to randomly generate the high level signals of the comparator over the whole conversion period [18]. From the figure, which uses exactly the same building blocks, it can be seen that the most significant bits of the digital buffer and the least significant bits of the ramp signal generated by the counter are compared. As a result of this operation the output pulse of the comparator is better randomized during the conversion period T. Around the middle range of the converter the frequency components are shifted to a higher

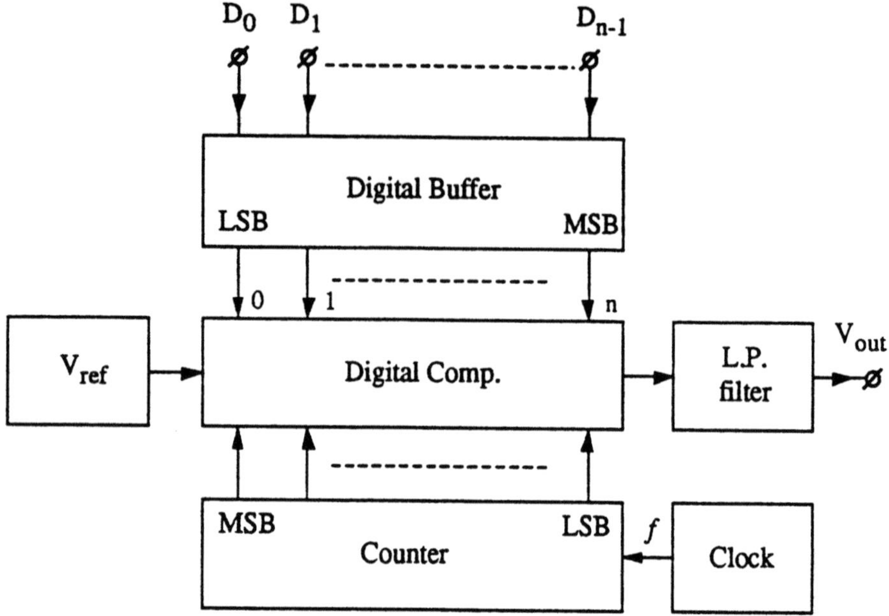

Figure 6.2 : Reverse comparing pulse-width modulating D/A converter

frequency band. As a result the requirements of the output filter can be reduced. Table 6.1 shows an example of this randomizing for a 3-bit code 110.

From the table the randomizing of the comparator output in the case of the reverse bit-weight connected counter can be easily distinguished.

6.3   Integrating D/A converters

In Figure 6.3 an example of a single slope D/A converter is shown [62]. The system consists of digital input latches that store the input information, a digital comparator, and a counter operating from a crystal controlled oscillator. In this part of the circuitry the input data are converted into a time signal which is the output of the digital comparator. The comparator output signal switches the reference current I to the integrator which converts the current signal into an analog output voltage across the capacitor C. At the end of the integration cycle the sample-and-hold using an operational amplifier and feedback elements R_1, R_2, and C_2 samples the final analog value of the converter and holds the signal during the time the next conversion is performed. Note that during the hold mode of the sample-and-hold amplifier the resistor R_2 remains in parallel with the hold capacitor C_2. A first order

Buffer	Reverse Counter	Normal Counter	Reverse Comparator	Normal Comparator
	0 0 0	0 0 0	S	S
	0 0 1	1 0 0	L	S
	0 1 0	0 1 0	S	S
1 1 0	0 1 1	1 1 0	L	E
	1 0 0	0 0 1	S	L
	1 0 1	1 0 1	L	L
	1 1 0	0 1 1	E	L
	1 1 1	1 1 1	L	L

Table 6.1 : Comparison Table

Notation:

S = Smaller

L = Larger

E = Equal.

output filtering is combined with the hold operation in this system. Finally the output signal can be applied to a low-pass filter that reconstructs the analog output signal. At the start of a conversion the integrator is reset by closing switch S_3. The sample-and-hold amplifier remains in the hold mode. When an input data word is loaded into the buffer, the reset switch S_3 is opened and the reference current I starts charging the capacitor C during the time the counter value is smaller than the buffer value (S_1 is closed). At the moment the comparator detects a counter value which equals the buffer value, then the reference current is switched off (S_1 is opened), and the integrator remains at the converted analog value. The sample-and-hold amplifier samples the output signal of the integrator (switch S_4 is closed). When the next start conversion comes, the sample-and-hold amplifier is switched in the hold mode (switch S_4 is opened), storing the analog information during the time the next digital-to-analog conversion is performed. The conversion speed of this systems depends on the number of bits that must be converted and the maximum clock speed that can be applied to the system.

$$T_{conversion} \approx \frac{2^n}{f_{clock}}. \tag{6.3}$$

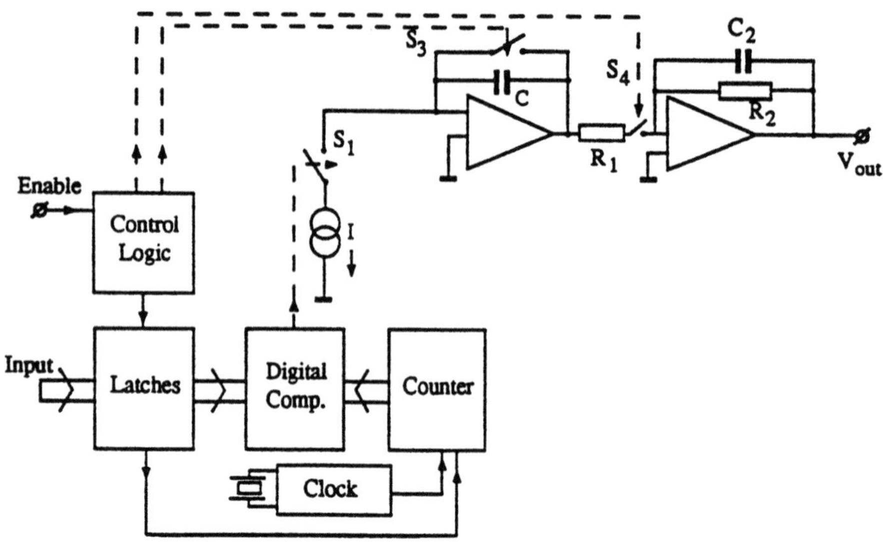

Figure 6.3 : Single-slope integrating D/A converter

Here f_{clock} is the frequency of the crystal clock.

To increase speed the input data word can be split-up into a coarse and fine value. At the same time the reference currents are split-up into two weighted values which correspond with the splitting of the coarse and fine digital values [30].

In Figure 6.4 a simplified form is shown. In this system the number of clock pulses to obtain a full coarse and fine counter is reduced. In a 16-bit system using an 8-bit coarse and an 8-bit fine split-up a 256 times larger conversion speed can be obtained compared to the circuit of Figure 6.3 having a 16-bit resolution.

A conversion starts with the reset of the integrator by closing switch S_3. At the same time the coarse and fine digital data are loaded into the input latches. Then switch S_3 is opened and the counter, which is reset to zero, starts counting. As long as the counter data are smaller than the input data, the comparator outputs are closing switches S_1 and S_2, respectively. The reference currents I and $\frac{I}{256}$ are charging the integrator capacitor C. This charging stops at the moment the contents of the counter is larger than the course or the fine input data. Finally, the analog data are sampled by

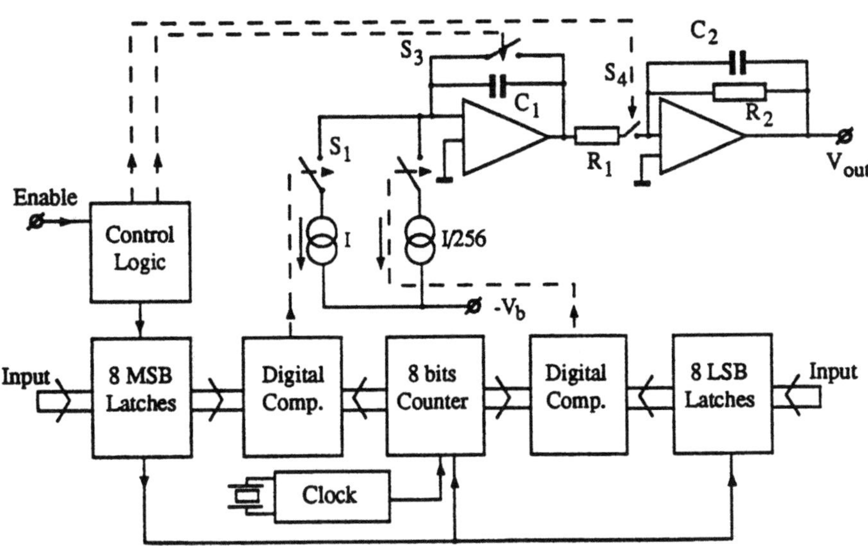

Figure 6.4 : Dual-ramp D/A converter system

the succeeding sample-and-hold amplifier consisting of an operational amplifier with feedback elements R_1, R_2, and C_2. The sample-and-hold circuit is switched into to hold mode during the time a new sample is loaded and converted into an analog value. Then this new value is sampled, giving the next output signal. A low-pass filter can be used to reconstruct the analog signal.

A disadvantage of single integrator systems is that full scale accuracy is dependent on the reference current value I and the integrator capacitor C. Furthermore the accuracy with which the switches can be operated determines the linearity of the system. Dynamic performance is determined by the performance of the sample-and-hold amplifier.

6.4   Current weighting using ladder networks

Different systems using ladder networks to obtain an accurate current weighting will be discussed. These ladder networks are simple to implement in integrated circuits. Furthermore, the very well matched value and thermal tracking of components on the same die improve the overall performance of the system.

6.4.1 *R-2R* ladder network

An *R-2R* ladder network with terminating transistors to generate binary weighted currents is shown in Figure 6.5. The good matching and excellent thermal tracking of integrated resistors is the main design criterion of this circuit. The circuit consists of equal resistors with a value R. The $2R$ resis-

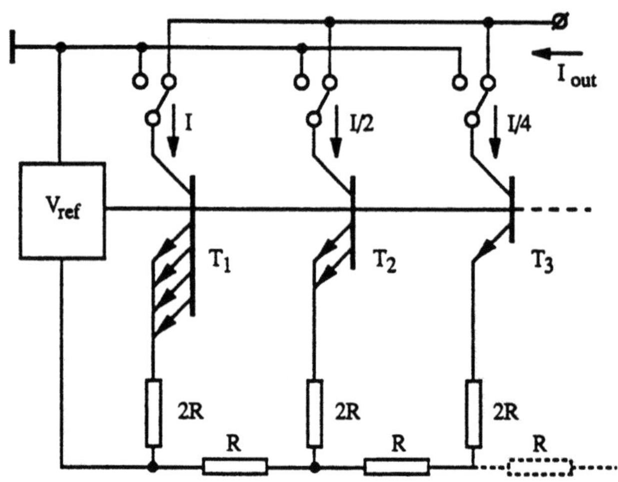

Figure 6.5 : R-2R ladder network D/A converter system

tor is formed by connecting two resistors in series to obtain the best resistor matching. Because the output currents are binary weighted, the base-emitter voltages of the terminating transistors will decrease if no special measures are taken. From Figure 6.5 it can be seen that the transistor sizes decrease by a factor two every time the current is reduced. An equal voltage at the emitters of the current source transistors is needed for an accurate operation of the *R-2R* ladder network. The division with such a network depends on a divide by two voltage division every time a section is added. At the time a low bit current value is reached, the voltage drop across the resistors is small compared to the base-emitter voltage of the transistors. Therefore, the current density per transistor must be equal, which results in the scaling of the emitter sizes of the transistors.

Some practical information about resistor matching is given in Table 6.2. The matching data are given for three types of resistors and for different widths W of 10μ and 40μ. Although these data change at the moment a technology improves, this table is given for the reader to obtain a feeling

Fabrication process	Matching tolerance			
	$\sigma(\%)$		mean(%)	
	10μ	40μ	10μ	40μ
Diffusion	0.44	0.22	-0.10	0.07
Thin film	0.24	0.11	-0.10	-0.06
Ion implant	0.34	0.12	0.05	0.05

Table 6.2 : Resistor matching data

about the problems of resistor matching. For more information see reference [64]. Thin film resistors are the most attractive for applications in integrated converters because of the low temperature coefficient and high linearity of these resistors. Furthermore, with laser trimming techniques these resistors can be accurately trimmed to the required value. A disadvantage of the thin film resistors is the need for additional process steps to incorporate these elements on the same die.

6.4.2 Resistor weighting current network

In Figure 6.6 an example of a network is shown that uses a binary resistor weighting combined with emitter scaling to obtain an accurately binary-weighted current network. The voltage drop across the current determining resistors remains about the same for all generated output currents. In this way the accuracy with which currents can be generated fully depends on the resistor weighting and less on the transistor matching. However, the total amount of resistance increases to nearly $2^n R$ for the LSB current. Here R is the resistor value for the MSB bit current generation. The die size even more increases compared to the circuit given in Figure 6.5. Because of the large total resistor value needed to construct this type of converter it is less suitable for implementation in an integrated circuit.

6.4.3 Equal currents output ladder network

In Figure 6.7 an example of a system using equal currents and an R-$2R$ resistor output network giving the binary weighting is shown. The basic idea behind this system is that for every technology there exists a transistor size and current density for which an optimum switching performance is

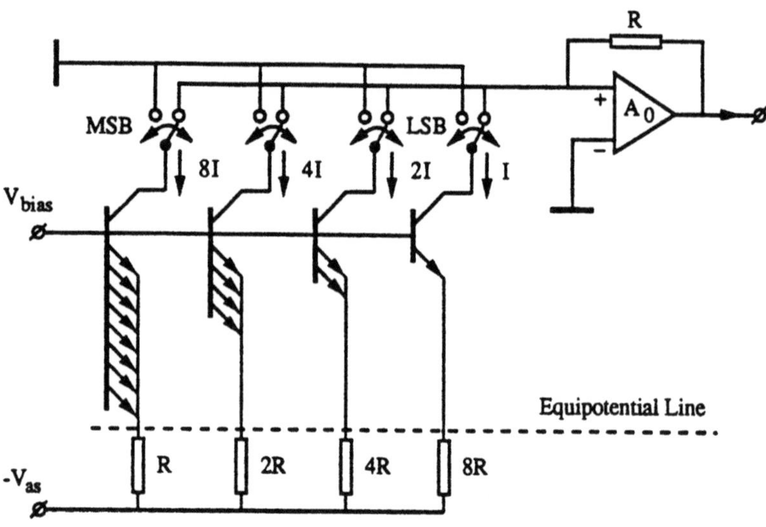

Figure 6.6 : Binary-weighted current network using resistor weighting

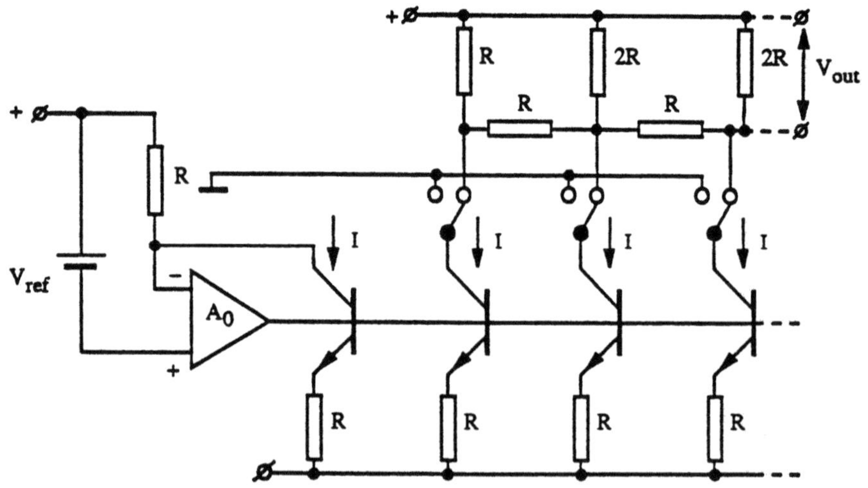

Figure 6.7 : High-speed equal current binary weighted resistor network

obtained. Therefore equal currents are accurately generated in the lower part of the circuit shown in Figure 6.7 using equal resistors R. Furthermore the equal current values need only equal transistors. The output capacitance of this network is small and equal to a single transistor capacitance. To accurately couple the reference source V_{ref} with the current value I an extra current source is added in a feedback loop with an operational amplifier. This operational amplifier adjusts the base voltages of the current transistors in such a way that the current I is equal to:

$$I = \frac{V_{ref}}{R'}.\tag{6.4}$$

In high-speed converter systems the switching speed is an important parameter. Therefore a network with equal currents is used in combination with
an output R-$2R$ resistor network to obtain the binary weighting. Mostly
the impedance of the R-$2R$ network is designed for 75 Ohm or 50 Ohm,
depending on the characteristic impedance of the system the converter has
to work with. The digital input information sets the switches in the required
position. The current I is applied to an R-$2R$ network to obtain a binary
voltage weighting depending on the weight of every input bit. With a low
value for R e.g., $R=$ 75 Ohm, the delay through the total R-$2R$ network is
so small that a small acquisition time of the total system is obtained. A disadvantage of this system is that the number of accurately matched elements
is increased. This has an influence on the yield of these systems, especially
when the resolution of the system is increased to 10 or 12 bits.

6.4.4   Data interleaved D/A converter

In very high-speed D/A converter applications the speed of the bit switches
and the D/A converter system shown in Figure 6.7 is not the limiting factor
[24,25,26,27]. Input data latches and decoding circuitry limit the maximum
throughput rate of the converter. Especially in low-glitch designs the three
to five most significant bits are encoded into a thermometer code, using
equal currents in a part of the circuit. This segment encoding improves
monotonicity of a design by increasing the output values with equal steps.
Furthermore, such a code minimizes in offset binary-coded converters the
glitch energy when small (video) signals are decoded. In this case not the
MSB but a much smaller current value is switched, resulting in smaller
glitches. The improvement in glitch reduction depends on the number of
currents used in the segmentation.

Note that for the generation of the lower bits again a system with equal
currents and a binary weighted output ladder network is used. Depending
on the technology used for the implementation of the circuit, an optimum
between three to five segmented bits with five to seven weighted bits is found
for an overall resolution between 8 to 12 bits. To increase the speed of the
digital system a multiplexing of the input data channel is used. In Figure

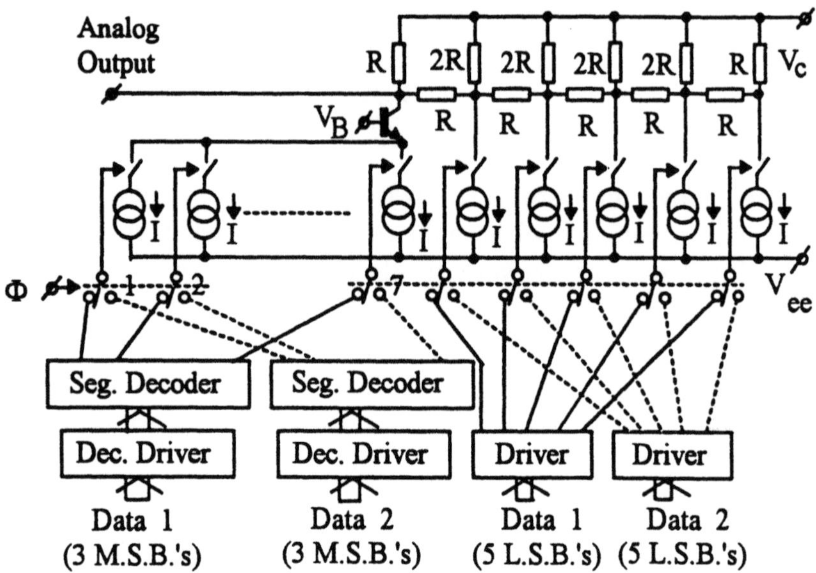

Figure 6.8 : input data interleaved D/A converter

6.8 an example of such a system is shown. As shown, the input data latches
and the decoding circuitry are doubled to allow more time for decoding. A
multiplexer switches the latched digital input to the bit switches. An increase
in throughput rate of the D/A converter is obtained. Furthermore, the data
of the latches remain fixed during the conversion. A clean control voltage
is applied to the switches in contrary with data coming from a master-slave
flip-flop. In a bipolar technology speeds up to 1GHz are possible.

6.4.5   Two-step current division network

By using a two-step current division principle, a number of problems from
the above discussed systems can be overcome. In Figure 6.9 an example of an
8-bit converter is shown. In the four most significant bit current generation
part of the system, an R-$2R$ current network with emitter scaling is used to
obtain the optimum accuracy. The most significant bit current is repeated
to form with the reference source a feedback loop that controls the base
voltages of the system in such a way that an accurate relation between the
output current I and the reference source is obtained. The current value $\frac{I}{8}$
is repeated. This current is applied to an emitter scaled fine divider network
that performs the 4-bit fine division. Note that the current value $\frac{I}{128}$ is
two times applied to make the total tail current equal to $\frac{I}{8}$. This second

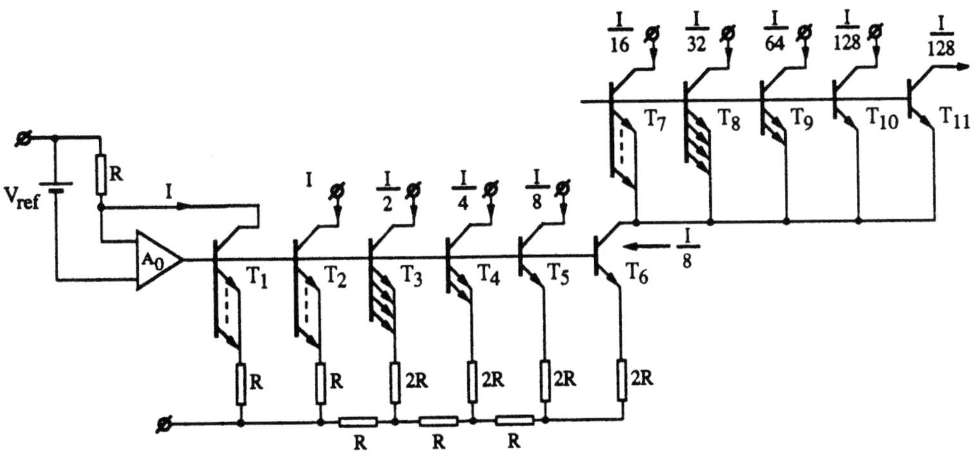

Figure 6.9 : Two-step current division network

current value is not needed for the binary-weighted network and is therefore connected to a supply voltage. Base current is lost in the second divider unit. In some cases this can introduce intolerable errors in the weighting function. To compensate for base current losses, the size of the second transistor that generated a current $\frac{I}{128}$ can be decreased in such a way that a compensation for the loss in base currents of the fine divider part is obtained. Such a compensation gives only a limited compensation possibility and must be analyzed for worst-case device tolerances and extreme temperature ranges. As can be seen from this system, an optimum in die size and amount of elements is obtained.

6.4.6 Base dropping R-$2R$ network with equal sized transistors

An interesting alternative exists by using a base dropping voltage to compensate for the binary emitter size weighting of the current source transistors [19,20,15]. As is known from device physics, the difference in base-emitter voltage of bipolar transistors with a current ratio of 1 to 2 and an equal emitter size is equal to:

$$\Delta V_{be} = \frac{kT}{q} \ln 2. \tag{6.5}$$

In Figure 6.10 an example is shown. In this system an R-$2R$ network is used to generate the accurately binary weighted currents. Between the bases of the current source transistors, which apply the binary-weighted currents to the bit switches, a resistor with a value $\frac{r}{2}$ is applied. Across this resistor a

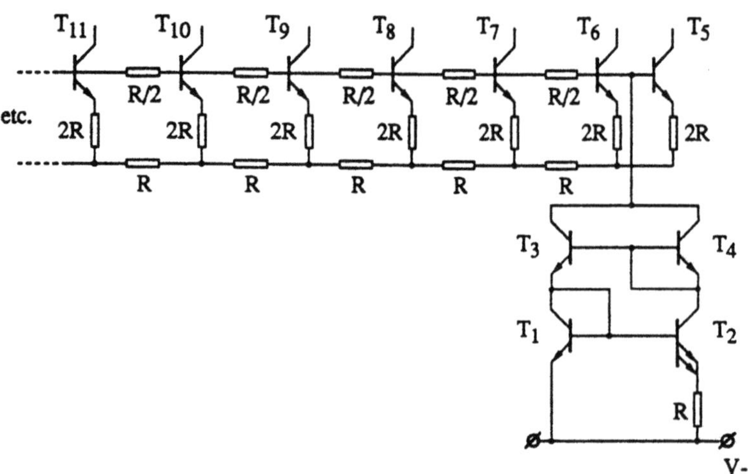

Figure 6.10 : Base dropping voltage compensation technique

voltage is generated which is equal to: $\frac{kT}{q}\ln 2$. For this purpose a current source is added that generates a current equal to

$$I_r = \frac{2kT}{rq}\ln 2. \tag{6.6}$$

The current source that generates the current I_r consists of a PNP current mirror with a mirror gain of 1 and an NPN current mirror with a device size ratio of 1 to 2 and an emitter degeneration resistor r. The NPN 1 to 2 device size scaling is exactly what is needed in a D/A system using a 1 to 2 current ratio. Across the resistor r a voltage equal to

$$I_r = \frac{2kT}{rq}\ln 2. \tag{6.7}$$

The current mirror action in the current stabilizer doubles the output current. Furthermore, equation 6.7 shows that the temperature coefficient of the current through the resistor r is proportional to the absolute temperature T. In Chapter 9 a more detailed analysis of these type of current sources will be given.

6.4.7　10-bit binary-weighted converter system

In Figure 6.11 an implementation of a 10-bit binary-weighted current converter network is shown. From Figure 6.11 it can be seen that a combination

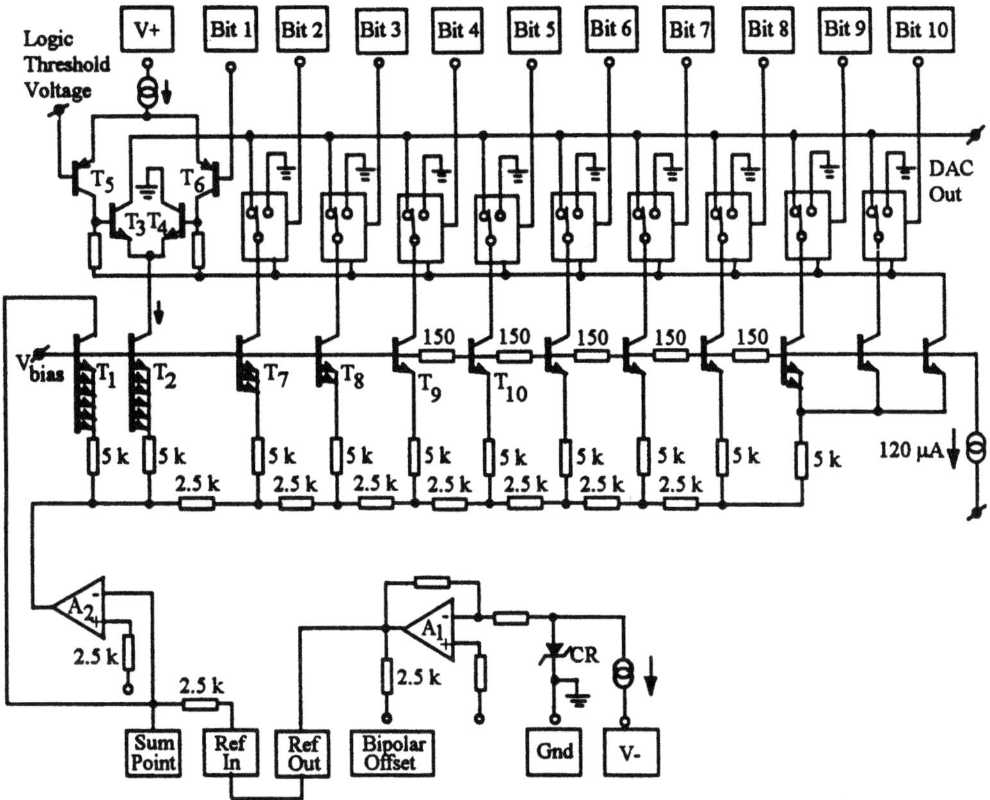

Figure 6.11 : 10-bit binary-weighted converter system

of techniques is used to generate binary-weighted currents. In the most significant bit part (4 bits) an R-$2R$ ladder network with emitter scaling is used. The next 5 bits use the base dropping voltage technique to compensate for the emitter scaling of the current source transistors. Finally the last two bits use emitter scaling. Note that the LSB bit currents are generated twice. The current source of 120 μA generates a voltage of about 18 mV across the 150 Ohm base dropping resistors. Furthermore, the most significant bit is copied again and compared with the reference source. In this system the bases of the current network are connected to a bias terminal and the operational amplifier A_2 controls the voltage across the R-$2R$ network in such a way that the MSB current value is equal to the reference current value flowing through the resistor of 2.5 k connected from *Ref In* to the inverting input of amplifier A_2. The zener diode CR_1 generates a reference voltage of about 7.5 V, which via operational amplifier A_1, is applied to the converter. As shown in a part of the circuit diagram the bit switches consists of differential pairs, at the MSB current value called T_3, T_4. These differential pairs are driven from lateral PNP transistors T_5, T_6 in the MSB example. One side of all the switches is connected to the logic threshold level. The other side can be directly driven from TTL logic levels or CMOS levels. Lateral PNP transistors have the advantage that the base-emitter reverse breakdown voltage is practically equal to the collector-base breakdown voltage of an NPN transistor. Therefore, lateral PNP transistors can be reverse biased with voltages of at least 10 Volts without destroying the devices. The output terminal of this D/A converter must be connected to an inverting output operational amplifier. Feedback resistors of 5 K are available as current to voltage elements having the same temperature coefficients as the resistors used in the binary weighted current network.

6.4.8 Binary-weighted current divider using device scaling

Although matching of active elements is limited in the range of 1% to 2% for two equal transistors, it is possible to use the law of the large numbers to improve the matching. In Figure 6.12 an example of a 10-bit binary-weighted current network using emitter scaling is shown. To obtain the bit weighting

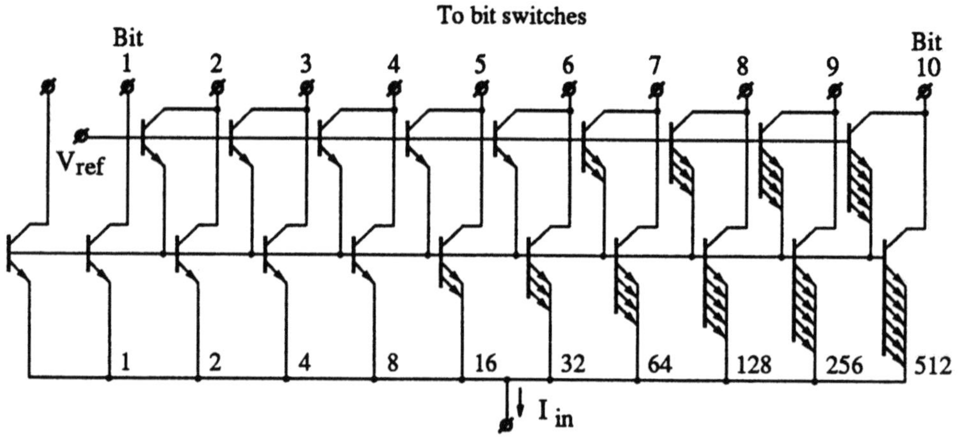

Figure 6.12 : Binary weighted current network using emitter scaling

for the most significant bit 512, transistors are randomly connected in parallel in the network [2]. The matching with respect to the ideal number of 512 is improved by $\sqrt{512}$ or about 22 times. This means that a 1% matching in this case decreases to a .05% matching error. The MSB-1 bit uses 256 transistors in parallel, which improves the matching about 11 times. We obtain finally about 0.1% matching. Going through the current network shows that the matching of every smaller bit decreases. In a binary-weighted network this is more or less allowed because the influence of the LSB bit on the overall integral linearity is much less than the MSB bit. Base current compensation in this system is used by weighting the total base current of the divider using a second simplified emitter scaled current divider. Individually the weighted base current is added to the individual bits to obtain an accurate division of the total tail current I_{in}. The advantage of using emitter scaling is that over a large input current range an accurate weighting is possible. In practice an integral linearity better than $\frac{1}{2}$ LSB is obtained over an input current range from a μA to a few mA. In Figure 6.13 a binary-weighted network using MOS devices is shown [33]. In the MOS system again 1024 devices

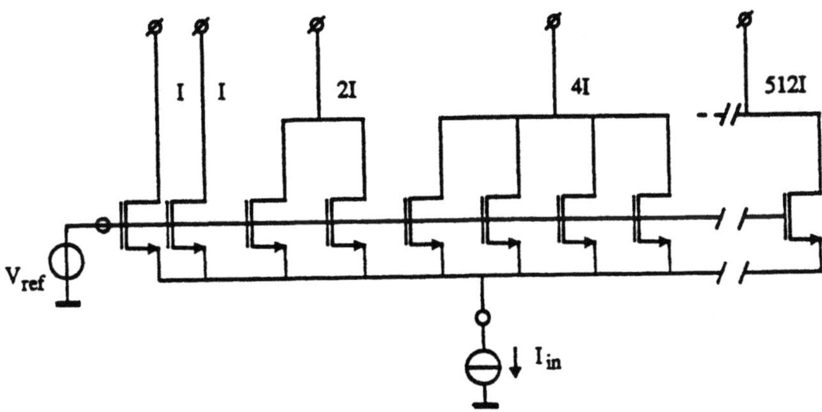

Figure 6.13 : Binary current weighting using MOS devices

are used to generate a binary-weighted current division. The connection of devices to generate the MSB current value is made in such a way that a good averaging concerning the matching of the devices is obtained as was the case with the bipolar implementation of this circuit type. In Figure 6.14 the integral non- linearity of an MOS binary-weighted current network as a function of the tail current is shown. Analyzing the measurement results shows that a 10-bit binary-weighted MOS network can be designed having $\frac{1}{2}$ LSB integral nonlinearity over a tail current value from 1 μA to 20 mA.

6.4.9 MOS ladder network converter system

In a special MOS technology thin film process steps to construct the *R*-2*R* ladder network are used in combination with CMOS (complementary P- and N-MOS devices) switches to design a D/A converter. Thin film resistors show a better matching than poly silicon resistors which are available in a

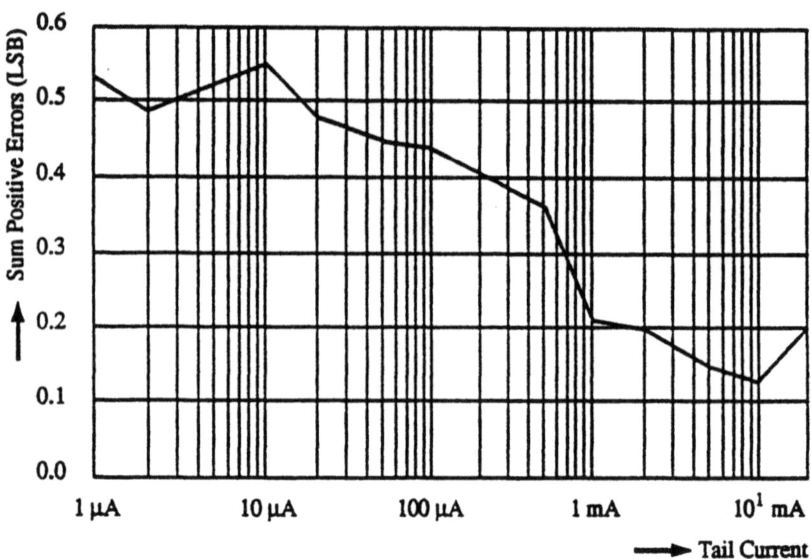

Figure 6.14 : Integral non-linearity measurement result

MOS technology. Moreover, thin film devices can be trimmed to improve the accuracy of the converter. In Figure 6.15 a basic circuit diagram of a 10-bit D/A converter is shown. As can be seen from Figure 6.15, the ON

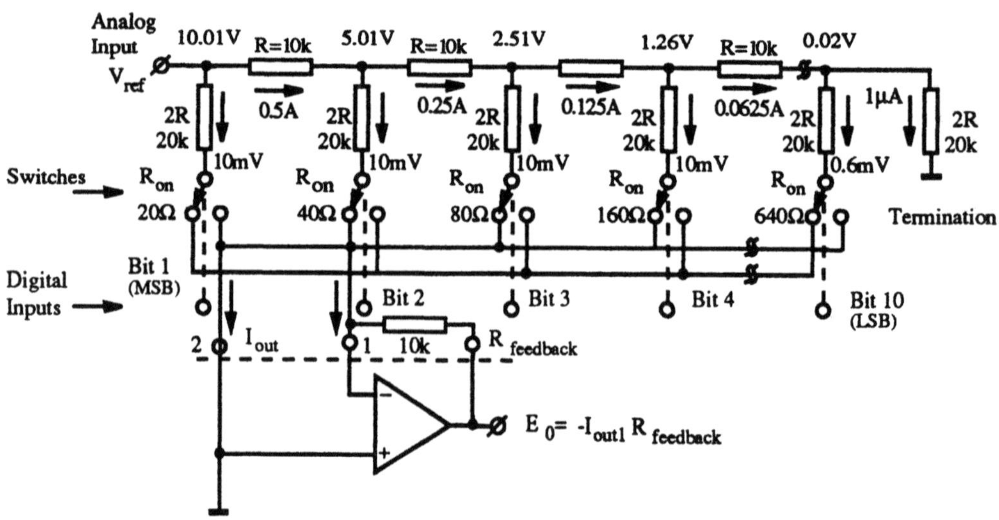

Figure 6.15 : MOS ladder network D/A converter system

resistance of the MOS bit switches is scaled to obtain a good accuracy of the converter. The reference voltage is applied at the analog input terminal. In this special situation a maximum of 10 Volts is used as reference. The resistor values are chosen depending on the size of the MOS switches and the ON resistance of these switches. A special application of this system is found if the analog input voltage is modulated. In that case a multiplying D/A converter structure is obtained. Such a construction can be used to increase the maximum dynamic range of the system or for digitally controlling an ac signal that can be used as analog reference voltage. The output signal of the ladder network is a current. This current is converted into a voltage using an external operational amplifier with a feedback resistor $R_{feedback}$. This feedback resistor is included with the ladder resistor network to obtain the same temperature coefficient and the same matching. Note, furthermore, that an extra termination resistor is included to obtain the required network loading. Using thin film resistor trimming techniques the linearity of this converter can be increased to 12 bits.

6.4.10 Weighted capacitor converter system

In a MOS technology capacitors with MOS switches perform an identical operation as resistors and transistors in a bipolar technology. In Figure 6.16 an example of a binary weighted capacitor D/A converter system is shown [28,31,29]. The system consists of n binary weighted capacitors, an

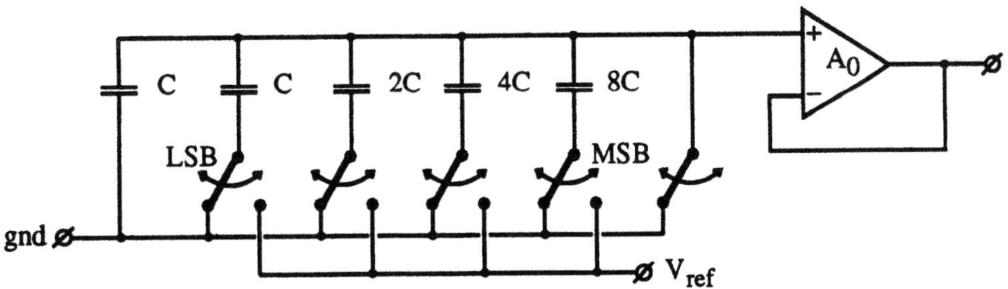

Figure 6.16 : Binary-weighted capacitor D/A converter

additional capacitor with the unit capacitance, an operational amplifier connected as a follower, and a set of switches that can connect the weighted capacitors to the reference voltage V_{ref}. At the start of the conversion all capacitors are discharged as shown in the switch configuration of Figure 6.16. Then all the capacitors are connected to the reference voltage to perform a precharge condition. In this case the unit capacitor C is still short-circuited to ground. When the conversion starts, every capacitor, depending on the digital input information, is connected to ground (gnd) or remains in the reference position. During the conversion the charge is redistributed over the capacitors and a binary-weighted D/A conversion is obtained for the generated output voltage.

Suppose that the total capacitance value in the system is equal to: C_{total}, then the unit capacitance C is equal to:

$$C = \frac{C_{total}}{2^N}. \tag{6.8}$$

The value of the capacitor connected to bit n, C_n becomes:

$$C_n = 2^{N-1} \times C. \tag{6.9}$$

The output voltage of the capacitor network can now be expressed in terms of capacitor values and the corresponding digital input code D_i. A calculation of the output voltage gives:

$$V_{out} = \frac{V_{ref}}{C_{total}} \sum_{i=1}^{N} C_i D_i. \tag{6.10}$$

This equation can be simplified into:

$$V_{out} = \frac{V_{ref}}{2^N} \sum_{i=1}^{N} D_i * 2^{i-1}. \tag{6.11}$$

This equation shows the binary weighting that is obtained by the charge redistribution technique. These techniques can be successfully applied in 10 to 12 bits D/A converters in a MOS technology.

6.4.11   Some remarks about the ladder converter systems

In a 10-bit system, for example, the largest transistor has a size which is $2^{N-1} = 512$ times the size of the transistor in the least significant bit. Such a ratio in transistor size requires a large die size. Moreover, the output capacitance of the most significant bit current source is large due to the parallel connection of 512 transistors. Such a large capacitance has a drawback for the switching of this current to the output terminal. As a result it is difficult to operate all switches at the same speed. This switching at the exact same moment is required to obtain a small output glitch. Mostly a reference source is added to accurately determine the value of the MSB current of the converter. Technology studies in a bipolar or CMOS implementation of these types of converters show that without trimming an accurate converter of up to 10 bits can be designed. In that case this converter does not need any trimming to obtain the required linearity and monotonicity specification. Modifying the design in a multiple equal current generation and switching of two or more of the MSBs improves the monotonicity of the converter. Furthermore, the accuracy is increased by using more elements in parallel to generate the high current values. The law of the numbers increases the weighting accuracy by the square root of the number of elements used. To increase the accuracy of the R-$2R$ system a trimming procedure is needed. With trimming it is possible to obtain the high accuracy. However, the system becomes sensitive to material stresses and therefore needs to be trimmed after the die has been mounted in its encapsulation. An example of a trimmed high-resolution converter is given in reference [65]. Such procedures are fairly expensive. To overcome this problem, a different system approach will be used.

6.5 Monotonic by design network systems

In this section attention will be paid to special circuit configurations which by construction give a monotonic behavior of the converter.

6.5.1 Current weighting operation

Sometimes it is not necessary to obtain a converter with a full integral linearity specification, but it is enough to guarantee monotonicity of the system. In that case a different design technique can be adopted [21]. Monotonicity of a converter is obtained if with an increasing digital input signal an increase in analog output signal can be guaranteed. In such a case always a current is added and not switched off and replaced by a larger current value. In Figure 6.17 an example of a current based *monotonic by design* converter type is shown. In the most significant 3 bits of the converter equal currents are used. The binary digital input signal uses a segment decoder to obtain for six of the eight switches a three-position condition. The next 5 bits use an R-$2R$ network with emitter scaling to obtain an accurate current division. In the third 3-bit weighting network, emitter scaling is used. The implementation of an R-$2R$ network does not improve the weighting accuracy because

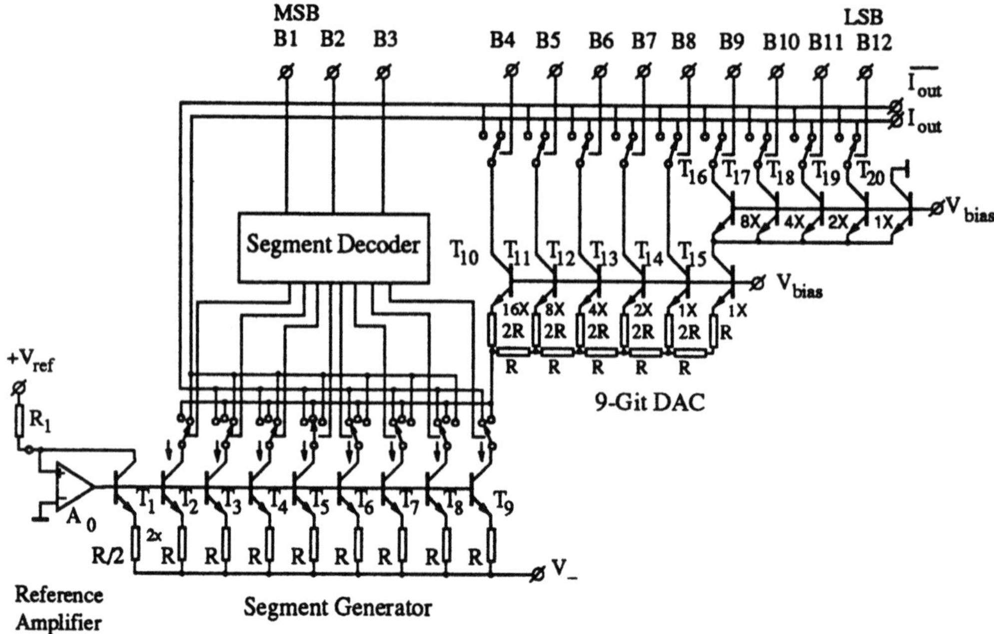

Figure 6.17 : Monotonic current based D/A converter system

of the small value of the current and the limited value resistors can be given in an integrated circuit. The operation of the system can be explained by supposing that a sawtooth like digital input code is applied at the input of the converter. First the current I_0 is switched to the 9-bit current weighting network. This current is switched to the output node I_{out} by the bit switches B_4 to B_{12}. At the moment the output current I_{out} must be larger than I_0, then the bit current I_0 is switched directly to the output terminal I_{out} by the segment current switches. At the same time the current I_1 is switched to the 9-bit current network to perform the next quantization steps until the output currents exceeds $I_1 + I_0$. At that moment the current I_1 is switched directly to the output terminal I_{out} too and the current I_2 is switched to the 9-bit current network. Using this procedure *monotonicity* of the converter can be guaranteed. Integral linearity basically is not guaranteed to be less than $\frac{1}{2}$ LSB.

6.5.2   Voltage division operation

An example of a 16-bit monotonic by design voltage division converter is shown in Figure 6.18 [22]. As can be seen from this figure, the system con-

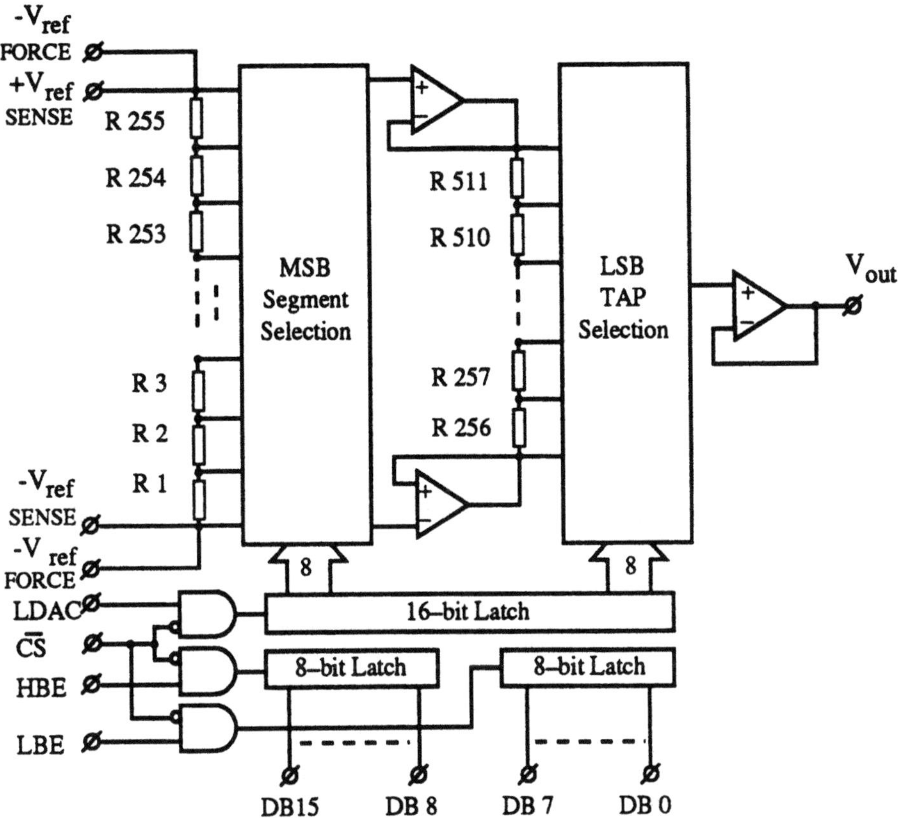

Figure 6.18 : Voltage division monotonic converter

sists of two resistor strings with 256 taps using resistors R_1 to R_{255} for the coarse divider part and R_{256} to R_{511} for the fine divider part. Two operational amplifiers connected in the follower mode apply the tap voltage of the coarse section across the total resistor string of the fine divider part. The MSB segment selection block consists of two sets of switches that are driven by the segment decoder. In the LSB tap selection block a set of switches is used to select a resistor tap and apply the output voltage to the output buffer. This output buffer consists of an operational amplifier in follower mode. A special operation of the switches in the MSB segment encoder is used to make the system independent of the offset voltage of the voltage followers. In Figure 6.19 a detailed circuit diagram of the circuit including the switch drivers is shown. From the figure it can be seen that one side of the even numbered coarse switches is connected to voltage follower A_2 while one side of the odd numbered switches is connected to voltage follower A_1. An

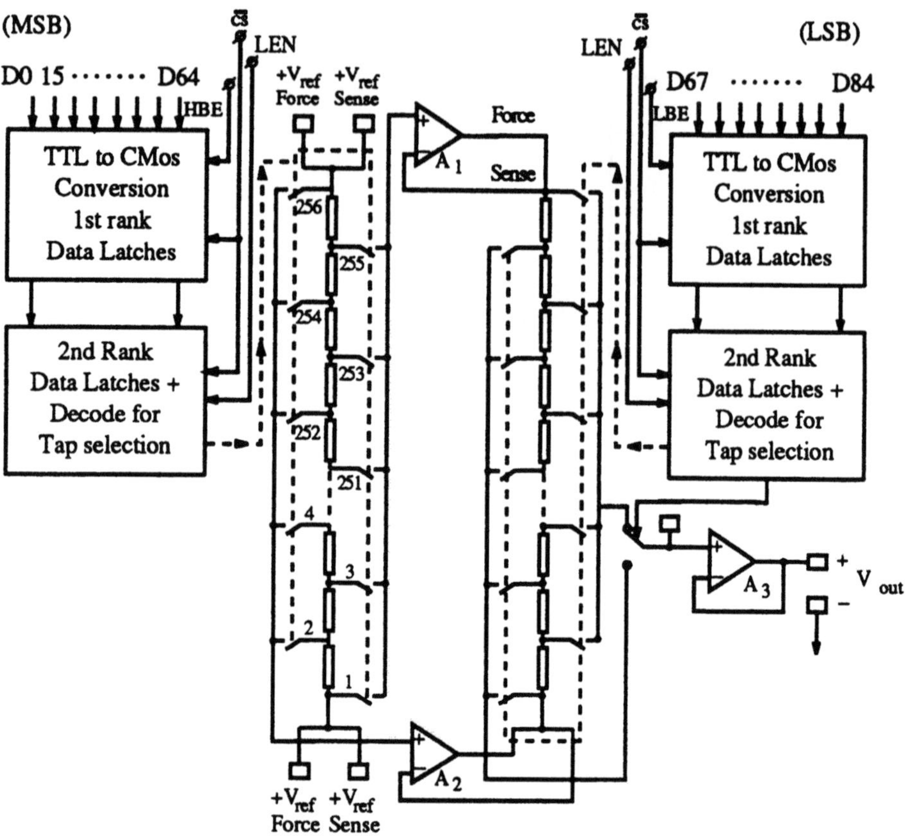

Figure 6.19 : Detailed circuit diagram of monotonic voltage converter

identical connection is obtained for the switches of the fine voltage divider. However, the output lines of the odd- and even-numbered fine switches are connected to an additional output switch. This decoding is used to simplify the logic circuit part.

The special connection used in the coarse voltage divider allows the analog signal to "roll-off" against the coarse divider without jumping from one position into another. In this way a continuous output voltage change can be guaranteed independent of the offset voltage of the voltage followers A_1 and A_2. By using this system, it is possible to construct a 16-bit guaranteed monotonic converter without needing the high division accuracy to obtain the integral nonlinearity of $\frac{1}{2}$ LSB.

6.5.3 Dual-ladder 10-bit D/A converter

A dual-ladder resistor string is used to implement a 10-bit high-speed linear
CMOS D/A converter [23]. In Figure 6.20 the basic dual-ladder implementation is shown. The coarse ladder consists of two rows, each built-up from

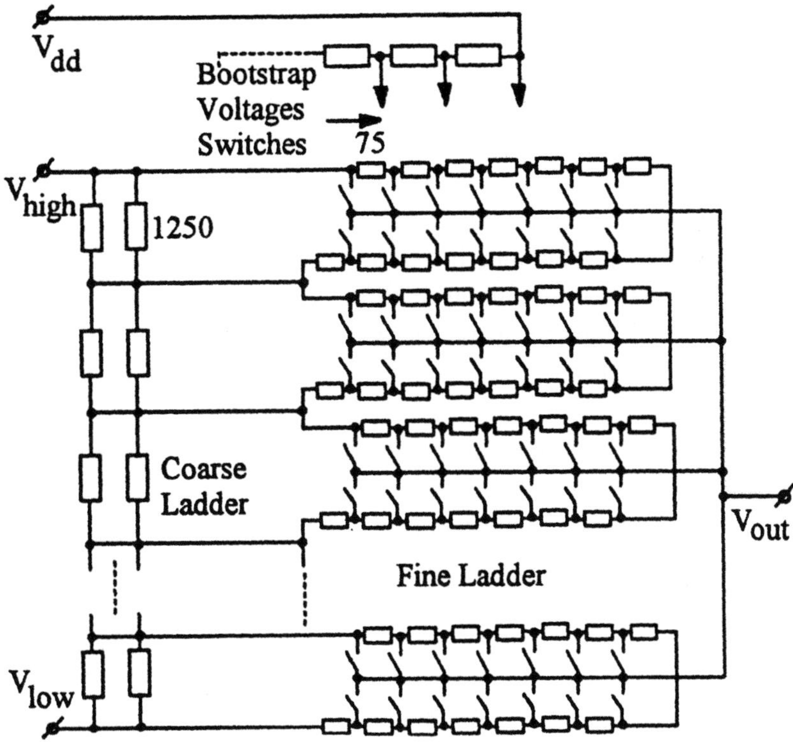

Figure 6.20 : Basic dual-ladder system

large size 250 Ω resistors which are connected anti-parallel to eliminate the
first-order gradient. The coarse ladder in this way determines 16 accurate
tap voltages and is responsible for the integral nonlinearity of the converter.
A 1024-resistor fine ladder is arranged in a 32-by-32 matrix. Every 64th
tap of this fine ladder is connected to the coarse ladder taps. As a result
of this configuration the resistivity of a fine ladder resistor can be chosen
at 75 Ω. By closing one of the 1024 switches an analog output voltage is
obtained at the output. From Figure 6.20 it is easily seen that the output
impedance varies as a function of the closed switch position in the network.
To compensate for a variable output of the system compensation resistors
are inserted. The insert of Figure 6.20 shows how these compensation re-

sistors are added, depending on the switch position. In the insert two fine rows are shown together with the output resistor row shown on the top. As a result of this operation, the output impedance at every position in the fine ladder network is made constant. No extra distortion or position dependent delay is found by this construction.

However, the ON resistance of MOS switches depends on the "drive" voltage that is applied to the device. As a result, the ON resistance of an MOS switch at the bottom part of the ladder is different from one at the top of the ladder. To overcome this problem an extra ladder is included in the system. This ladder voltage is used to bootstrap the drive voltage of the MOS switching devices. In this way equal drive voltages are applied to nearly all switching devices resulting in equal ON resistances. The coarse ladder can be designed with poly silicon or diffused resistor types, while in the fine ladder poly silicon resistors are preferred because of the low parasitic capacitance of these resistors.

In Figure 6.21 the block diagram of the 10-bit D/A converter is shown. The input digital data is split-up into two five bit parts which are applied to a 5-to-32 decoder. The 5-to-32 decoding is performed in two steps: a predecoder converts into 10 lines that control 32 three-input NOR gates of which one gate is activated. The two decoders are placed on two sides of the system. To minimize the glitches, the two sets of decoded signal lines are latched by the main clock before running horizontally and vertically over the matrix. In this matrix, the 1024 AND gates perform the final decoding and drive the addressed switch. This switch connects the ladder tap to the output line. In the output line a multiplexer is used, connecting only one output rail to the output terminal, while keeping the other rails at the corresponding middle tap voltage. This scheme reduces the load capacitance and minimizes the (dis)charging of the matrix output rails. At the output terminal an output buffer amplifier is connected to drive a 75 Ω output terminal. In Figure 6.22 a circuit diagram of this output buffer is shown. The amplifier consists of a folded cascode followed by a Miller compensated common source PMOS. Miller compensation is used to stabilize this unity gain buffer stage. An extra 75 Ω on chip resistor is used to prevent shorting of the output amplifier by the bond-pad capacitor at high frequencies. The 75 Ω resistor can be controlled within about ±10 % so the gain variation is for video applications within acceptable values. In Table 6.3 the D/A converter data are shown. The dual ladder system results in a high-performance D/A converter with an excellent

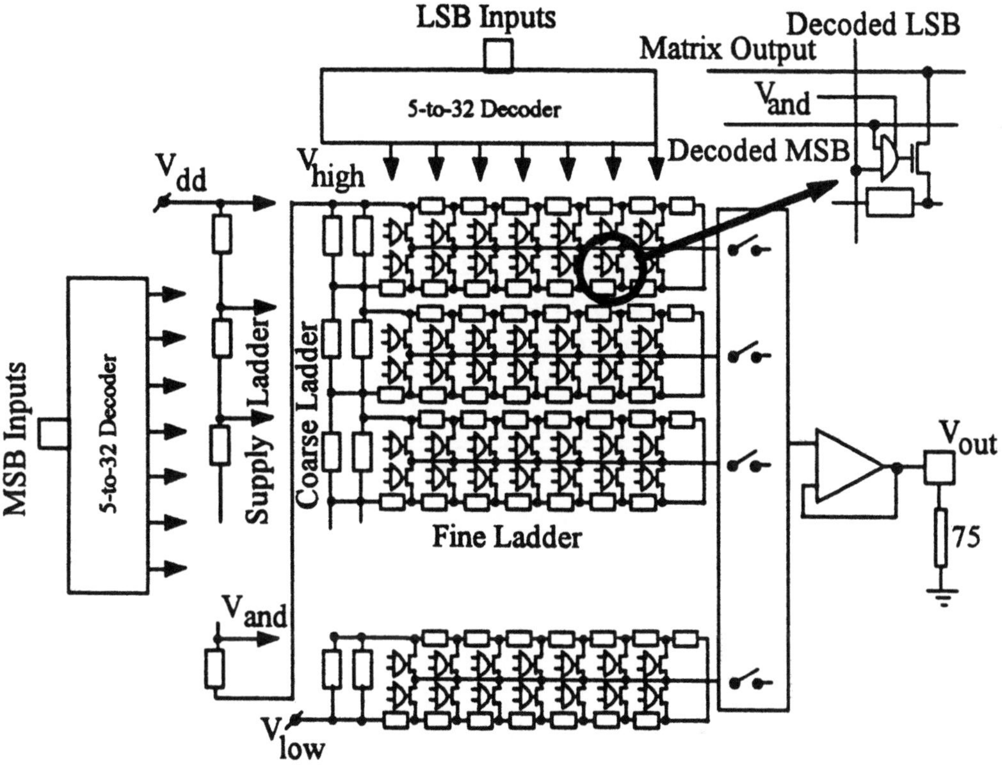

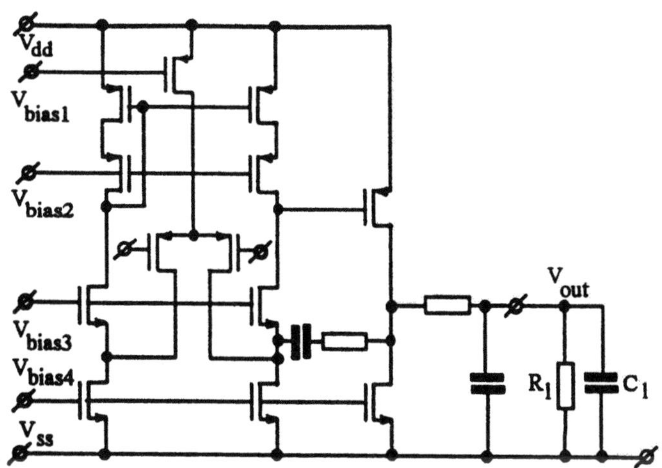

Figure 6.21 : Block diagram of 10-bit D/A converter

Figure 6.22 : Output buffer amplifier

Process	1.6 μm CMOS
DC resolution	10 bit
Differential non-linearity	< 0.1 LSB
Integral non-linearity	$\pm$ 0.35 LSB
Glitch energy	100 pVs
Rise/Fall time (10% $-$ 90%)	6 ns
Signal bandwidth (-1 dB)	19 Mhz
Nominal supply voltage	5 V
Output voltage over 75 Ω	1 V
Power (50 MHz, 75 Ω)	65 mW
Die size	2.5 mm^2

Table 6.3 : 10-bit D/A converter data

differential nonlinearity and a good glitch specification.

6.6   Self calibrating D/A converter system

If D/A converters are not continuously is use, a calibration procedure can be used to eliminate inaccuracies of elements [73]. In Figure 6.23 an example of a D/A converter system that uses a self-calibration cycle is shown. The system consists of a main D/A converter and a sub-D/A converter used to eliminate the errors in the main D/A converter. During the calibration cycle an analog ramp is generated using a ramp generator. The output of the main D/A converter is compared with the ramp signal by the comparator. The correction logic generates a digital correction signal that is stored in the RAM unit. The RAM data control the sub-D/A converter to correct the output signal of the total D/A converter function to obtain the full $\frac{1}{2}$ LSB integral nonlinearity. In Figure 6.24 the calibration operation is shown. At the moment the main D/A output signal does differ from the ideal D/A converter output, which is generated by the ramp signal source, then a counting operation starts that controls the output of the sub D/A converter to add a small correction value to the main D/A converter output to obtain the ideal output value. The correction value is stored for every weighting value of the main D/A converter. This means that in a 14-bit system 14 correction values of 6 to 8 bits must be stored to correct the main

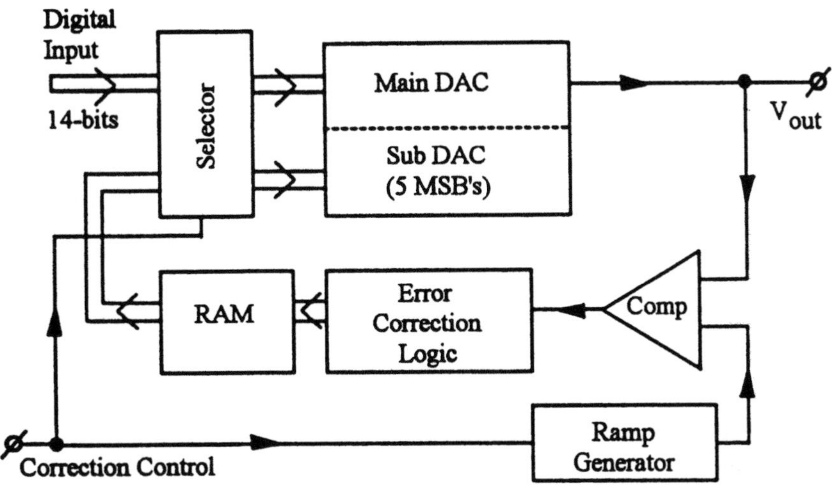

Figure 6.23 : Self-calibrating D/A converter system

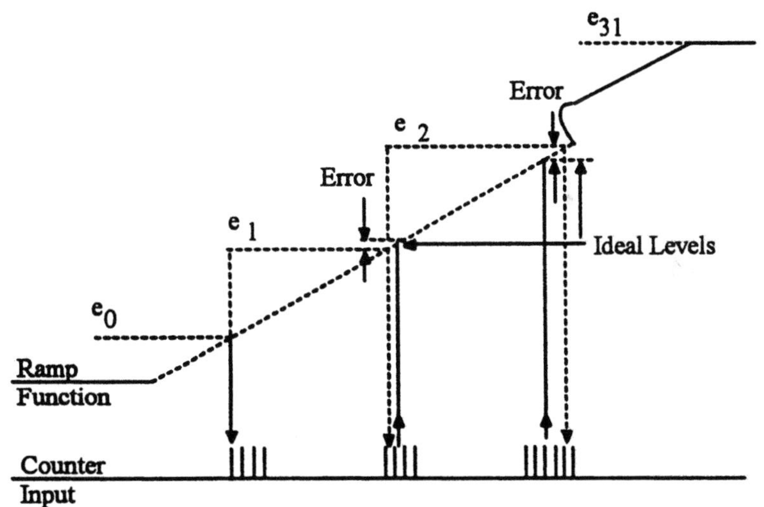

Figure 6.24 : D/A correction cycle

D/A converter output. When a code is applied to the D/A converter, then the correction value is called from the RAM and applied to the sub D/A to correct the output. Note that during calibration the D/A converter cannot be used for conversion. In some applications this can be a disadvantage. To overcome this problem a different method will be used. Self-calibration techniques of converters have become fairly popular during the last few years. In references [67] and [74] examples of these techniques are given, applied in a D/A and an A/D converter.

6.7   Dynamic Element Matching

In this section a system will be described that combines a good passive division with a dynamic time accurate interchanging operation to obtain an element independent high-accuracy current division [63]. Practical implementations of the system will be shown in a bipolar technology. However, the basic technique can be used in CMOS or GaAs technologies, too. An optimum performance is expected in a BiCMOS technology. The designer in this technology can optimize his/her design with respect to matching, noise, and minimum supply voltage.

6.7.1   Basic dynamic divider scheme

A simplified circuit diagram of the basic dynamic current divider is shown in Figure 6.25. The circuit consists of a passive current divider and a set

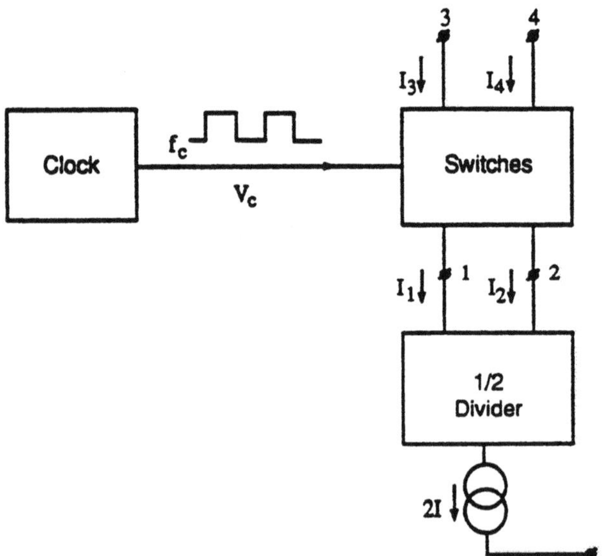

Figure 6.25 : Basic dynamic current divider

of switches driven by the clock generator f. The passive divider divides the total current $2I$ into two nearly equal parts: $I_1 = I + \Delta I$, $I_2 = I - \Delta I$. The currents I_1 and I_2 are now interchanged during equal time intervals with respect to the output terminals 3 and 4. At these terminals currents then flow whose average values are exactly equal and have a dc value I. In Figure 6.26 the currents as a function of time are shown. The figure shows that a small ripple current with a value of $2\Delta I$ and frequency f is also present at

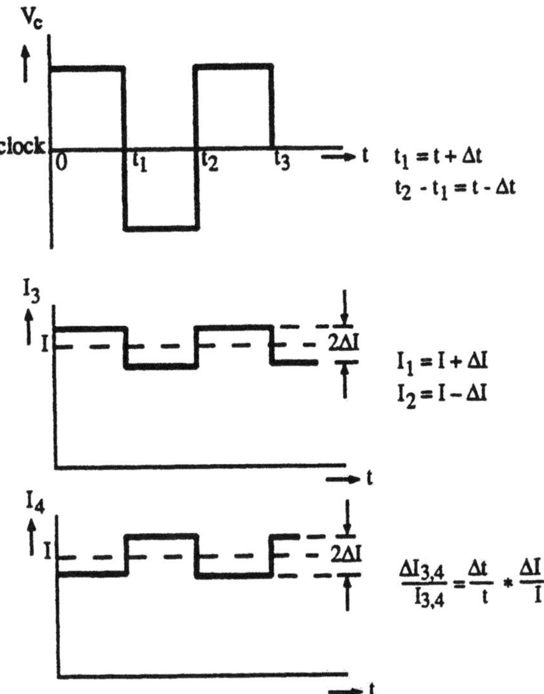

Figure 6.26 : Currents as a function of time in the dynamic current divider

the output terminals. This ripple is a measure of the matching accuracy of
the passive divider. With a simple low-pass filter this ripple can be removed
from the output current and, as a result, accurate output currents with a
value of I are obtained. In principle, a current divider with an exact division
ratio of one-to-two is obtained with this system. Moreover, if the matching
of the passive divider stage changed during operation, for example, because
of a slowly changing temperature, then the accuracy of the system would
not change. The only variation which occurs is the increase or decrease of
the ripple current value depending on the direction in which the matching
in the passive divider changes. In the previous analysis an accurate time
division by two is supposed. If there is a slight difference in the timing of
the two clock cycles, then a high-accuracy is still possible. Suppose that the
clock time periods differ by a value Δt, then we obtain:

$$t_1 = t + \Delta t, \tag{6.12}$$

and

$$t_2 = t - \Delta t. \tag{6.13}$$

Also, using the already defined values for I_1 and I_2

$$I_1 = I + \Delta I, \tag{6.14}$$

and

$$I_2 = I - \Delta I. \tag{6.15}$$

The current differences ΔI are equal because the sum of the two currents must be equal to $2I$, which is the starting current of the divider stage. During the first phase of the clock cycle the output currents become:

$$I_{out11} = (t + \Delta t) \times (I + \Delta I) \tag{6.16}$$

and

$$I_{out21} = (t + \Delta t) \times (I - \Delta I). \tag{6.17}$$

In the second phase of the clock cycle the output currents change into:

$$I_{out12} = (t - \Delta t) \times (I - \Delta I) \tag{6.18}$$

and

$$I_{out22} = (t - \Delta t) \times (I + \Delta I). \tag{6.19}$$

Now at every output terminal the currents must be added and averaged over the total clock time $2t$. We obtain:

$$\frac{I_{out11} + I_{out12}}{2t} = I(1 + \frac{\Delta t \times \Delta I}{t \times I}) \tag{6.20}$$

$$\frac{I_{out21} + I_{out22}}{2t} = I(1 - \frac{\Delta t \times \Delta I}{t \times I}). \tag{6.21}$$

As is seen from equations 6.20 and 6.21 the final accuracy in this system is determined by the product of two small errors. In practice it is not difficult to make $\frac{\Delta t}{t} \leq 0.1\%$ and the matching of the passive divider can easily be made smaller than 1% so $\frac{\Delta I}{I} \leq 1\%$. An overall accuracy better than 10^{-5} is obtained without using extremely difficult circuitry or elements which are difficult to match. The ripple which is present at the output of the divider scheme is a measure of the matching accuracy in the passive divider. This method which obtains a high accuracy without using highly matched and highly accurately trimmed elements is called *Dynamic Element Matching*. The value of the ripple can be reduced by optimizing the matching characteristics of the passive divide-by-two stage. A small ripple on the output values of the circuit allows the use of a simple low-pass filter network to reject this ripple below the quantization noise of the system. Usually such a network consists of a single *RC* stage.

Generally speaking, *Dynamic Element Matching* can be used to increase the accuracy of a divider scheme when the network can be split up into a number of nearly equal elements. The accuracy improvement is obtained by a continuously cyclic interchanging of the elements with respect to the respective output terminals. The output signals must be averaged over the total interchanging time period. At least an order of magnitude improvement in accuracy is obtained in this way. Furthermore, this system is independent of element aging and does not need highly accurate elements.

In reference [66] an example is shown that does not use binary weighting of current. Instead a decimal weighting is used to generate reference currents for resistor measurements.

6.7.2 Practical dynamic divider circuit

In Figure 6.27 a circuit diagram of a practical divider in a bipolar technology is shown. The passive divide-by-two current division is performed by transistors T_1, T_2. Two cross-coupled differential stages (T_3 to T_{10}) interchange the currents I_1 and I_2 for equal time intervals between the output terminals 3 and 4. The already discussed improvement in division accuracy with respect to the basic current divider T_1, T_2 is now obtained. The base currents of the Darlington switches limit the division accuracy. The *only* criterion determining this overall accuracy for the *whole* circuit is a high current gain of the Darlington circuits (e.g., $\beta^2 > 10^4$ and with $\frac{\Delta\beta}{\beta} \leq 0.1$ a practical limit better than 10^{-5} is obtained). If the current gain of a standard process is not high enough, then an extra processing step to obtain high-gain, low breakdown transistors can be used. The value of the output ripple current depends on the matching of the current mirror T_1, T_2. A problem in this current mirror is the base current of transistors T_1 and T_2, which is added to the current flowing through transistor T_2. This reduces the matching accuracy in the system to about $\frac{2}{\beta}$. An improvement in matching accuracy is obtained by inserting emitter degeneration resistors R_{E1} and R_{E2} as shown in Figure 6.27. An exact compensation of the base currents is obtained by inserting a resistor with a value of $2 \times R_{E2}$ in series with the base of transistor T_2. This results in the same effect as increasing the value of the emitter resistor R_{E2} by a value $2\frac{R_{E2}}{\beta}$. The voltage drop across the resistors is about

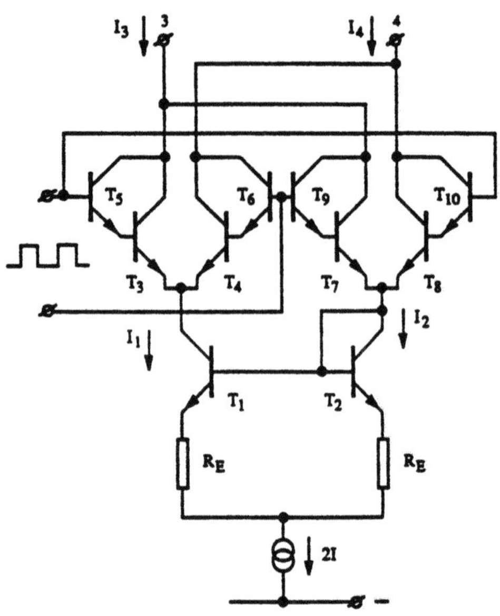

Figure 6.27 : Practical dynamic divider scheme

5 to 10 times the value of $\frac{kT}{q}$, giving a value between 125 mV and 250 mV. The minimum supply voltage drop across the circuit to operate under linear condition is about $3V_{be}$ or between 1.8 and 2.5 V.

In a CMOS implementation of the practical divider circuit from Figure 6.27 the Darlington switches are replaced by single MOS devices. Depending on the output impedance of the passive current mirror divider, these switches can be operated in the triode region. The voltage drop across these switches can be made very small. As a result a total divider needs between one and two MOS threshold voltages. Furthermore, PMOS switched current mirrors can be used to minimize the total supply voltage of a complete current network.

6.7.3   Two-bit dynamic current divider scheme

In the circuit of Figure 6.27 per dynamic divider stage only one accurately matched bit current is generated. When such stages are cascaded, the supply voltages will increase to an impractically high voltage level. To overcome this problem a 2-bit system will be used. The basic system diagram is shown in Figure 6.28. In this system the passive divider is extended to divide a

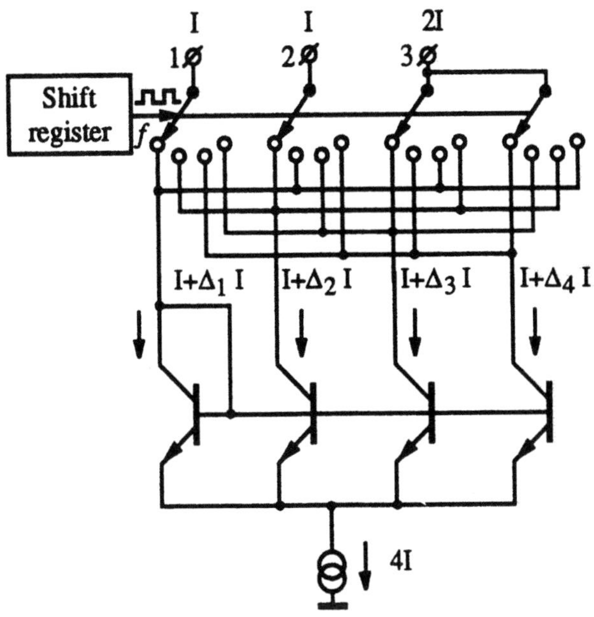

Figure 6.28 : 2-bit current divider scheme

current $4I$ into four nearly equal currents: $I_1 = I + \Delta_1 I$, $I_2 = I + \Delta_2 I$, $I_3 = I + \Delta_3 I$, and $I_4 = I + \Delta_4 I$. Note that $\Delta_1 + \Delta_2 + \Delta_3 + \Delta_4 = 0$, because the total current which is divided into four nearly equal parts is $4I$. These currents are now fed into a switching network that interchanges all currents during equal time intervals with respect to the output terminals 1, 2, and the combined terminal 3. At these outputs averaged currents with values I, I, and $2I$ are obtained. In Figure 6.29 the output currents as a function of time are shown. From Figure 6.29 it can be seen that the currents with a value I have the same frequency as the clock f, while the current with a value $2I$ has a ripple with a frequency $\frac{f}{2}$. Timing errors have the same influence as in the one-bit divider scheme. In Figure 6.30 a circuit diagram of a practical divider circuit is shown. It consists again of a passive divide by four stage using transistors T_1, T_2, T_3, and T_4, with the emitter degeneration resistors R. This stage divides the current $4I$ into four nearly equal values of I. These currents are applied to the interchanging system consisting of Darlington switches to minimize base current losses that might reduce the overall accuracy of the system. In Figure 6.30 the bases of the passive divider are connected to a reference voltage V_{ref}. Usually this reference voltage is the collector terminal of, for example, transistor T_1. In this way a current

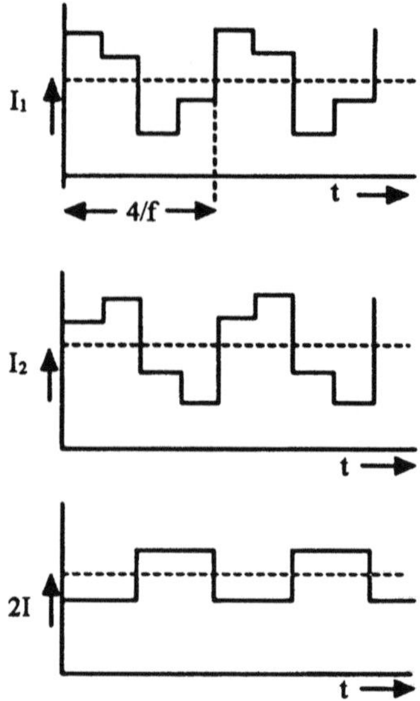

Figure 6.29 : Output currents as a function of time for the 2-bit divider system

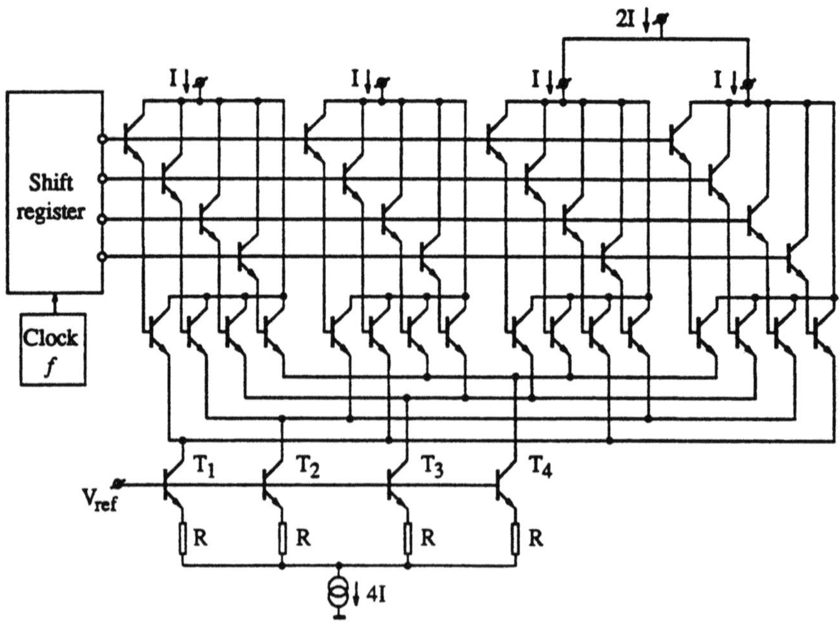

Figure 6.30 : Practical 2-bit dynamic current divider

mirror circuit with four output signals is obtained. Furthermore, such a system does not lose base currents due to the final current gain of the bipolar transistors used in this system. By tuning one of the emitter resistors a more accurate division in this divide-by-four stage is obtained. The same method as used in Figure 6.28 can be used again for transistor T_1.

A four-stage shift register provides the time-accurate signals for the interchanging of the currents. In this circuit a high time division accuracy is easily obtained. The only additional criterion to improve the accuracy of the system is a high current gain of the bipolar transistors in the Darlington switches.

6.7.4 High-speed Darlington switching stages

It is known in general that Darlington stages do not switch accurately and rapidly. In Figure 6.31 a single Darlington stage consisting of Transistor T_1 and Transistor T_2 is shown. The parameters which determine the switch-

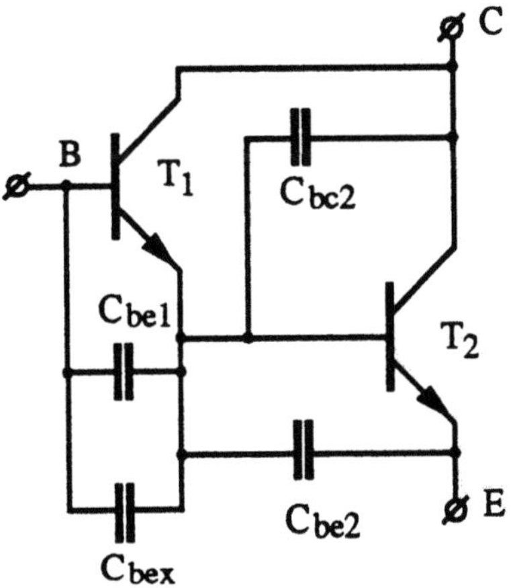

Figure 6.31 : Basic Darlington switch configuration

ing performance at high frequencies are shown as capacitors C_{be1}, C_{be2}, and C_{bc2}. Capacitor C_{bex} is an extra capacitor. The function of this capacitor will be explained later. Assume at first that the value of $C_{bex} = 0$. When the switch is conducting and is intended to be switched off, then due to the loading of transistor T_1 with a capacitance of $C_{bc2} + C_{be2} > C_{be1}$ transistor T_1 will be switched off much faster and with a larger voltage variation than the voltage swing across T_2. The capacitive division of the switching voltage is not balanced equally across the transistors $T1$ and T_2. A much smaller variation of the switching voltage across T_2 is obtained in comparison with T_1. Furthermore, due to the forward biasing of transistor T_2 at a much higher level than T_1, the base-emitter capacitance C_{be2} of T_2 is much larger than C_{be1} of T_1. As a result, the capacitor unbalance is increased for high-speed switching signals. To overcome this problem, an extra capacitor C_{bex} is added to the base-emitter capacitance of transistor T_1. The value of this capacitor must be chosen in such a way that an equal switching voltage variation across transistor T_1 and transistor T_2 is obtained. By using an accurate transistor model, a simulation can show the exact value of C_{bex}. In a practical application, however, this capacitor is rather difficult to construct and to apply in an integrated circuit layout. Therefore, a device size-dependent model implementation is used in which the size of transistor T_1 with respect to transistor T_2 can be increased. In this way the capacitor matching and implementation are done by properly adjusting the sizes of transistors T_1 with respect to T_2. In Figure 6.32 an example of a Darlington switch layout is shown.

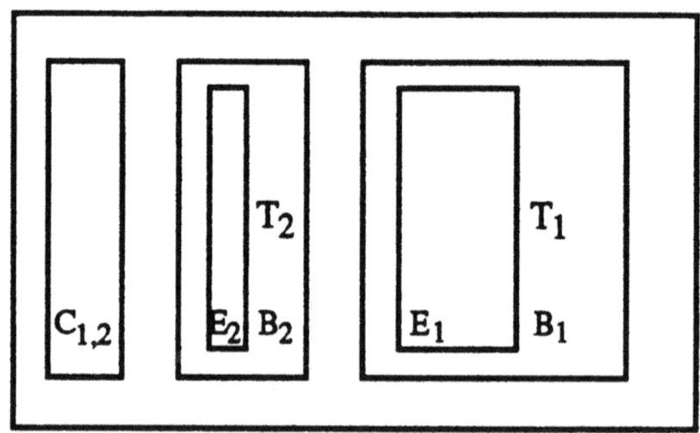

Figure 6.32 : Compensated Darlington switch layout

6.7.5   Dynamic current mirror circuit

In some applications it is useful to obtain an accurately matched current mirror. A circuit diagram which uses *Dynamic Element Matching* to obtain a device matching independent accurate current mirror is shown in Figure 6.33 The basic current mirror consists of transistors T_1 and T_2 with emitter degeneration resistors R_E to increase the matching accuracy. Darlington stages T_3 to T_{10} interchange the currents of T_1 and T_2 with respect to the sum and output terminal of the mirror. Darlington stages T_{12} and T_{13} with the level-shifting diode T_{11} form the current-source feedback loop. In the previous circuit diagrams a connection between the collector and the base of a transistor was used to construct a diode which, in cooperation with a parallel connected transistor, forms a current mirror. Two cases of operation

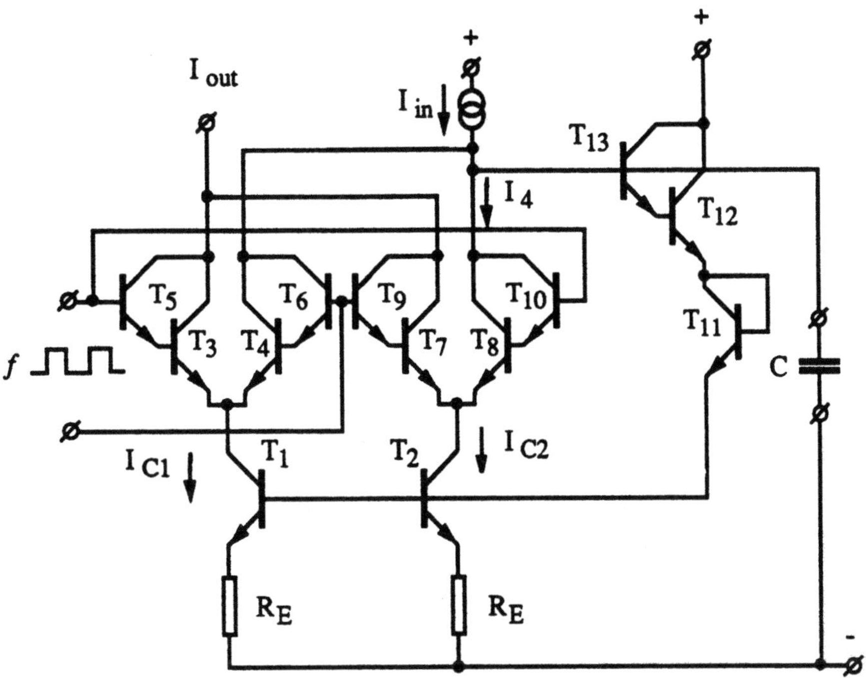

Figure 6.33 : Dynamic current mirror circuit diagram

can now be distinguished.

1. An averaging capacitor is connected across the diode connection of the
   current mirror. This means that the base of transistor T_{13} is decoupled
   to ground with the capacitor shown as C. The average value of the
   collector currents I_{c1} and I_{c2} shown as I_4 is made equal to I_{in}. Then
   the average output current I_{out} is the average value of I_4. As a result,
   an accurate current mirror $I_{out} = I_{in}$ is obtained.

2. The capacitor C is omitted. During the first half period of the clock
   phase the current I_{c2} is controlled in such a way that $I_{c2} = I_{in}$. By
   supposing an error Δ between the transistors T_1 and T_2, an output
   current equal to $I_{out} = I_{in} \times (1 + \Delta)$ is obtained. During the second
   phase of the clock cycle, I_{c1} and I_{c2} are interchanged, resulting in an
   output current $I_{out} = \frac{I_{in}}{1+\Delta}$. When Δ is made small, by obtaining a
   good matching between the transistors T_1 and T_2 then the division by
   $(1 + \Delta)$ can be approximated by a finite series expansion. As a result

we obtain:

$$I_{out} \cong I_{in}(1 - \Delta + \Delta^2). \tag{6.22}$$

After averaging the results over the total clock period we obtain:

$$I_{out} \cong I_{in}(1 + \frac{1}{2}\Delta^2). \tag{6.23}$$

Inserting a value for $\Delta \leq 0.5\%$ the mirroring error term is below 10^{-5}. The current mirror without the capacitor C shows a less accurate performance than the one with a large capacitor C.

6.7.6   Binary-weighted accurate current network

A binary-weighted current network is formed by cascading current division elements. In Figure 6.34 a simplified block diagram using single current divider stages is shown. In the first divider stage a combination with a ref-

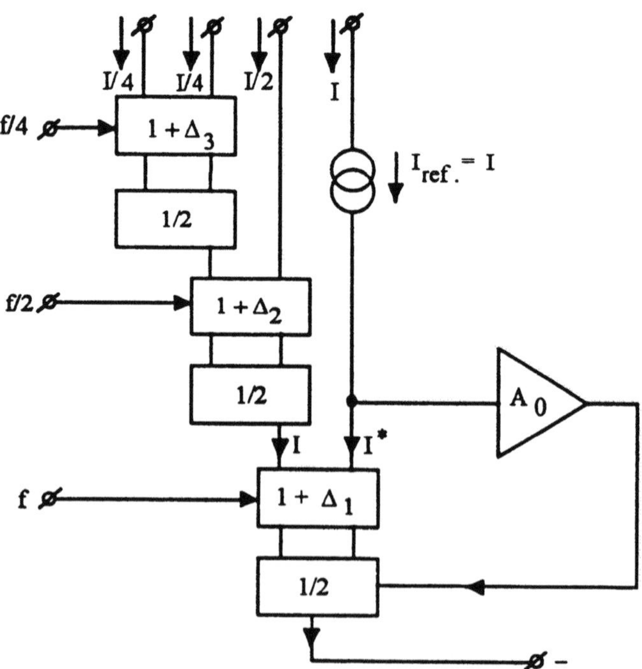

Figure 6.34 : Binary-weighted current network

erence current source I_{ref} and a current amplifier A_0 is made to obtain an accurate current mirror.

Now two possibilities exist with respect to the interchanging frequency of the cascaded current divider stages.

1. Every following divider stage is operated at half or double the switching frequency of the first divider stage.

2. Every divider stage obtains the same interchanging frequency.

6.7.7 Binary-weighted current network with divided interchanging clock

In this case digital clock circuitry that accurately divides the interchanging clock frequency by two is needed. Supposing that we lower the interchanging frequency every time a current is divided by two, then we will not have any interaction between the individual current dividers concerning the finite matching accuracy of every divider. In Figure 6.35 the currents in the first and second divider stage as a function of time are shown. Suppose that the

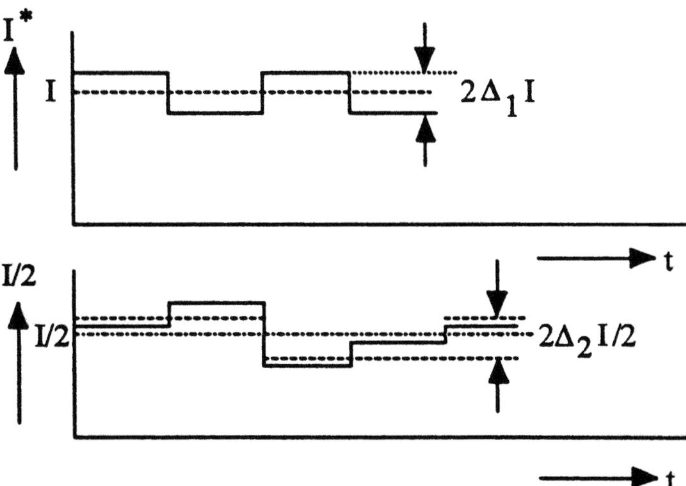

Figure 6.35 : Output currents of first two current dividers as a function of time

output current of the first divider I^* shows the inaccuracy Δ_1 as shown in the top part of Figure 6.35. In the output current $\frac{I}{2}$ of the second divider stage we can recognize the error Δ_2 with a frequency $\frac{f}{2}$ and the error Δ_1 of the first stage with a frequency f in the lower part of Figure 6.35. During a half period of the clock $\frac{t}{2}$ the average value of the current $\frac{I}{2}$ does not contain an error term originating from the first divider stage of frequency f. The same is valid for the second half clock period of the frequency $\frac{f}{2}$. As a result of this operation no interaction of the division accuracy of the first stage on the second stage is found. If the clock frequency division is continued, the same arguments can be used for the next stage, and so on. An independent operation of the individual stages in the total divider chain is obtained. A disadvantage of the division of the interchanging clock frequency by factors of two is the large increase in digital circuitry that is needed to accurately drive the individual divider stages. Moreover, the ripple frequency decreases with decreasing current value. This means that the filtering capacitors that are used in the passive ripple filter increase and might need impractically large values.

6.7.8 Binary-weighted current network using equal interchanging clock frequencies

When all divider stages are operated with the same interchanging clock, it is expected that in the error analysis interactions between the individual divider stages will occur. Timing errors in this case are equal for all the divider stages. These timing errors can therefore be separated from the division errors in the individual stages.

Suppose the error in the first stage is denoted as Δ_1, then the total error of the first stage including the timing error can be written as:

$$I^* = I_{ref}(1 + \Delta_1 \times \frac{\Delta t}{t}). \tag{6.24}$$

In the expression given in equation 6.24 the filtering capacitor as described with regard to the accurate current mirror circuit is ignored. The average value of the output current of the second stage can be calculated:

$$\frac{I}{2} = I_{ref}\frac{(1 + \Delta_1)(1 + \Delta_2)(t + \Delta t) + (1 - \Delta_1)(1 - \Delta_2)(t - \Delta t)}{4t}. \tag{6.25}$$

This long formula can be simplified into:

$$\frac{I}{2} = \frac{I_{ref}}{2}(1 + \Delta_1\Delta_2 + (\Delta_1 + \Delta_2)\frac{\Delta t}{t}). \tag{6.26}$$

The average value of the output current of the third stage is again determined. The simplified result is shown in the following equation:

$$\frac{I}{4} = \frac{I_{ref}}{4}(1 - \Delta_1\Delta_2 + \Delta_1\Delta_3 - \Delta_2\Delta_3 + (\Delta_1 - \Delta_2 + \Delta_3)\frac{\Delta t}{t}). \qquad (6.27)$$

If the error terms of the individual stages are made small ($\Delta_{1 \to n} \leq 0.5\%$), then the influence of the interactions between the individual divider stages on the overall accuracy of the binary-weighted current network can be kept very small. To increase the resolution of the active divider stages, a 2-bit-per-switching-level configuration can be used advantageously.

6.7.9 14- and 16-bit binary network examples

An example of a 14-bit binary-weighted current network using 2-bit-per-level divider stages is shown in Figure 6.36. In this system 5 active divider stages are cascaded followed by a 4-bit passive divider using emitter scaling of transistors. In the 14-bit system a choice between 5 active and a 4-bit passive

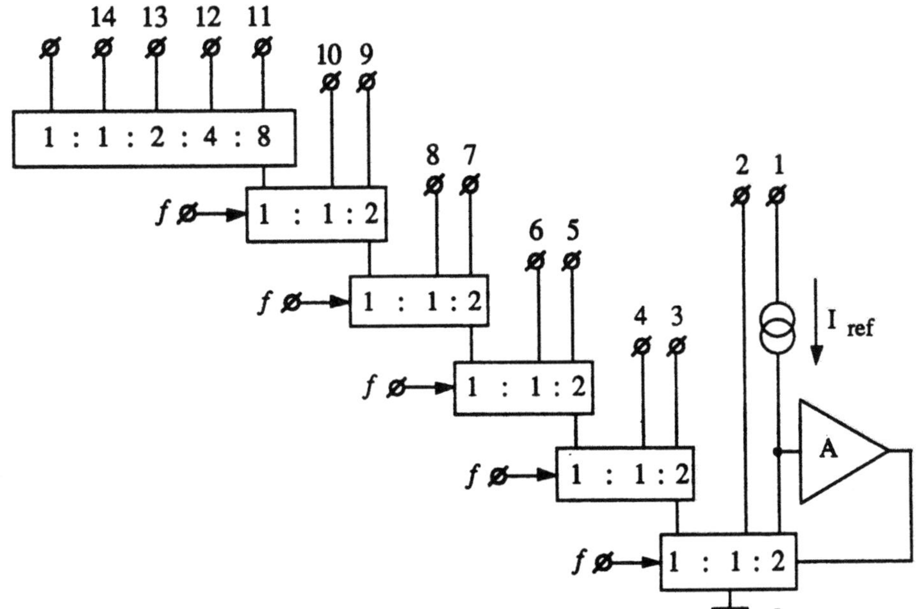

Figure 6.36 : 14-bit binary-weighted current network

divider has been made to obtain a high circuit yield. The disadvantage of the large number of cascaded active stages is the large supply voltage needed ($V_{supply} > 15$ V). To overcome this drawback the number of active divider stages can be reduced. In Figure 6.37 a 16-bit binary-weighted network is shown. Here 3 active divider stages are cascaded, followed by a 10-bit passive

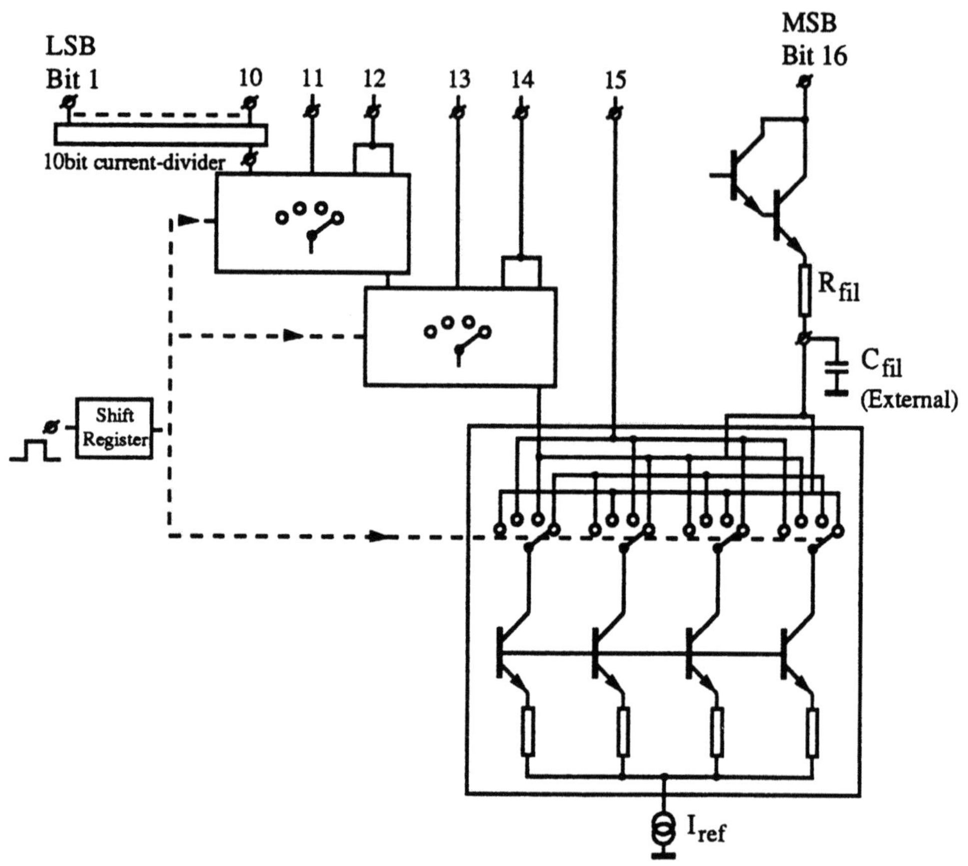

Figure 6.37 : 16-bit binary-weighted current network

divider using emitter scaling. The number of external filtering capacitors is reduced to 7 in this architecture. To obtain a high matching accuracy in the passive divider the 1024 transistors used are randomly interconnected. In this way the accuracy increases with the square root of the number of transistors connected in parallel. This means for the MSB that the practical matching between transistors of about 1% is increased by $\sqrt{512} \approx 23$. This results in a matching accuracy of 0.05 %. Tests on practical dividers built-up from 1024 transistors to obtain a 10-bit divider showed that these circuits can be designed with a high circuit yield. The supply voltage of the total circuit can be reduced to 12 V. In Figure 6.38 a 10-bit binary-weighted current divider using emitter scaling and base-current compensation is shown.

Note that all the bases of the transistors are connected to obtain information about the total base current which would have been lost when no additional circuitry is added. To obtain an accurate base-current compensation a second scaling of the base currents is used with a second passive divider stage, again using emitter scaling to improve the division accuracy. In the case of base-current compensation the averaging effect improves the base current compensation, again resulting in a high-accuracy base-current-independent passive division network.

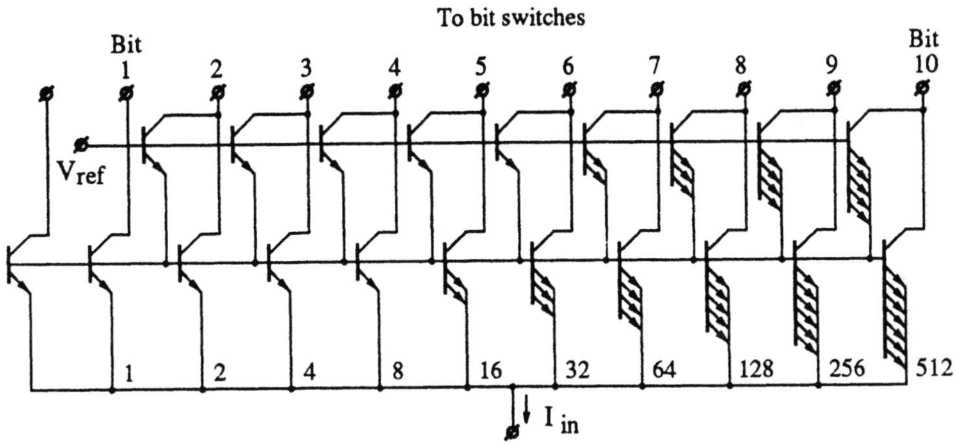

Figure 6.38 : 10-bit binary-weighted current divider using base current compensation

6.7.10 Filtering and switching

In Figure 6.39 the filtering and switching operation of the bit currents of the active divider stages is shown. A first-order filtering operation using the elements R_1C_1, R_2C_2 is performed to remove the ripple current from the final accurate bit current. The external capacitors C_1 and C_2 are added to the chip for this purpose. Additional Darlington cascode stages (T_3, T_4 and T_5, T_6) isolate the filter operation from the switching of the binary-weighted bit currents. Furthermore, the individual filtering of the bit currents minimizes the noise on the bit output currents, resulting in a high signal-to-noise ratio of the total converter. Bit switching is performed with a diode transistor configuration (T_1, D_1, and T_2, D_2). This switching configuration obtains a very high switching accuracy in steady state because no currents are lost by the final current gain of a bipolar transistor. A diode conducts all the applied

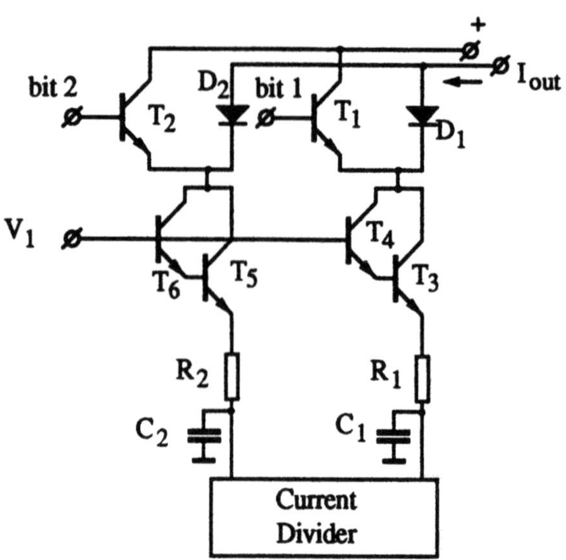

Figure 6.39 : Filtering and switching of bit currents

input current to the output terminal. One problem, however, is found with this diode-transistor switch. The switching speed at low bit current levels is reduced because of the large capacitive loading in the collector leads of the Darlington stages (e.g., T_3, T_4). This capacitance must be discharged before the bit current is conducted to the output through the diode D_1. Because the ON switching of the bit current needs a discharging of the collector capacitance of T_3, T_4 this situation shows the worst-case switching condition. OFF switching of the bit current is performed at a much higher speed because transistor T_1 can be switched on very fast and the collector capacitance is charged very quickly.

6.7.11 Compensated bit switch

To overcome the speed problem at low current values, an additional switching stage is added to the circuit diagram of Figure 6.39. The objective of this additional switching stage is to maintain the same voltage levels in the switching branch of the circuit. In this way parasitic capacitances do not need to be charged or discharged. In Figure 6.40 the compensated switch is shown. To obtain the compensation effect an extra differential pair T_7, T_8 with the resistor R are added to the circuit from Figure 6.39. The differential amplifier T_7, T_8 has a tail current with value I_{comp}. The value of this

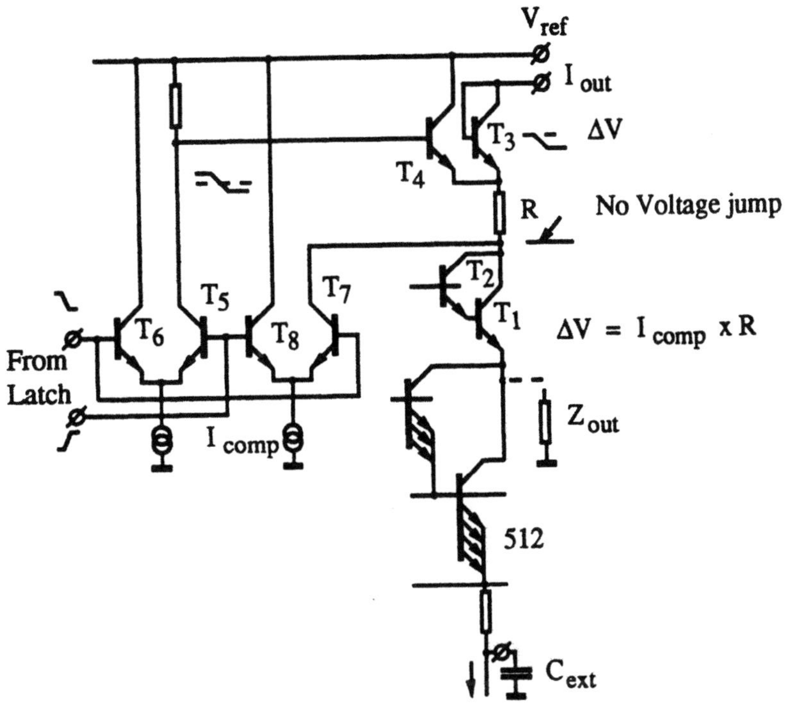

Figure 6.40 : Compensated diode-transistor bit-switch

current is chosen in such a way that:

$$I_{comp} \times R = \Delta V. \tag{6.28}$$

In this equation the value of ΔV is equal to the voltage swing which is applied to the diode-transistor switch to get a high switching accuracy. In this case this means that the on-off ratio of the switch must be better than 1 to 10^{-6} over temperature and supply voltage ranges for which the 16-bit D/A converter is designed. The operation of the compensation loop is as follows: Suppose first that the bit current is switched to the output terminal of the converter. This means that the diode T_3 is conducting and the compensation current is flowing through transistor T_8 to the supply line. The voltage level across the diode T_3 and the compensation resistor R is preset. At the moment the bit current must be switched off, the diode T_3 is switched off, and transistor T_4 starts conducting because the base voltage of this transistor now equals the supply voltage V_{ref}. At the same time transistor T_7 starts conduction and a current will flow through the resistor R and transistor T_4. The value of the resistor R with the compensation current I_{comp} is chosen in such a way that the voltage jump in the collectors of T_1 and T_2 is made zero. The voltage swing at the collector of T_3 is kept at a constant level by an operational amplifier acting as a current-to-voltage converter. As a result of this action, the parasitic capacitor in the collectors of T_1 and T_2 does not change in voltage level. No charging current is subtracted from the bit current which is switched on or off. A faster operation of the switch is obtained. This faster switching also reduces the possibility of glitches occurring at the output of the converter.

6.7.12   Output current-to-voltage converter

The diode-transistor switch shows a big advantage in terms of accuracy of switching. One drawback, however, is the coupling of the output voltage level with the switching voltage level applied to the transistor part of the switch. Due to the diode coupling there is a direct connection between the output voltage and the voltage at the emitter of the diode. Voltage variations at this point reduce the effective switching voltage, which might result in a loss in switching accuracy. To overcome this problem an inverting operational amplifier must be connected at the output of the switch. The voltage level around which the switching occurs can now be accurately determined. As a result, the voltage swing which is applied to the switch can be optimized with respect to accuracy, switching speed and, minimum glitch output.

6.7.13   14-bit D/A parallel converter

A complete circuit diagram of a 14-bit D/A converter with parallel data inputs is shown in Figure 6.41. This system forms the core of the commercial available audio converter TDA 1540. Note that in the reference input circuit a temperature-compensated current reference source is used in a current mirror circuit to apply this current to the binary current divider network and that the other terminal of the current reference is used as the MSB current of the converter. This current is switched directly by the MSB current switch. A master-slave flip-flop is used to accurately generate time moments for the dynamic matching operation in the circuit. This flip-flop is driven by an emitter-coupled multivibrator to generate clock pulses for the time-dividing flip-flop. To avoid feed-through of the digital data information to the analog output terminal of the converter, a serial input data system is needed.

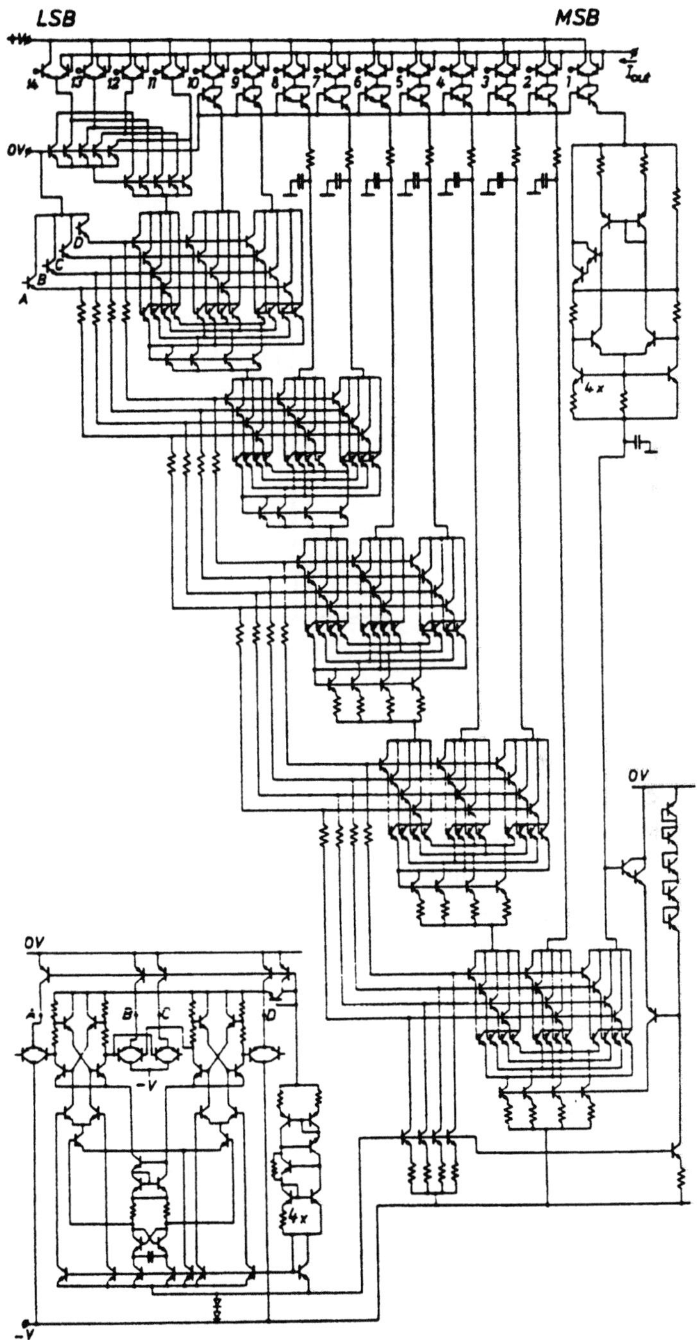

Figure 6.41 : Complete 14-bit D/A converter circuit diagram with parallel data input

6.7.14   16-bit dual D/A converter system

In Figure 6.42 the system diagram of the dual 16-bit D/A converter TDA
1541 is shown. The digital TTL-compatible input data are applied to a shift

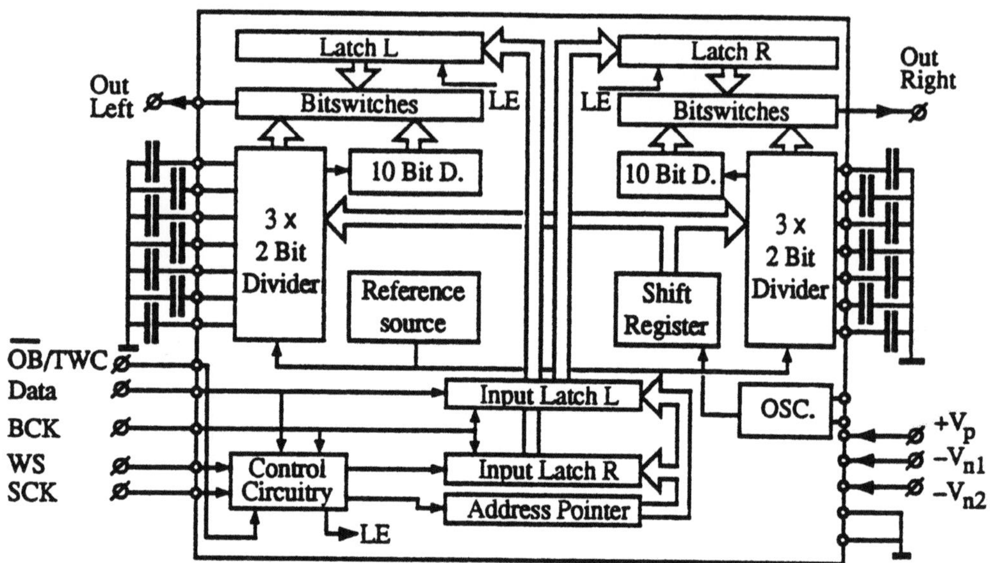

Figure 6.42 : Dual 16-bit D/A converter system diagram

register which transforms the serial input data into parallel format, which
is latched and applied to the D/A converter switches. The data transfer
between the integrated circuits is according to the Inter IC Signal standard
(I^2S). In this system a 6-bit active division is combined with a 10-bit passive division to obtain a 16-bit accurate current network per D/A converter
section. The number of external components can be minimized in this way.

6.7.15   16-bit converter data

In Table 6.4 the performance data of the dual 16-bit D/A converter TDA
1541 are given. Additional measured data are shown in Figure 6.43 and
Figure 6.44. These figures show the signal-to-noise plus distortion as a function of amplitude and of frequency. As a result, it can be said that using
Dynamic Element Matching very high performance D/A converters can
be built with a dynamic performance close to the theoretically attainable
maximum. In Figure 6.45 a die photograph of a 16-bit dual D/A converter
is shown. Both converters are made using a bipolar process with double
layer metal.

Resolution	16 bits
F.S output current	4.0 mA
Temperature coefficient	200 ppm/degree C
Integral nonlinearity (INL)	1.0 LSB maximum
Differential nonlinearity (DNL)	1.0 LSB maximum
Settling time to 1 LSB	1.0 μsec maximum
Channel separation	90 dB minimal
Harmonic distortion	-96 dB typical
Temperature range	-20 to +70 degree C
Power dissipation	800 mW typical
Supply voltages	+5, -5, -15 V
Package	28 pin plastic DIL
Chip dimensions	$3.8 \times 5.43\ mm^2$

Table 6.4 : 16-bit D/A converter data

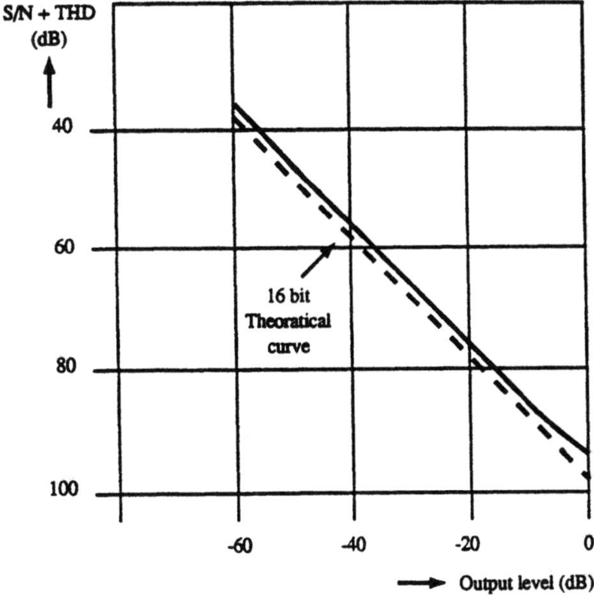

Figure 6.43 : Signal-to-noise plus distortion as a function of amplitude

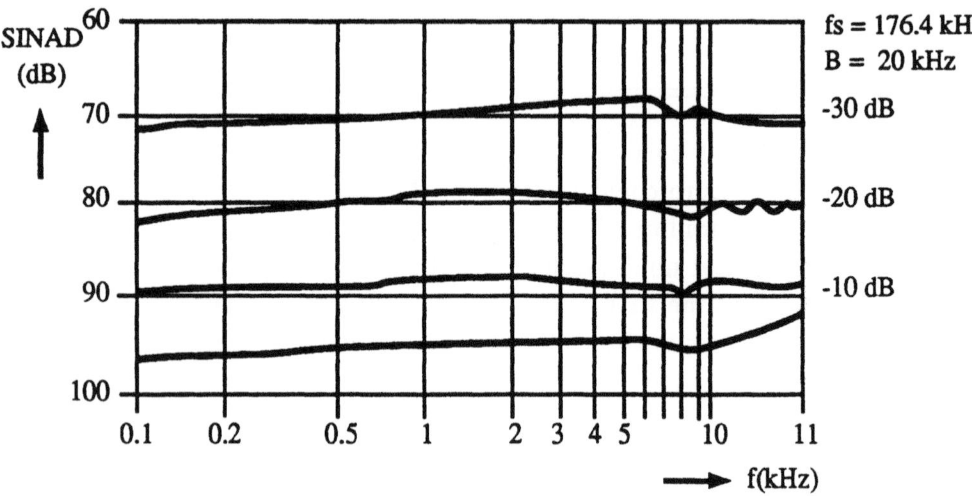

Figure 6.44 : Signal-to-noise plus distortion as a function of frequency

Figure 6.45 : Die photograph of a dual 16-bit D/A converter

6.8   Current calibration principle

In an MOS system it is possible to use a charge storage principle in an accurate current calibration system [67,68,69,70,71]. In Figure 6.46 the basic operation of the current calibration system is shown. The figure shows the

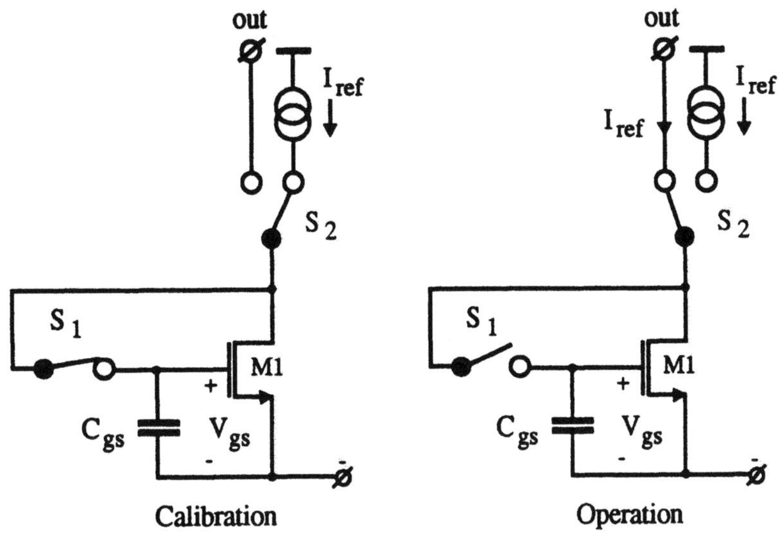

Figure 6.46 : Current calibration principle

calibration and the operational cycle. During calibration of the MOS current source, the MOS device M_1 is connected as a diode by closing switch S_1. The current I_{ref} is applied to the system and because of the diode connection of M_1, the gate-source voltage V_{gs} is adjusted in such a way that the drain current is made equal to I_{ref}. After the current has been calibrated to the reference value I_{ref}, the switch S_1 is opened and the gate-source voltage of the transistor M_1 remains at the calibration value. The output switch S_2 is switched to the output terminal. At that moment a current I_{ref} will start to flow through the output terminal. As long as the capacitor C_{gs} is not discharged, the drain current remains at the value I_{ref}. In a practical configuration, however, the capacitor is discharged because of the drain-substrate leakage current of the switch S_1. Moreover, the charge feed-through of the switch S_1, in case this switch is switched off, is added to the charge in C_{gs}. This means that the output current is not exactly equal to the calibration current value I_{ref}. In Figure 6.47 the two dominant error sources are shown. The leakage current of the source-to-substrate diode of transistor M_2 discharges the capacitor C_{gs}, while the charge feed-through of this switch is added or subtracted from the charge on the capacitor C_{gs}. These error sources result in variations of the gate-source voltage of M_1 which results in output current variations between two repeating calibration cycles. In Figure 6.48 the drain-source current of M_1 as a function of time is shown.

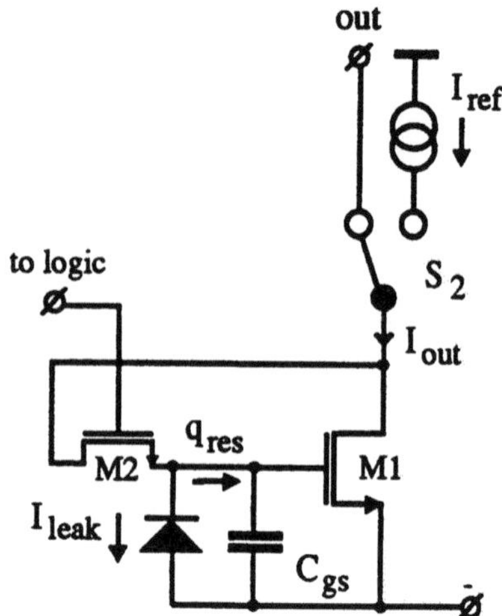

Figure 6.47 : Two dominant error sources

During the time the current is compared with the reference current I_{ref} the

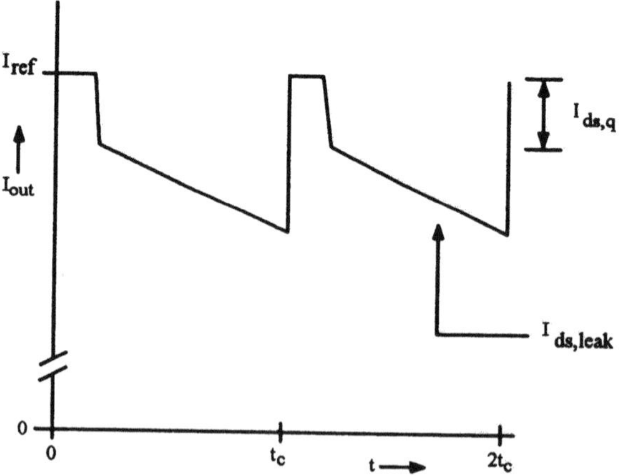

Figure 6.48 : Drain current of calibrated device as a function of time

same drain current is found. At the moment the switch S_1 is opened, the feed-through charge is subtracted from the charge on the capacitor C_{gs}. As a result, a decrease in output current is obtained. This is shown with the steep decrease in current just after switching of S_1. Then the leakage current discharges the capacitor resulting in a roughly linear decrease in output current. If at time t_c the calibration cycle starts again, the current is adjusted to I_{ref} and the cycle repeats.

6.8.1 Improved current calibration principle

To overcome some of the problems encountered with the system shown in Figure 6.47, the calibration is applied to the error current value only. The improved system is shown in Figure 6.49. The basic system is equal to

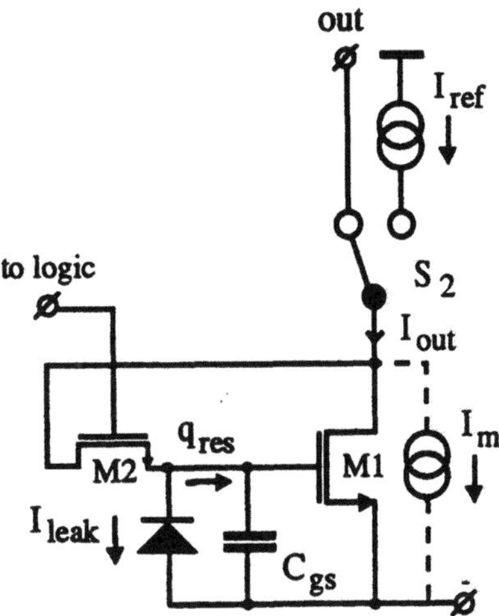

Figure 6.49 : Improved current calibration principle

the circuit shown in Figure 6.47. However, a current source I_m is added to the system. The current value of I_m is close to the value of I_{ref}. The difference between I_{ref} and I_m is stored in M_1 during the calibration cycle. This means that $I_m < I_{ref}$. In a practical case the difference current that can be stored in M_1 is between 0.1 and 0.05 of I_{ref}. Using this system the $\frac{W}{L}$ ratio of transistor M_1 can be chosen to minimize the g_m of the device. Moreover, due to this device choice a large V_{gs} voltage is needed to drive the device. Charge feed-through and leakage current influences can at least be reduced with a factor ten. As a result the calibrated current is much more accurate than was the case in the former system solution.

6.8.2 Continuous current calibration system

A system that uses a continuous current calibration is shown in Figure 6.50. The systems consists of a $N + 1$ -bit shift register and $N + 1$ current sources. The $N + 1$ current source is called the "spare" source. The output of this system is N calibrated currents. The operation of the system is as follows. The $N + 1$ stage shift register determines which current of the network is switched to the calibration source to be adjusted to the reference value I_{ref}.

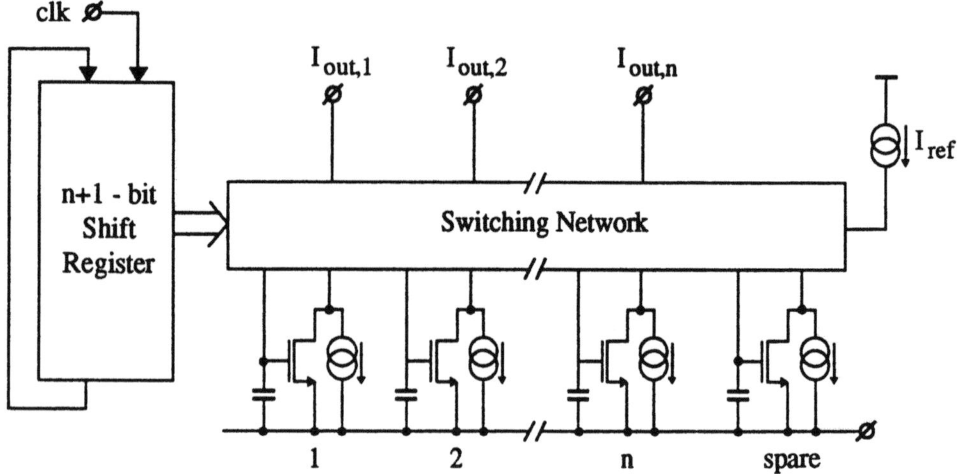

Figure 6.50 : Continuous current calibration system

Successively every current is compared with the reference source I_{ref} and then inserted back into the system. The switching network performs the necessary switching operation to perform the calibration of every current source in the system. By using $N + 1$ current sources no time is lost during calibration, because the current which is calibrated is replaced by the spare current source. In this way a continuous calibrated current network is obtained.

6.8.3 Practical current calibration implementation

In Figure 6.51 an example of a practical current calibration stage is shown. In the circuit transistor M_4 supplies the current I_m from the previous circuit. This is the main current source. Transistor M_1 holds the calibration current which is added to obtain an accurate value of I_{ref} of the total output current. The switches M_5, M_6, and M_7 perform the switching from calibration to operation of the current cell. When transistor M_5 is switched on, the switch M_7 is switched on as well and the spare current is applied to the bit switches in the D/A converter part. At the same time switch M_2 is on and transistor M_8 connects the drain of M_1 with the gate. Note that a current I_{bias} is added to the reference current I_{ref}. This bias current is forward biasing transistor M_8 and is subtracted from this reference current by the current source I_{bias}. In this way a good speed of transistor M_8 is obtained under all conditions. Transistor M_3 is added to the system to compensate for the

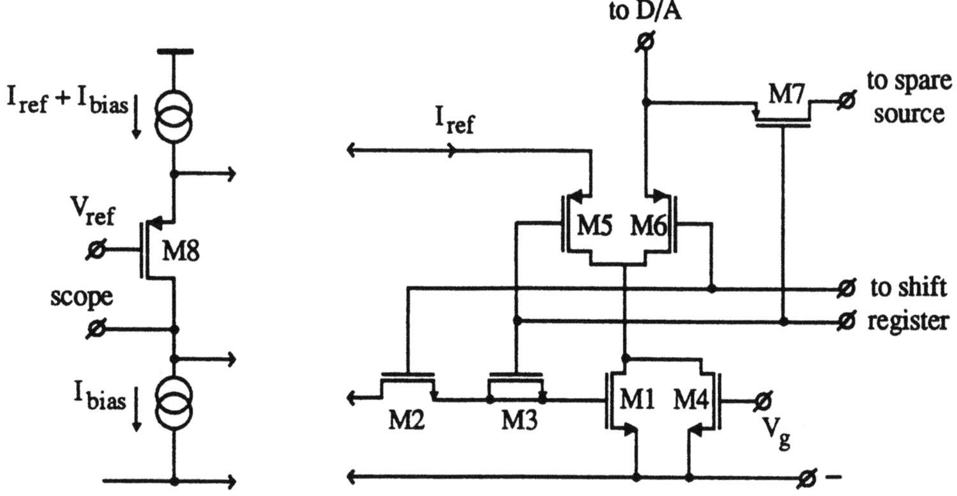

Figure 6.51 : Practical current calibration circuit

charge feed-through of M_2. Therefore, the gate of M_3 is connected to the inverse control voltage applied to M_2.

6.8.4   16-bit D/A converter system

An example of an MOS calibrated 16-bit D/A converter system is shown in Figure 6.52. The system consists of a 6-bit segmented current calibrated network to generate the six most significant bits. Current "64" is applied to a 10-bit binary weighted network using source scaling. In this way a 16-bit binary weighted current network is obtained. The digital input data which must be converted into an analog value are stored in the data register. The outputs of this data register controls the bit switches that switch the weighted currents to the output to obtain the converted analog value. The output current of the converter is converted into a voltage using an operational amplifier with a resistive feedback. To minimize the glitches that occur during the switching from calibration into output current generation of the MSB calibrated currents, a deglitcher can be implemented before the total output current is applied to the current to voltage converting operational amplifier. Another possibility exists by operating the deglitcher and calibration cycle at the same time the digital data applied to the switches are refreshed. A synchronization between shift register clock and input data clock is required.

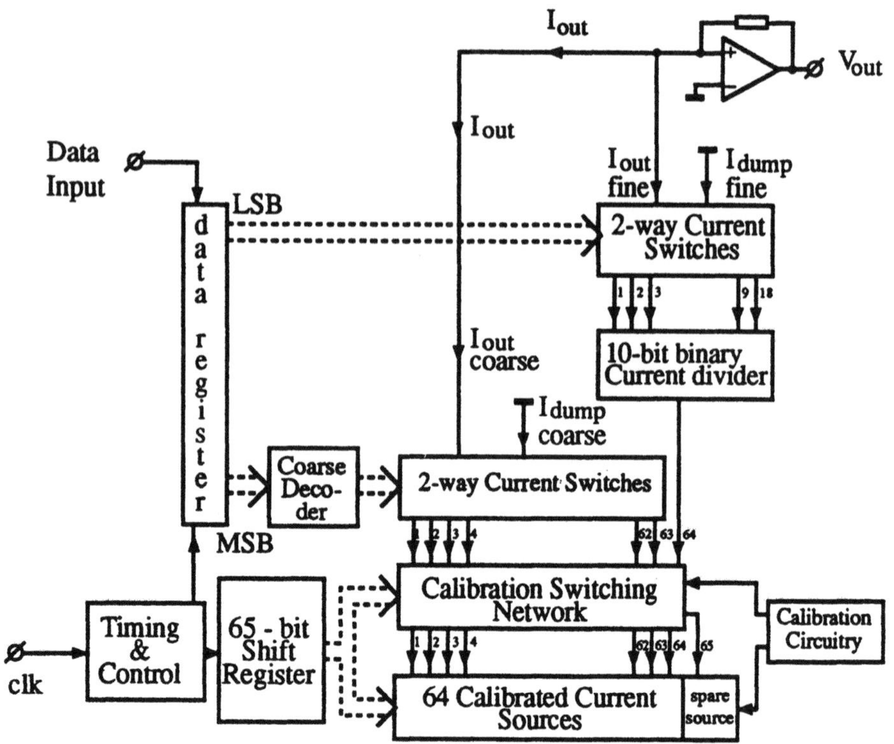

Figure 6.52 : 16-bit current calibrated D/A converter system

6.8.5 Integral nonlinearity measurement

In Figure 6.53 the result of the integral nonlinearity measurement of the system shown in Figure 6.52 is shown. Due to the segmented construction of the converter, *monotonicity* is guaranteed while the integral nonlinearity in this case equals ± 1 LSB. The curve characteristic shows a strong dependence of the calibration accuracy on the position of the calibrated weighted current values. Layout improvements are possible to obtain a full 16-bit accuracy.

6.8.6 Dynamic performance measurement

In Figure 6.54 the measurement result of signal-to-noise plus distortion is shown as a function of the output amplitude. The measurement result shows that the dynamic performance is close to the expected theoretical value.

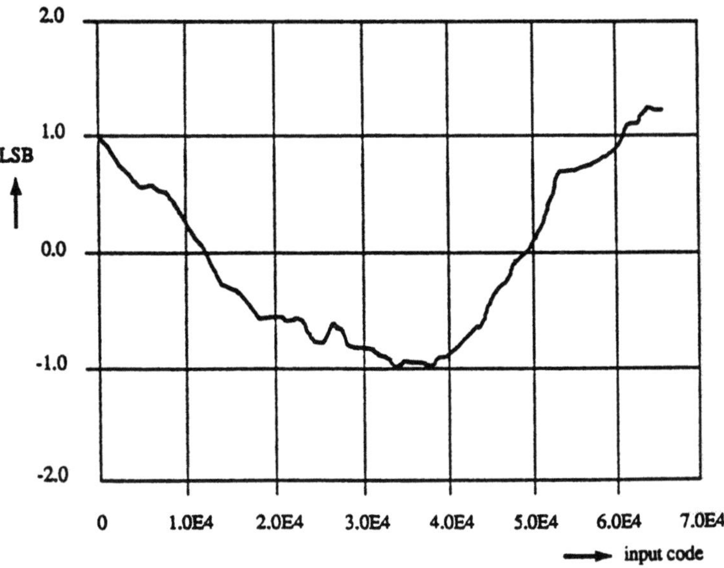

Figure 6.53 : Integral nonlinearity measurement result

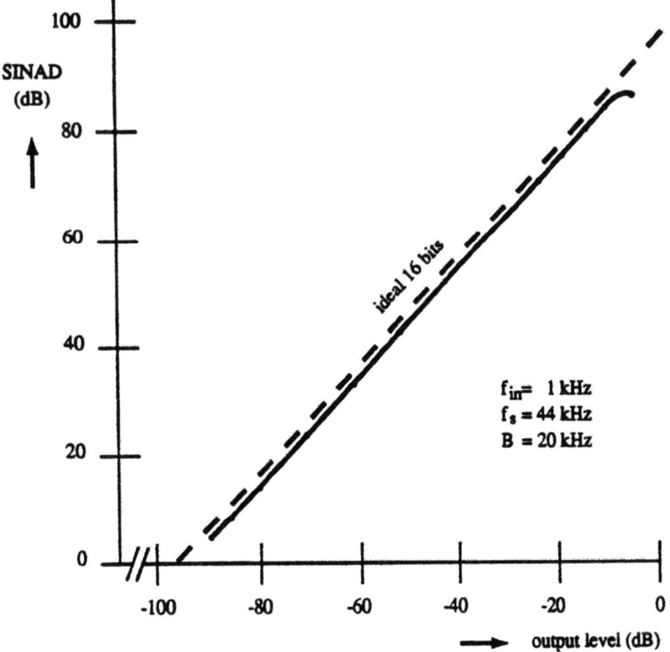

Figure 6.54 : Signal-to-noise plus distortion as a function of amplitude

6.8.7   D/A converter specifications

In Table 6.5 the specifications of the complete MOS 16-bit D/A converter
are shown.

Resolution	16 bits
Dynamic range	94 dB
S/(N + THD) at 0 dB	92 dB
S/(N + THD) at -10 dB	84 dB
Supply voltage range	3 to 5 V
Power dissipation	20 mW at 5 V
Temperature range	-10 to 70 degree C
Process	1.6 μm CMOS
Active chip area	3 mm^2

Table 6.5 : 16-bit D/A converter specifications

6.9   Conclusion

In this chapter an overview of basic circuits has been presented. Systems
with resolutions up to 10 bits do not require a calibration system. In monotonic designs resolutions up to 16 bits are possible. The only requirement is
that with an increasing input code the output at least increases. This can
be obtained by special circuit implementations that have a limited absolute
accuracy.

At the moment the full specification for the converter is needed, calibration
or trimming procedures are required. A special system is presented that
makes it possible to divide currents with a very high accuracy without needing accurate elements. The method presented uses a combination of a passive
divider with a dynamic interchanging method to improve the final accuracy.
After removing the error ripple by simply using low-pass filter structures
D/A converters with accuracies from 14 to 18 bits can be designed. The
circuits presented show a high accuracy and, by optimizing the dynamic behavior of the systems, an overall performance is obtained that is close to the
theoretically possible behavior. In practice the *Dynamic Element Matching*

system can be extended into generally applicable systems to obtain high-accuracy integer ratios of currents. A second method, which uses a continuous calibration principle of a multiple of equal reference currents, is very suitable for implementations in a CMOS technology. This system does not need external filtering. Furthermore, the use of an additional (spare) current makes a continuous calibration possible during the operation of the converter in a system.

A third self-calibration method makes a circuit less sensitive to component variations; however, during warming up of the system a re-calibration might be needed. During the calibration cycle the system is not able to produce an output signal. This might be a large drawback for many systems.

Chapter 7

High-accuracy A/D converters

7.1   Introduction

High-resolution monolithic A/D converters are subject to growing interest due to the rapidly expanding market for digital signal processing systems. The introduction of digital audio recording equipment such as the Digital Compact Cassette (DCC) players requires resolutions of 16 to 18 bits. Monolithic converters with such a high linearity are difficult to design and require special circuit configurations. When a low conversion speed is needed, integrating types of converters can be used. In integrating types of high-resolution A/D converters basically the analog input signal is converted into a time which is proportional to the input signal. Time is measured using a counter with an accurate clock. These systems are relatively slow because of the counting operation in the time-to-number conversion cycle. A speed improvement is obtained by using a coarse and fine conversion cycle in the time-to-number counting operation. A well-known analog-to-digital converter based on this system is the dual slope converter. This converter is mostly used in digital voltmeters.

In fast and highly accurate A/D converters, the successive approximation method is commonly used. Accuracy and linearity in this system are determined by the D/A converter, while the conversion speed depends on the comparator response time and the settling time of the D/A converter. In a successive approximation system the analog input signal is approximated by the step-by-step built-up analog output voltage of the D/A converter, starting with the most significant bit. To obtain the high accuracy for the D/A converter needed to construct a 14- to 16-bit A/D converter, *Dynamic Element Matching* is used. In the 14-bit A/D converter, which will be discussed in the following sections, the 14-bit D/A shown in Figure 6.41 is used. Due to the construction of the bit switches, which in the high-accuracy part consist of a diode-transistor configuration, the output voltage swing at the D/A current output must be small. This voltage swing, called output voltage compliance, reduces the bit switching accuracy of the D/A converter. To avoid problems in this system, a special comparator operation is needed. In general-purpose A/D converters, a general nonlinear comparator circuit with high gain around the zero-crossing level is used. Such comparators generally have a certain voltage compliance which is above the range needed for the D/A converter from Figure 6.41. Therefore, a wide-band, high-speed operational amplifier with diode clamps is applied in the inverting mode. The voltage compliance in this case remains below a few millie volts. This is accurate enough to obtain the full accuracy of the D/A converter.

The ease with which a hold operation can be constructed allows an A/D converter implementation using a cyclic converter algorithm. In the cyclic converter the number of components is drastically reduced and consists of two sample-and-hold amplifier circuits with an accurate 2 times amplifier stage and a subtracter circuit. Per conversion step the remaining signal is compared with a reference signal. If the remainder is larger than the reference signal, a subtraction of the reference signal from the remainder is performed. The error signal what is then generated is amplified by two and compared with the reference signal again. This operation is repeated until the total number of bits what can be converted is obtained. Finally self-calibration schemes are shown, which use an additional cycle to calibrate the complete converter. During calibration, however, it is not possible to use the converter effectively in the system.

7.2 Single slope A/D converter system

In Figure 7.1 a block diagram of a single slope A/D converter is shown. The circuit consists of a reset-table integrator, which generates the accurate reference ramp signal, a comparator, and a counter. The input signal is applied to one input of the comparator. At the moment the conversion starts,

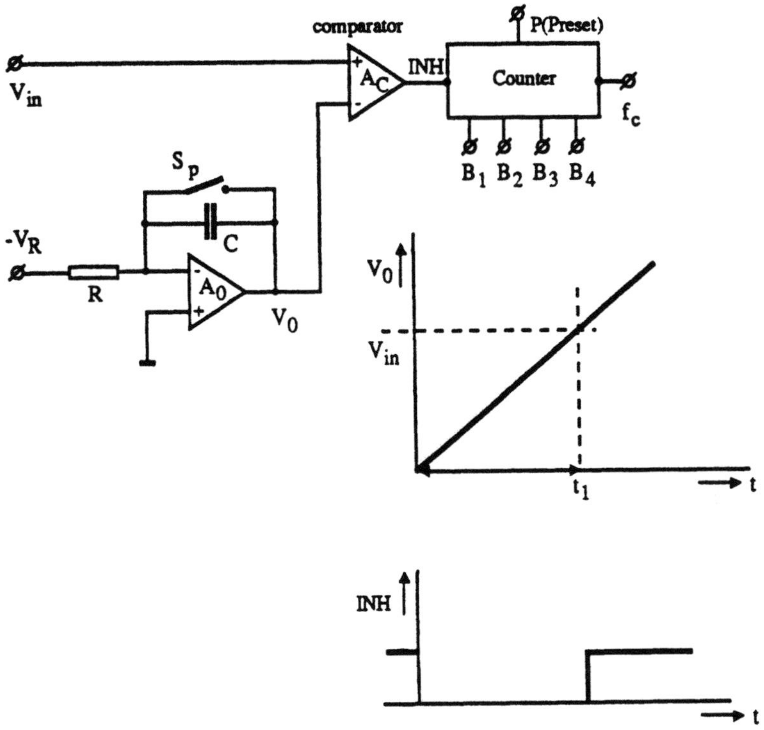

Figure 7.1 : Single-slope A/D converter system

the counter is set at zero and the integrator is reset by closing switch S_p. When a positive input signal V_{in} is applied, the integrator starts generating the ramp function. In the meantime a gate is opened that applies counting signals to the counter. At the moment the output signal of the integrator equals the input signal, the gate is closed and the counter stops. The analog input signal is converted into a time that is measured by counting clock pulses during that time. An accurate time-to-number conversion is obtained. The accuracy of the system is determined by the clock generator, the RC time constant of the integrator, and the reference source V_R. A simple calculation shows that the time to which an input signal is converted is equal to:

$$t_1 = RC\frac{V_{in}}{V_R}.$$

(7.1)

The digital output value then becomes:

$$N_{digital} = t_1 \times f_{clock}.$$

(7.2)

Offset of the comparator can be canceled by measuring the offset with a zero input signal. This offset number can be used to preset the counter. In this way an automatic offset compensation is obtained.

7.3 Dual-slope A/D converter system

To overcome a number of the accuracy problems encountered with the system from Figure 7.1 a dual-slope system has been designed. A system diagram is shown in Figure 7.2. The system consists of an input switch, an integrator

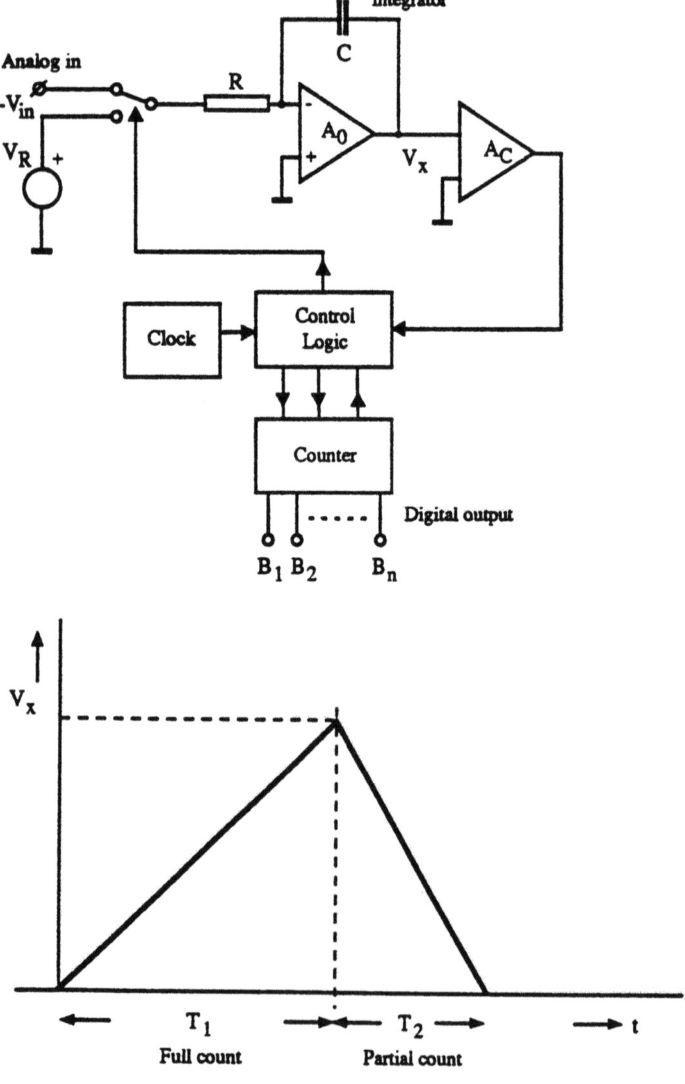

Figure 7.2 : Dual slope A/D converter system

with a comparator, a clock generator with control logic, and a counter. The operation of the system is as follows. Starting from a reseted integrator, the input signal V_{in} is integrated during a time t_1 which corresponds with a full count of the counter. Then the input is switched to the reference voltage V_R having the opposite sign compared to the input signal. The integrator is now discharged. During the discharge time pulses are counted. Counting stops when the comparator detects zero. As a result, the counts in the counter represent the digital value of the input signal. A simple calculation shows:

$$V_{in} = V_R \times \frac{T_2}{T_1}.\tag{7.3}$$

Here T_2 is the time during which the integrator is discharged from the integrated input signal to zero. As is shown in equation 7.3 the clock is not critical; only the ratios between the charge and discharge times is important. A disadvantage of this system is the low conversion speed if a high resolution is required. In digital voltmeters these systems are very popular.

7.4 Dual-ramp single-slope A/D converter system

To decrease the conversion time of the single-slope A/D converter as shown in Figure 7.1 a dual-ramp system has been designed. The block diagram of the dual ramp converter is shown in Figure 7.3 [30]. The system consists of an inverting sample-and-hold amplifier with feedback resistors R and hold capacitor C, two reference current sources with current values I and $\frac{I}{256}$, comparator $comp_1$ with threshold voltage V_t, and comparator $comp_2$ which control the coarse and fine counting operation using the control logic function, a clock generator, and a coarse and fine counter. At the start of conversion, the counters are set to zero and the switch S_3 is closed. Switches S_1 and S_2 are open, so no current flows into the sample-and-hold integrator. Closing switch S_3 causes the operational amplifier to act as an inverter charging the hold capacitor C. At the moment switch S_3 is opened, the input signal is sampled and held on the capacitor C. Then switch S_1 is closed and the reference current I starts discharging the capacitor until the output signal of the integrator reached the threshold voltage V_t. During the discharge time pulses are counted in the MSB counter which is able to store a maximum of 255 pulses. However, the threshold voltage V_t applied at comparator $comp_1$ is larger than the voltage that can be obtained during a full count

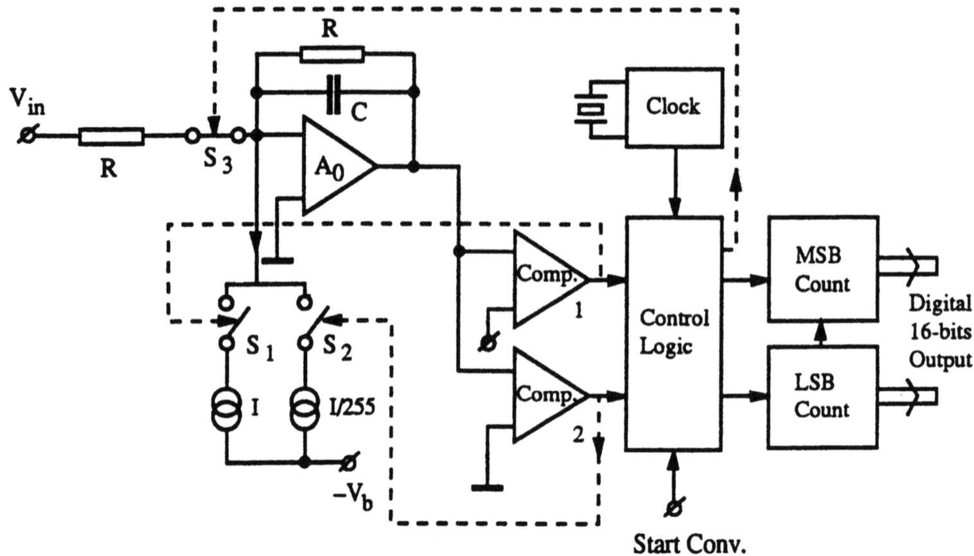

Figure 7.3 : Dual-ramp single-slope A/D converter system

of the fine counter (255 pulses) and integrating a current $\frac{I}{256}$ on the capacitor C. This threshold voltage is not at all critical, as will be explained later.

After comparator $comp_1$ detects the threshold voltage V_t, switch S_1 is opened and switch S_2 is closed. At the same time the fine counter starts counting clock pulses until comparator $comp_2$ detects zero. If the number of clock pulses applied to the fine counter is larger than the number it can store (in this case 255 pulses), then a carry is generated which is applied to the coarse counter and increases the count of this counter with one count. In this way an automatic adjustment of the threshold voltage V_t is obtained. At the output of this system a 16-bit digital number is obtained that corresponds with the analog input signal. The speed of this system is increased with a factor 256 divided by 2 equals 128 times. Using a very high counting clock, it is possible to obtain a 16-bit A/D conversion with a conversion speed of about 44 kHz.

In Figure 7.4 the output signal of the sample-and-hold amplifier/integrator as a function of time is shown. The coarse and fine discharging period are clearly distinguishable in the figure.

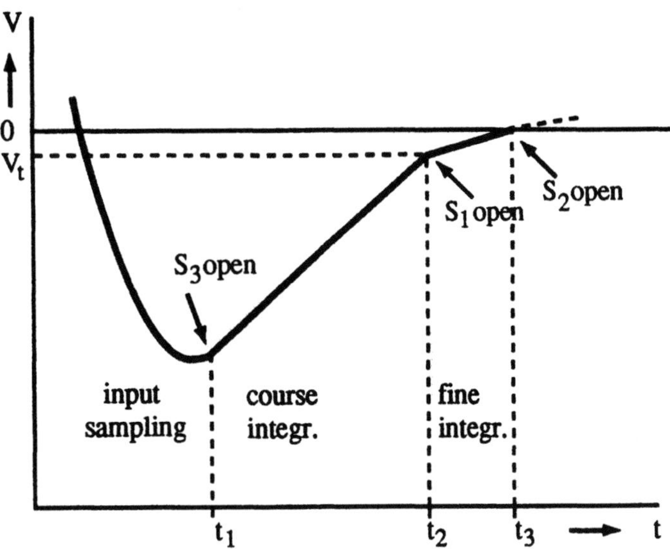

Figure 7.4 : Output signal of the sample-and-hold/integrator amplifier

7.4.1 Accuracy analysis of the dual ramp A/D converter

In Figure 7.5 the output signal of the sample-and-hold amplifier/integrator is shown. Note that the coarse current value is equal to $I + \frac{I}{255}$ and the

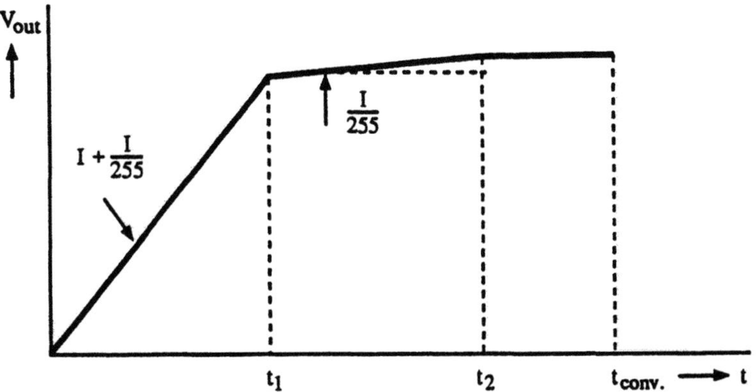

Figure 7.5 : Styled output signal of the sample-and-hold/integrator amplifier

fine current value is equal to $\frac{I}{255}$. This construction is used because a large current can be switched better at high speeds than a small current. This means that $\frac{I}{255}$ is always switched to the integrator. The ratio between the currents is 1 to 256 which is required in this system.

Suppose that the current $\frac{I}{255}$ has an accuracy of $(1 - \delta)$; then after a full count of the fine integrator the error between the coarse current and the full fine integrated current value must be smaller than k LSB or:

$$I + \frac{I}{255} \times (1 - \delta) - 256 \times \frac{I}{255} \times (1 - \delta) \leq k \times \frac{I}{255}, \qquad (7.4)$$

or

$$\delta \leq \frac{k}{255}. \qquad (7.5)$$

In a 16-bits system a matching between the coarse and fine discharge currents better than 0.2 % is needed to obtain $\frac{1}{2}$ LSB differential linearity.

The accuracy with which the currents must be switched can be determined. Suppose that the clock time $t_0 = \frac{1}{f_{clock}}$, then with a time uncertainty in switching the current I of δt we obtain in case the fine full-scale integrated value must be smaller than k LSB error:

$$I \times (t_0 + \delta t) + \frac{I}{255} \times t_0 - 256 \times \frac{I}{255} \times t_0 \leq k \times t_0 \times \frac{I}{255}, \qquad (7.6)$$

or

$$\delta t \leq \frac{k}{255} t_0. \qquad (7.7)$$

With $t_0 = 40$ nsec and $k = \frac{1}{2}$ LSB we get $\delta t = 80$ psec.

7.5 Successive approximation converter system

The basic architecture of a successive approximation analog-to-digital converter is shown in Figure 7.6 The basic converter consists of a comparator stage (Comp.), the successive approximation register (SAR) and the digital-to-analog converter. A sample-and-hold and an anti-alias filter are added to limit the maximum analog input frequency and to convert the continuous time input signal into a discrete time signal. At the beginning of the conversion the MSB is switched on and the input signal is compared to the output signal of the D/A converter. When the input signal is larger than the output signal of the D/A converter, then the MSB remains on and the next bit is switched on and a comparison will be performed. A bit by bit operations is performed in this system to bring the D/A output signal within 1 LSB to the time discrete input signal. In the lower part of fig. 7.6 the conversion procedure as a function of bit weighting is shown. The output value

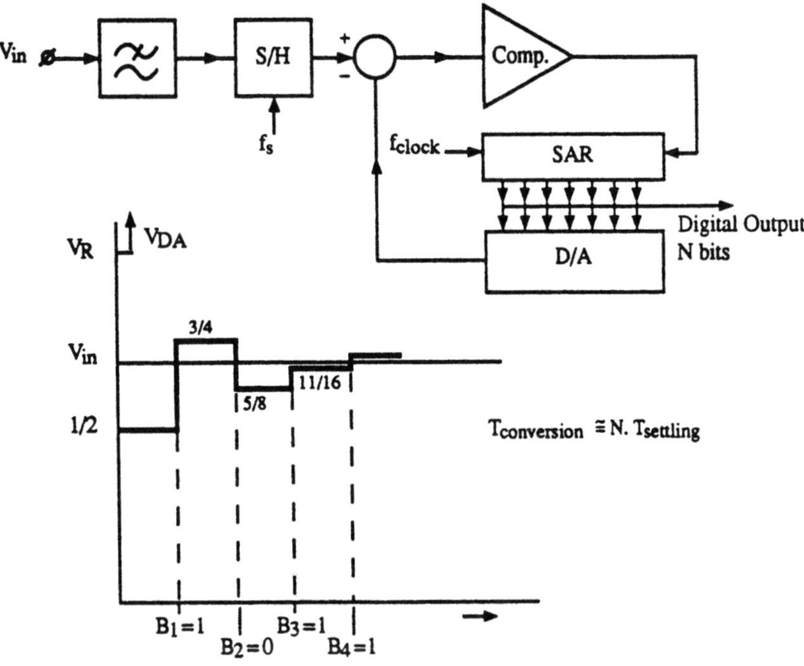

Figure 7.6 : Basic successive approximation A/D converter

in the figure equals 1011. A complete conversion in this system requires N switching and comparison operations to convert the input signal into a N-bit digital output value. The conversion time equals:

$$T_{conversion} = N.T_{settling}. \qquad (7.8)$$

The settling time is defined as the time required to settle within $\frac{1}{2}$ LSB of the D/A converter. The linearity and accuracy of this system depends on the D/A converter.

7.5.1 Practical successive approximation A/D converter

A block diagram of a practical A/D converter [3] system is shown in Figure 7.7. The most important parts are the successive approximation logic, the D/A converter with the reference current source, the subtracter-comparator circuit using an operational amplifier to convert the analog input signal into a current, and the internal clock generator with the control logic. Digital input and outputs are TTL compatible. In the internal digital part current mode logic (CML) is used for speed. Because of the differential operation of

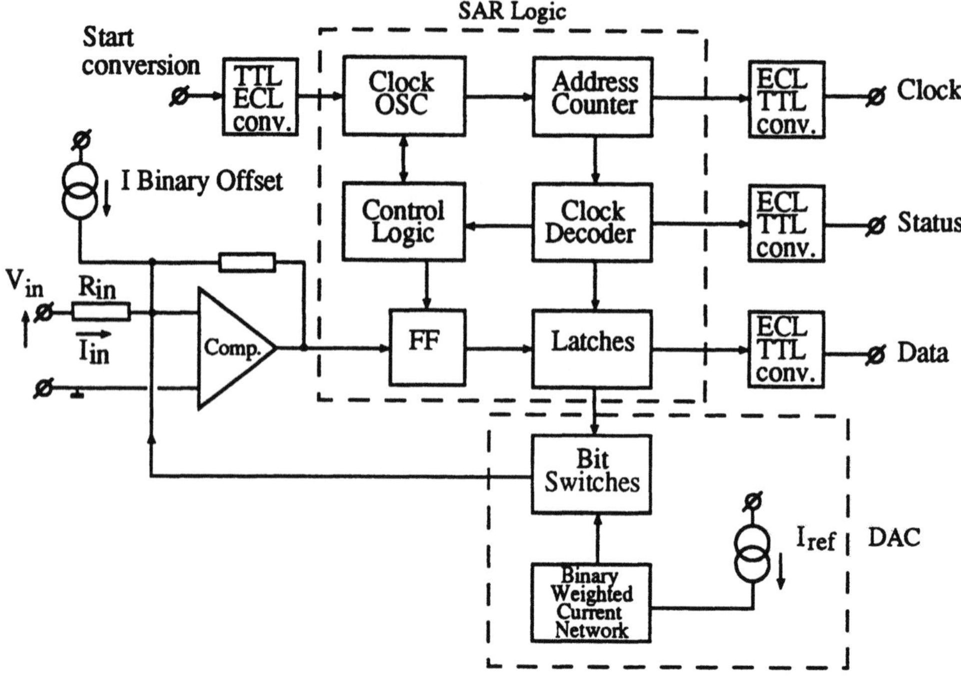

Figure 7.7 : Block diagram of the A/D converter system

the current mode logic low interference of the switching signals on the power supply lines is obtained. At the input a TTL-to-CML logic converter stage is incorporated, while at the output CML-to-TTL levels converters with output buffers are used. A scheme with addressable latches is used in the successive approximation register to minimize the number of components. Output data flow of the converter is in a serial mode to minimize the number of circuit pins and to reduce the noise generated by the TTL output signals. The start conversion pulse activates the internal clock and control logic for one conversion cycle. A constant "trial-and-decision" time period is used, except for the most significant bit which has a 50 percent longer decision time. A special output signal, "enable status," is available during the time in which the conversion is performed. With this signal an external sample-and-hold function can be switched from sample into hold mode. The aperture jitter of the enable pulse, which switches the sample-and-hold amplifier from sample into hold mode, is kept below the theoretical limit defined by the resolution and the maximum signal frequency. Then conversion starts. Because of the voltage-to-current input mode configuration, the input voltage signal is converted into a current by the resistor R_{in}. An extra offset binary current equal to the most significant bit value is available at the input of the converter to allow bipolar signals to be converted into offset binary output code.

All circuitry to operate the A/D converter is on chip except the input sample-and-hold amplifier, filtering capacitors to remove the ripple from the interchanging network and gain-setting resistors.

7.5.2 Comparator-subtracter circuit

As already said, a high-speed operational amplifier in the inverting mode is used for subtraction of the D/A converter output current from the analog input current. A simplified circuit diagram is shown in Figure 7.8. A wide

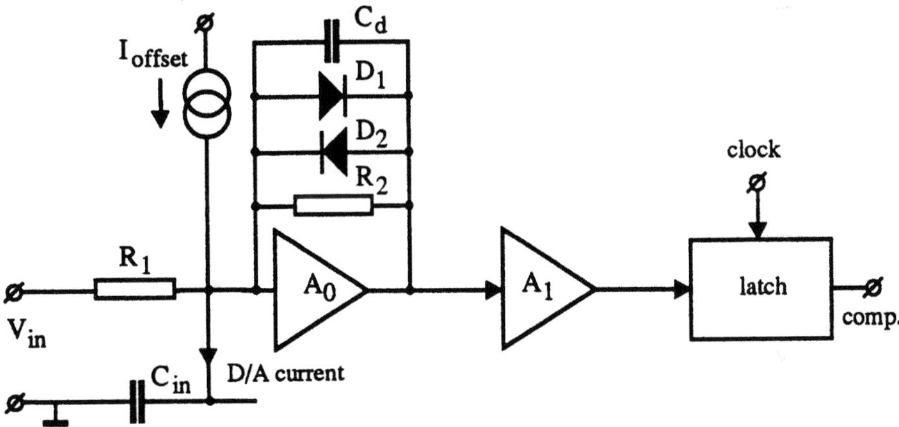

Figure 7.8 : Comparator-subtracter circuit diagram

band feed-forward coupled operational amplifier with a unity-gain-frequency-compensated bandwidth of 75 MHz is used. A high-value feedback resistor R_2 is used to have a large gain around the zero crossing of the signal. Diodes D_1 and D_2 are connected in parallel with the resistor R_2 to prevent large output voltage swings, which maybe even larger than the applied supply voltage of the converter, being generated at the output. With these diodes, the input voltage excursion remains very small, thus having no influence on the D/A converter performance and the input voltage-to-current transformation. Speed limitations in the system are due to the diode capacitance C_d across the feedback resistor R_2 and the input capacitance C_{in}. The output capacitance of the D/A converter in this case is the dominating contributor to the input capacitance C_{in}. In practice using a standard bipolar process, a value of approximately 20 pF for C_{in} is found. The comparator bandwidth

with a diode capacitance C_d of 2 pF is limited to about 4 MHz. To make
the bandwidth of the total system only dependent on the feedback and loading elements of the operational amplifier a minimum compensated unity-gain
bandwidth of 40 MHz is required. With a practical designed limit of 75 MHz
this demand is easily fulfilled. The output signal of the operational amplifier
is further amplified and then latched into a current mode logic master-slave
flip-flop. This flip-flop converts the comparison information into current
mode logic levels, which are applied to the successive approximation logic
to perform the complete conversion cycle. A total average settling time of
the D/A converter currents of 400 nsec to $\pm\frac{1}{4}$ LSB is obtained with this
comparator system. The low noise of the filtered bit currents of the D/A
converter and a low noise design of the operational amplifier keep the total
noise in the system below $\frac{1}{3}$ LSB, which is small enough to obtain a good
dynamic performance of the converter.

The low offset voltage of the operational amplifier (below 0.5 mV) does not
need an extra offset trim of the comparator with sufficiently large input
voltage. An extra current which is equal to the most significant bit current
value is available at the input of the operational amplifier to allow a bipolar
signal operation of the A/D converter.

7.5.3   Complete practical A/D converter

The total A/D converter is implemented using a standard bipolar technology
and needs a chip size of 3.5x4.4 mm^2. Double-layer metallization simplifies
circuit layout and improves performance. A die photograph of the chip is
shown in Figure 7.9. In the layout special guard rings are added to separate
accurate and sensitive analog parts from the digital part of the circuitry.
Supply connections for the analog and the digital circuit part of the system
are separated to avoid interference and noise inductions due to the operation
of the circuit. To operate the converter, a sample-and-hold amplifier, a
number of resistors, and some filtering capacitors are needed.

7.5.4   Measurements

An A/D-D/A converter loop has been built to obtain quick and reasonably
accurate information about conversion time, linearity, signal-to-noise ratio,
distortion, and so on, of the A/D converter. During dynamic measurements,
a sample-and-hold module with sufficiently high performance is used. Analog

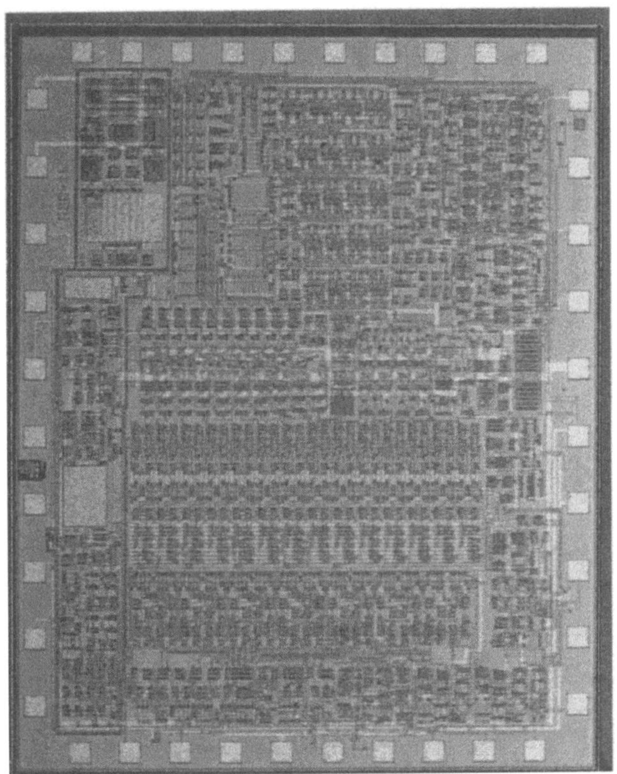

Figure 7.9 : Die photograph of the 14-bit A/D converter

low-pass filters have enough stop-band attenuation ($\approx 100\ dB$) to allow accurate measurements. In Figure 7.10 the block diagram of the test set-up with the specified measurement instruments is shown. In Figure 7.11 the total signal-to-noise plus distortion ratio as a function of conversion time is shown. During the measurement procedure of this curve, only the conversion time is changed. All other system parameters are maintained at the basic settings. From Figure 7.11 a conversion time of 7 μ sec is obtained. A loss in S/N with respect to long conversion times of 1 dB is used as a test criterion. In Figure 7.12 the signal-to-noise as a function of amplitude is shown, while Figure 7.13 the signal-to-noise as a function of

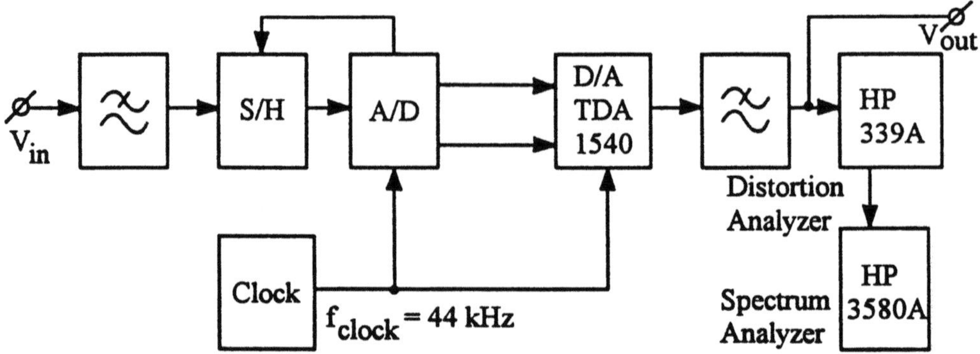

Figure 7.10 : A/D converter measurement test set-up

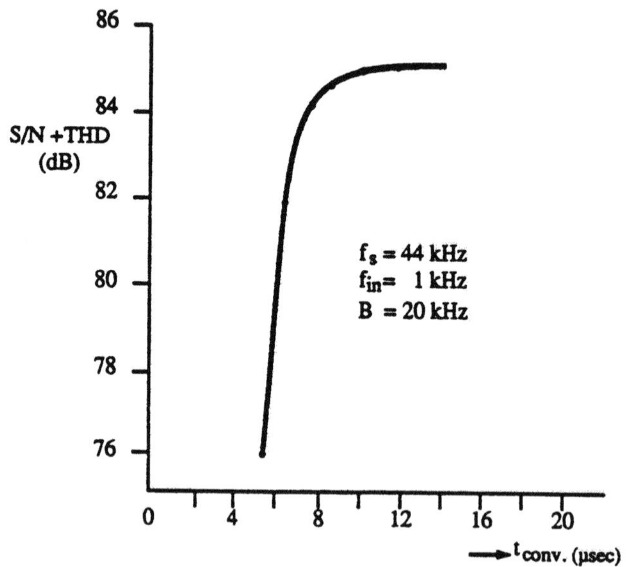

Figure 7.11 : S/N plus distortion as a function of conversion time

frequency for various amplitude levels is plotted.   Additional information
about the performance of the A/D converter chip is shown in Table 7.1.

7.6   Algorithmic A/D converter

A block diagram of a cyclic (algorithmic) A/D converter is shown in Figure
7.14 [35,87,88]. In a cyclic system the output code is generated in a serial
way. In every cycle the comparator, subtracter, and multiply by two amplifier are used. A very compact converter system is obtained at the cost

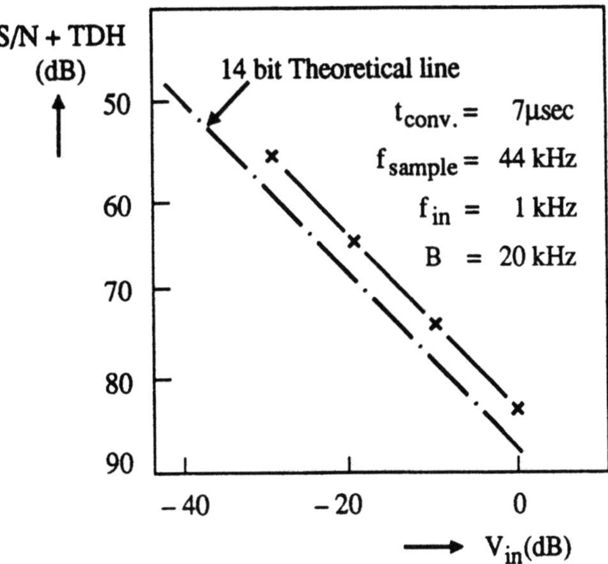

Figure 7.12 : S/N plus distortion as a function of amplitude

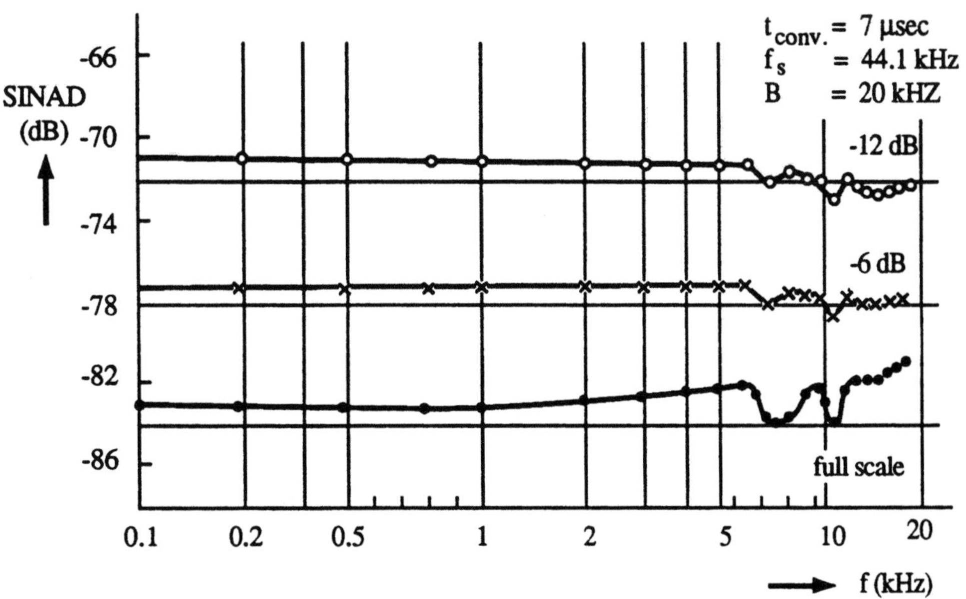

Figure 7.13 : S/N plus distortion as a function of frequency

of a larger conversion time (p times the time of one cycle operation). The system consists of a times two differential amplifier with a built-in hold function, a summer/subtracter circuit, a comparator, a reference source, and a

Resolution	14 bits
Integral nonlinearity (INL	± 1/4 LSB
Analog input	± 2 mA
Minimum conversion time	7 μsec
Logic levels	TTL compatible
Output data	Serial
Supply voltages	+5, -5, -17 V
Power dissipation	450 mW
Temperature Coefficient of	
Reference source	0.5 ppm per degree C Δ T = 100^0
S/(N + THD)	84 dB, f_{sample} = 44.1 kHz,
	bandwidth = 20 kHz
Distortion	-96 dB, f_{in} = 1 kHz
Package	40 pins DIL
Chip dimensions	3.5 $\times$ 4.4 mm^2

Table 7.1 : 14-bit A/D converter data

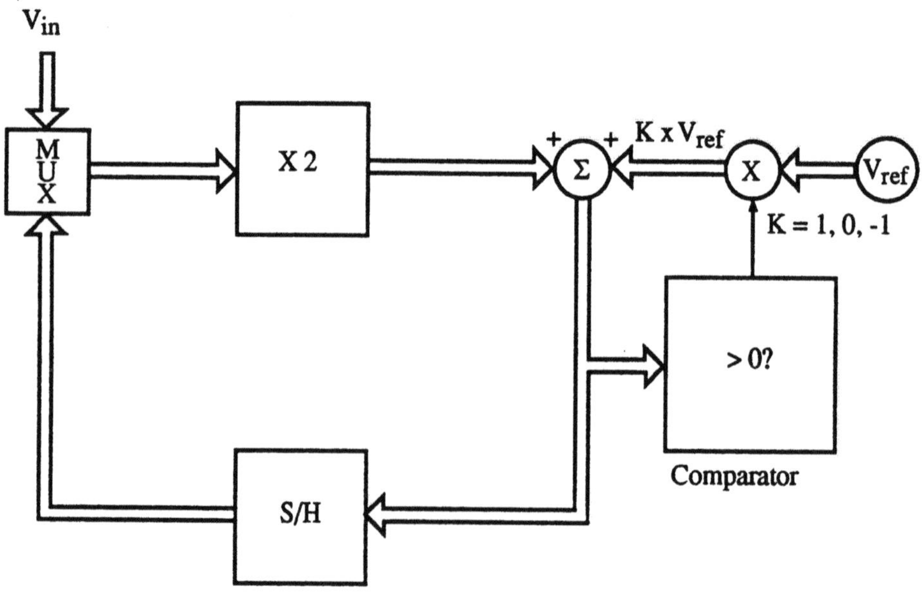

Figure 7.14 : Algorithmic A/D converter system

sample-and-hold amplifier. The multiplexer at the input switches the input of the sample-and-hold amplifier between the input terminal and the output of the summer/subtracter. The conversion starts by sampling the input signal V_{in} and amplifying this signal with an accurate factor of two. The signal is then applied to the comparator which subtracts or adds the reference voltage V_{ref} to the two times amplified input signal. A reference source operation is performed when the comparator detects a signal larger than zero. The positive error signal is transferred to the sample-and-hold amplifier. In the meantime the multiplexer has been switched to the output of the sample-and-hold amplifier. The remaining signal is amplified by the two times amplifier and a reference operation is performed again. This cyclic comparison, subtraction, and signal transfer are repeated, depending on the number of bits with which the converter is operating. A serial code appears at the output of the comparator. The only accurate element in the system is the times two amplifier and the subtracter/summator circuit with the reference source.

In Figure 7.15 a detailed operation of the accurate times two amplifier system is shown. The operation of the system is based on the addition of two sampled signals on two capacitors. In Figure 7.15a the input differential signal is sampled on the capacitors C_1 and C_2, while at the same time the input offset voltage is sampled on capacitors C_3 and C_4. In Figure 7.15b the input signal charge is transferred from capacitors C_1 and C_2 to capacitors C_3 and C_4. The second input sample is taken as shown in Figure 7.15c. Capacitors C_3 and C_4 are disconnected and therefore do not get any charge. The input sample is stored on capacitors C_1 and C_2. The repeated addition of the two samples is shown in Figure 7.15d. The voltage across C_3 is added to the voltage across C_1 and gives a doubling of the input signal. An identical operation is found for the voltages across C_4 and C_2. The sampling of the signal accurately stores the input voltage on a capacitor, while due to the charge transfer it can be shown that an offset correction occurs. Furthermore, the system is nearly independent of the capacitor mismatching. The differential operation of the system compensates for charge feed-through errors due to the switching operation that is performed in this system. A larger dynamic range can be obtained, while, furthermore, such a system is less sensitive to supply voltage variations. In Figure 7.16 the complete analog part of the A/D converter is shown. Amplifier A_1 with the surrounding switches performs the accurate times two amplification with the subtraction or addition of the reference value. Note that by inverting the reference voltage to capac-

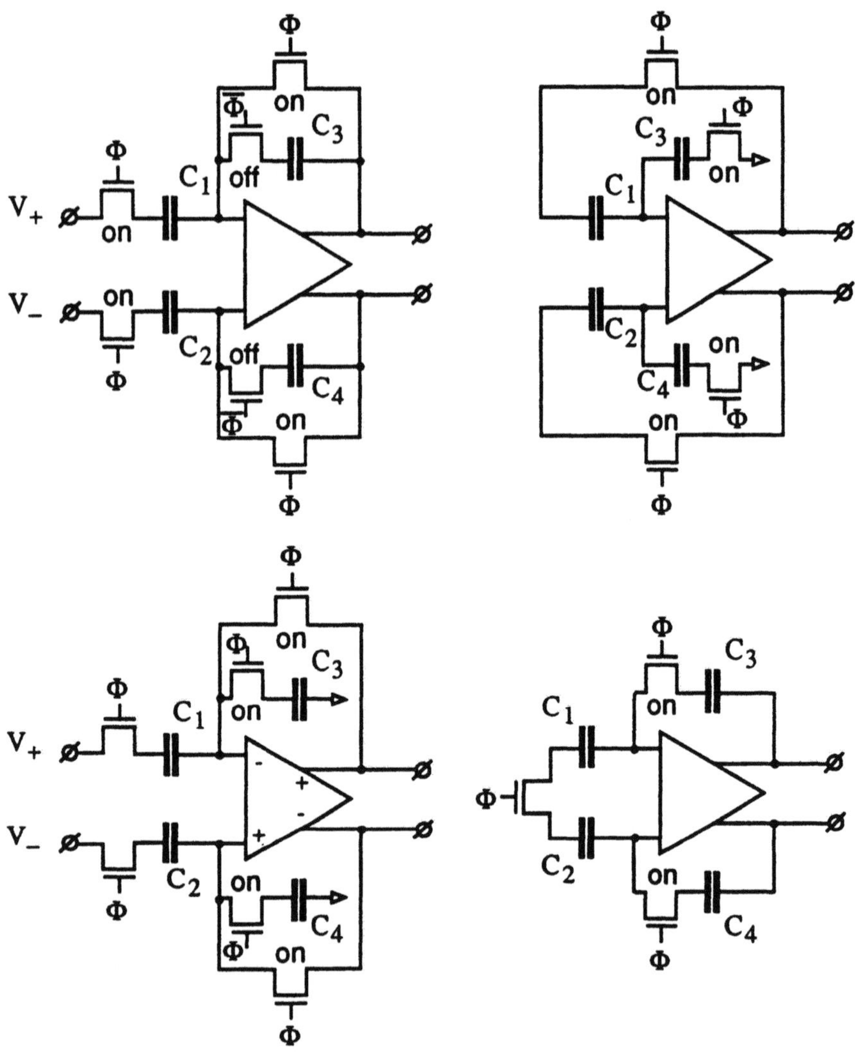

Figure 7.15 : Detailed operation of the accurate times two amplifier

itor C_5 or capacitor C_6, a subtraction or addition can be performed without loosing accuracy. Amplifier A_2 with capacitors C_7 and C_8 perform an off-set independent sample-and-hold amplifier function. Capacitors C_9 and C_{10} with the amplifier perform an accurate comparison with zero. The output signal of the comparator is stored in the output latch and can be transferred as a series output signal of the converter. A single-ended sampling of the input signal is used. No amplification by two is obtained during the sampling of the input signal.

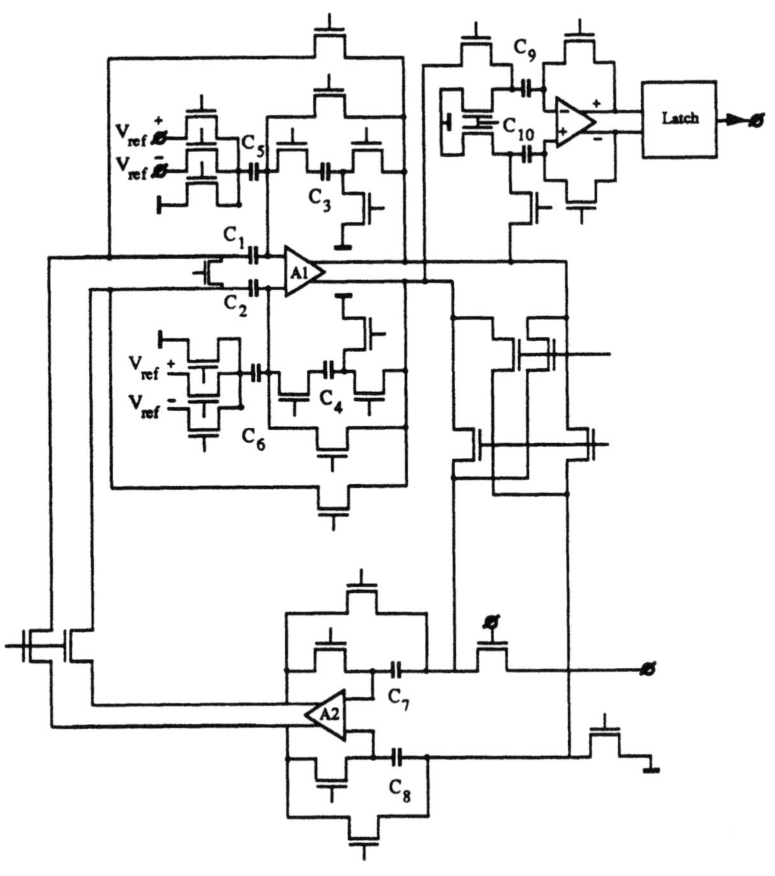

Figure 7.16 : Complete analog part of the A/D converter

7.7 Cyclic Redundant Signed Digit A/D converter

A cyclic converter that is less dependent on component matching and comparator offset voltages uses the Redundant Signed Digit Coding [72]. The main advantage of this coding that it gives the option to perform a subtraction as well as an addition. The RSD coding is well known from digital multiplier coding using special algorithms (Booth). Because of the subtraction operation decision, errors introduced by offset in the comparators can be corrected for. In Figure 7.17 the basic implementation of an RSD algorithm is shown. The system consists of a sample-and-hold amplifier, an accurate time two amplifier, a switchable adder-subtracter circuit, and two comparators. The comparison levels P and Q are arbitrary but can best be chosen at half the reference voltage ($P = -Q = \frac{1}{2}V_{ref}$).

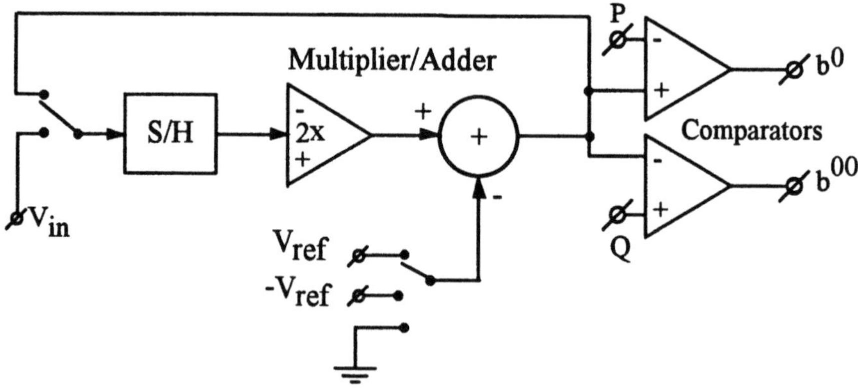

Figure 7.17 : Basic RSD converter implementation

b^0	b^{00}	b
1	0	+1
0	0	0
0	1	-1

Table 7.2 : Signed Digit coding

The operation of the comparison is shown in Table 7.2. Starting with a sample of the input signal V_{in} in the sample-and-hold amplifier, this signal is amplified by two and compared with the levels P and Q. At the moment the signal is between of the levels P and Q corresponding to b = 0, then no addition or subtraction of the reference voltage is performed. The signal is amplified by two and again a comparison is performed. At the moment b = -1, then the reference voltage is subtracted and the remainder is amplified by two again to perform a new bit value determination. At the moment b = +1, then a reference voltage addition is performed. The remainder is multiplied by two again and a new comparison takes place. This cycle is repeated until the required amount of serial bits is obtained. Because of this restricted comparison method with the ability to add or subtract a reference voltage from the partial remainders, a comparator offset independent operation is found. The serial output code containing 0, +1 and -1 bits can be converted into a binary-weighted code by the appropriate subtraction operation of the -1 valued bits. In Figure 7.18 a switched capacitor

Signed bit b	V_A	V_B	Transfer function
-1	V_{ref}	Gnd	$\hat{V}_x = 2\,V_x + V_{ref}$
0	Gnd	Gnd	$\hat{V}_x = 2\,V_x$
+1	Gnd	V_{ref}	$\hat{V}_x = 2\,V_x - V_{ref}$

Table 7.3 : Switching and operation schemes of converter

implementation of the system is shown. The circuit consists of an input

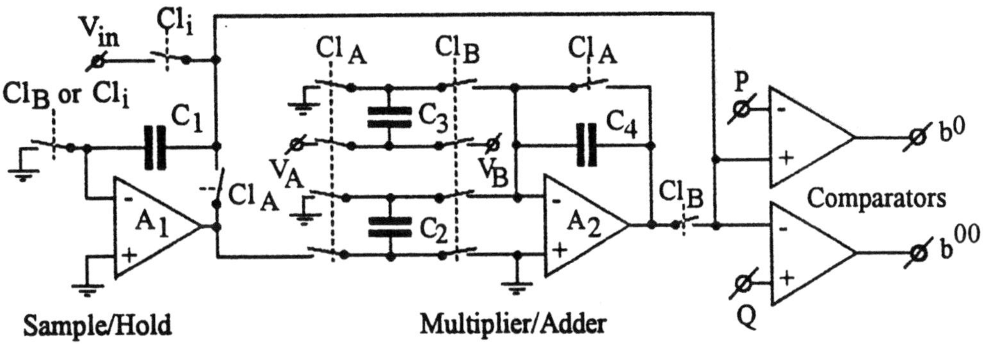

Figure 7.18 : Switched capacitor RSD converter implementation

sample-and-hold amplifier, a multiplier/adder circuit, and the two comparators with appropriate decision levels. In the multiplier/adder circuit a times two operation is performed. Note that only capacitor C_2 has a value 2C while all other capacitors have a value C. The operation starts by closing switch Φ_I and sampling the input signal. Then by sequentially closing switches Φ_A and Φ_B, respectively, the cyclic conversion takes place. Depending on the output of the signed bit b, which is generated according to Table 7.2, the voltages applied to the terminals V_A and V_B are applied according to Table 7.3. The output bits of the comparators are again converted into a binary code by the appropriate subtraction of the negative values from the output signal. The shown implementation is a simplified schematic of the preferred embodiment of the system. Especially in the switched capacitor CMOS implementation, a differential operation of the circuit is required to minimize the switch feed-through on the overall performance. The only important factor in the system is the times two amplifier.

7.8   Self-calibrating capacitor A/D converter

In Figure 7.19 a block diagram of a self-calibrating A/D converter is shown
[74]. The system consists of a 10-bit weighted capacitor D/A converter com-

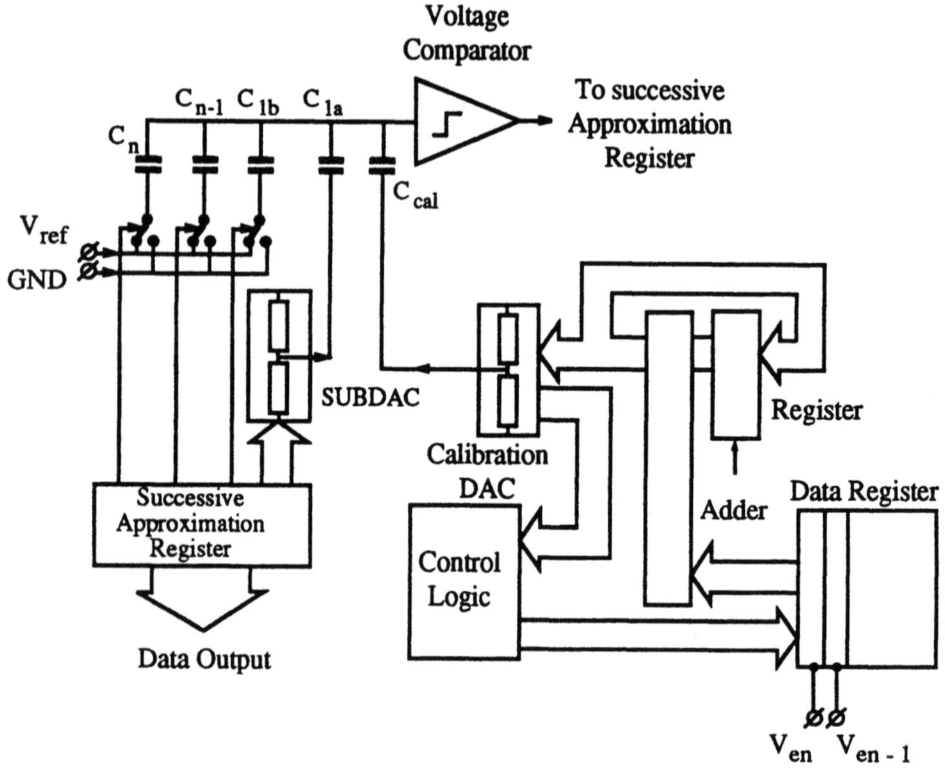

Figure 7.19 : Self-calibrating A/D converter

bined with a 5-bit resistive divider sub-D/A converter. A 7-bit calibration
D/A converter is used to calibrate the errors from the binary-weighted capacitor D/A converter and the resistive divider sub-D/A converter. The
accuracy of a 10-bit binary-weighted D/A converter is so good that only the
five most significant capacitor values need to be calibrated to obtain a 15- to
16-bit overall accuracy. The calibration of the main D/A converter plus the
sub-D/A converter is performed by comparing every major carry transition
of the main D/A converter. The calibration principle is shown in Figure
7.20. In the first calibration cycle as shown in Figure 7.20(a) the most significant capacitor C_{512} is compared with the sum of the residual capacitor

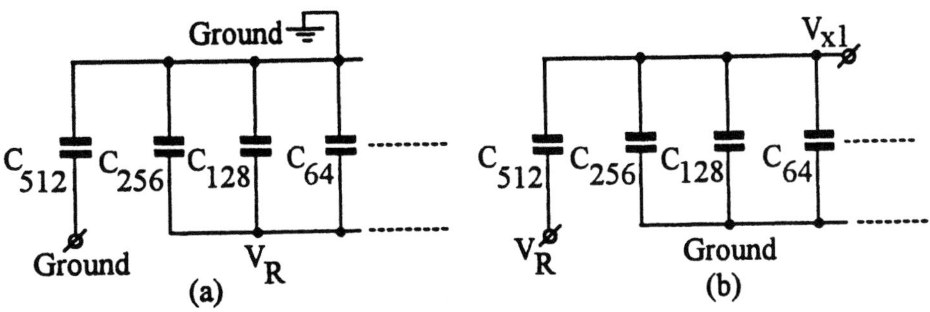

Figure 7.20 : C-2C calibration principle

values $C_{residual}$.

$$C_{residual} = C_{256} + C_{128} + C_{64} + ...C_1 + C_1. \tag{7.9}$$

In the ideal case the value of the capacitor $C_{residual}$ is equal to C_{512}. The deviation from the ideal value can be measured by using a charge redistribution technique. This is performed by reversing the reference voltage across the capacitor network. In the top Figure 7.20 the reference voltage is applied to the bottom plates of the residual capacitors $C_{256}, C_{128}, ..C_1$ and the voltage across C_{512} is made zero and this capacitor is connected to ground. In the lower Figure 7.20 ground (GND) and V_R are exchanged. As a result the voltage at terminal V_{x1} in case of ideal matching must be zero. With mismatches in capacitor values a voltage at V_{x1} appears that is a linear measure for the capacitor mismatch. With use of a successive approximation conversion cycle the voltage V_{x1} is measured and the data are stored in a random access memory (RAM). The sub-D/A converter is used during this error search. During subsequent cycles the remaining capacitors of the capacitor D/A converter are calibrated and the errors are stored in the RAM. To overcome problems with noise during calibration, a number of error cycles are averaged. Furthermore, a longer word length in the error correction system is used to avoid the surpassing of the accumulated correction errors of the half LSB linearity error. Usually two extra bits are sufficient. When the calibration cycle is finished, then the normal successive approximation conversion is performed. Every time a bit value is "true," then the calibration signal accompanying this bit is called from the memory and via the calibration D/A converter the correction signal is added. In this way it is possible to extend the linearity of a converter into the 16- to 18-bits linearity range. At predetermined time intervals the calibration of the system is repeated and new data are stored in the calibration data register. The only disadvantage of this system is that during calibration the A/D converter is not able to operate. In certain systems this cannot be tolerated.

7.9 Conclusion

In this chapter some basic circuits have been presented. Systems with low conversion speed can use integrating circuit to obtain a high-accuracy voltage-to-time conversion. This time is converted into a digital number by simply counting pulses. Accuracies up to 20 bits are possible. A good compromise between speed and accuracy is obtained by the successive approximation system. Maximum resolution is determined by the bandwidth and the noise generated in the comparator circuits. A practical limitation is about 16 bits with 100 kHz conversion rate. The use of multi-level comparator increases the speed at the cost of power dissipation and die size. The ease of implementation of sample-and-hold amplifiers results in simple algorithmic analog-to-digital converters. The most difficult part in such a system is the accuracy with which the times two amplification with addition or subtraction is performed. A special Redundant Signed Digit algorithm does not needs a low-offset requirement for the comparators.

Finally, self-calibration systems make a circuit less sensitive to component variations; however, during warming up of the system a re-calibration might be needed. A very high accuracy, between 16 to 18 bits, is possible with such a calibration system. However, during this calibration cycle the system is not able to produce an output signal. This might be a large drawback for many systems applications.

Chapter 8

Sample-and-hold amplifiers

8.1 Introduction

In this Chapter sample-and-hold amplifiers are discussed. In most cases practical implementations of these amplifiers are track-and-hold amplifiers. During the sampling mode the input signal is "tracked" and a sample is taken at the moment the system is switched into the hold mode. The terminology sample-and-hold amplifier will be used in this chapter although most systems are track-and-hold amplifiers.

A sample-and-hold amplifier is a crucial part in a high-resolution A/D converter system. Using a very simple model of a sample-and-hold amplifier, a number of specifications will be analyzed and the optimum switch configuration will be defined. Next, different circuit configuration options for sample-and-hold amplifiers will be discussed. An important factor is charge feed-through when a sample-and-hold amplifier is switched from the track mode into the hold mode. Furthermore, this charge feed-through must be signal level independent to avoid distortion. Single ended circuit implementations are limited by charge feed-through. An optimum circuit configuration is obtained with a differential system implementation. In an MOS technology switch charge feed-through can be minimized while in a bipolar technology a droop rate compensation is obtained in case no JFETS or MOS devices are available for implementation in the hold amplifier. In a bipolar system, however, the implementation of a power switch is a problem. An example of a switch implemented as a class-B output stage will be shown. This switch can be optimized to have a very low charge feed-through that is very competitive with MOS implementations. Overall A/D converter system performance, such as dynamic range, distortion, and noise, is largely dependent on the sample-and-hold amplifier. Examples of differential sample-and-hold circuits in MOS and bipolar technologies will be discussed. Furthermore an example of a high performance single ended implementation will be shown.

Monolithic versions of sample-and-hold amplifiers are limited available. A practical example with a 12-bit performance is described in reference [75]. This design does not have the performance required in a digital audio system. Special optimized versions that can, for example, be used in high-dynamic-range digital audio systems with sufficiently low distortion figures are not widely available on the market. A bipolar version of a sample-and-hold amplifier that meets the high demands for digital audio is described in this chapter. See reference [4] for more information. The circuit uses high-speed operational amplifiers combined with a specially designed switchable class-B output stage to perform the power switching operation. This switch is optimized to have a low track-to-hold step.

8.2   Basic sample-and-hold configuration

The simplest form of a non-inverting sample-and-hold circuit consists of a switch S with ON resistance R_s and the hold capacitor C_H. At the input of the circuit a voltage source V_{in} is applied. A circuit with basically an infinite input impedance must be used to sense the output voltage V_{out} available across the hold capacitor. In this simple system this "hold" mode amplifier is not shown. The information is stored as charge on the capacitor C_H. To avoid voltage droop a low leakage capacitor must be used with a small dielectric adsorption. The dielectric adsorption can result in voltage changes during the hold mode. As an example, a fully discharged capacitor may still show a voltage across the terminals after the short circuit switch, used to discharge the capacitor, is opened. The basic system is shown in Figure 8.1. This simple circuit can be used to determine some of the most important limitations of a sample-and-hold system.

8.2.1   Signal bandwidth

The small signal bandwidth of the system is determined by the ON resistance of the switch R_s optionally enlarged with the output impedance of the signal

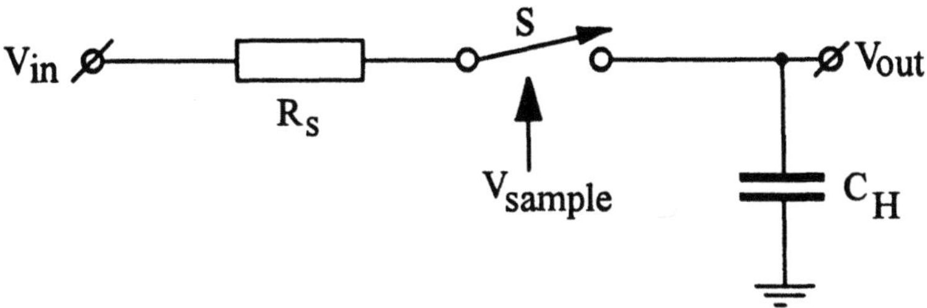

Figure 8.1 : Basic sample-and-hold configuration

source driving this system. We obtain for the small signal bandwidth:

$$f_{-3\ dB} = \frac{1}{2\pi R_s C_H}.$$ (8.1)

Suppose that the maximum input signal voltage the system can operate with is V_{max} then the large signal bandwidth is equal to the small signal bandwidth when the current through the switch and the current of the input driver stage is equal to:

$$I_{max} = \frac{V_{max}}{R_s}.$$ (8.2)

When the input driver cannot supply enough current or the ON resistance of the switch R_s is dependent of the signal, then a smaller large signal bandwidth will be obtained. Mostly the switch resistance R_s increases when the input signal reaches the maximum supply voltage of the system. The smaller current results in a "slewing" of the signal across the capacitor during the track mode. The maximum rate of change during the "slewing" is equal to:

$$\frac{dV}{dt} = \frac{I_{source}}{C_H}.$$ (8.3)

Here I_{source} is the maximum current the source can apply to charge the capacitor C_H.

8.2.2 Acquisition time

The acquisition time of the sample-and-hold circuit can be calculated by applying a full-scale step to the system. The output signal becomes:

$$V_{out} = V_{in}(1 - e^{\frac{-t}{R_s C_H}})$$ (8.4)

Accuracy	Time constants $R_s C_H$
10 %	2.3
1 %	4.6
.1 %	6.9
.01 %	9.2
.001 %	11.5
.0001 %	13.8

Table 8.1 : Accuracy as a function of acquisition time

Equation 8.4 is very easy to analyze in regard to the final settling. In Table 8.1 the settling time as a function of the final accuracy is shown. The results of Table 8.1 are valid as long as the input signal source can deliver the maximum current defined by equation 8.2. When the signal source or the switch cannot deliver such a large current, then the acquisition time will increase. The acquisition time in this situation is equal to the "slew" time plus the settling time during the non-slew operation of the system.

8.2.3   Aperture time accuracy

In the circuit shown in Figure 8.1 the switch S is controlled by the sample pulse. In an ideal case this pulse has infinitesimal small fall times. However, in a practical application a finite rise and fall time of the sampling pulse is found. When we assume, furthermore, that the switching from track to hold occurs at the moment the analog input signal and the sample signal are equal, then an alteration of the sample moment is found. This sample moment alteration is previously defined as aperture uncertainty time. In Figure 8.2 this phenomenon is shown as Δt_s. As a result of this signal-dependent sampling, the aperture time of the sample-and-hold circuit is not constant. Sampling of the input signal is not performed at equidistant time intervals, resulting in distortion after the signal is reconstructed at equal time intervals. This problem can be overcome by bootstrapping the switch control signal with the analog input signal. The switch is always controlled at equidistant time intervals. Another solution is obtained by changing the system implementation in such a way that the switch is always activated around a fixed bias point, which preferably is ground level. A

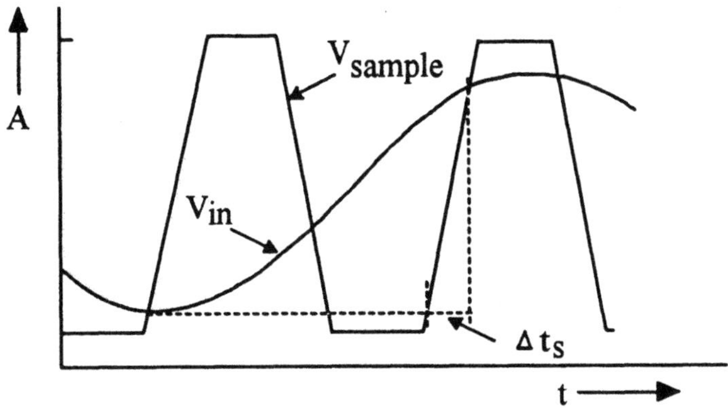

Figure 8.2 : Signal-dependent sampling moment

further advantage of this switch operation is found in a signal independent feed-through of the control signal on the "hold" signal. Furthermore, the switch performance is not influenced by the back-bias of the substrate which is the case in a bootstrapped system configuration.

8.2.4 Sample and hold mode errors

In this simple example an ideal switch is used. As a result, no track mode offset is introduced. Furthermore, at the moment the switch is opened no hold mode error is found. When practical switches are used, the following will be found:

1. Bipolar technology
   During track mode a switch offset can be introduced resulting in a track mode offset.
   Track-to-hold step is obtained by feed-through of the sampling pulse and possible charge stored in the device.

2. MOS technology
   No offset during track mode. A MOS device in the triode region behaves like a resistor. In a switch the MOS device is mostly used in the triode region.
   Track-to-hold step introduced by feed-through of the sampling pulse and channel charge in the device. The size of the step depends on the speed the switch is turned off.

8.2.5   MOS switch charge injection

To a designer it is important to know what effects are causing the charge feed-through and charge injection of MOS devices when used as a switch. In Figure 8.3 a simplified model of a sample-and-hold is shown [109,110]. The system consists of the input signal source V_i with source resistance R_i

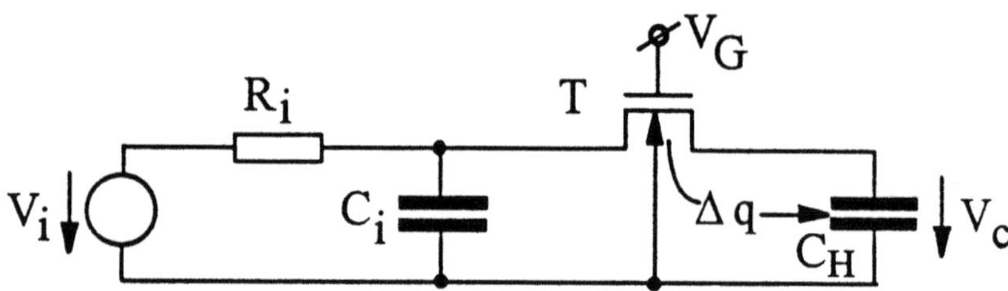

Figure 8.3 : Simplified model of a MOS sample-and-hold

and source capacitance C_i, the MOS switch T, and the hold capacitor C_H. C_i includes the junction capacitance of the drain or source of the device. Turning ON of the switch is obtained when the gate voltage V_g equals V_{gon}. Switching OFF occurs when the gate voltage is below the threshold voltage V_{th} of the device. In this case sub-threshold effects are neglected.

At the moment a device is turned off, then the gate charge Q_g must be released. The amount ΔQ_g that flows to the hold capacitor C_H causes the charge feed-through, resulting in a track-to-hold step $\Delta V_H = \frac{\Delta Q_H}{C_H}$. In [109] is has been shown that the amount of charge that flows to the substrate can be neglected in most practical cases.

To obtain an accurate sample-and-hold system the switch-off time of the MOS transistor is much smaller than the time constant R_iC_i. In this case a small aperture time is obtained. This assumption simplifies the analysis of the problem. In Figure 8.4 a distributed model for the MOS device embedded as a switch in the sample-and-hold system is shown. The total gate capacitance C_{GC} of the device is built-up from a distributed gate capacitance (total value WLC_{ox}) and two overlap capacitances C_{ov}. In a symmetrical device both overlap capacitors are equal. W is the width of the device and L is the length. Furthermore, in this model g_{ch} is the channel conductance of the device. The channel conductance g_{ch} can be expressed in device parameters resulting in:

$$g_{ch} = \beta(V_g - V_{Te}),$$

(8.5)

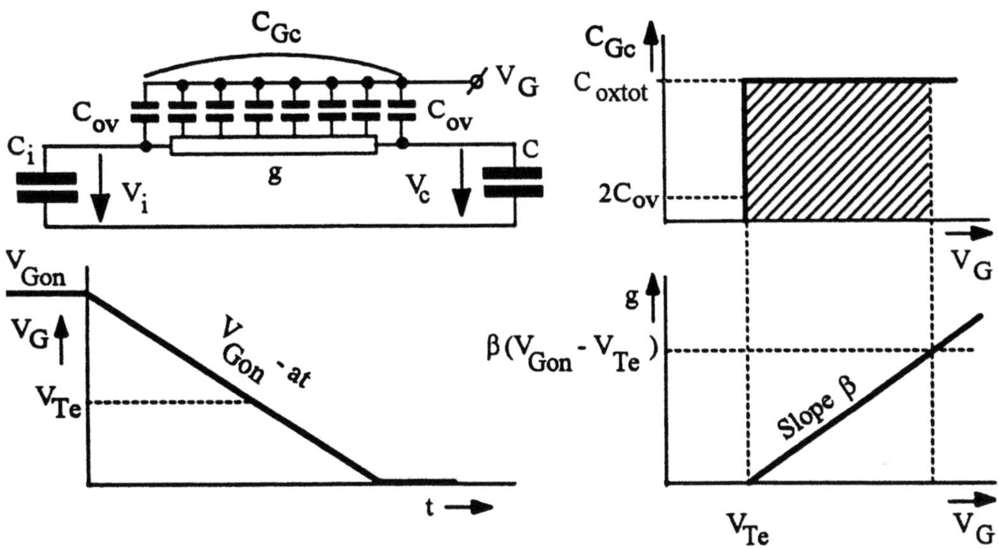

Figure 8.4 : Distributed MOS sample-and-hold model

with

$$\beta = \mu C_{ox}\frac{W}{L} \tag{8.6}$$

and

$$V_{Te} = V_{T0} + bV_{CH_n}. \tag{8.7}$$

Here V_{Te} is the effective threshold voltage of the device. This threshold voltage is influenced by the bulk modulation effect b and the voltage across the hold capacitor V_{CH_n} during the track mode. When the switch is ON, then the total gate capacitance C_{GC} is equal to $WLC_{ox} + 2\,C_{ov}$.

At the moment the switch is OFF, the total gate capacitance C_{GC} becomes $2\,C_{ov}$, because g = 0.

Assume, furthermore, that the gate voltage linearly decreases with time starting from the ON voltage V_{GON}:

$$V_G = V_{GON} - at. \tag{8.8}$$

The slope a of the gate control pulse can be expressed in fall time t_{fall} (10 % to 90 %) and maximum amplitude V_{max} or:

$$a = \frac{0.8V_{max}}{t_{fall}}. \tag{8.9}$$

Inserting 8.9 into equation 8.8 gives:

$$V_G = V_{GON} - \frac{0.8V_{max}}{t_{fall}}t. \tag{8.10}$$

Suppose that due to charge feed-through on the hold capacitor the voltage change ΔV_{CH} is small with respect to $V_{GON} - V_{Te}$, then the potential at each end of the channel conductance g may be considered equal. When, furthermore, it is assumed that the voltage drop across the channel is neglectable even during the decrease of the gate voltage, then the model of Figure 8.4 can be decomposed into the model shown in Figure 8.5 The validity of

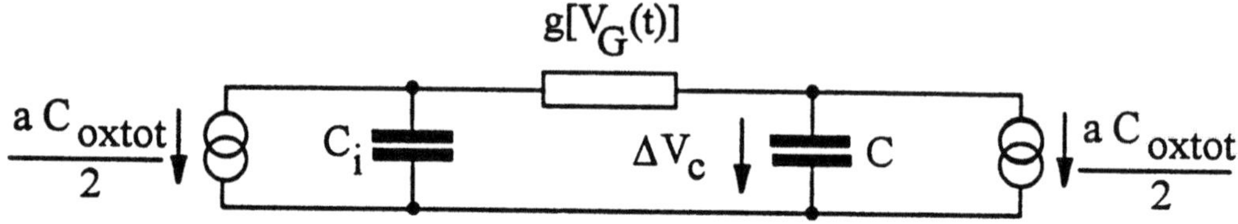

Figure 8.5 : Decomposed sample-and-hold charge injection model

this model has been proven in [109] and [110]. The linear decrease of the gate voltage V_G from V_{GON} across C_{GC} is thus equivalent to current sources $\frac{aC_{GC}}{2}$ flowing at both ends of the channel. Now the following differential equation holds for the model of Figure 8.5 including 8.5 and 8.8:

$$\frac{dU}{dT} = (T - B)[(1 + \frac{C}{C_i})U + 2\frac{C}{C_i}T] - 1. \tag{8.11}$$

The following normalized variables are introduced:

$$U = \frac{\Delta V_{CH}}{\frac{C_{ox}}{2}\sqrt{\frac{a}{bC}}} \qquad \text{(normalized voltage error)} \tag{8.12}$$

$$T = \frac{t}{\sqrt{\frac{C}{a\beta}}} \qquad \text{(normalized time)} \tag{8.13}$$

and

$$B = (V_{GON} - V_{Te})\sqrt{\frac{\beta}{aC}} \qquad \text{(switching parameter).} \tag{8.14}$$

For U = 0 and T = 0, the numerical integration of 8.11 from T = 0 to T = B yields the value for U until the transistor is switched off ($V_G = V_{Te}$).

The value of $\frac{\Delta Q_H}{Q_H}$ can then easily be deducted from U. The value of $\frac{\Delta Q_H}{Q_H}$ is plotted in Figure 8.6 as a function of the switching parameter B with the ratio $\frac{C}{C_i}$ as a parameter. The first design choice a designer can make is:

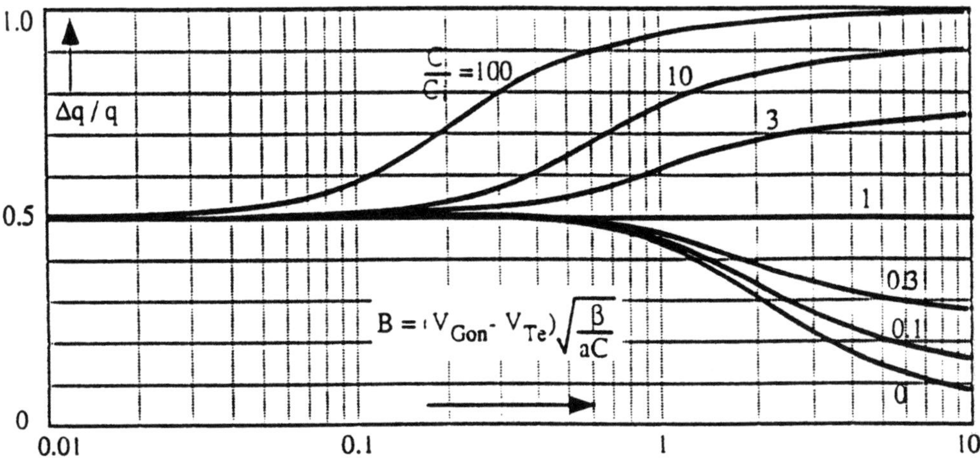

Figure 8.6 : Hold charge injection plot

$$\frac{C}{C_i} \ll 1 \qquad \text{and} \qquad B \gg 1. \tag{8.15}$$

This choice results in most of the charge to flow back onto C_i. The large B is obtained by using a small fall time for the sampling pulse. At the moment the device is switched off, a residual effect due to the overlap capacitance C_{ov} remains. This overlap results in a small charge injection on the hold capacitor C_H. A minimum over-drive of the gate voltage below V_{Te} minimizes this charge injection.

A second design choice is to choose a symmetrical (mostly differential) circuit configuration:

$$C = C_i. \tag{8.16}$$

A fully symmetrical device with equal overlap capacitances is needed. As a result of this operation, half the charge is stored on the input capacitance C_i and the other half is stored on the hold capacitor C_H. A compensation of the charge feed-through is possible by introducing at both ends of the switches dummies that are half the size of the switching device. In Figure 8.7 this compensation scheme is shown. The half-sized dummy switches have

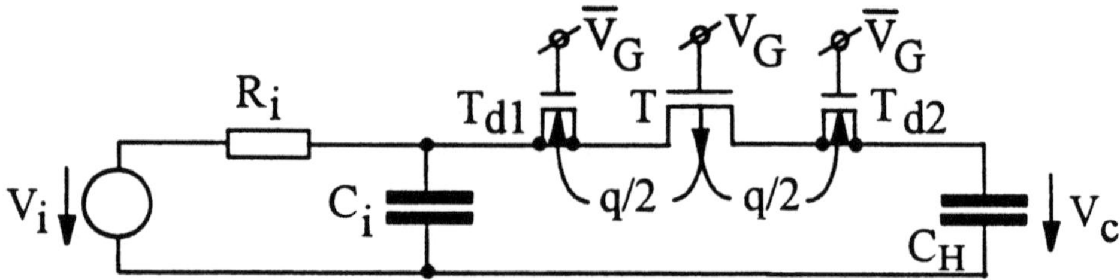

Figure 8.7 : Charge-injection compensation with half-sized dummy switches

an opposite clock signal with respect to the main switch. As a result, half the charge will flow through the dummy switches, thus compensating the charge injection into C_i and C_H. Basically under this condition the injection is not strongly dependent on the variable B. However, mismatches in the circuitry have less influence on the charge division at the moment a slow pulse (small B) is used. In section 8.5.4 an example of a full differential implementation of a sample-and-hold amplifier in MOS technology is shown.

8.2.6   Noise in sample-and-hold circuits

The noise in the system originates from the ON resistance of the switch and the source resistance of the input source. Both resistances are incorporated into the resistor R_s shown in Figure 8.1. The noise generated by this resistor is equal to:

$$\bar{e}^2 = 4kTR_s\delta f. \tag{8.17}$$

The total amount of noise generated across the capacitor C_H at the output of the sample-and-hold circuit is found by integration of the following equation:

$$\bar{e}^2_{total} = 4kTR_s \int_0^\infty \frac{1}{1 + (\omega R_s C_H)^2} df. \tag{8.18}$$

Working out this equation we obtain:

$$\bar{e}^2_{total} = \frac{kT}{C_H}. \tag{8.19}$$

From equation 8.19 it is found that the noise does not depend on the bandwidth but is determined by the value of the hold capacitor C_H. To obtain the given noise calculation results some practical assumptions have been made.

At first the acquisition time of a sample-and-hold amplifier must be small compared to the sampling time. This assumption results in the fact that switching the system from the hold mode into the track mode, only a short time is needed before the signal is settled into the "steady state mode." As a result of this settling, the input signal is applied to a low-pass filter consisting of $R_s C_H$. When the sampling operation is performed, this operation is performed on the filtered input signal. At the moment signals such as noise with a much larger bandwidth than half the sampling frequency are applied, then these signals are folded back into the base band. In case of noise, the amount of noise over half the sampling frequency is reduced, but due to the folding operation the total amount of noise increases with the same factor as the reduction over half the sample frequency band. As a result the total amount of noise is equal to the wide-band noise as is given by formula 8.19.

Using power definitions, the result of equation 8.19 can be correlated to the signal power using:

$$P_{signal} = \frac{1}{2} C_H V_{in}^2 \tag{8.20}$$

$$P_{noise} = \frac{1}{2} C_H \left(\frac{kT}{C_H} \right) = \frac{1}{2} kT. \tag{8.21}$$

V_{in} is the amplitude of the input sine wave.

The signal-to-noise ratio becomes:

$$S/N_{power} = \frac{P_{signal}}{kT/2}. \tag{8.22}$$

Equation 8.22 shows that for a defined signal-to-noise ratio in the system a minimum signal power is required.

At the moment a "hold" amplifier is added to the system, then the noise of this hold-amplifier must be added to the noise given by equation 8.19. This hold-mode noise is time continuous and the noise bandwidth is equal to $\frac{\pi}{2}$ times the small signal bandwidth of this hold amplifier.

8.3 Generalized non-inverting configurations

In this section different circuit implementations are shown based on the basic circuit from Figure 8.1. Mostly these circuit implementations make the system independent of the output impedance of the input signal source and show a low-output impedance to drive a low-impedance load resistor.

8.3.1　Double-buffered sample-and-hold circuit

In Figure 8.8 a double-buffered system is shown. The system consists of two

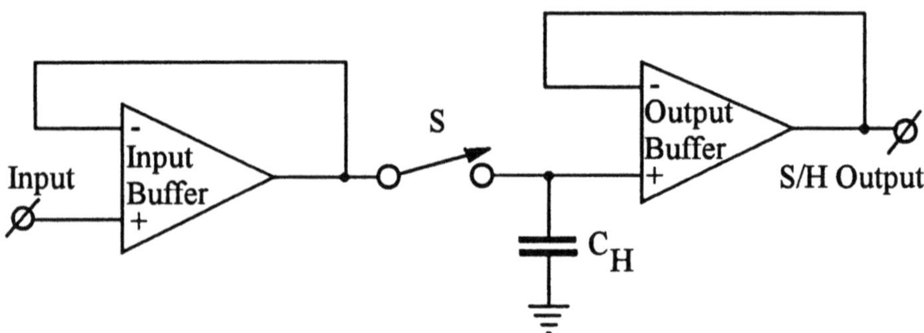

Figure 8.8 : Double-buffered sample-and-hold circuit

unity gain operational amplifiers that act as buffer, a switch S and the hold capacitor C_H. The input current of the output buffer amplifier must be very low to obtain a low droop rate during hold mode. The input buffer is used to supply a large charging or discharging current to the hold capacitor. The overall accuracy of the system is determined by the gain of the operational amplifiers and distortion of the system is determined by the linearity of these amplifiers. Offsets of both amplifiers will add, while charge feed-through of the switch determines the track-to-hold step. Stability of this system in determined by the individual amplifier stabilities and is not influenced by the system configuration.

8.3.2　Feedback improved sample-and-hold circuit

To overcome the accuracy problem of the system shown in Figure 8.8, the gain of the input amplifier-buffer is used to fix the overall accuracy. In Figure 8.9 the basic system configuration is shown. In this system the output buffer amplifier is still in the follower mode, but the input amplifier is used in a feedback mode to obtain during track mode a high overall accuracy. The offset of the output buffer is canceled out by this feedback configuration. Switch offsets during track mode also cancel. The total offset during track mode is only determined by the input amplifier. Common mode rejection of the input amplifier stage must be high to avoid errors. In the hold mode operation of the system the charge feed-through of the switch results in a track-to-hold step. Aperture uncertainty is still determined by the rise or fall time of the sampling clock due to the changing signal levels at the switch.

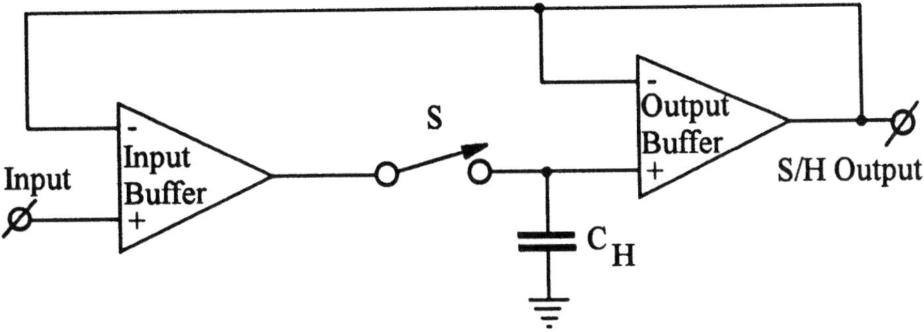

Figure 8.9 : Feedback improved sample-and-hold circuit

Furthermore, additional circuitry is needed to switch off the input amplifier from the input signal source. This is not drawn in Figure 8.9, but will be discussed in practical implementations. System stability is obtained when the unity gain bandwidth of the input feedback amplifier is much smaller than the unity gain of the output buffer amplifier. Furthermore, care must be taken to obtain a large enough phase margin for the total system. A small gain or phase margin results in overshoot during the signal acquisition of the system.

8.3.3 Integrating sample-and-hold circuit

The preferable switch configuration can be obtained by using the gains of both amplifiers in feedback configurations. In Figure 8.10 the circuit configuration is shown [108]. In this system the output buffer is replaced by

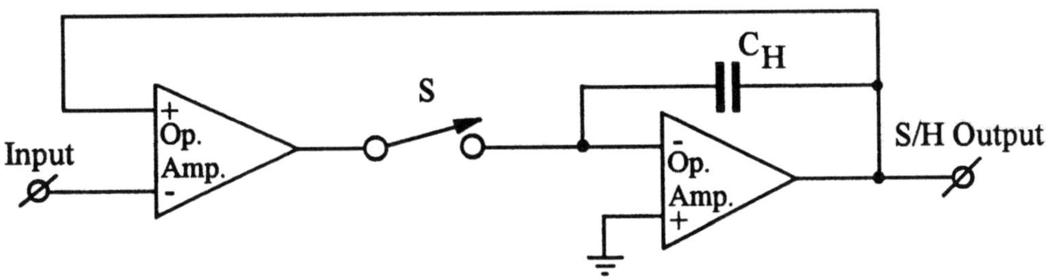

Figure 8.10 : Integrating sample-and-hold circuit

an integrator function. The hold capacitor is connected between the output and the negative input terminal of the output amplifier. The switch drive is around the level of the non-inverting integrator amplifier terminal. Therefore, the switching moment and charge feed-through can be controlled very well. The optimum in switching accuracy can be obtained. The input amplifier is again connected in a feedback loop to increase the overall accuracy of the system. Note, however, that the input buffer is connected as a non-inverting amplifier. This connection is needed because the integrator already introduces an inversion in the loop. Stability of this system is more complex. A more detailed analysis will be given during the discussion of a practical circuit implementation. The only important factor concerning the overall accuracy of the system is the common mode rejection, especially at high input frequencies, of the input amplifier. Again special circuitry is needed to stabilize the input amplifier during the hold mode.

8.3.4   Switched capacitor sample-and-hold circuit

A special switched capacitor sample-and-hold circuit configuration is shown in Figure 8.11. A minimal system consists of three switches, S_1, S_2 and S_3,

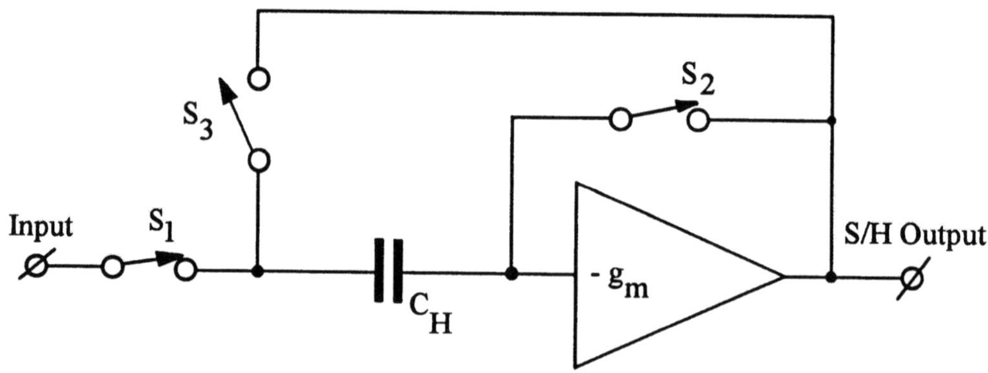

Figure 8.11 : Switched capacitor sample-and-hold circuit

a hold capacitor C_H, and a transconductance amplifier g_m. During the track mode the transconductance amplifier is in a full feedback mode, resulting in a low input impedance $(1/g_m)$. The capacitor is charged to the input voltage using switch S_1 and switch S_2. Bandwidth of this system during track mode is determined by:

$$f_{-3dB} = \frac{1}{2\pi(1/g_m + R_s)C_H}.$$ (8.23)

In this equation R_s includes the "on" resistance of switch S_1 and the signal source impedance.

When switches S_1 and S_2 are opened and switch S_3 is closed, the circuit switches into the hold mode. The stored signal value on the capacitor C_H is connected between the output and the input of the transconductance amplifier. As a result, nearly the same sampled output signal appears at the output terminal. Amplifier offset is automatically canceled. Only a small input signal variation occurs due to the generation of the output signal by the transconductance amplifier. When the gain of this amplifier is large, only a small variation is found. Charge feed-through and aperture uncertainty are determined by the switches. A very simple and compact unit can be designed in this way.

8.3.5   Bi-CMOS non-inverting mode S/H example

As a high-speed, simple-to-implement switch, a diode bridge can be used. The basic circuit implementation of the sample bridge is shown in Figure 8.12 [85]. The system consists of four as a bridge connected diodes, a set

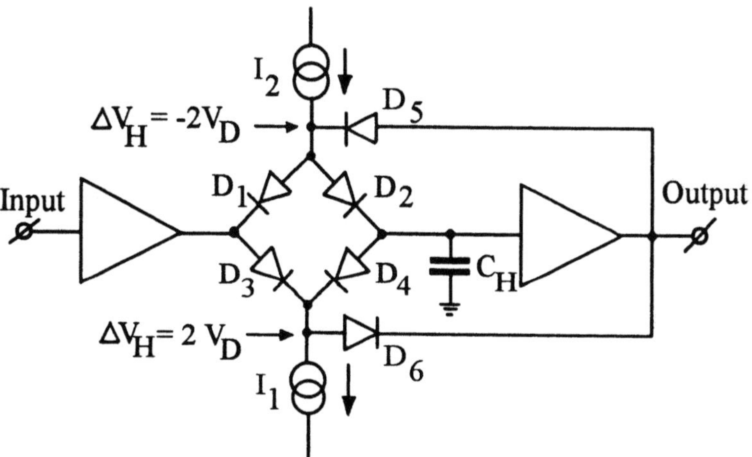

Figure 8.12 : Sampling bridge circuit implementation

of current sources that are switched to the bridge to make it active, a hold capacitor C_H, and an output buffer amplifier. When the currents I_1 and I_2 are switched to the bridge, then the diodes in the bridge are forward biased. With equal diodes and only a small signal current flowing through the bridge diodes, then the voltage level at the input of the circuit is equal to the voltage across the capacitor. A proper biasing of the diode bridge compared to the charging or discharging current of the hold capacitor results in a small distortion of the sample-and-hold amplifier. At the moment the current through the bridge is switched off, the impedance of the diodes becomes very high and the input signal is sampled. The currents I_1 and I_2 are reversed, resulting in a forward biasing of the diodes D_5 and D_6. These diodes are "bootstrapped" with the hold mode output voltage to maintain a constant "off" voltage of the diode bridge equal to about $2V_{diode}$. Capacitive feed-through due to the reverse biased bridge diodes between the input and the hold capacitor is reduced as well. During this "off" mode of the bridge circuit a "large" $(2V_{diode})$ input signal can be applied without disturbing the voltage across the capacitor. Analyzing the circuit shows that under all conditions always a diode is reverse biased. When diodes with low reverse bias currents are used, a low droop rate due to the switch is obtained. In ultra high-speed applications mostly Schottky diodes are used in the sampling bridge, while in a discrete circuit solution the biasing of the bridge is performed with a pulse transformer. A well-known example of this diode bridge is used in sampling oscilloscopes. A practical Bi-CMOS circuit implementation of a diode bridge switch is shown in Figure 8.13. In this system the sample bridge is controlled by a differential amplifier and the biasing sources PM_9 and PM_{10}. At the moment transistor T_2 is conducting, a current I_0 is flowing through the diode bridge. During this time the circuit is in the track mode. At the moment transistor T_1 is conducting, then the diode bridge is reverse biased. Diodes D_5 and D_6 are used to prevent transistor T_1 from bottoming. The voltage level is kept at one diode voltage above the hold signal and one diode voltage below the hold signal. At the output of the system a PMOS differential buffer is used having a very low input bias current. Droop rate, is therefore, low. Bipolar emitter followers in the buffer stage allow a large output current to be supplied to the external load.

8.4   Generalized inverting configurations

A general inverting mode sample-and-hold amplifier configuration is shown in Figure 8.14. The system uses two operational amplifiers (A_1, A_2), two switches (S_1, S_2), a hold capacitor C_H, and two feedback resistors R. The "hold" part of the system is formed by amplifier A_2 with the hold capacitor C_H. During "track" mode the resistors R perform an inversion of the input signal and apply the sampled signal across the hold capacitor C_H. Amplifier A_1 decouples the capacitor C_H from the feedback resistor R, thus increasing

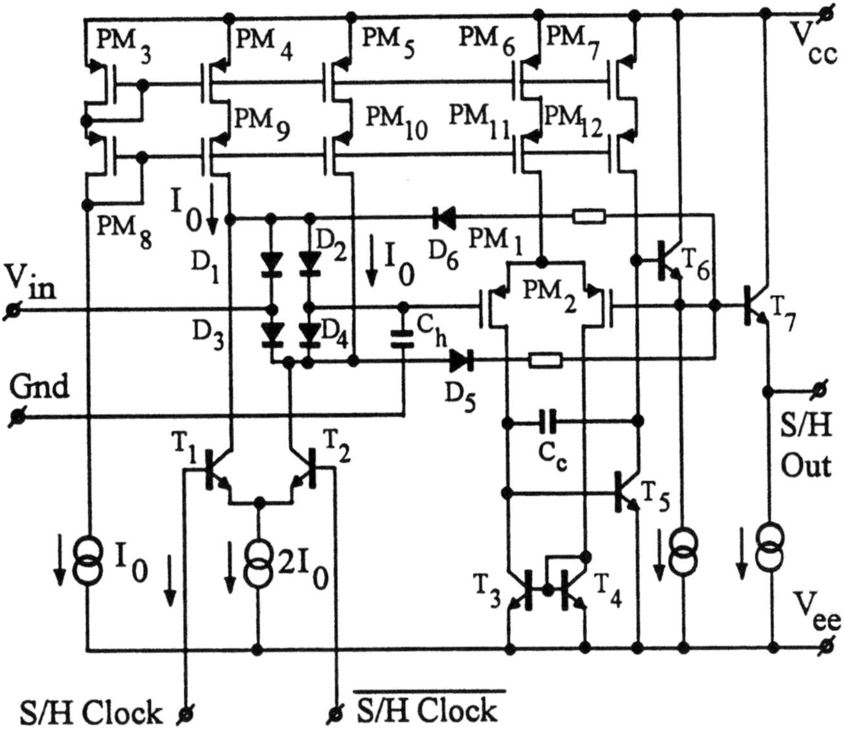

Figure 8.13 : Bi-CMOS S/H circuit implementation

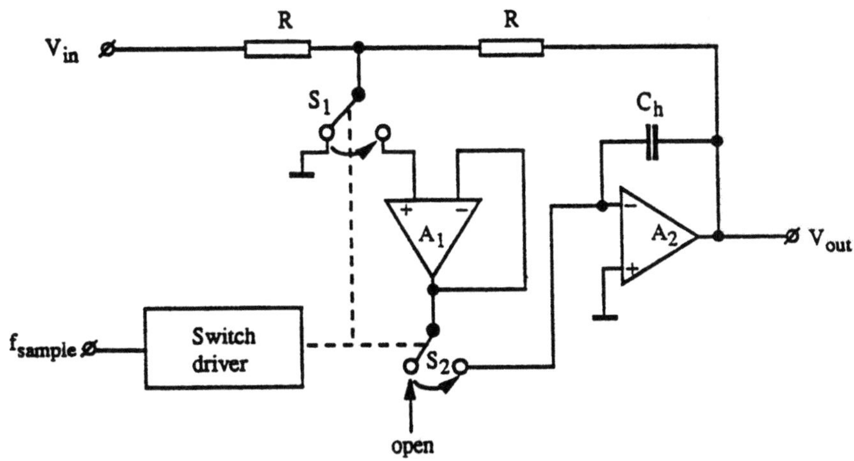

Figure 8.14 : Inverting sample-and-hold amplifier

the bandwidth of the system. This amplifier can also supply a large current to charge or discharge the hold capacitor C_H. In this way a small acquisition time of the system is obtained. Switch S_1 is used to minimize feed-through of the input signal during hold mode. A low signal feed-through is obtained in this way. One disadvantage of this system is that the gain of amplifier A_1 is not used to improve the overall performance of the system. Furthermore, the offset voltages of both amplifiers are additive, as is the voltage drop across the "on" resistance of the switch S_2. This might result in a larger total offset voltage and a larger distortion of the system. In the system that is discussed in the following sections, some of these drawbacks will be resolved, resulting in a high-performance system.

8.4.1   Modified sample-and-hold amplifier

The modified sample-and-hold amplifier that uses the gain of amplifier A_1 to reduce offset and distortion is shown in Figure 8.15. [4]. The circuit

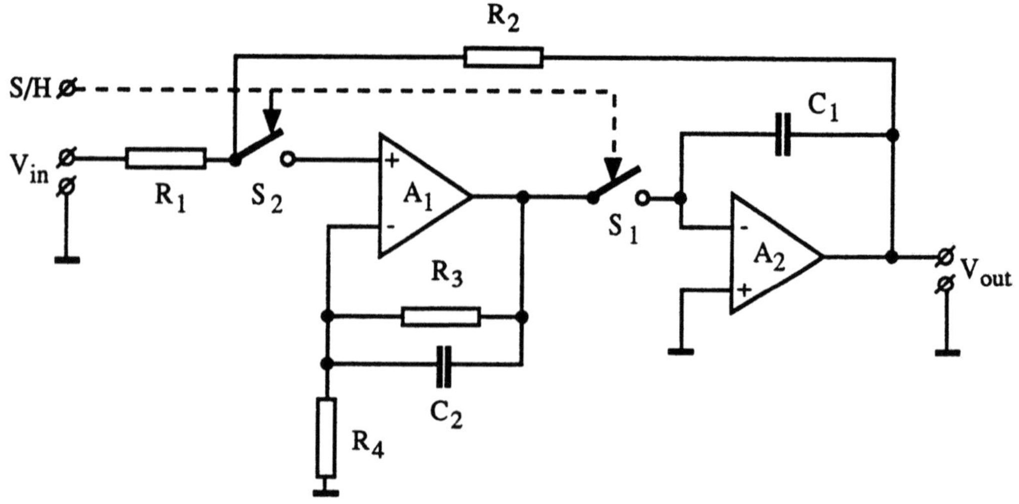

Figure 8.15 : Modified sample-and-hold amplifier

consists of a wide-band operational amplifier A_1 with feedback network R_3, C_2, R_4, and an ultra-low distortion amplifier A_2 with J-FET input devices. Overall gain is determined by the resistors R_1 and R_2. To obtain a high accuracy (for example, $\frac{1}{4}$ LSB of 16 bits), the open-loop gain of the cascaded amplifiers is kept flat over the audio band (20 kHz). To come up with a practical circuit solution a second-order roll-off of the total open-loop gain is the only possibility. At the point where the open-loop frequency characteristic crosses the feedback characteristic, this roll-off is changed into a first-order roll-off to obtain a stable system. The total open-loop amplitude response as a function of frequency is shown in Figure 8.16. The gain of

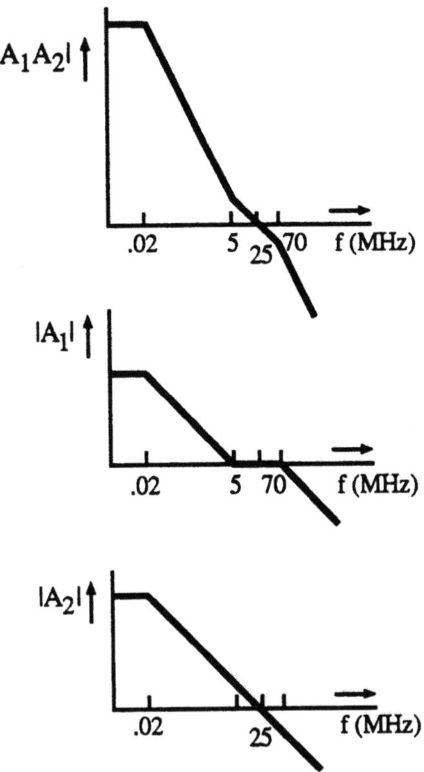

Figure 8.16 : Open-loop amplitude response curves

amplifier A_1 is set at 250 using the feedback network R_3, C_2, and R_4. The gain of this amplifier is set at 1 in the frequency range from 5 MHz to 70 MHz. The unity gain bandwidth of this amplifier is 70 MHz. Furthermore, this amplifier is connected in the non-inverting mode. The open-loop gain of amplifier A_2 is set at 1000, using an internal feedback loop. The unity gain bandwidth of this amplifier is 25 MHz. As a result of this operation, a total open-loop gain of about 2.5×10^5, which is flat over the audio band, is obtained.

In the operational amplifiers a special feed-forward frequency compensation technique is used.

8.5   Differential sample-and-hold configurations

To improve the dynamic performance of sample-and-hold amplifiers, especially concerning parameters like droop rate and switch feed-through, it is
possible to use a differential system operation. In differential systems only
the difference between the (small) errors are important. In practical systems an improvement of at least one order of magnitude is possible. Such
improvements are usually unobtainable by using compensation techniques
with dummy switches and inverted control signals. Furthermore, leakage
currents of capacitors or input currents of amplifiers during the hold mode
cancel for a large amount. The only disadvantage of a differential operation
is the increase of the noise level. However, this disadvantage disappears at
the moment the maximum signal level is increased by a factor two. Dynamic
range under this condition improves with 6 dB.

8.5.1   Bipolar differential sample-and-hold

In Figure 8.17 the principle of a differential sample-and-hold circuit is shown
[102]. The system consists of a differential amplifier with a differential gain

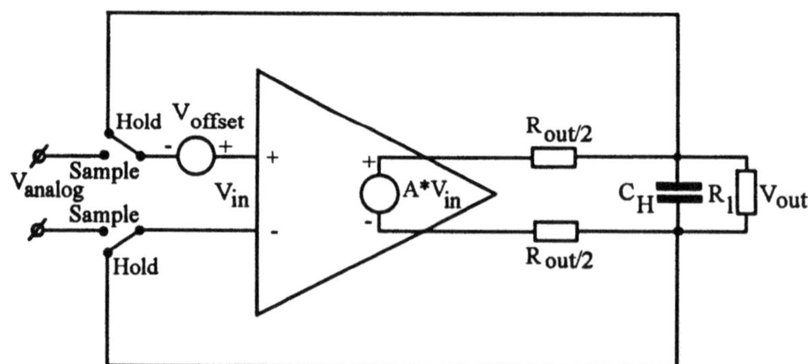

Figure 8.17 : Bipolar differential sample-and-hold circuit

as close as possible to one, a set of switches performing the operation mode,
and the hold capacitor C_H loaded with a resistor R_l. During the sample
mode, the input signal is nearly exactly reproduced across the hold capacitor.
The low output impedance $R_{out/2}$ of the differential amplifier reduces the
acquisition time of the system to $T_{acquisition} = R_{out} \times C_H$. In this case it is
supposed that $R_l \gg R_{out}$. In the hold mode the input signal is disconnected
from the circuit and the output voltage across the hold capacitor C_H is
applied at the input, thus reducing the voltage across the output resistors

$R_{out/2}$. As a result, the loading of the hold capacitor is reduced, giving a small droop rate. Suppose that the gain of the differential amplifier only slightly differs from 1, then putting:

$$\frac{V_0}{V_{in}} = A = 1 + \Delta A. \tag{8.24}$$

Here V_0 is the output voltage of the amplifier.

Inserting $V_{in} = V_{out}$ into equation 8.24 then V_0 becomes:

$$V_0 = V_{out}(1 + \Delta A). \tag{8.25}$$

Using equation the voltage across the resistors $R_{out/2}$ then becomes:

$$V_{Rout} = \Delta A V_{out}. \tag{8.26}$$

Due to the reduction of the voltage across the resistors $R_{out/2}$ with a factor $\frac{1}{\Delta A}$ the droop rate becomes:

$$V_{droop} = \frac{dV_{out}}{dt} = \frac{\Delta A}{R_{out}C_H}V_{out}. \tag{8.27}$$

At the moment an amplifier with a gain of exactly 1 is used, the droop rate of the system is reduced to zero.

In practical systems, however, an offset voltage is usually present. This offset voltage is not reduced. A constant charge or discharge current equal to $\frac{V_{offset}}{R_{out}}$ will flow causing a minimum droop rate:

$$V_{droopminimum} = \frac{V_{offset}}{R_{out}C_H}. \tag{8.28}$$

The charging of the hold capacitor with an impedance R_l can be incorporated in the above given equations by replacing R_{out} by $R_{out} // R_l$. In a practical circuit a gain error $\Delta A = 10^{-3}$ can be obtained resulting in a charge/discharge resistance ratio of 1000. This value is not large enough and can be increased by switching off the output stages of the operational amplifier. At that moment the input bias currents of the amplifier are discharging the hold capacitor. The charge/discharge ratio in a bipolar circuit can be increased to 10^6. A simplified circuit diagram of a practical sample-and-hold amplifier is shown in Figure 8.18 The differential input voltage $A_n, \bar{A}_n$ is sup-

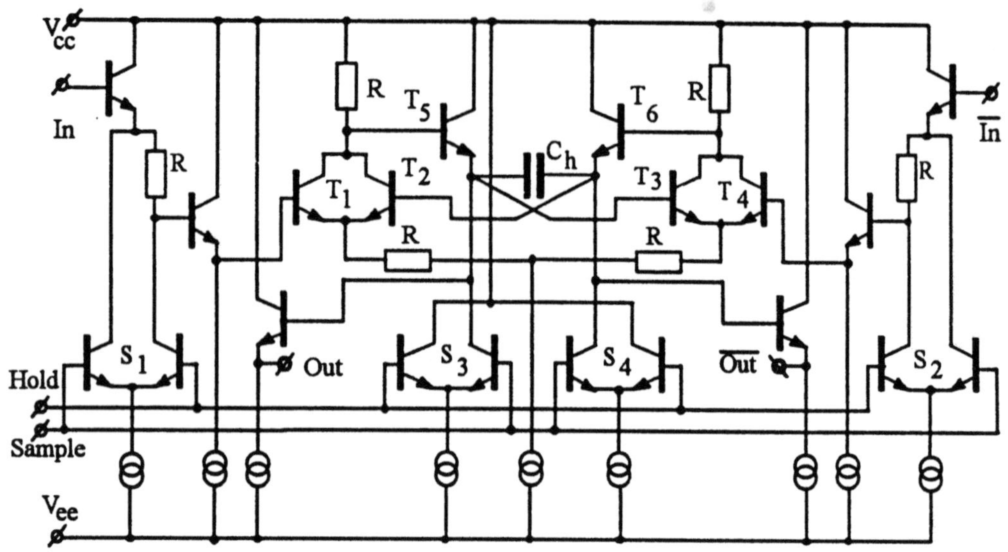

Figure 8.18 : Practical sample-and-hold circuit diagram

posed to be symmetrical around the reference voltage that can be ground. This input voltage passes switchable level shifts introduced across the resistors R_1. During the track mode the voltage drop across the resistors R_1 is zero. At the bases of transistors T_1 and T_4 this level shifted analog input voltage appears. The biasing of the circuit is performed in such a way that during track mode transistors T_2 and T_3 are switched off. The feedback loop from output to input terminals is therefore disconnected. Emitter followers T_5 and T_6 can apply large charge or discharge currents to the hold capacitor C_H. The input signal is tracked across the capacitor with a large accuracy. Switches S_3 and S_4 bias the output emitter followers with a large current, resulting in a small output impedance of followers T_5 and T_6. Switching the circuit into the hold mode introduces a voltage drop across the resistors R_1. This voltage drop switches off transistors T_1 and T_4. At the same time the feedback loop is closed by switching on transistors T_2 and T_3. The voltage across the hold capacitor is now applied to the input of the hold amplifier consisting of T_2, T_5 with the resistors R and T_3, T_6 with the resistors R connected as an accurate one times differential amplifier. At the same time the biasing of the output emitter followers T_5 and T_6 is reduced to zero. The base currents of transistors T_2 and T_3 are the only currents left discharging the hold capacitor. A low droop rate is possible, especially when these transistors consist of so-called Darlington stages. The output buffers, which in the circuit diagram are directly connected to the hold capacitor, can be included in this Darlington stage as well. Due to the differential operation, the difference in input bias currents of transistors T_2 and T_3 results in the droop of the hold signal while the common part of the bias currents is supplied by the buffer transistors T_5 and T_6. As a result of whole this operation, a sample-and-hold amplifier can be designed that uses only NPN devices and resistors.

8.5.2 Low-power bipolar differential sample-and-hold

The basic system implementation of a low-power high-speed bipolar sample-and-hold amplifier is shown in Figure 8.19 [103]. The system consists of a

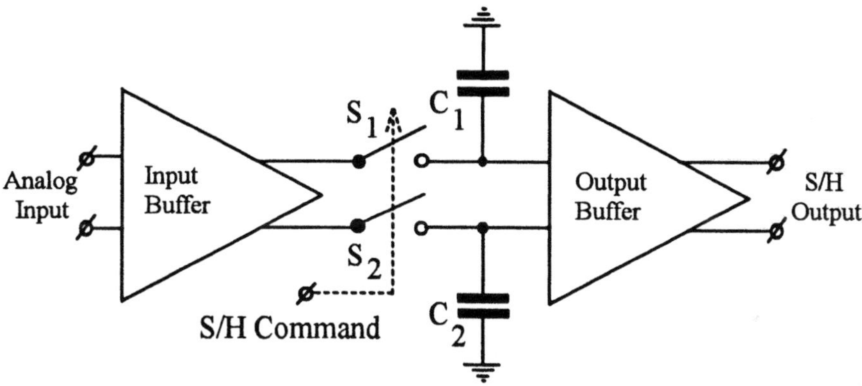

Figure 8.19 : Differential low-power high-speed sample-and-hold system

differential input buffer amplifier with a voltage gain of about one, a differential output buffer with again a voltage gain of about one, two hold capacitors C_1 and C_2, and two switches S_1 and S_2. In the output buffer amplifier, no MOS or JFet devices with an ultra low input current are used. This means that at the moment the system switches from the track mode into the hold mode an input bias current starts to discharge the two hold capacitors. When equal capacitors and equal input bias currents are assumed, then the difference between the voltages across the two hold capacitors remains constant. A good common mode rejection ratio and low and equal input bias currents for the buffer amplifier are needed. The maximum hold time of this system is determined by the input bias currents of the output buffer amplifier and the value of the hold capacitors. Furthermore, when switches with a low and equal charge feed-through are used, the track-to-hold step, which is finally found at the output, nearly vanishes in the differential output signal.

In Figure 8.20 a practical implementation of the sample-and-hold amplifier is shown. The input buffer amplifier consists of the differential amplifier T_1, T_2

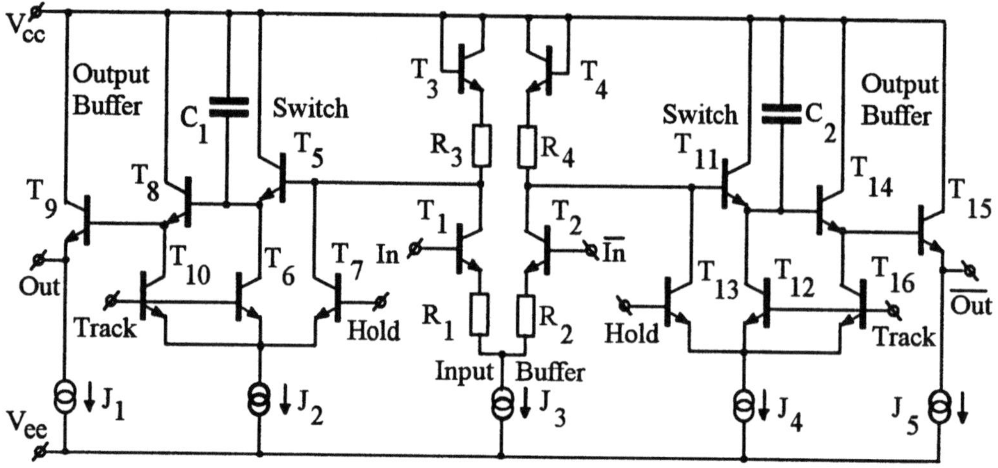

Figure 8.20 : Practical low-power circuit implementation

with emitter degeneration resistors R_1, R_2. The input voltage is converted into a current having a nonlinear transfer function. To compensate for this nonlinearity an equivalent collector load is applied consisting of T_3, T_4, and the resistors R_3 , R_4. With all equal elements an exact compensation for the nonlinear behavior of the system is obtained. As a result, an exact replica of the input differential voltage appears in the collectors of T_1, T_2. The switches consist of transistors T_5, T_6, T_7 and T_{11}, T_{12} ,T_{13} with the tail current sources J_2 and J_4, respectively. During the track mode terminals T are high compared to terminals marked H, half of the currents J_2 and J_4 flow through T_6, T_5 and T_{12}, T_{11}, respectively. Transistors T_5 and T_{11} perform a follower operation and are charging or discharging the hold capacitors C_1 and C_2. The differential operation of the system results in only odd harmonics introduced by the limited linearity of the followers T_5 and T_{11}. The hold mode is obtained at the moment the terminals H become high with respect to T. The currents J_2 and J_4 are switched through T_7 and T_{13} to the collector loads of T_2 and T_1, respectively. As a result of this operation, the base-emitter voltages of transistors T_5 and T_{11} are reversed, switching off the follower operation. The input signal is stored on the hold capacitors C_1 and C_2 at this moment. The output buffers T_8, T_9 and T_{14}, T_{15} apply the held information to the output terminals. At the same time the currents through transistors T_8 and T_{14} are switched off, resulting in a reduction of the base currents of T_5 and T_{11}. The discharge currents of the hold capacitors are largely reduced, resulting in a lower droop rate. A good matching between transistors T_8 and T_{14} is needed to obtain equal base currents for these devices.

At high frequencies the hold-mode feed-through is determined by the ratio of the base-emitter capacitance of T_5 and T_{11} and the hold capacitance $C_1 = C_2 = C_H$. We obtain:

$$A_{feedthrough} = \frac{C_{be.T_5}}{C_H + C_{be.T5}}. \qquad (8.29)$$

The hold-mode feed-through can be reduced by adding cross-coupled compensation capacitors as shown in Figure 8.21. The capacitors C_{FF} are cou-

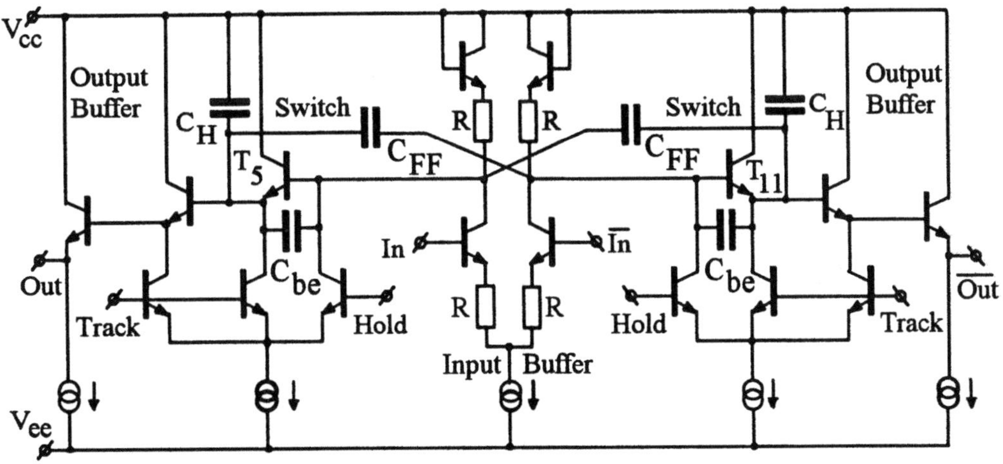

Figure 8.21 : Hold-mode feed-through compensation

pled to the hold capacitors from the opposite signal points as shown in Figure 8.21. As a result the hold-mode feed-through reduces to:

$$A_{compensated} = \frac{C_{be.T_5}}{C_H + C_{be.T_5}}.(1 - \frac{C_{FF}}{C_{be.T5}}). \qquad (8.30)$$

A complete cancellation is obtained at the moment the base-emitter capacitance of T_5 and T_{11} are equal to C_{FF}. A circuit implementation that gives nearly identical capacitance values is shown in Figure 8.22

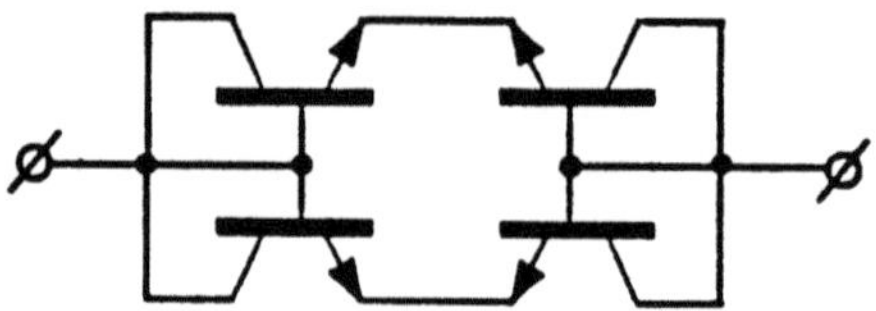

Figure 8.22 : Hold-mode compensation capacitor implementation

Process	Bipolar 3 GHz f_t
Input range (peak-to-peak)	1 V
Low-frequency distortion	< -70 dB
Differential droop rate	2.5 mV/ms
Sample rate frequency range	0.3 - 200 MHz
10-bit full Nyquist sample rate	120 Msample/s
Hold-mode feed-through	< -60 dB
Hold capacitance	10 pF
Supply voltage	-5 V
Power consumption	40 mW
Chip size	0.25 mm^2

Table 8.2 : Sample-and-hold amplifier data

8.5.3   Practical implementation

The circuit has been implemented in a 3 GHz f_t, 2×9 μm^2 minimum emitter size, bipolar process. A differential-to-single-ended converter completes the circuit. This converter is used for measuring purposes and can apply single-ended signals to single-ended analog-to-digital converters. Table 8.5.3 shows the main characteristics of the sample-and-hold amplifier. The differential structure of the sample-and-hold amplifier results in a low-power, high performance circuit needing no MOS or JFet devices in the hold amplifier.

8.5.4   MOS differential sample-and-hold

In Figure 8.23 a basic example of a MOS differential sample-and-hold circuit is shown [106,105]. In principle the first approach consists of two circuits of Figure 8.11 in parallel. Charge feed-through introduced by the sampling switches cancels when a differential output signal is sensed by the analog-

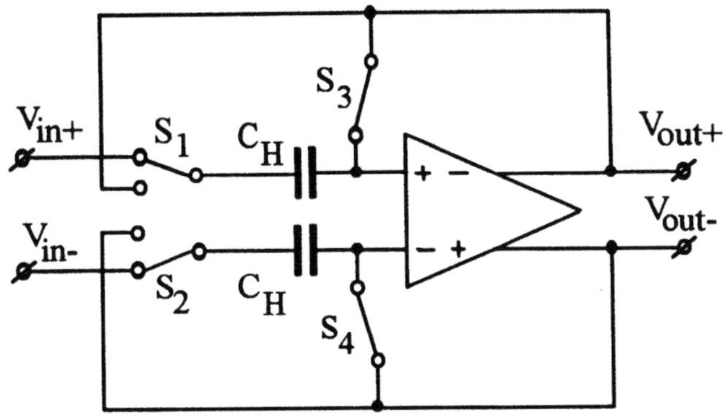

Figure 8.23 : MOS differential sample-and-hold circuit

to-digital converter. A disadvantage of this system is the variation of the input signal from track-to-hold caused by the variation in output voltage during the track and the hold mode. This problem can be overcome by modifying the system into the circuit shown in Figure 8.24. During the

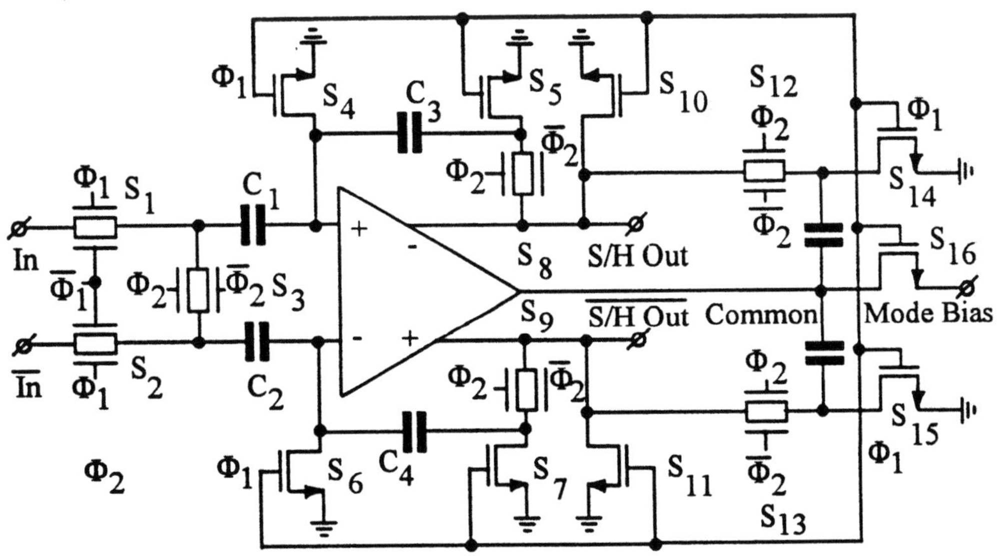

Figure 8.24 : Improved differential sample-and-hold circuit

sampling mode the clock Φ_1 is high closing the accompanying switches. The input signal is sampled on the input capacitors C_1 and C_2. During this phase the output of the differential amplifier is shorted to prevent over-drive of the amplifier. An analog sample is taken at the moment the clock phase Φ_2 is made high. When switches S_1 and S_2 are opened, charge feed-through is stored on the capacitors C_1 and C_2. Furthermore, the opening of switches S_4, S_5 and S_6, S_7 results in feed-through effects that are common mode for hold capacitors C_3 and C_4. Only differences are stored that result from circuit un-symmetries. At the moment switch S_3 is closed, the opposite feed-through charge on capacitors C_1 and C_2 cancels out. As a result, an accurate sample charge is transferred onto capacitors C_3 and C_4. At the same time the high clock Φ_2 closes switches S_8, S_9, S_{12}, and S_{13}. During the high level of Φ_2 the sampled signal is available at the output of the system. A high-performance low-feed-through sample-and-hold amplifier is obtained.

8.6 Bipolar inverting mode S/H example

The circuit diagram of Figure 8.15 requires operational amplifiers for implementation. These operational amplifiers are designed to have a good first-order frequency response with a large phase margin and a maximum attainable bandwidth [4]. Special frequency compensation techniques are used to obtain an optimum bandwidth using a standard bipolar technology. The different methods of frequency compensation are discussed.

8.6.1 Miller integrator frequency-compensation

When operational amplifiers consist of more than one stage, a frequency compensation mechanism is needed to perform a first-order overall amplitude frequency response. In most cases a so-called "Miller Integrator" compensation technique is used. A basic circuit diagram of this compensation system is shown in Figure 8.25 [104]. The system consists of two amplifier

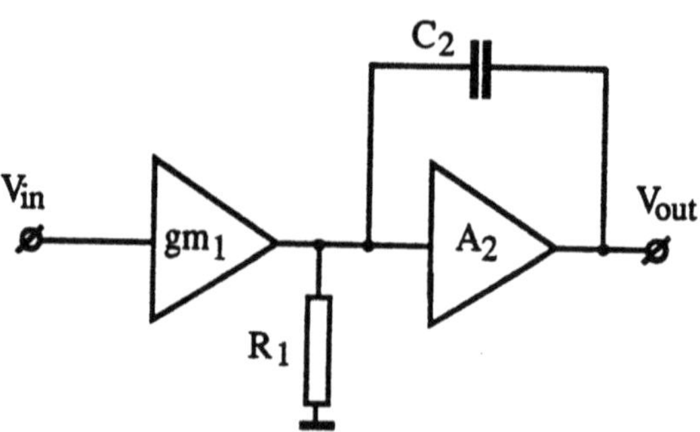

Figure 8.25 : Miller integrator frequency-compensation system

stages. The first stage is a transconductance stage with value gm_1. The second stage is a Miller integrator that converts the output current of the first stage into an output voltage across the compensation capacitor C_2. In standard bipolar processes the input transconductance stage consists of a conglomerate of NPN and PNP transistors to combine a dc level shift with the transconductance operation. These PNP transistors are built in a lateral manner and thus have a small transition frequency in comparison with NPN devices. As a result of this construction, the overall frequency response of the operational amplifier does not show a first-order amplitude frequency response. The lateral PNP transistors add an extra time constant, which results in the dashed line shown in Figure 8.26. A_{20} is the dc gain of the amplifier A_2. In a practical solution using this frequency compensa-

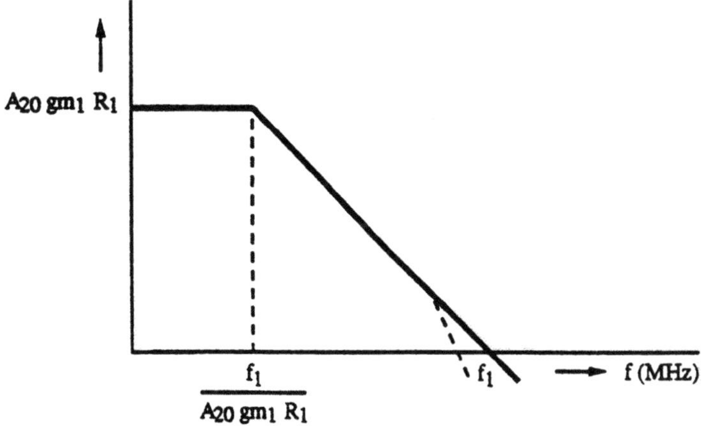

Figure 8.26 : Frequency response of a Miller compensated operational amplifier

tion method a first-order compensated unity gain bandwidth of maximum 5 MHz is possible. However, this bandwidth is not large enough to obtain a building block for a digital audio sample-and-hold amplifier. Therefore, a special compensation method will be used, as described in the next section.

8.6.2 Feed-forward wide-band frequency-compensation

A block diagram of the basic feed-forward frequency compensation technique is shown in Figure 8.27. The circuit consists of a voltage amplifier A_1 followed by a transconductance amplifier with transconductance of gm_2 and loaded with the output impedance consisting of R_L. If no extra measures have been taken the amplitude frequency response of the system shows at high frequencies a second-order roll-off. In stable feedback systems using operational amplifiers a second-order roll-off is not allowed. To overcome

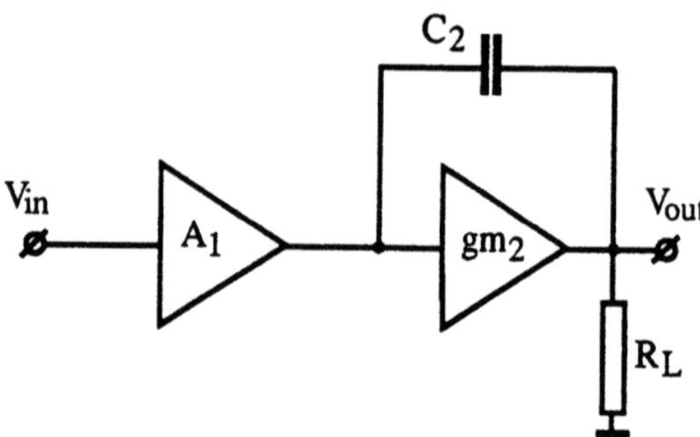

Figure 8.27 : Basic feed-forward frequency compensation technique

this problem, a so-called frequency compensation technique is used which reduces the second-order roll-off at high frequencies to a first-order roll-off. Frequency-compensation techniques result mostly in a much smaller compensated bandwidth than is possible in the basic cascaded two-stage amplifier. The design challenge is to obtain a circuit construction that allows the largest compensated bandwidth possible in a two-stage system.

In the circuit shown in Figure 8.27 frequency compensation is obtained using the capacitor C_2. The operation of this compensation technique can be explained as follows. At low frequencies the influence of the capacitor on the transfer function of the total amplifier can be ignored. A maximum open-loop gain of $A_{openloop} = A_{10}.gm_2.R_L$ is found. Here A_{10} is the dc gain of amplifier A_1. When the signal frequency increases, the capacitor C_2 short-circuits the transconductance stage gm_2 and applies the output signal of amplifier A_1 directly to the output terminal. In this way a first-order amplitude response equal to the high-frequency response of amplifier A_1 is obtained. The unity gain bandwidth depends on the maximum bandwidth of amplifier stage A_1. Note also that the phase of the output signal of A_1 and the phase of the output signal of the transconductance stage are the same. Due to the low output impedance of A_1, no instability can occur in the transconductance stage gm_2 with the capacitor C_2 as a feedback element.

In Figure 8.28 the amplitude frequency response with an exact frequency compensation is shown. A simple calculation gives the following equation for the transfer function with $f_1 = \frac{1}{\tau_1}$ where f_1 is the unity gain bandwidth

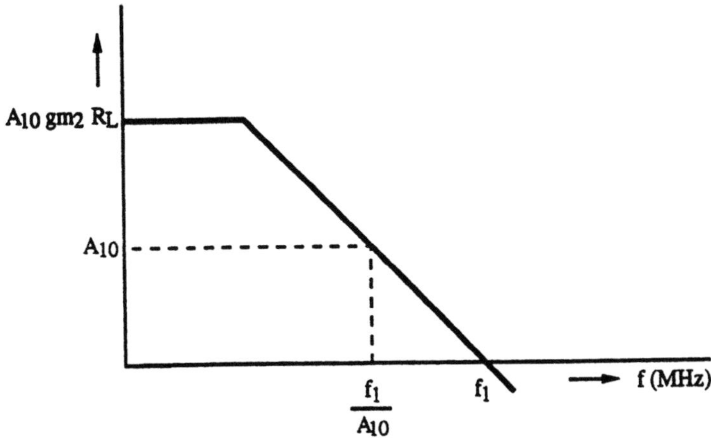

Figure 8.28 : Amplitude frequency response with an exact frequency compensation

of amplifier A_1:

$$\frac{V_{out}}{V_{in}} = \frac{A_{10}gm_2R_L(1 + pC_2/gm_2)}{(1 + pA_{10}\tau_1)(1 + pR_LC_2)}. \tag{8.31}$$

An exact frequency compensation is obtained when:

$$C_2/gm_2 = A_{10}\tau_1 \, or \, C_2 = gm_2A_{10}\tau_1. \tag{8.32}$$

Then:

$$\frac{V_{out}}{V_{in}} = \frac{A_{10}gm_2R_L}{1 + pR_LC_2}. \tag{8.33}$$

In practice, capacitor values ranging from 5 to 15 pF are used, and a bandwidth from 10 to 100 MHz with a large phase margin at unity gain can be obtained. This circuit, however, does not give the lowest distortion. The reason for distortion is found in the nonlinear capacitance at the output of the transconductance stage. The main part of this capacitance is the parasitic substrate capacitance of the devices used in this amplifier stage. This nonlinear capacitance forms along with the capacitor C_2 a nonlinear divider at high frequencies, resulting in signal distortion. A modification of the basic amplifier structure is shown in the following section. First a practical implementation of the simple system is given.

8.6.3 Practical compensated amplifier

A simplified circuit diagram of a practical feed-forward frequency-compensated operational amplifier is shown in Figure 8.29. In the input stage with transistors T_1 and T_2 a resistive load consisting of resistors R_1 and R_2 is used

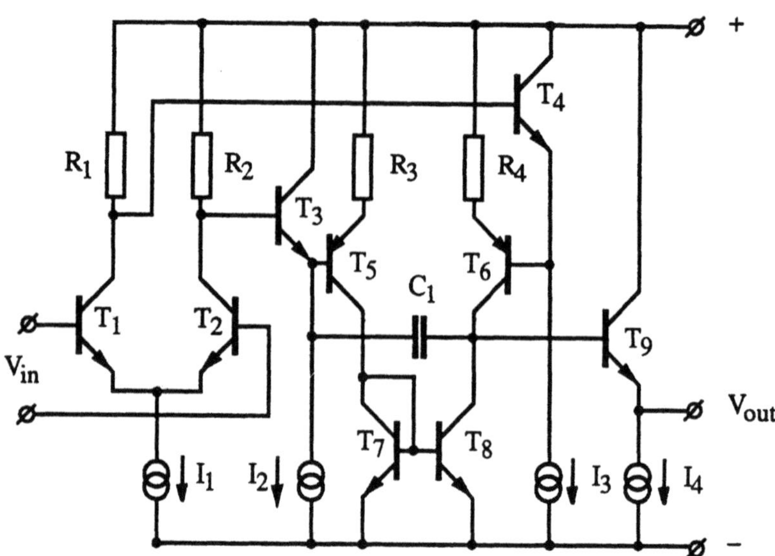

Figure 8.29 : Simplified circuit diagram of a feed-forward frequency-compensated operational amplifier

to obtain a low-noise performance. Emitter followers T_3 and T_4 are further incorporated to obtain the low output impedance needed for frequency compensation. The lateral PNP transistors T_5 and T_6 with the emitter degeneration resistors R_3 and R_4 and an active load consisting of T_7 and T_8 form the transconductance stage.

The capacitor C_1 short-circuiting the transconductance stage at high frequencies gives the necessary frequency compensation. An additional advantage of this method of compensation is the low output noise of the total system at frequencies above the unity gain bandwidth of the amplifier. The emitter follower T_9, which is normally a class-B output amplifier, matches the amplifier output impedance to the externally applied load.

Measurement results of amplitude and phase on a practical integrated amplifier are shown in Figure 8.30. A phase margin of 45° at unity gain frequency is found. The bump in the amplitude curve is due to the combination of a low transition frequency of the lateral PNP transistors and the high takeover frequency of the capacitive feed-forward compensation, thus leaving a void in between. This high frequency has been chosen to avoid slow-settling parts in the pulse response of the amplifier due to inaccuracy in the frequency

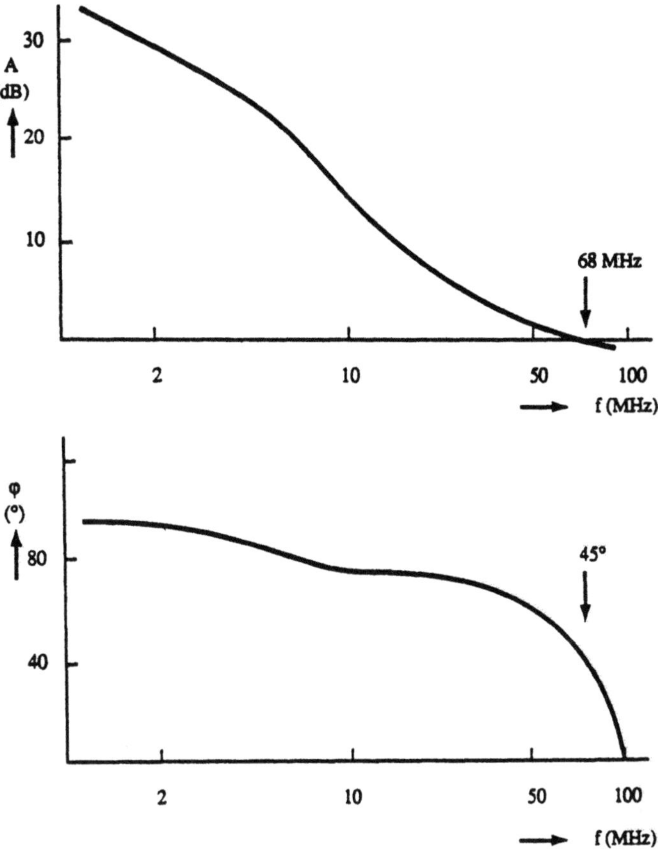

Figure 8.30 : Amplitude and phase measurement results of a practical amplifier

compensation elements. In the final sample-and-hold application that uses this amplifier, a negligible effect on the total performance is found.

8.6.4   Low-distortion frequency-compensation

The block diagram of the low-distortion feed-forward frequency compensation is shown in Figure 8.31. The system consists of a voltage amplifier A_1 followed by a transconductance stage gm_2 with load resistor R_L and the output voltage amplifier A_3. Frequency compensation is performed using capacitors C_2 and C_3. The operation of the circuit can be explained as follows. At low signal frequencies the influence of the frequency-compensation capacitors C_2 and C_3 on the transfer function of the amplifier can be neglected. The overall dc voltage gain then becomes: $A_{openloop} = A_{10}gm_2R_LA_{30}$. Since

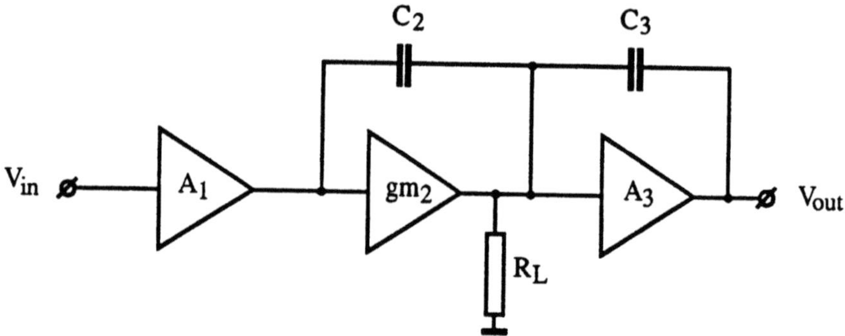

Figure 8.31 : Low-distortion frequency-compensation diagram

the amplifier A_3 with the capacitor C_3 operates as a Miller integrator, the output current of the transconductance stage gm_2 is converted into a voltage across C_3. At the same time, however, the capacitor C_2 short-circuits the transconductance stage and forms with C_3 and A_3 an inverter stage with a gain equal to the ratio between C_2 and C_3. Due to the Miller integrator, a low distortion is obtained because the maximum output voltage swing has to be generated only at the output of A_3. The frequency response of an exactly frequency compensated operational amplifier is shown in Figure 8.32. The

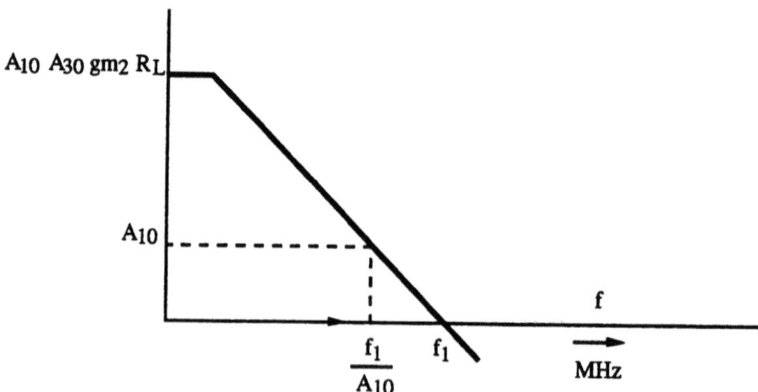

Figure 8.32 : Amplitude frequency response of an exactly compensated amplifier

transfer function of the amplifier can be calculated and the result of this calculation is:

$$\frac{V_{out}}{V_{in}} = \frac{A_{10}A_{30}gm_2R_L(1 + pC_2/gm_2)}{(1 + pA_{10}\tau_1)(1 + pR_L(C_2 + (1 + A_{30})C_3))}. \tag{8.34}$$

An exact frequency compensation is obtained if:

$$A_{10}\tau_1 = C_2/gm_2 \quad \text{or} \quad C_2 = A_{10}gm_2\tau_1. \tag{8.35}$$

In that case the transfer function of the operational amplifier becomes:

$$\frac{V_{out}}{V_{in}} = \frac{A_{10}A_{30}gm_2R_L}{1 + pR_LC_2(1 + (1 + A_{30})C_3/C_2)}. \tag{8.36}$$

Mismatches in frequency compensation elements results in slow settling tails during transient operation. This phenomenon is not analyzed here.

When the value of the capacitor C_2 is made equal to the value of C_3, then an optimum in overall bandwidth is found.

8.6.5 Practical low-distortion amplifier

The simplified circuit diagram of a low-distortion feed-forward frequency-compensated operational amplifier is shown in Figure 8.33. Compared with

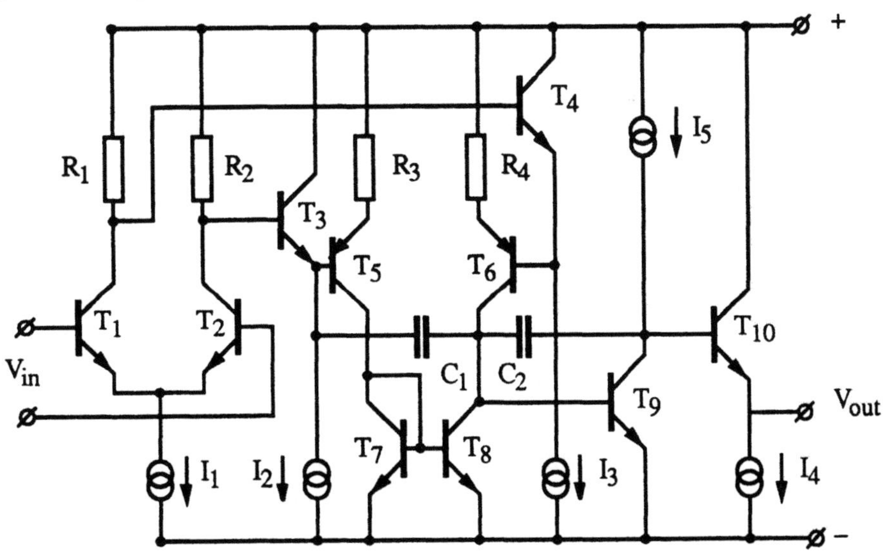

Figure 8.33 : Practical low-distortion operational amplifier circuit

the circuit diagram of Figure 8.29, the Miller amplifier stage consisting of transistor T_9 and capacitor C_2 is added. Amplitude and phase measurements of a practical integrated amplifier are shown in Figure 8.34. In this amplifier the phase margin at a unity gain frequency of 27 MHz is 35°. A practically ideal first-order amplitude frequency response is obtained in this circuit.

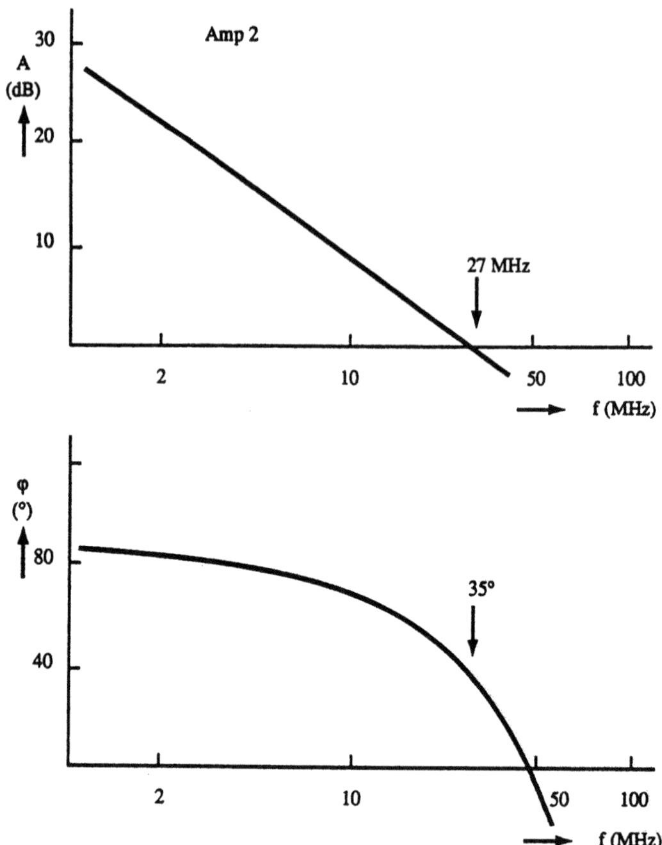

Figure 8.34 : Amplitude and phase measurements of a low-distortion operational amplifier

8.6.6　Class-B output stage

In wide-band operational amplifiers wide-band output stages are required. Standard complementary NPN-PNP output stages cannot be used because of the frequency limitations found in the lateral PNP transistors. An all-NPN class-B output stage must be designed to overcome this problem. An example of such a circuit is shown in Figure 8.35. From this Figure it can be seen that transistors T_1 and T_3 are the output devices that deliver the output signal to the external load. Most class-B stages contain a control loop that adjusts the quiescent current and applies the proper signals to the bases of transistors T_1 and T_3. For this purpose, transistors T_2, T_4, and T_5 are added. The bias current flowing through the output transistors is sensed

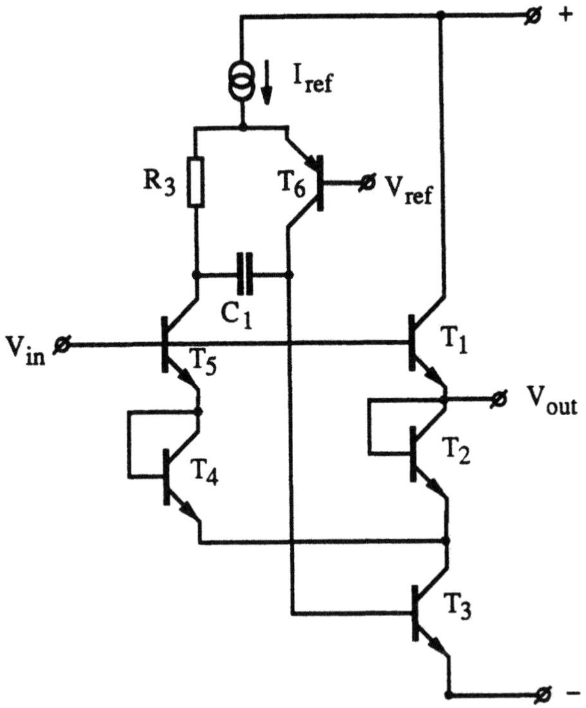

Figure 8.35 : All NPN class-B output stage

by transistors T_1 and T_2. The sum of the base-emitter voltages of T_1 and T_2 is now compared with the sum of the base-emitter voltages across T_4 and T_5. As a result, the collector current of T_5 is a measure of the bias current flowing through T_1, T_2, and T_3. The collector current of T_5 is compared to the reference current I_{ref}, and the lateral PNP transistor T_6 supplies base current to transistor T_3 until a stable operating point is obtained.

The class-B circuit operation is obtained as follows. Suppose an input source is connected to the V_{in} terminal and a load resistor R_L is connected to the output terminal. With a positive-going input signal, T_1 acts as an emitter follower that supplies current to the load resistor. The increase in collector current corresponds to an increase in the base-emitter voltage of transistor T_1. Using the control condition, which was needed to obtain a stable bias current stabilization:

$$V_{be1} + V_{be2} = V_{be4} + V_{be5}, \tag{8.37}$$

it can be shown that the increase of V_{be1}, results in an increase of V_{be4} and

V_{be5}, thus giving an increase in the collector currents of T_4 and T_5. More current is subtracted from the reference current I_{ref}, causing a decrease in the collector currents of T_6 and T_3.

When a negative-going input signal is applied, the collector current of T_1 decreases. At the same time, the collector current of T_5 decreases and more current is applied to T_6. As a result, the base current of T_3 increases resulting in an increase in the collector current of this transistor. The output signal is brought down by transistor T_3 to the negative supply voltage.

At high frequencies the frequency response of the PNP transistor is eliminated by a feed-forward coupling using R_3 and C_1. A bandwidth up to 100 MHz is obtained with this output stage. It can be designed to supply output currents with a maximum value of 100 mA.

8.6.7 Switchable class-B output stage

A class-B output stage which can be switched on and off is shown in Figure 8.36. To obtain a switching operation in the class-B output stage a differential amplifier stage consisting of transistor T_7 and T_8 and the resistors R_1 and R_2 are added to the circuit diagram of Figure 8.35. The operation of the circuit can be explained as follows. When transistor T_7 is switched on and the voltage drop across R_1 is zero, then the normal operation of the class-B stage as explained in connection with Figure 8.35 is obtained. The circuit is switched into the off mode when transistor T_8 is conducting. Across resistor R_1 a voltage drop equal to a diode voltage is generated. The resistor R_2 reduces the voltage drop across transistor T_5 to zero. In this way, the base-emitter voltages of transistors T_1 and T_2 are made zero and no conductance is possible. The transistor loop consisting of T_3, T_4, and T_6, however, remains active irrespective of the switching condition. The leakage current of the switch can be made very low, while capacitive feed-through is compensated by equal voltage swings across the base-emitter junctions of transistors T_1 and T_2. The voltage compliance of this switch, however, is small. The reverse biasing of the switching transistors equals about one diode voltage. Therefore, only a voltage compliance between 200 mV and 500 mV is allowed at the output of the switch. To overcome this problem the switch is operated at, for example, an inverting input of an operational amplifier. It is known that at these input points a small voltage variation, well below the required range by the switch, is found. With special circuit con-

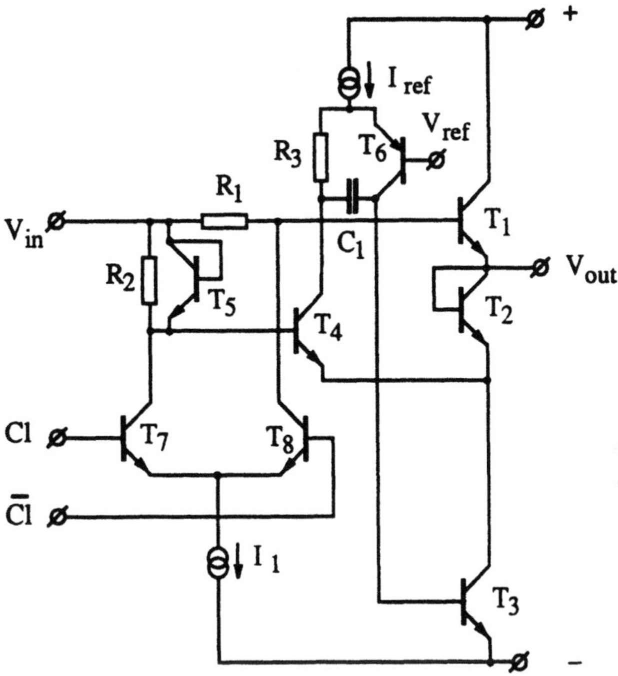

Figure 8.36 : Switchable class-B output stage

figurations, however, it is possible to operate this switch in a non-inverting sample-and-hold mode.

8.6.8   Complete sample-and-hold amplifier

The complete sample-and-hold amplifier is incorporated on a 1.5x2.5 mm^2 chip using a standard bipolar technology with double-layer interconnect. In this chip all capacitors used in the circuit consist of the capacitance between the two interconnect metal layers. A very linear capacitance-voltage characteristic is obtained in this way. Because of this construction these capacitors require a large part of the chip area. With a special process option using thin oxide/nitride layers, a much larger capacitance per unit area can be obtained. In such a process the size of the compensation capacitors can be reduced and the "Hold" capacitor can also be put on the chip.

8.6.9   Measurements

The result of a signal-to-noise + THD measurement as a function of amplitude is shown in Figure 8.37. A high-performance A/D and D/A loop is used to perform this measurement. A second method to perform this measurement is the use of a second sub-sampled sample-and-hold amplifier to obtain a beat frequency. The acquisition time measurement of the circuit

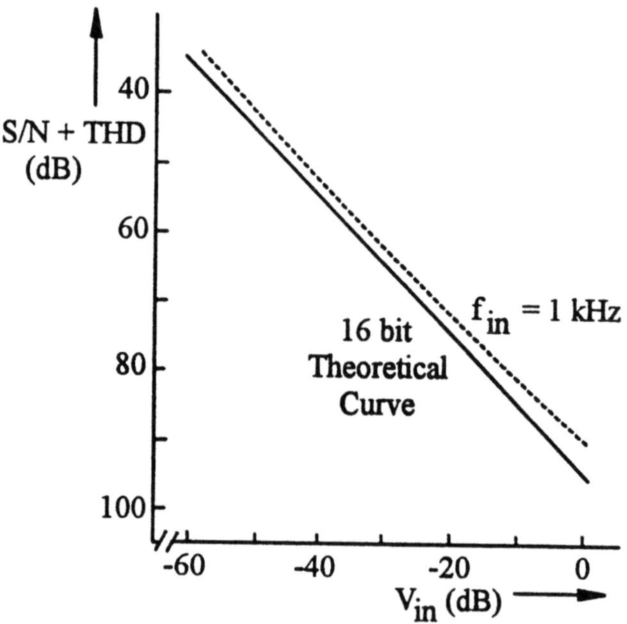

Figure 8.37 : Signal-to-noise + THD measurement as a function of signal amplitude

is performed by using a full-scale input signal of 19 kHz. The sampling frequency is kept constant, while the hold time is varied during this measurement. The measurement results are shown in Figure 8.38. A photograph of the hold-track-hold operation of the circuit is shown in Figure 8.39. The input frequency is taken at one quarter of the sampling frequency to obtain a stable picture. In Table 8.3 the measurement results of the total system are shown. A die photograph is shown in Figure 8.40.

8.7   Conclusion

Different system implementations that can be applied for high-performance or high-speed sample-and-hold amplifiers have been analyzed. In MOS applications a differential solution seems to be the best concerning switch feedthrough. In bipolar applications the differential solutions show a big advantage in the hold amplifier. An order of magnitude lower droop rate can be

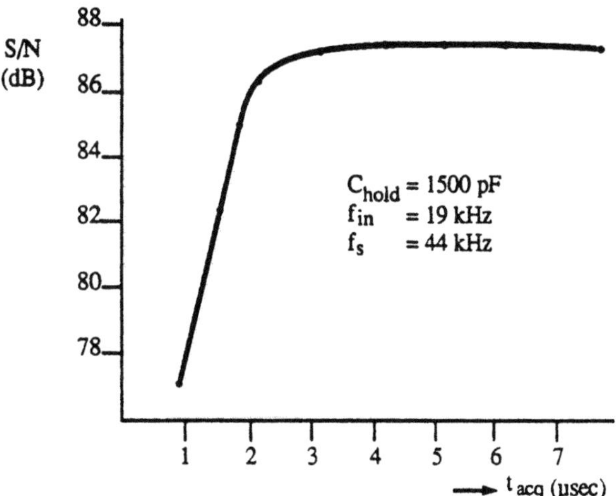

Figure 8.38 : Acquisition time measurement result

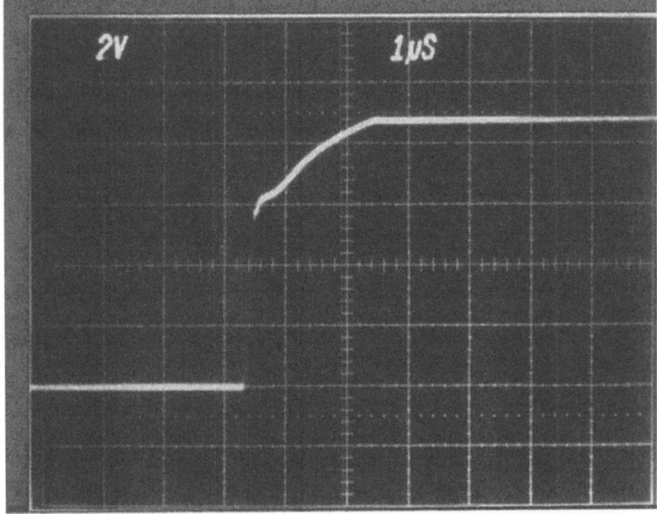

Figure 8.39 : Hold-track-hold operation with an input signal of one-quarter of the sample frequency

obtained in this system. Even harmonic distortion is reduced in a differential system for both technologies.

The highest sampling rates are still obtained using a diode bridge as a sampling switch. In such a case accuracy trade off against speed.

Using an all-NPN switchable class-B output stage and special feed-forward

Acquisition time	1.5 μsec to 0.001 %
Aperture uncertainty	< 0.5 nsec
Small signal bandwidth	8 MHz
Slew rate	100 V/μsec
Gain	-1 V/V
Distortion	< 0.001 %
S/N ratio	> 90 dB
Hold offset	< 2 mV
Droop rate	5 V/sec
Digital input	TTL compatible
Power supply	+5, -5 V
	27 mA
Chip size	1.5 $times$ 2.5 mm^2

Table 8.3 : Sample-and-hold amplifier measurement data

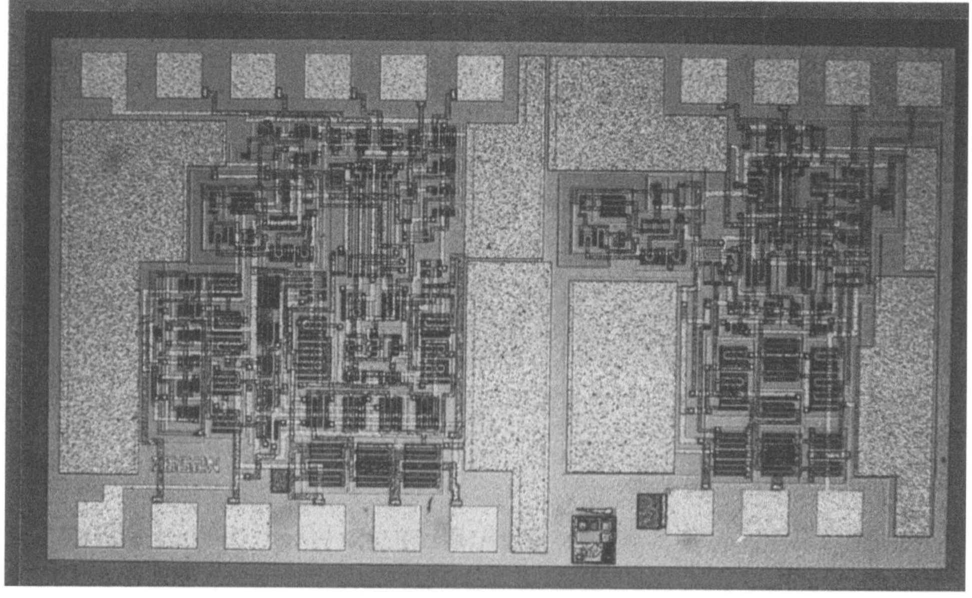

Figure 8.40 : Die photograph of sample-and-hold amplifier

frequency compensation techniques for multistage operational amplifiers, a
high-performance sample-and-hold amplifier can be designed for 16-bit dig-

ital audio systems. A further advantage of the frequency compensation method used is the low output noise of the compensated amplifiers for frequencies above the unity gain bandwidth of the amplifiers. The application of resistive loads in the input stages in comparison with active loads using current mirrors reduces the noise in the amplifier system increasing the dynamic range. Today, however, bi-MOS technologies, which combine bipolar devices with MOS devices, can be used advantageously in a sample-and-hold amplifier. The relatively large gain in the first sampling amplifier stage reduces the inherently large low-frequency noise of the MOS devices used in the hold amplifier. Furthermore, the lateral PNP transistors that are used in level shifts or biasing current sources can be replaced by P-MOS devices. These MOS devices show a better frequency characteristic than the lateral PNP transistors.

Chapter 9

Voltage and current reference sources

9.1 Introduction

In A/D and D/A converters the full-scale value is determined by the reference
source. A low noise and low temperature coefficient of the output signal
of the reference source is very important for high-resolution, high-accuracy
converters. A well-known device for stabilizing a reference voltage is a zener
diode. In integrated circuits, however, the zener diode can cause problems
with the reliability of the circuit. In modern technologies it is not always
possible to reverse-bias the emitter-base junction of a transistor to obtain
a zener diode operation. The yield of circuits is reduced by reverse-biasing
transistors. Today's reference sources are built using the band-gap voltage
of silicon as a low-temperature dependent reference voltage. In this chapter
different circuits will be described that use the band-gap principle to stabilize
a voltage or a current. Examples of band-gap reference voltages are given in
references [76,77,79].

9.2 Zener diode reference voltage

One of the simplest and most used methods to obtain a stable voltage source
is the zener diode. The operation of such a diode is for an important factor
based on the avalanche multiplication of carriers in a reverse biased junction.
A drawback of the avalanche multiplication effect is the relatively large noise
of the device. Discrete devices can be optimized for a minimum temperature

coefficient and a minimum noise.

9.2.1   Reverse-biased base-emitter zener junction

In an integrated circuit the reverse-biased base-emitter junction can be used
as a zener reference diode. In Figure 9.1 the basic circuit configuration
and the IC implementation are shown. From the figure is it seen that the

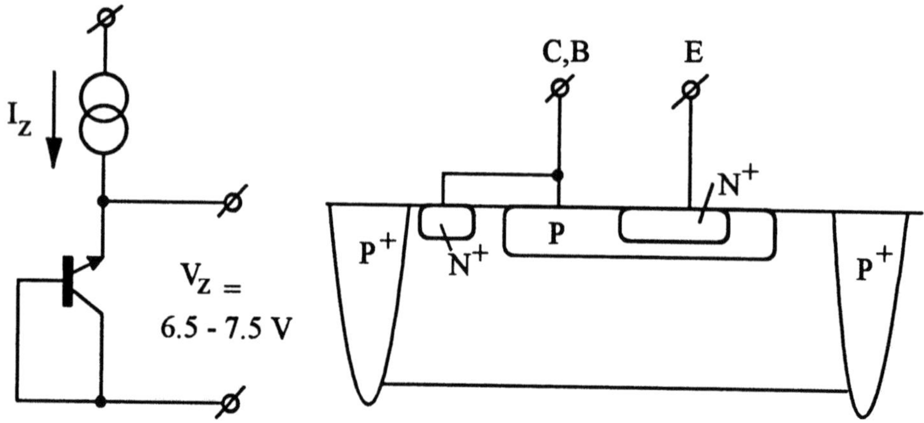

Figure 9.1 : Reverse-biased emitter-base zener junction

collector-base junction is short-circuited to avoid parasitic effects. The zener
voltage of the device depends on the process, and in practice voltages between 6.5 and 7.5 V are found. As a result of this voltage spread, the
temperature coefficient of the zener voltage varies from about zero at 6 V
to about $+5$ mV/^{0}C at 8 V. Integrated zeners usually have a temperature
coefficient of about 2 mV/^{0}C. The best biasing condition of such a device is
a constant current source I_z. A reliable operation of the device is obtained
with a diode current of less than 1 mA. Furthermore, long-term stability
of the output voltage depends on the surface conditions between the emitter and the base areas. The temperature coefficient of the device can be
improved by using an additional bipolar transistor.

9.2.2   Zener diode with extra transistor

To improve the performance of a single zener diode and to compensate for
the positive temperature coefficient, an extra bipolar transistor is added to
the circuit. In Figure 9.2 the improved circuit is shown. The base-emitter
voltage of the bipolar transistor is connected in series with the zener voltage

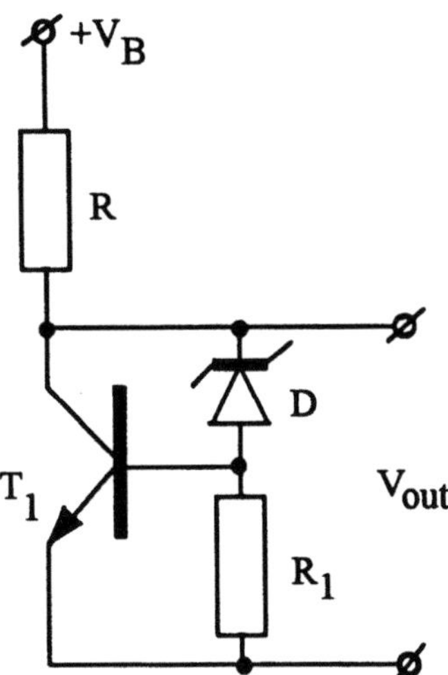

Figure 9.2 : Zener diode with extra transistor

of the reference device. Furthermore, the current through the zener diode is stabilized by the voltage drop across the base-emitter voltage of the transistor, generating a nearly constant current through resistor R_1. The negative temperature coefficient of the base-emitter voltage of the transistor compensates for the positive temperature coefficient of the zener diode, resulting in a practically temperature independent output voltage. The resistor R biases the total system. The output impedance of the total system is reduced by the gain of the transistor. It can be shown that:

$$R_{out} = \frac{1}{g_m(1 + 1/\beta) + 1/R_1} \approx \frac{1}{g_m}. \tag{9.1}$$

The approximation is only valid when $g_m R_1 \gg 1$.

Long-term stability of this device is still dependent on the base-emitter surface conditions. A very good long-term stability can be obtained by modifying the zener diode construction.

9.2.3 Buried zener diode construction

To improve the long-term stability of the zener output voltage, a buried construction of the zener diode is used. The breakdown in such a device occurs fully in the bulk material. The surface conditions of the process on the long-term zener voltage are thus fully eliminated. In Figure 9.3 the basic

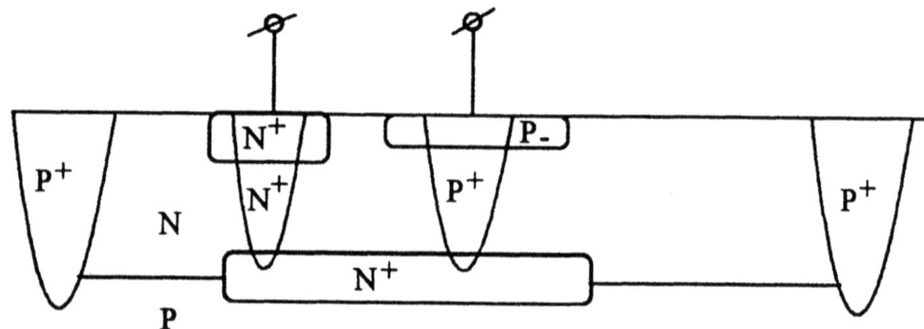

Figure 9.3 : Basic buried zener diode construction

construction of such a diode is shown. In the figure it is shown that the zener operation is obtained between the P^+ diffusion and the buried N^+ diffusion in the buried material. Such a diode shows a good long-term performance but has a disadvantage, which is that due to process variations between the P^+ and the N^+ concentrations, the breakdown voltage can vary between 8 and 12 V. In such a case a large positive temperature coefficient is found as well. A different structure with much better control of the parameters is required.

9.2.4   Buried zener diode with additional implantation

To allow an adjustment of the breakdown voltage of the zener diode in the bulk material, an additional implantation is used. In Figure 9.4 the basic construction of such a diode is shown. The figure shows the basic

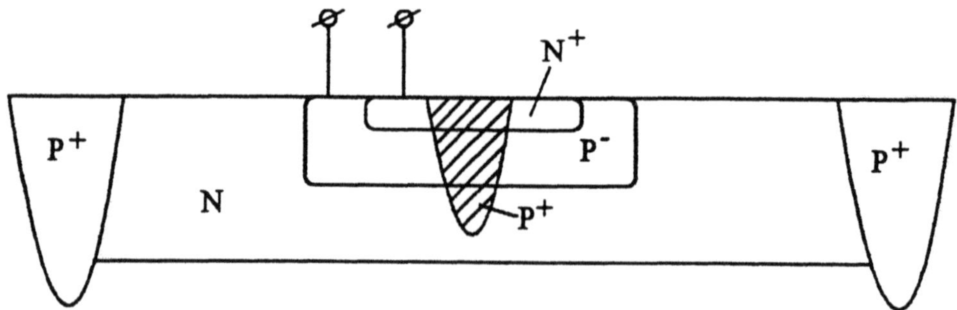

Figure 9.4 : Buried zener with extra implantation

construction of an NPN device with the exception that an extra P^+ implant is applied before the emitter diffusion is applied. The extra P^+ implant is completely surrounded by the N^+ emitter diffusion, resulting in a breakdown completely in the bulk material. By changing the dopant of the additional P^+ implant the zener voltage can be adjusted to the desired value. Usually a voltage in the order of 6.3 V is a good optimum. Long-term drifts in the order of 10 ppm/^{0}C are possible over 1000 hours. Temperature coefficients in the order of 5 ppm/^{0}C with compensation circuitry can be obtained in an optimized design.

9.3   Base-emitter voltage used as a reference

The base-emitter voltage of a bipolar transistor can be used as a very simple reference voltage. A simple practical example of such a stabilizer is shown in Figure 9.5. The circuit consists of two transistors T_1, T_2 and two resistors

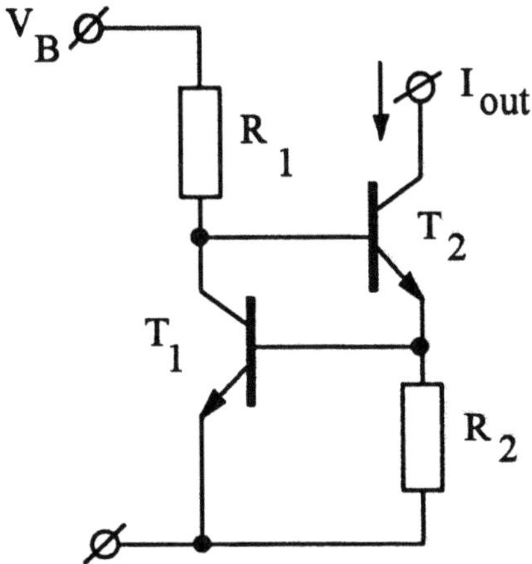

Figure 9.5 : Base-emitter voltage stabilizer example

R_1, R_2. As a stabilized voltage the base-emitter voltage of transistor T_1 is used. The voltage stabilization is obtained because of the logarithmic characteristic between the base-emitter voltage of a transistor and the collector current:

$$V_{be} = \frac{kT}{q} \ln(\frac{I_c}{i_0}) \tag{9.2}$$

where i_0 is the base-emitter reverse current of the bipolar transistor.

Transistor T_1 with the resistor R stabilizes the base-emitter voltage of T_1. Transistor T_2 with the resistor R_2 converts this stabilized voltage into an output current flowing through the collector of T_2. Transistor T_2 then acts as a current source with a limited stabilization factor. A simplified calculation of the output current gives:

$$I_{out} = \frac{V_{BE1}}{R_2} = \frac{kT}{qR_2}\ln(\frac{V_B}{i_0R}). \tag{9.3}$$

Equation 9.3 is valid when $V_B \gg V_{BE1} + V_{BE2}$.

The stabilization factor of the circuit for variation in the supply voltage V_B can be found after differentiation of equation 9.3 with respect to V_B:

$$\frac{1}{I_{out}}\frac{dI_{out}}{dV_B} = 1/(V_B\ln(\frac{V_B}{i_0R})) = \frac{kT/q}{V_BV_{BE1}}. \tag{9.4}$$

A high stabilization factor is obtained if a large supply voltage is used. The temperature coefficient of the output current can be calculated using a simplified temperature dependence of i_0:

$$i_0 = Ce^{-\frac{qV_g}{kT}}. \tag{9.5}$$

In this simplified formula the constant C is supposed to be temperature independent and V_g is the band-gap voltage of silicon.

The temperature coefficient of the output current is calculated by differentiating equation 9.3. The result becomes:

$$\frac{1}{I_{out}}\frac{dI_{out}}{dT} = \frac{1}{T}\frac{I_{out}R_2 - V_g}{I_{out}R_2} = \frac{1}{T}\frac{V_{BE1} - V_g}{V_{BE1}}. \tag{9.6}$$

By inserting practical values in equation 9.6, we obtain with:

$V_{BE1} \approx 0.6$ V and $V_g = 1.2$ V:

$$\frac{1}{I_{out}}\frac{dI_{out}}{dT} \approx -\frac{1}{T}. \tag{9.7}$$

A negative temperature coefficient of the output current is obtained when a base-emitter voltage is used as a reference.

9.3.1 Improved base-emitter voltage stabilizer

The stabilization factor of the circuit given in Figure 9.5 can be improved. A simplified circuit diagram of the improved system is shown in Figure 9.6.

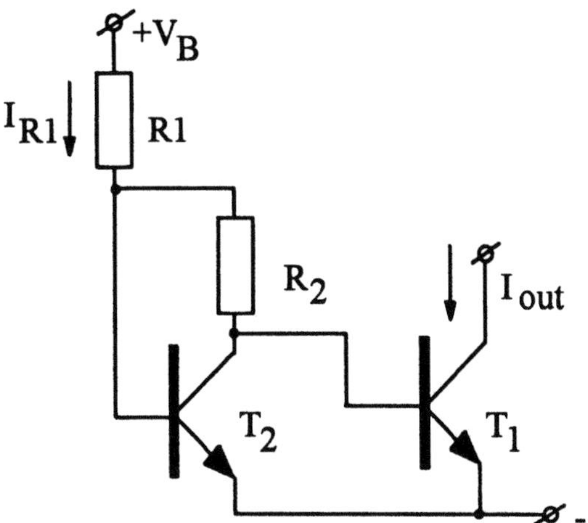

Figure 9.6 : Improved base-emitter voltage stabilizer

Basically this system consists of a bipolar current mirror consisting of transistor T_2 connected as a diode and transistor T_1 acting as the output current device. Between the collector and the base of transistor T_2 an extra resistor R_2 is connected. The purpose of this resistor is to compensate for the increase in the base emitter voltage of T_2 with increasing supply voltage V_b. With an increasing voltage V_b the voltage drop across the resistor R_2 increases, and this voltage drop is subtracted from the increased base-emitter voltage of transistor T_2. The output voltage applied to the base-emitter junction of transistor T_1 in this way is nearly constant, resulting in a constant output current.

Suppose a current $I_R \approx V_b/R_1$ flows through the collector of transistor T_2, then the output current of the circuit can be calculated from:

$$I_R R_2 = \frac{kT}{q} \ln\left(\frac{i_{01} I_R}{i_{02} I_{out}}\right). \tag{9.8}$$

The value for R_2 can then be calculated to make I_{out} independent of the supply voltage (I_R) so:

$$\frac{dI_{out}}{dI_R} = \frac{I_{out}}{I_R}\left(\frac{kT}{q} - I_R R_2\right)\frac{kT}{q} = 0. \tag{9.9}$$

This results in:

$$I_R R_2 = \frac{kT}{q}. \tag{9.10}$$

Substituting this value into equation 9.9, the following relation is found:

$$\frac{I_R}{I_{out}}\frac{i_{01}}{i_{02}} = e \tag{9.11}$$

$$I_{out} = I_R\frac{i_{01}}{i_{02}}\frac{1}{e} \tag{9.12}$$

where $e = 2.73..$.

In case we want an output current equal to the input current, the ratio between transistors T_1 and T_2 must be equal to e or:

$$i_{01} = ei_{02}. \tag{9.13}$$

In practice a value of 3 is easy to draw in an ic layout. A plot of the output current as a function of the input voltage (or current I_R) is shown in Figure 9.7. The temperature coefficient of the output current is found after

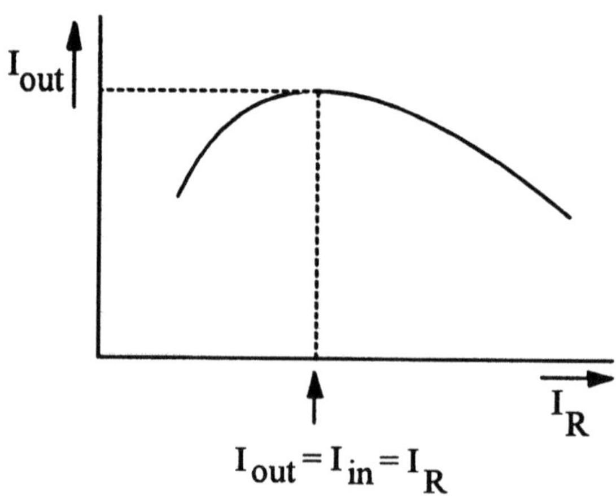

Figure 9.7 : Output current as a function of the input voltage

differentiating equation 9.8 to temperature. We obtain:

$$\frac{1}{I_{out}}\frac{dI_{out}}{dT} = \frac{1}{T}\frac{I_R R_2}{kT/q}. \tag{9.14}$$

By inserting the optimum value for the voltage across R_2, we obtain:

$$\frac{1}{I_{out}}\frac{dI_{out}}{dT} = \frac{1}{T}. \tag{9.15}$$

In this case a positive temperature coefficient for the output current is obtained.

9.4 All-NPN band-gap voltage reference source

In Figure 9.8 a practical example of a voltage reference source that uses only NPN transistors is shown. The circuit consists of a cross-coupled quad

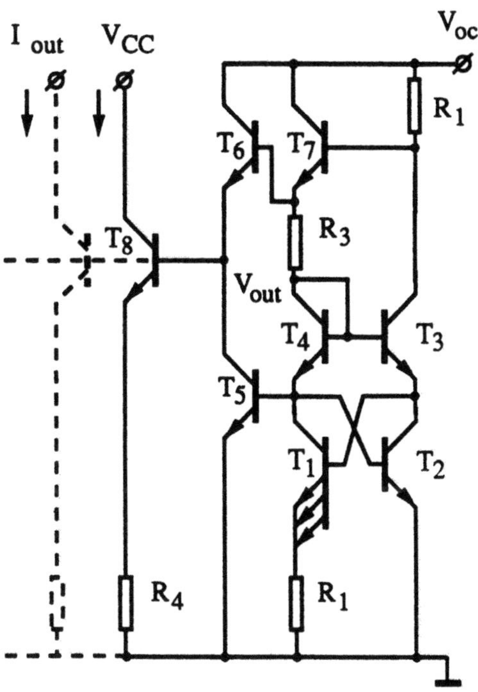

Figure 9.8 : All-NPN band-gap voltage reference source

transistor unit T_1, T_2, T_3, and T_4, and the current-converting resistor R_1. Transistor T_1 has a p times larger emitter area than the other three transistors in the quad. Via resistor R_2 a current from the supply source is applied to the circuit. Resistor R_3 is added to obtain an adjustable output voltage. Transistor T_7 acts as a buffer device and results in a smaller current variation through resistor R_2 with supply voltage variations (V_{cc}). Transistors T_6 and T_5 are added to obtain a compensation for supply voltage dependence of the output voltage V_{out}. The operation of the system will be explained as follows. The current flowing through transistors T_4, T_1, and R_1 can be expressed in the following parameter of the quad unit:

$$I_{R1} = \frac{kT}{qR_1} \ln(p). \tag{9.16}$$

Equation 9.16 shows that basically the current is supply-voltage-independent; however, the output voltage at the collector of T_4 shows a variation that contains a supply voltage dependence. This can be explained by the fact that the output voltage at the collector of T_4 equals $V_{be4} + V_{be2}$. The current through transistor T_4 is well stabilized at a value given by equation 9.16 so the base-emitter voltage of this transistor is independent of the supply voltage variation. The current through transistor T_2, however, varies with the supply voltage. Therefore, the base-emitter voltage of this transistor shows a supply-voltage-dependent part. To the output voltage at the collector of T_4 a value equal to:

$$V_{R3} = \frac{R_3}{R_1}\frac{kT}{q}\ln(p) \tag{9.17}$$

is added. As can be seen from equations 9.16 and 9.17, the temperature dependence of the current I_{R1} is linear with T, and thus the voltage drop across R_3 increases linearly with increasing temperature.

Now transistors T_5 and T_6 are added. Transistor T_5 is connected in parallel with transistor T_2 and forms a current mirror. As a result, the current through T_5 varies exactly in the same way as the current through transistor T_2. This current flows through transistor T_6, too, and therefore the voltage variation of the base-emitter voltage of T_6 is identical to the base-emitter voltage variation of transistor T_2. The voltage variation across transistor T_6 is subtracted from the voltage developed across T_2 and T_4. Therefore, the voltage variation at the emitter of transistor T_6 is zero. At the output terminal marked V_{out}, a voltage equal to the base-emitter voltage of T_4 plus the voltage drop across resistor R_3 is obtained.

$$V_{out} = V_{be2} + V_{be4} + V_{R3} - V_{be6}. \tag{9.18}$$

Making

$$V_{be2} = V_{be6} \tag{9.19}$$

we obtain:

$$V_{out} = V_{be2} + \frac{R_3}{R_1}\frac{kT}{q}\ln(p). \tag{9.20}$$

When this output voltage is adjusted to the value given in equation 9.32, a temperature dependence equal to a band-gap voltage source is obtained. This circuit does not contain PNP transistors and is therefore very suitable for use in integrated circuits that use special high-frequency processes (See reference [80]). In these processes mostly lateral PNP transistors with poor dc and high-frequency performance are available. The temperature dependence of the resistors R_1 and R_3 is canceled provided that the temperature coefficients of the resistors R_1 and R_3 are equal and have a good thermal tracking.

9.5 General purpose current stabilizer

A general purpose current stabilizer consisting of a PNP current mirror with a current transfer ratio of 1 and a current dependent current mirror consisting of an NPN current mirror is shown in Figure 9.9. The PNP current

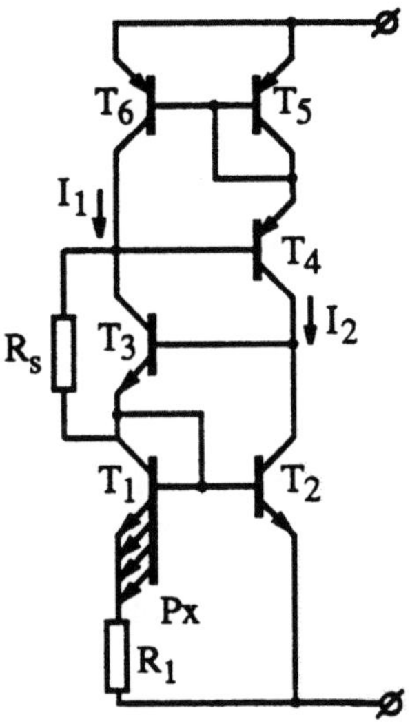

Figure 9.9 : General purpose current stabilizer

mirror with a current independent transfer ratio consists of transistors T_4, T_5, and T_6. The NPN current mirror with a current dependent transfer ratio is formed by transistors T_3, T_2, and T_1 which has a p times larger emitter area than transistor T_2. The resistor R_1 is inserted to adjust at the stabilizing current value the ratio of the NPN current mirror to the transfer ratio of the PNP current mirror which equals 1. As a result, the current through R_1 becomes:

$$I_{R1} = \frac{kT}{qR_1} \ln p. \tag{9.21}$$

The total output current of this stabilizer $I_t = 2\,I_{R1}$. Furthermore, a positive temperature coefficient of the total output current is obtained.

A drawback of these types of current stabilizers is found that a zero current value is a solution of the equations too. The stabilizers have a starting problem. To avoid this problem the resistor R_s is inserted. This resistor prevents that zero is a solution. When a supply voltage is connected to the stabilizer terminals, then through the diode part of the PNP current source and the NPN diode consisting of T_1 with R_1 in series a current will always flow. As a result of this current zero can not be a solution anymore.

9.6 Basic band-gap reference voltage source

The basic band-gap voltage stabilizer is shown in Figure 9.10. It consists of a

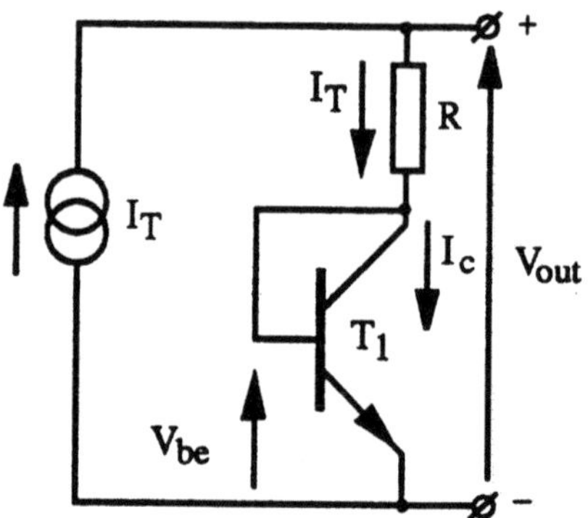

Figure 9.10 : Basic band-gap voltage reference source

current source with a well-determined temperature relation, a resistor R and a transistor which is switched as a diode. The output voltage of the system is generated across the resistor R and the transistor T_1. The temperature dependence of the current source I_T is given by:

$$\frac{1}{I_T}\frac{dI_T}{dT} = \frac{1}{T}. \qquad (9.22)$$

The output voltage of the system which is made nearly temperature independent is given by:

$$V_{out} = RI_T + V_{be}. \qquad (9.23)$$

In the following simplified analysis we assume that the base current of the transistor can be ignored with respect to the collector current. This simplification does not change the results considerably and makes it much easier to understand the operation of the system. The basic idea behind the band-gap voltage stabilization is the following: With increasing temperature it is known that the base-emitter voltage of a transistor decreases. At the same time, the positive temperature coefficient of the current source I_T generates a voltage across resistor R that increases linearly with temperature. The output voltage of the circuit can be adjusted to a value that is nearly equal to the band-gap voltage of silicon. The decrease in base-emitter voltage in that case is compensated by the increase in the voltage across the resistor R.

It is known that for a bipolar transistor the following equation is valid for the relation between collector current and base-emitter voltage:

$$I_c = I_0(e^{\frac{q}{kT}V_{be}} - 1). \tag{9.24}$$

I_0 is the base-emitter reverse current. The temperature relation for I_0 can be expressed in the following parameters:

$$I_0 = CT^n e^{-\frac{qV_g}{kT}}. \tag{9.25}$$

In equation 9.25 the value of V_g is given by the band-gap voltage of silicon which is equal to 1.208 V. In silicon n is approximately 1.4 and C is a constant depending on the size of the device.

If we assume that $e^{\frac{q}{kT}V_{be}} \gg 1$, then we can approximate equation 9.24 and equation 9.25 by:

$$I_c = CT^n e^{\frac{q}{kT}(V_{be}-V_g)}. \tag{9.26}$$

After differentiation of equation 9.26 with respect to temperature we obtain:

$$dI_c = CnT^{n-1} e^{\frac{q}{kT}(V_{be}-V_g)} dT -$$
$$CT^n e^{\frac{q}{kT}(V_{be}-V_g)} \frac{q(V_{be} - V_g)}{kT} \frac{dT}{T} +$$
$$CT^n \frac{q}{kT} e^{\frac{q}{kT}(V_{be}-V_g)} dV_{be}. \tag{9.27}$$

This equation can be simplified using the expression given in equation 9.26. After the insertion we obtain:

$$dI_c = \frac{n}{T} I_c dT - \frac{q}{kT} I_c \frac{V_{be} - V_g}{T} dT + \frac{q}{kT} I_c dV_{be}. \tag{9.28}$$

With a negligible base current of the transistor, the collector current variation now equals the variation of the current source I_T. By inserting equation 9.22 into equation 9.28 we end up with the following relation for the temperature dependence of the base-emitter voltage of the transistor:

$$\frac{dV_{be}}{dT} = \frac{k}{q}(1 - n) + \frac{V_{be} - V_g}{T}. \tag{9.29}$$

The temperature relation of the output voltage V_{out} can be expressed in the following terms, assuming that R is temperature-independent:

$$\frac{dV_{out}}{dT} = R\frac{dI_T}{dT} + \frac{dV_{be}}{dT}. \tag{9.30}$$

Inserting equation 9.22 and equation 9.29 into equation 9.30 results in, after rearrangement of the terms:

$$\frac{dV_{out}}{dT} = \frac{I_T R}{T} + \frac{k}{q}(1 - n) + \frac{V_{be} - V_g}{T}. \tag{9.31}$$

The output voltage of the reference source can be adjusted in such a way that at $T = T_0$ the temperature coefficient of the reference source $\frac{dV_{out}}{dT} = 0$. The following value for the output voltage at $T = T_0$ is obtained:

$$V_{out}(T = T_0) = V_{beT_0} + I_{T0}R = V_g - \frac{kT_0}{q}(1 - n). \tag{9.32}$$

The remarkable point is now obtained that at $T = T_0$ the temperature coefficient of the output voltage is zero. The temperature dependence of the output voltage can be calculated. By substituting equation 9.32 into equation 9.31, the following result is obtained:

$$\frac{dV_{out}}{dT} = \frac{k}{q}(1 - n)(1 - \frac{T}{T_0}). \tag{9.33}$$

This equation shows that there is a change in sign of the temperature coefficient of the output voltage around $T = T_0$.

After integration of equation 9.33 we obtain an expression for the output voltage of the band-gap reference source as a function of temperature T. The integration and substitution of the boundary conditions result in:

$$V_{out}(T) = V_g - \frac{k}{q}T(1 - n)(1 - \ln\frac{T}{T_0}). \tag{9.34}$$

The expression given in equation 9.34 represents a parabolic temperature dependence around $T = T_0$. The temperature dependence of the reference source as expressed by equation 9.34 is calculated, and the result is shown in Figure 9.11.

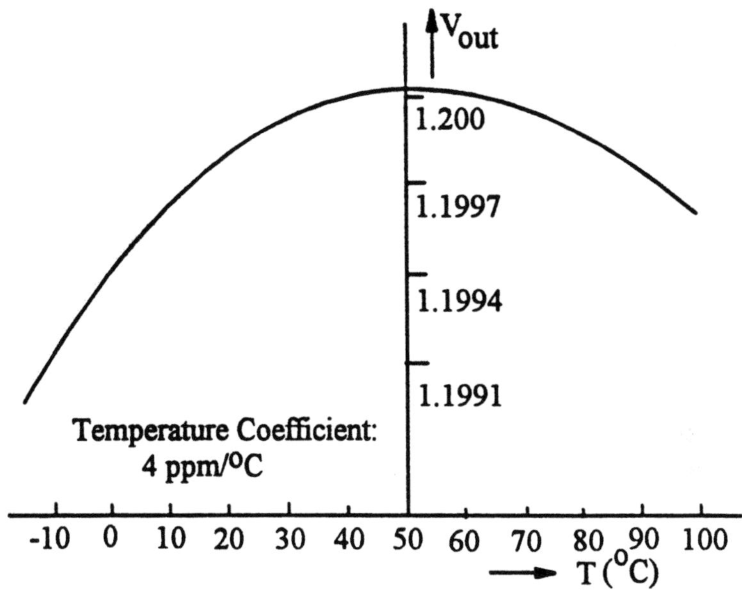

Figure 9.11 : Temperature dependence of a band-gap reference source

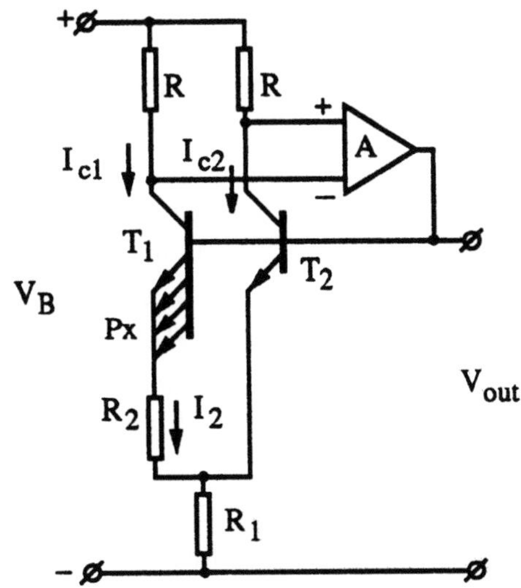

Figure 9.12 : Practical band-gap reference source

9.6.1 Practical band-gap voltage source

A practical circuit implementation of a voltage source is shown in Figure
9.12 [78]. The circuit consists of NPN transistors T_1 and T_2 with an emitter
ratio of p, resistor $R2$, an operational amplifier, two equal resistors R that

set the current ratio flowing through the NPN transistors T_1 and T_2 to one, and an additional resistor $R1$. The operational amplifier controls the base voltage of the transistors T_1, T_2 in such a way that equal currents will flow through T_1 and T_2. Equal currents will also flow through the resistors R. The operational amplifier replaces in this way the PNP current mirror from the circuit shown in Figure 9.9. Furthermore, a much higher accuracy in the current ratio is possible, resulting in a more accurate output current. The output voltage is generated across the base-emitter voltage of T_2 plus the voltage across resistor R_1. The voltage across the resistor R_1 is linearly increasing with temperature, while the base-emitter voltage of transistor T_2 nearly linearly decreases with temperature. A cancellation of both temperature effects can be obtained with a proper adjustment of the resistor R_1. When the output voltage is adjusted to:

$$V_{out}(T = T_0) = V_g + (n - 1)\frac{kT_0}{q}, \tag{9.35}$$

then the parabolic temperature dependence of the band-gap source is found.

9.7 Band-gap reference current source

A simplified circuit diagram of a reference current source is shown in Figure 9.13. It consists of transistors T_1, T_2, with the voltage-to-current converting resistor R_1. An operational amplifier A_0 with the resistors R controls the current in transistors T_1 and T_2 in such a way that equal currents will flow through these transistors. The resistor R_2 is added for temperature compensation and gives an identical temperature dependence, as is the case with the band-gap reference voltage source. This will be explained later.

Transistor T_1 has a p times larger emitter area than transistor T_2. With equal collector currents flowing in T_1 and T_2, the following expression for the current flowing through R_1 is found:

$$I_{R1} = \frac{kT}{qR}\ln(p). \tag{9.36}$$

As can be seen from equation 9.36, the currents through T_1 and T_2 are linearly dependent on the temperature T.

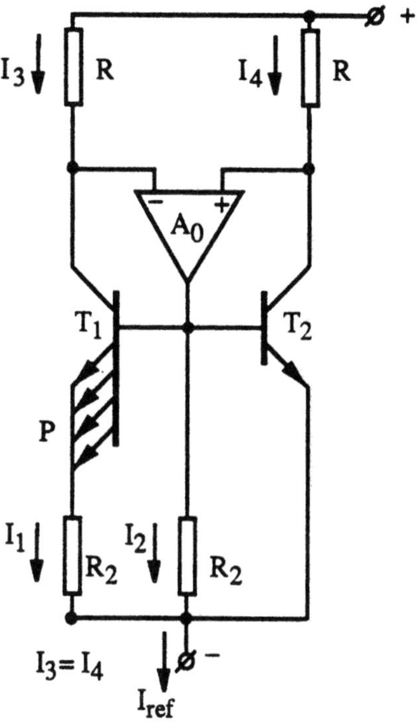

Figure 9.13 : Basic reference current source

Now resistor R_2 is added to the circuit. The current through R_2 then becomes:

$$I_{R2} = \frac{V_{be2}}{R_2}.$$ (9.37)

With increasing temperature it is known that the base emitter voltage of a transistor decreases. As a result, the current through resistor R_2 will decrease. As already stated, the currents through T_1 and T_2 will increase with increasing temperature. The resistor value of R_2 can be chosen in such a way that the increase in current flowing through T_1 and T_2 is canceled by the decrease in the current flowing through R_2. In this way a temperature independent current source is obtained. Furthermore, the operational amplifier A_0 can be constructed in such a way that it uses only currents that flow through R_2, T_1 and T_2.

The value of R_2 for which an exact compensation of the temperature coefficients of the currents occur can be calculated. The result of this calculation, which is identical to the calculation given in the basic band-gap voltage reference source, is shown. We obtain:

$$R_2 = (V_g + \frac{kT_0}{q}(n-1))/I_{out}. \tag{9.38}$$

Moreover, the generated current reference I_{out} shows the same temperature dependence as the band-gap reference voltage. The output current of this reference source becomes:

$$I_{out}(T) = \frac{V_g}{R_2} + \frac{k}{q}T(n-1)\frac{1}{R_2}(1 - \ln\frac{T}{T_0}). \tag{9.39}$$

9.7.1 Practical reference current source

In Figure 9.14 an example of a practical implementation of a reference current source is shown. In this system the operational amplifier is replaced

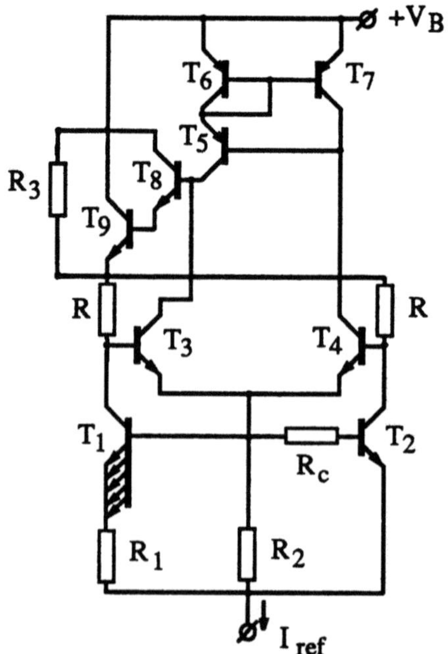

Figure 9.14 : Practical reference current source

by a differential pair T_3 and T_4 with a PNP current mirror consisting of T_5, T_6, and T_7. A Darlington stage, T_8 and T_9, controls the current through the resistors R. An additional resistor R_3 is added to the circuit to obtain a "starting condition" under all circumstances. The resistor R_3 supplies current into the circuit and in this way prevents "zero current" from being a solution to the equations as well. The zero solution is not a desired solution. R_C is an extra temperature-compensation resistor that is process-dependent and can give an extra second-order temperature compensation. The advantage of this type of current source is that the current that flows through the "bottom" is the same as the current that flows in the "top" of the circuit. Therefore, this current source can be included in the reference loop of an A/D or D/A converter, to generate at one side the most significant bit current, while at the other side this current can be divided in the binary division stage to obtain a binary-weighted current network.

9.7.2 Second-order temperature compensation

Up until now, all described reference sources show a parabolic temperature dependence. In high-accuracy converters this temperature dependence is too great to keep the absolute accuracy within the size of the least significant bit. An additional temperature compensation is therefore needed to overcome this problem. In Figure 9.15 an example of a reference current source with a second-order temperature compensation is shown. The main stabilizer core consists of transistors T_1 and T_2 which have a p times larger emitter area than transistors T_3 and T_4. Resistor R_1 is used to convert the base-emitter voltage difference into a current. The operational amplifier OA with the resistors R forces equal currents to flow through transistors T_1, T_2, T_3, T_4, and the resistor R_1. The resistors R_2, R_3 and transistor T_5 are added to obtain the band-gap current stabilizer with second-order temperature compensation. The output current I_{out} in the bottom part of the circuit is identical to the current that flows in the top part of the circuit. The relation between the circuit parameters and the temperature dependence of the reference current source will be calculated. The following equations will be used:

$$I_{out} = 2I_1 + I_2 + I_3 \tag{9.40}$$

$$I_1 R_1 = 2\frac{k}{q}T \ln p \tag{9.41}$$

$$I_2 R_2 = 2\frac{k}{q}T \ln \frac{I_1}{I_0} \tag{9.42}$$

$$I_3 R_3 = 2\frac{k}{q}T \ln \frac{I_1}{I_0} - \frac{k}{q}T \ln \frac{I_3}{I_0}. \tag{9.43}$$

During the calculation we suppose that all transistors are equal, that is, all emitter-base reverse currents I_0 are equal. Furthermore, equations 9.24 and 9.25 will be used to calculate the temperature dependence of the current

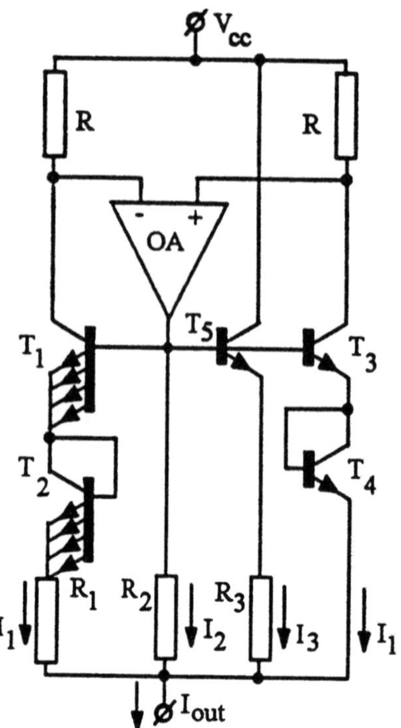

Figure 9.15 : Current reference source with second-order temperature compensation

source.

To obtain analytical expressions for the values of resistors R_1, R_2, and R_3 we have to solve the equations with the requirement that at $T = T_0$, $\frac{dI_{out}}{dT} = 0$. After differentiation and substitution we obtain:

$$\frac{dI_{out}}{dT} = \frac{I_{out}}{T} - \frac{V_g}{T}\left(\frac{2}{R_2} + \frac{1}{R_3}\right) -$$
$$\frac{k}{q}(n-1)\left(\frac{2}{R_2} - \frac{1}{R_3}\frac{V_g - I_3 R_3(n-1)}{\left(\frac{k}{q}T + I_3 R_3\right)(n-1)}\right). \tag{9.44}$$

In this equation a linear temperature relation is shown in the first part, while in the second part a nonlinear temperature dependence is shown. This nonlinear temperature relation introduces the already shown parabolic temperature dependence of the output current. However, in working through the equations a relation between the resistors R_2 and R_3 is found which can be adjusted in such a way that the parabolic temperature relation is canceled out. When at $T = T_0$ the nonlinear term is made zero we obtain for the ratio between R_2 and R_3:

$$\frac{R_2}{R_3} = \frac{2(n-1)(I_{30}R_3 + \frac{k}{q}T_0}{V_g - I_{30}R_3(n-1)}).$$
(9.45)

I_{30} is the current value of I_3 at $T = T_0$. At the same time the output current of the circuit at $T = T_0$ can be found from the following equation:

$$I_{out} = V_g(\frac{2}{R_2} + \frac{1}{R_3}).$$
(9.46)

This result shows a basically "ideal" band-gap reference source. The voltage drop across resistor R_3 which is equal to $I_3 R_3$ is close to the value of a base-emitter voltage of a transistor. This value is much larger than $\frac{k}{q}T$. Therefore equation 9.45 can be approximated by:

$$\frac{R_2}{R_3} \cong \frac{2}{\frac{V_g}{(n-1)I_{30}R_3} - 1}.$$
(9.47)

This ratio between R_2 and R_3 is obtained at the temperature $T = T_0$. The output current of the total system can be obtained by integrating equation 9.44. This is not possible without applying some simplifications. Therefore, we suppose that for temperatures that do not differ too much from $T = T_0$ the nonlinear term

$$\frac{k}{q}(n-1)(\frac{2}{R_2} - \frac{1}{R_3}\frac{V_g - I_3 R_3(n-1)}{(\frac{k}{q}T + I_3 R_3)(n-1)})$$
(9.48)

is a constant. The output current then becomes:

$$I_{out}(T = T_0) = V_g(\frac{2}{R_2} + \frac{1}{R_3}) + \frac{k}{q}(n-1) \times$$
$$(\frac{2}{R_2} - \frac{1}{R_3}\frac{V_g - I_{30}R_3(n-1)}{(\frac{k}{q}T_0 + I_{30}R_3)(n-1)})(1 - \ln\frac{T}{T_0}).$$
(9.49)

When the value of the resistor R_3 is made infinite, a well-known circuit configuration as shown in Figure 9.8 is obtained. The output current in the case of $R_3 \rightarrow \infty$ becomes:

$$I_{out}(T) = V_g\frac{2}{R_2} + \frac{k}{q}T(n-1)\frac{2}{R_2}(1 - \ln\frac{T}{T_0}).$$
(9.50)

Because two base-emitter voltages are connected in series, a factor 2 in output current value with respect to the value predicted by equation 9.39 is obtained.

A second important limit is obtained at the moment the value of the resistor $R_2 \to \infty$. In that case a parabolic temperature which is opposite with respect to the curve shown in Figure 9.11 is obtained. The output current in this case becomes:

$$I_{out}(T) = V_g \frac{1}{R_3} - \frac{k}{q}T(n-1)\frac{1}{R_3}(1 - \ln \frac{T}{T_0}). \tag{9.51}$$

When at $T = T_0$ the ratio between R_2 and R_3 is chosen as given by equation 9.47, then an exact relation for the output current is not available. The estimate shows that the cancellation of the parabolic temperature relation must occur. A computer analysis of the basic equation gives the best results.

9.7.3   Reference current source measurements

A practical current source as shown in Figure 9.15 is integrated in an A/D converter chip. Measurements have been performed to verify the formulas given above. The very low temperature dependence of this reference source makes it difficult to obtain accurate measurement results because the stability of this source is in the same order as the stability of the reference sources used in digital voltmeters. In Figure 9.16 the measurement results with an optimum resistor setting, and the two extremes are shown. It is clear that the above equations predict the temperature dependence of the reference source rather well. A detailed measurement result, which shows the optimum compensation situation of the reference source, is shown in Figure 9.17. The result shown in Figure 9.17 indicates that over a temperature change from -20° to + 85° C, a maximum change of ±50 ppm is attainable. Hysteresis effects over a temperature cycle are very small, < 10 ppm. A temperature coefficient of ±0.5 ppm/°C is obtained over the given temperature range. The output current of the reference source is 2 mA.

9.8   Noise of a band-gap reference current source

When reference current sources are applied in A/D converters, it is very important to know more about the noise behavior of this source. The bandwidth of comparators is much higher than the analog bandwidth that must

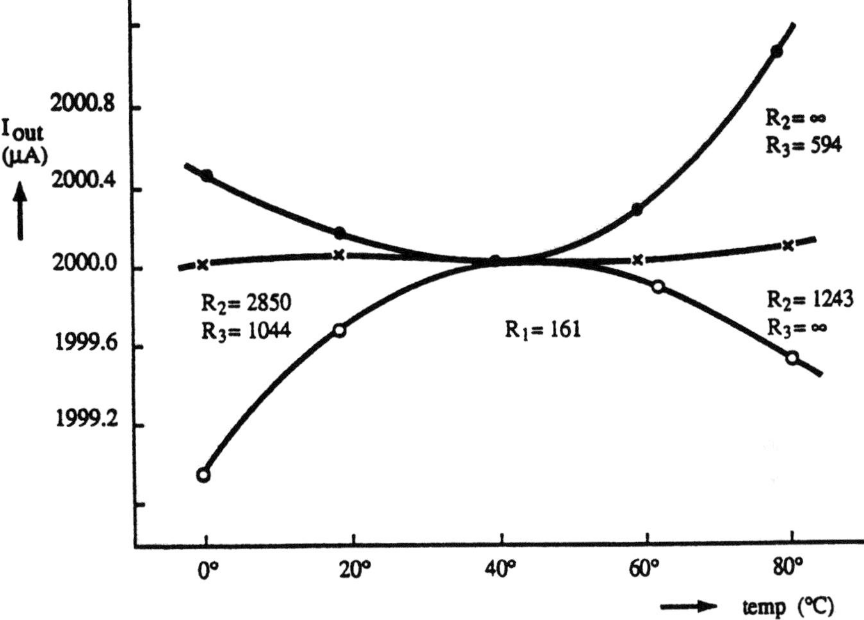

Figure 9.16 : Measurement results of a reference current source

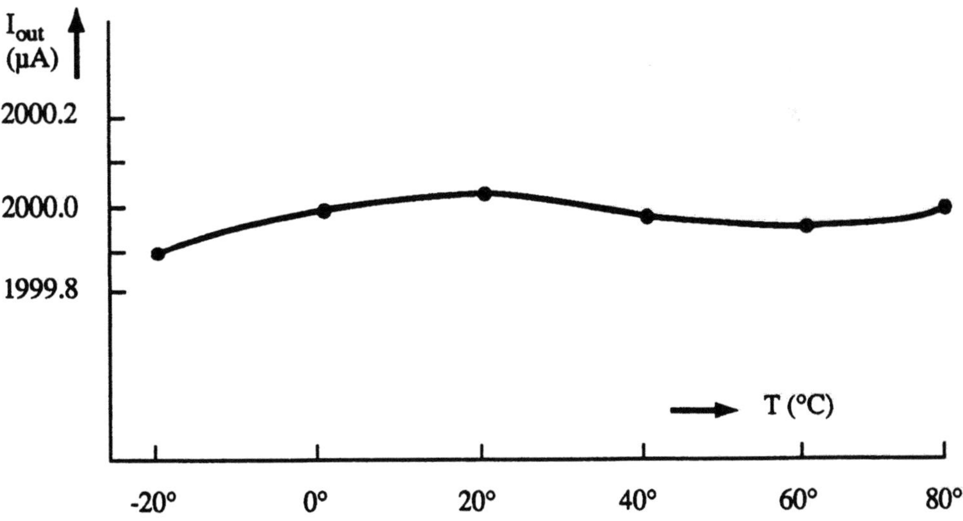

Figure 9.17 : Detailed measurement result of the reference current source

be converted. Therefore, in high-resolution A/D converters the reference source might limit the maximum resolution of the converter. To obtain an impression of the maximum resolution that can be obtained, the current noise of the reference source shown in Figure 9.13 is determined by computer simulation. When the value of the output current equals 1 mA, then the output current noise density with transistor sizes for which the internal base resistance $R_{bb} \ll \frac{1}{gm}$ tends to reach $I_{noise} = 30 \ pA/\sqrt{Hz}$. Supposing that the 1 mA output current is the value of the most significant bit of the converter, then an rms value for a sine wave that fits the full-scale value of the converter equals 0.7 mA. The signal-to-noise ratio of the current source can be calculated. We obtain:

$$S/N = 20 \log \frac{I_{signal}}{I_{noise}}. \tag{9.52}$$

Inserting the values given above results in: $S/N = 147$ dB $\sqrt{Hz}$. Over a bandwidth of 20 kHz a value of 104 dB is found.

The same calculation is performed for a current source built up according to Figure 9.15. The output current is adjusted to 1 mA. In this case the noise current equals 23 $pA/\sqrt{Hz}$. Signal-to-noise is calculated in the same way. The result becomes: $S/N = 150$ dB per $\sqrt{Hz}$, giving over a bandwidth of 20 kHz a value of 107 dB.

The doubling of the voltage drop across resistor R_1 improves the noise performance of the reference current source by 3 dB. A further decrease in noise could be obtained by increasing the voltage drop across resistor R_1. Such a configuration, however, leads to an impractical circuit solution because of the cascoding of a large amount of transistors requiring a large supply voltage.

From the noise analysis it is found that most noise contributions come from: resistor R_1, transistor T_2 and transistor T_1. Further noise additions are small and can be ignored. It must be noted furthermore that the equivalent input voltage noise of a transistor is:

$$e^2_{rms} = 4kT(R_{bb} + \frac{1}{2gm})\Delta f. \tag{9.53}$$

As a result an increase of the reference current by a factor two results in a decrease in noise current by $\sqrt{2}$ or 3 dB. The base resistors of the transistors must be small with respect to $\frac{1}{gm}$ to validate this statement. The signal-to-noise ratio of the system of Figure 9.15 amounts to 110 dB over 20 kHz bandwidth.

The noise bandwidth of the current source can be reduced by inserting a capacitor across the collector-base junction of transistor T_3. The noise reduces by a factor ten for frequencies above $f_{band} = \frac{1}{2\pi RC}$.

9.9 Conclusion

After the introduction of some basic voltage and current stabilizer circuits using special zener diodes, stabilizers with well-determined temperature coefficients are presented that can be used in several parts of a converter system. In integrated circuits accurate and stable reference sources can be designed. These reference sources use the band-gap of silicon as the reference voltage. This band-gap voltage is nearly temperature-independent. Therefore, reference sources that use a second-order temperature compensation with temperature coefficients below 1 ppm/^{0}C can be built. The noise performance of band-gap circuits is good, but improvements are still necessary to be able to apply these reference sources in wide band, high-resolution A/D converters. The signal-to-noise ratio of the described reference sources is adequate for applications in 16- to 18-bit D/A converters in digital audio systems without the need for additional filtering.

Chapter 10

Noise-shaping coding

10.1   Introduction

In this chapter noise-shaping techniques to improve the dynamic range of a system will be described. Noise-shaping can be very useful when speed can be exchanged with accuracy. The quantization errors in a noise-shaping system are removed from the signal band of interest. Mostly the suppressed quantization errors appear enlarged as out-of-band noise in the system. With a simple filter these errors are removed. An increased dynamic range of the coder is obtained. In digital systems word length can intelligently be reduced using a noise-shaping operation without losing dynamic range significantly. An ultimate in bit reduction is obtained when the noise-shaping operation reduces the number of bits to 1. Examples of such an operation is sigma-delta analog-to-digital conversion or noise-shaping digital-to-analog conversion based on single-bit word-lengths. The advantage of a 1-bit converter is the extreme linearity of such a device. A very good differential linearity is obtained with these converters. The most important design criteria for these converters will be given. At the moment the dynamic range of a system must be enlarged, but the maximum clock rate of the system cannot be increased because of technology limitations, then a multi-bit digital-to-analog converter can be used in the feedback loop. At that moment, however, the linearity of the digital-to-analog converter determines the linearity and the distortion in the system. To overcome this problem *Dynamic Element Matching* or *Continuous Current Calibration* techniques can be used to obtain the extreme linearity of the D/A converter without needing extra trimming steps. In the stability analysis of the noise-shaping coders the root-locus method can be successfully applied. Examples for first-, second-, and third-order coders will be treated. A coder is said to be stable when idling occurs at the highest possible frequency. In this case a maximum of oversampling ratio is obtained resulting in a maximum dynamic range. Mostly an idle pattern having a frequency around half the sampling frequency gives the best result. However, idle patterns at other frequencies depending on the input signal randomly generated are also allowed. In this case the quantization errors are randomized and appear as noise. Examples of designed 16- to 18-bits digital-to-analog converters will be shown [101,94,95].

10.2   Digital oversampling filtering

In this section an example of oversampling to increase the dynamic range of a digital-to-analog converter including a combined digital and analog output filtering will be described. The analysis method will be extended into the following sections for applications to noise-shaping coders.

10.2.1   Combined digital-analog D/A output filter

In Figure 10.1 a block diagram of a combined digital-analog output filter for audio applications including the filter response is shown [13]. The system consists of a Finite Impulse Response (FIR) digital filter followed by an oversampled D/A converter. A third-order Bessel filter with a nearly linear phase removes the signal bands around the high sampling frequency. The digital filter performs the steep signal filtering at the pass band edge, while the analog filter reduces the repeated spectrum of the input signal at a frequency that is equal to four times the sampling frequency. In this system the input word-length is 16 bits. The filter coefficients have a word-length of 12 bits. As a result, at the output of the filter a minimum word-length of 28 bits is obtained. By using an oversampling and noise-shaping operation, the output word-length is reduced to 14 bits without a significant loss in signal-to-noise ratio. In the next sections the noise-shaping operation will be explained. Noise-shaping is an intelligent reduction of the word-length in a digital signal processing system without significantly reducing the dynamic range.

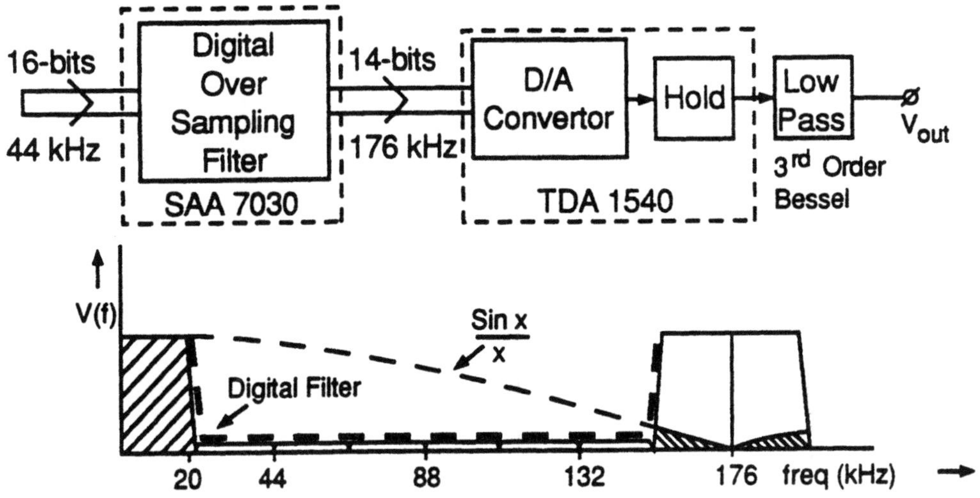

Figure 10.1 : Combined digital-analog low-pass output filter

10.2.2 Digital filter configuration

In Figure 10.2 a simplified block diagram of an oversampling digital low-pass filter is shown. The filter uses a 96-tap FIR structure to perform the low-pass filtering operation. The input word-length is 16 bits, while the coefficients have a 12-bit word-length. To obtain the oversampling operation, four sets of 24 coefficients having a word-length of 12 bits are used to calculate the output words at a four times higher sampling frequency. These words appear at the output with a 28-bit word-length at the higher sampling rate. At least 12 of the least significant bits do not contribute to the overall performance of the system. With a noise-shaping technique the number of significant output bits can be further reduced. The total dynamic range of the system in this case is only slightly reduced when a D/A converter with only 14 bits is connected to the output. However, to obtain the increased dynamic range of the system at the full output signal range, a 16-bit linearity of the 14-bit converter is needed. When the 14-bit converter gives a good 14-bit linearity performance, then the quantization noise is reduced and the distortion of the converter determines the dynamic range of the system. In offset-binary coding, glitches are important in this application.

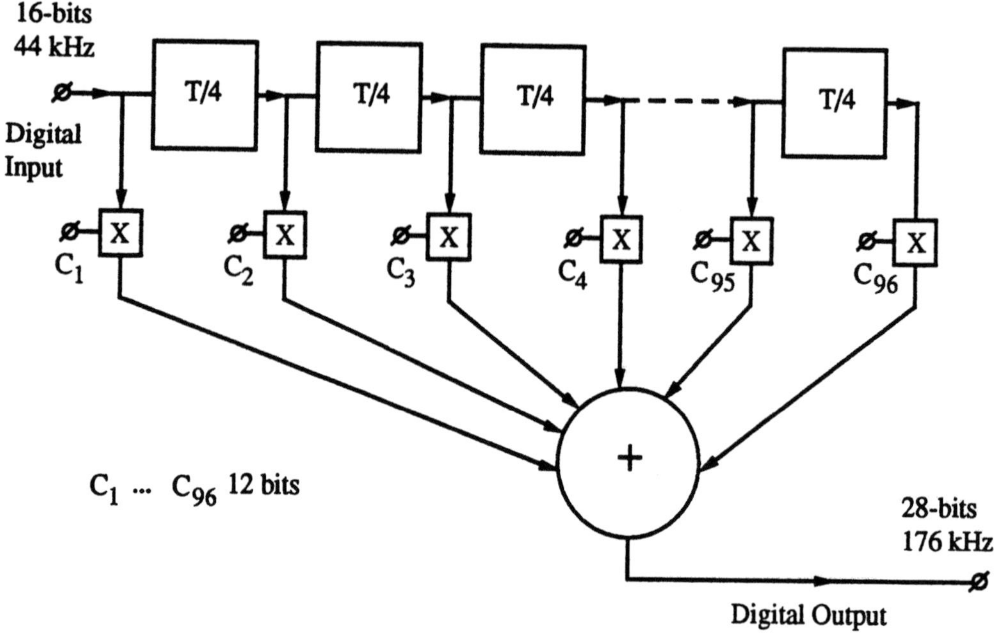

Figure 10.2 : Block diagram of an oversampling filter

10.2.3　Quantization errors

In Figure 10.3 the quantization error as a function of frequency is shown. The shaded area equals the quantization error over half the sampling frequency. Part (a) of the figure shows the quantization error of the input signal, while part (b) shows the effect of the four times oversampling operation. From Chapter 1 it is known that the noise density with a four times oversampling ratio is reduced with a factor four. Limiting the bandwidth of the system to $\frac{1}{8}f_s$ shows a four times reduction in the total quantization error (part (c) of the figure). As a result, an increase in dynamic range with 1 bit is obtained. A second operation is needed to improve further the dynamic range of the system in the signal band.

10.3　Noise-shaping coders

In this section noise-shaping techniques are described that can be applied to increase the dynamic range of digital-to-analog converters without increasing the number of bits. In case a large oversampling is used, the number of bits in the digital-to-analog converter can be decreased to 1 bit. These 1-bit coders

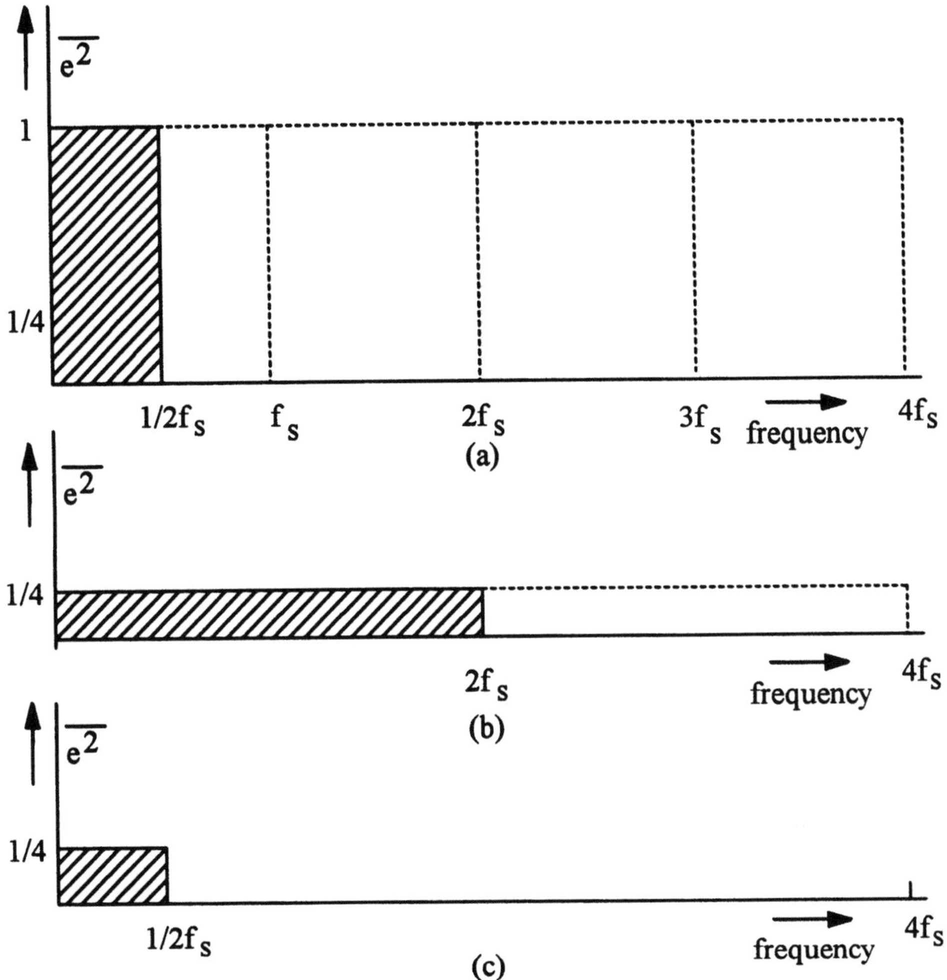

Figure 10.3 : (a), (b) and (c) Quantization errors as a function of oversampling ratio

will show a large dynamic range and have a very low distortion without needing accurate elements.

10.3.1 First-order noise-shaper

A circuit diagram of a noise-shaper is shown in Figure 10.4. At the input of the system a 28-bit input signal is applied. The word-length is reduced into a minimum number of significant bits using an intelligent reduction algorithm. As can be seen from Figure 10.4, the most significant 14 bits are applied to

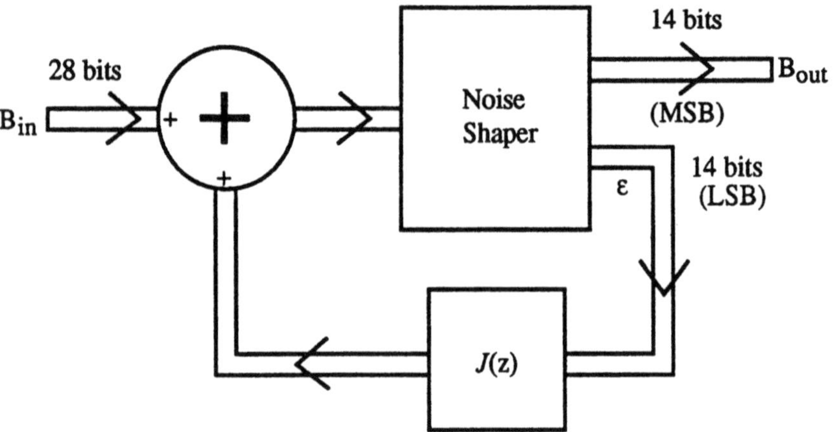

Figure 10.4 : General noise-shaper

the output that drives the D/A converter. The lower 14 bits, denoted with ε, are filtered by the error filter $J(z)$ and added to the input word. The simplest implementation of $J(z)$ is one clock delay so: $J(z) = z^{-1}$. In this way a first-order filtering operation will be performed. The error that is obtained by this operation can be calculated. A general expression for a noise-shaping function using Figure 10.4 yields:

$$B_{in}(z) + \varepsilon(z)J(z) = B_{out}(z) + \varepsilon(z). \tag{10.1}$$

After rearranging equation 10.1 the output signal can be expressed as:

$$B_{out}(z) = B_{in}(z) - \varepsilon(z)(1 - J(z)). \tag{10.2}$$

Normalizing equation 10.2 we obtain:

$$\frac{B_{out}(z)}{B_{in}(z)} = 1 - \frac{\varepsilon(z)}{B_{in}(z)}(1 - J(z)). \tag{10.3}$$

From equation 10.3 it is found that the signal transfer function is equal to 1, while the second part of the equation

$$\varepsilon(z)(1 - J(z)) \tag{10.4}$$

determines the error at the moment the output word length is reduced.

The term $(1 - J(z))$ is usually called the *noise transfer function* and determines the coloring of the output noise. Because the signal gain equals 1, this error can be related to the input signal as well. In case the word length of a system is reduced to N bits, then a value

$$\frac{\varepsilon}{B_{in}} = \frac{1}{2^N - 1} \approx 2^{-N} \qquad (10.5)$$

is obtained. The value of ε is thus equal to the LSB value of the N-bits output word. The noise-shaping operation reduces the error even more, as will be shown. From equation 10.2 it is seen that the spectral density of the error signal at the output of the system is determined by the filter operation $(1 - J(z))$.

To obtain zero error at $z = 1$ ($f_{in} = 0$ Hz) the term $(1 - J(1))$ must be zero. This requires a suitable choice for $J(z)$.

In Figure 10.5 a first order noise-shaper is shown. A first-order noise-shaper

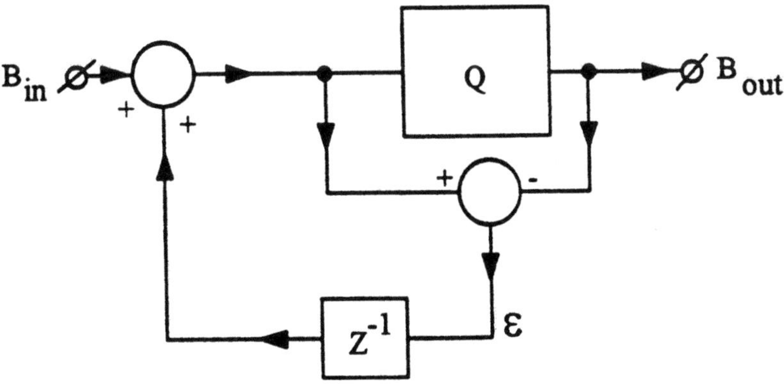

Figure 10.5 : First-order noise-shaper

uses a one clock delay (z^{-1}) for $J(z)$. Then we obtain:

$$B_{out}(z) = B_{in}(z) - \varepsilon(z)(1 - z^{-1}). \qquad (10.6)$$

The first order filter operation reduces the error $\varepsilon(z)$. Especially at low frequencies ($z \approx 1$) a large reduction is obtained. The total amount of quantization error is found by integrating $\varepsilon(z)$ over the signal bandwidth f_b. Supposing that the quantization error density is frequency independent in the band ($f = 0$ to $f = \frac{f_s}{2}$), then the result becomes:

$$e_{tot}^2 = \int_0^{\theta_1} \varepsilon^2 |\, 1 - z^{-1}\,|^2 d\theta. \qquad (10.7)$$

In this equation $| 1 - z^{-1} |$ is the amplitude characteristic of a first-order noise filter and $\theta_1 = \frac{2\pi f_b}{f_s}$.

Note that f_s is equal to the *output* sampling frequency!

Inserting $z = e^{j\theta}$ we obtain:

$$| 1 - z^{-1} |^2 = \frac{| e^{j\theta} - 1 |^2}{| e^{j\theta} |^2}. \tag{10.8}$$

Working out equation 10.8 the following result for the amplitude characteristic is obtained:

$$| 1 - z^{-1} |^2 = 2(1 - \cos \theta). \tag{10.9}$$

In Figure 10.6 the square of the amplitude response as a function of frequency is shown. Note that at half the sampling frequency the gain is 2 ($z = -1$). Inserting equation 10.9 into equation 10.7 and integrating the function results in:

$$e_{tot}^2 = 2(\theta_1 - \sin \theta_1)\varepsilon^2. \tag{10.10}$$

Without noise-shaping the total uniformly distributed noise over the band 0 to Θ_1 is equal to:

$$e_{uniform}^2 = \varepsilon^2 \int_0^{\theta_1} d\theta = \varepsilon^2 \theta_1. \tag{10.11}$$

In comparing the results of equations 10.10 and 10.11, the improvement with respect to noise and dynamic range of the system is obtained. Using F_1 as the dynamic range improvement factor we get:

$$F_1 = \frac{e_{tot}}{e_{uniform}} = \sqrt{2(1 - \frac{\sin \theta_1}{\theta_1})}. \tag{10.12}$$

In this equation $\theta_1 = \frac{2\pi f_b}{f_s}$.

The oversampling factor obtained by the digital filtering operation introduces an extra decrease in noise with a factor G. This factor can be expressed in terms of sample rate f_s and signal bandwidth f_b or:

$$G_1 = \sqrt{\frac{\theta_1}{\pi}} = \sqrt{\frac{2 f_b}{f_s}}. \tag{10.13}$$

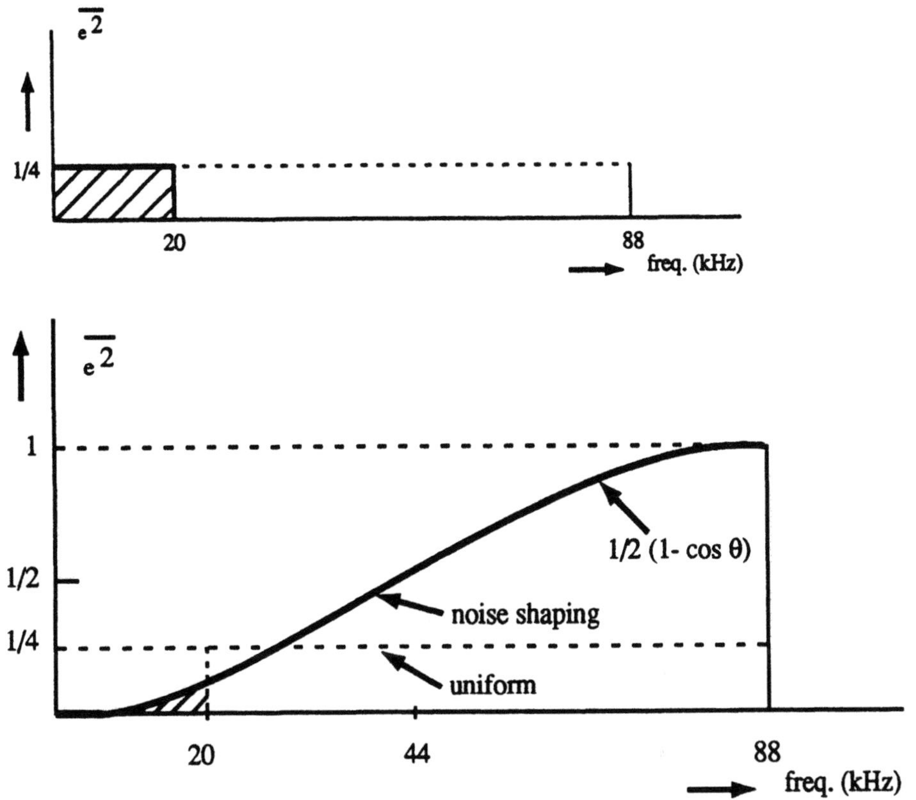

Figure 10.6 : Square of the filter amplitude response

The total improvement H_1 in the signal-to-noise ratio due to noise-shaping and oversampling becomes:

$$H_1 = \frac{1}{F_1 G_1} = \frac{1}{2\sqrt{\frac{f_b}{f_s}\left(1 - \frac{sin2\pi f_b/f_s}{2\pi f_b/f_s}\right)}}. \tag{10.14}$$

Inserting values for $f_s = 176$ kHz and $f_b = 20$ kHz, H_1 becomes 14.2 dB. The signal-to-noise ratio of a 14-bit D/A converter with oversampling filter and noise-shaping becomes $86.0 + 14.2 = 100.2$ dB. At the input of the filter an infinite word length is assumed. This is not true for a practical system. At the input of the system a 16-bit input word is applied. The input noise and the rounding noise of the noise-shaper must be added to obtain the total noise in the system. In this calculation it is supposed that the rounding errors are not correlated and behave like noise sources. As a result of the noise-shaping operation the signal-to-noise of the system becomes:

$$S/N_{system} = \frac{S/N_{16bits}}{\sqrt{1 + k^2}}. \tag{10.15}$$

k is the ratio between the signal-to-noise ratio of the 16-bits input word (98.1 dB) and the signal-to-noise ratio of the noise shaped oversampled 14-bit digital-to-analog converter (100.2 dB). With $k = 0.776$, the signal-to-noise ratio of the system becomes:

$$S/N_{system} = 96.0 \ dB. \tag{10.16}$$

The total operation of noise-shaping and a four times oversampling results in an increase of the dynamic range of the 14-bits D/A converter with nearly 2 bits. An excellent digital-to-analog conversion system including linear phase low-pass filtering with a dynamic range of nearly 16 bits, is obtained.

10.3.2 Second-order noise-shaper

At the moment the order of the noise-shaping filter is increased from a first-order into a second-order, then an improvement in dynamic range is expected. The filter function in the feedback loop can be determined using equation 10.2. We obtain:

$$1 - J(z) = (1 - z^{-1})^2. \tag{10.17}$$

In case of a second-order noise-shaper $J(z)$ becomes:

$$J(z) = z^{-1}(2 - z^{-1}). \tag{10.18}$$

In Figure 10.7 the implementation of equation 10.18 is shown. The improvement in dynamic range due to noise-shaping with a second-order noise-shaping function can be performed in the same manner as was done to get F_1 (10.12). After some calculations we obtain:

$$F_2 = \sqrt{6 + \frac{2 \sin \theta_1 \cos \theta_1 - 8 \sin \theta_1}{\theta_1}} \tag{10.19}$$

The improvement due to oversampling remains identical to 10.13 so:

$$G_2 = \sqrt{\frac{\theta_1}{\pi}} = \sqrt{\frac{2 f_b}{f_s}}. \tag{10.20}$$

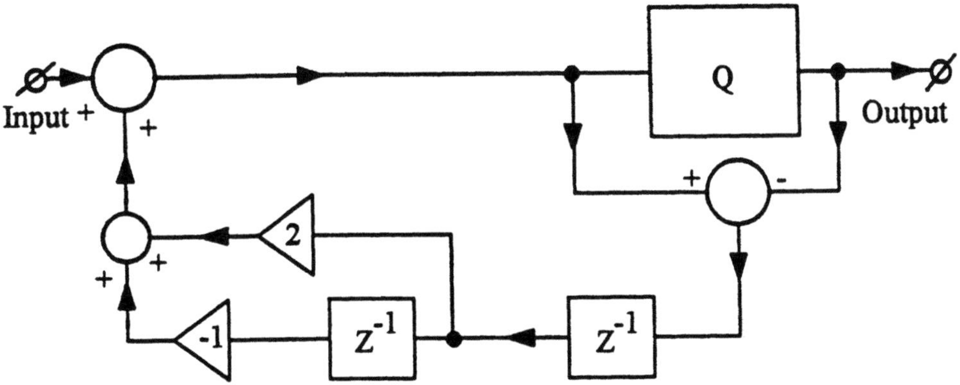

Figure 10.7 : Second-order filter implementation

The total improvement in dynamic range H_2 of a second-order system becomes:

$$H_2 = \frac{\sqrt{\pi}}{\sqrt{6\theta_1 + 2\sin\theta_1 \cos\theta_1 - 8\sin\theta_1}}. \qquad (10.21)$$

By inserting values for $f_s = 176$ kHz and $f_b = 20$ kHz, H_2 becomes 19.5 dB. In Figure 10.8 the noise-shaping operation as a function of frequency for a first- and second-order filter function are shown. Note that over the signal band of interest (0 to f_b) the total noise is reduced with the second-order filtering operation. The improvement in dynamic range when the noise-shaping filter order is increased from a first-order into a second-order equals 5.3 dB, which is close to 1 bit.

10.3.3   Third-order noise-shaper

The filter order of the noise-shaper can again be increased from second order into third order. However, problems with stability of the system can arise. Therefore it is important to discuss two possibilities:

1. The output number of bits is larger than 1 (multi-bit converter).

2. The output number of bits is limited to 1 (Sigma-delta converter).

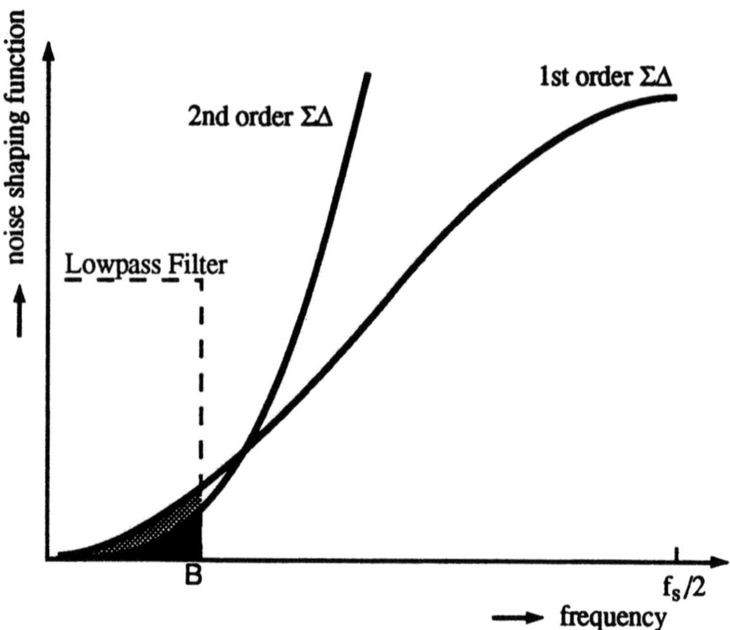

Figure 10.8 : First- and second-order noise-shaping functions

Case 1)

The number of output bits is larger than 1.
The noise transfer function in this case becomes:

$$1 - J(z) = \left(1 - z^{-1}\right)^3, \tag{10.22}$$

giving for the filter function:

$$J(z) = z^{-1}\left(3 - 3z^{-1} + z^{-2}\right). \tag{10.23}$$

The improvement in dynamic range due to noise-shaping with a third-order
noise-shaping function is calculated in the same manner as was done to get
F_1 (10.12). After some calculations we obtain:

$$F_3 = \sqrt{20 + \frac{-30\sin\theta_1 + 6\sin 2\theta_1 - \frac{2}{3}\sin 3\theta_1}{\theta_1}} \tag{10.24}$$

The improvement due to oversampling remains identical to 10.13, so:

$$G_3 = \sqrt{\frac{\theta_1}{\pi}} = \sqrt{\frac{2f_b}{f_s}}. \tag{10.25}$$

The total improvement of a third-order multi-bit noise-shaper becomes:

$$H_3 = \frac{\sqrt{\pi}}{\sqrt{20\theta_1 - 30\sin\theta_1 + 6\sin 2\theta_1 - \frac{2}{3}\sin 3\theta_1}}. \qquad (10.26)$$

By inserting values $f_b = 20$ kHz and $f_s = 176$ kHz we obtain $H_3 = 16.02$ or 24.1 dB.

Case 2)

The number of output bits is reduced to 1.

The loop gain in the system increases in such a way that the stability of the system must be analyzed. This stability analysis will be performed in section 11.6. To generalize the problem the approach used in reference [11] will be applied here. The noise-shaping function is changed into:

$$1 - J(z) = \left(\frac{1 - bz^{-1}}{1 - az^{-1}}\right)^n. \qquad (10.27)$$

In case of a third-order noise-shaper 10.27 changes into:

$$1 - J(z) = \left(\frac{1 - z^{-1}}{1 - az^{-1}}\right)^3. \qquad (10.28)$$

In reference [11] a stable operation of the system is obtained when $b = 1$ and $a = 0.5$. The transfer function for $J(z)$ becomes:

$$J(z) = z^{-1}\frac{1.5 - 2.25z^{-1} + 0.875z^{-2}}{(1 - 0.5z^{-1})^3}. \qquad (10.29)$$

In Figure 10.9 an implementation of equation 10.29 is shown. The improvement in noise performance due to a third-order noise shaper can be calculated again. However, because of the stability problem the implemented function is more difficult to integrate.

Using $b = 1$ but keeping the variable a in the equation, we obtain with:

$$T_{3a} = \frac{1}{a^3}$$

$$U_{3a} = \frac{2(1 + 5a + 10a^2 - 10a^3 - 5a^4 - a^5)\arctan\left[\left(\frac{a+1}{a-1}\right)\tan\frac{\theta_1}{2}\right]}{\theta_1 a^3(1 + a)^5}$$

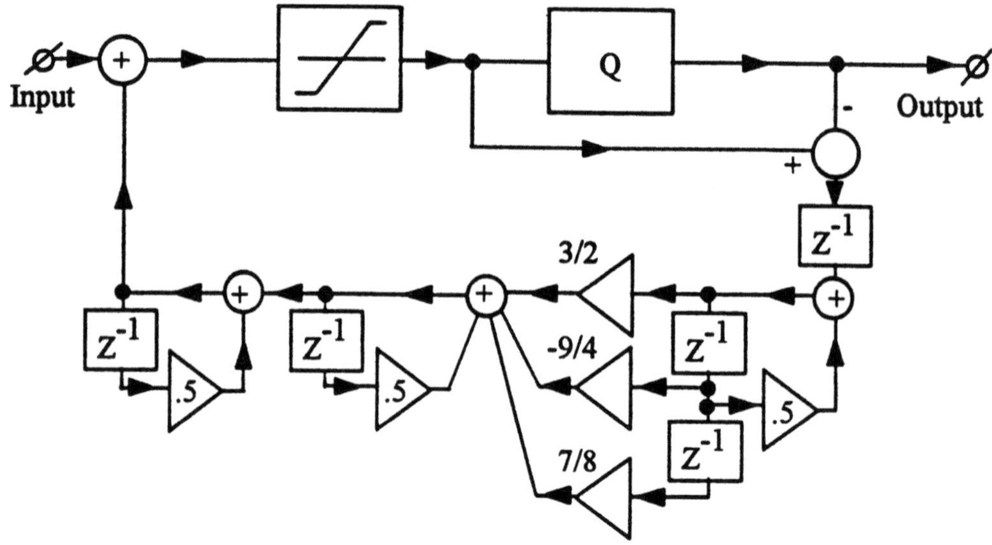

Figure 10.9 : Third-order filter implementation

$$V_{3a} = \frac{3(a-1)^2(1+4a+a^2)\sin\theta_1}{\theta_1 a^2(1+a)^4(1+a^2-2a\cos\theta_1)}$$

$$W_{3a} = \frac{(a-1)^4\sin\theta_1}{\theta_1 a^2(1+a)^2(1+a^2-2a\cos\theta_1)^2} \tag{10.30}$$

and $0 < a \leq 1$ for F_3:

$$F_{3a} = \sqrt{T_{3a} + U_{3a} + V_{3a} - W_{3a}}. \tag{10.31}$$

The notation (3a) is used to indicate that we are talking about a third-order noise-shaper having an extra stabilizing function $(1 - az^{-1})$ in the filter section. The total improvement in noise-shaping H_{3a} can be calculated, resulting in:

$$H_{3a} = \frac{\sqrt{\pi}}{\sqrt{\theta 1 (T_{3a} + U_{3a} + V_{3a} - W_{3a})}}. \tag{10.32}$$

Inserting values in 10.32 of $a = 0.5$, $f_b = 20$ kHz and $f_s = 176$ kHz, we obtain: $H_{3a} = 4.52$ or 13.1 dB. The expected improvement in signal-to-noise ratio is smaller then expected with a third-order noise-shaper. This is due to the term $(z - a)^3$ required to obtain a stable system.

10.3.4 Fourth-, fifth-, and sixth-order noise-shaper

Depending on the number of output bits used in the system, again a split-up has to be made into:

1. The number of output bits is larger then 1,

2. The number of output bits is 1.

Case 1)

The noise shaping function becomes:

$$1 - J(z) = (1 - z^{-1})^n \text{with n } = \text{ 4, 5, and 6.} \qquad (10.33)$$

The improvement in noise-shaping becomes:

$$F_n = \sqrt{\frac{\int_0^{\theta_1} [2(1 - \cos \theta)]^n \, d\theta}{\theta_1}}. \qquad (10.34)$$

Even with higher-order filter functions, the improvement due to oversampling remains equal to:

$$G_n = \sqrt{\frac{\theta_1}{\pi}}. \qquad (10.35)$$

The total improvement due to noise-shaping and oversampling generalizes into:

$$H_n = \frac{1}{H_n G_n} = \frac{\sqrt{\pi}}{\sqrt{\int_0^{\theta_1} [2(1 - \cos \theta)]^n \, d\theta}}. \qquad (10.36)$$

The analytical results of the integrals for n = 4 to 6 are not given in this section because of the complexity of the formulas.

In Figure 10.10 the improvement in signal-to-noise ratio of a multi-bit system as a function of oversampling ratio and filter order n are shown. From Figure 10.10 it is seen that at the moment the oversampling ratio is larger then 4 nearly a linear relation between oversampling and increase in signal-to-noise ratio is obtained. In Figure 10.11 a detailed part of Figure 10.10 with an oversampling ratio between 1 and 10 is shown. From Figure 10.11 it is found that oversampling ratios smaller than 4 do not give an increase in signal-to-noise ratio of the system when the order of the noise shaper is increased.

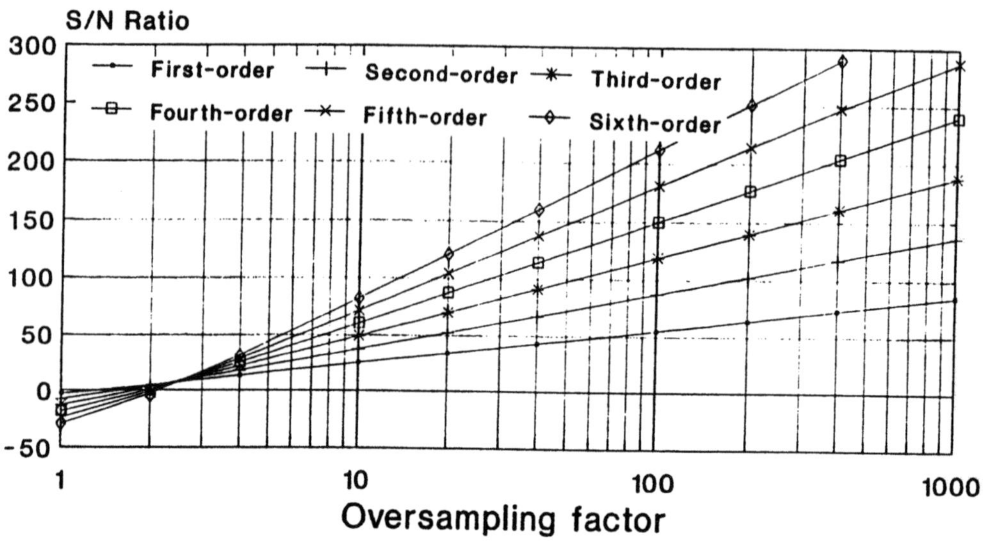

Figure 10.10 : Signal-to-noise improvement as a function of oversampling ratio and filter order in a multi-bit system

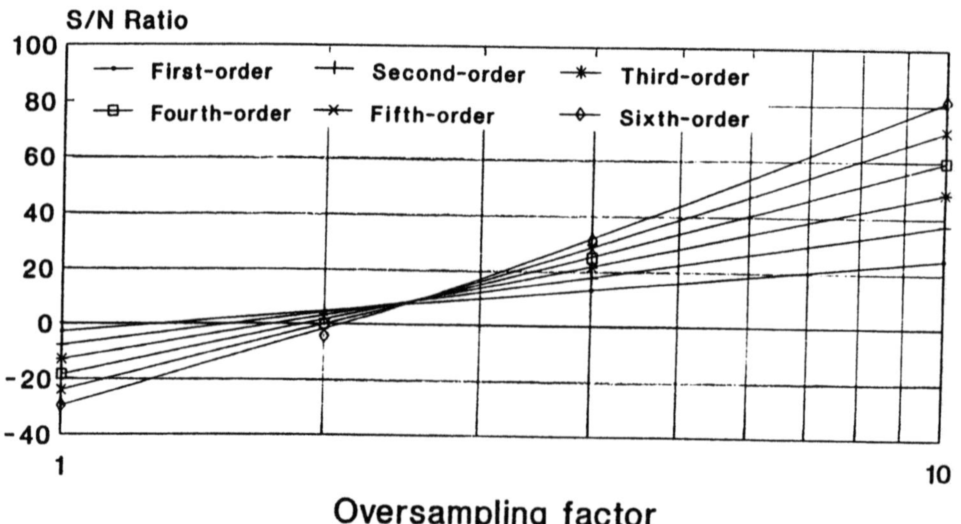

Figure 10.11 : Signal-to-noise improvement of multi-bit system with a 1 to 10 oversampling ratio

Case 2)

The noise-shaping function $1 - J(z)$ is again assumed to be:

$$1 - J(z) = \left(\frac{1 - z^{-1}}{1 - az^{-1}} \right)^{n}. \tag{10.37}$$

However, in cases for n = 4 to 6, the integrals are more difficult to obtain. By using advanced computer programs [89] it is possible to get analytical expressions for the noise-shaping improvement function H_n. These analytical expressions are quite large and are therefore not presented in this section. However, from a designer's viewpoint it is important to get an insight in the

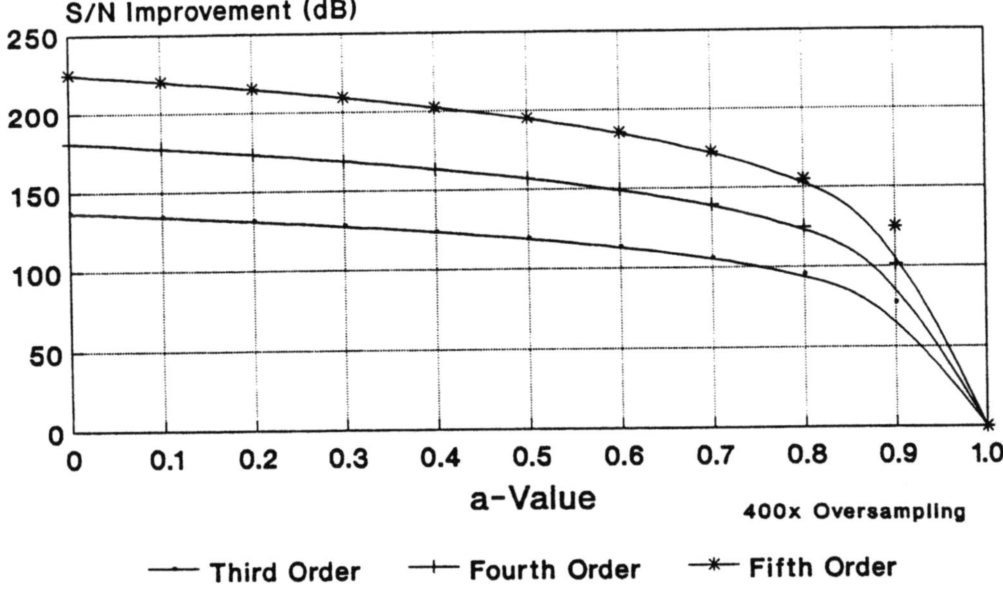

Figure 10.12 : Normalized noise-shaping improvement function

improvement in dynamic range or signal-to-noise ratio at the moment the order of the noise-shaping function is increased and at the same time a large oversampling ratio is used. To simplify the understanding of the results, the improvement factor H_n is normalized with respect to the oversampling factor G_n, so:

$$H_{normalized} = H_n \sqrt{\frac{\theta_1}{\pi}}. \tag{10.38}$$

As a result of the normalization operation the improvement reduces to 1 at the moment the factor $a = 1$. Depending on the order of the noise-shaper not all values of a result in a stable operation of the system. However, when the number of output bits is increased the value of a tends to decrease while maintaining stability of the system. In Figure 10.12 the improvement in dynamic range due to noise-shaping as a function of the value of a, the order of the noise-shaper from a third-order to a sixth-order noise-shaping function are shown. A large oversampling factor of 400 is used in this figure. Note, furthermore, that this figure is only valid for a 1-bit noise-shaper!

In practical systems the stability of the noise-shaping system must be determined resulting in minimum values for the factor a. In the following table 10.1 an indication of practical values for a as a function of filter order are given. These values are valid for a **1-bit output code!** It must be noted

n	a
3	0.54
4	0.66
5	0.73
6	0.81

Table 10.1 : Filter order and minimum stability value

that the presented values for a are indications for systems that show a good optimum between noise-shaping improvement, stability at large signals, and implementation complexity. In Figure 10.13 the improvement in signal-to-noise ratio due to oversampling and noise-shaping for a 1-bit system as a function of the filter order is shown. The presented results use values for the stabilization term according to Table 10.1. From Figure 10.13 it is found that for a filter order above 2 the increase in signal-to-noise ratio is only obtained when the oversampling ratio is large than about 40. Below an oversampling ratio of 40 a lower order filter gives a better improvement. To show this in more detail, Figure 10.14 gives the signal-to-noise improvement in case of an oversampling ratio between 10 and 100. Again from Figure 10.13, it is seen that for large oversampling ratios (more than 40) a linear increase in dynamic range with increasing oversampling ratio is obtained.

In the next section an approximation of the increment in signal-to-noise ratio as a function of filter order and oversampling ratio for a multi-bit system

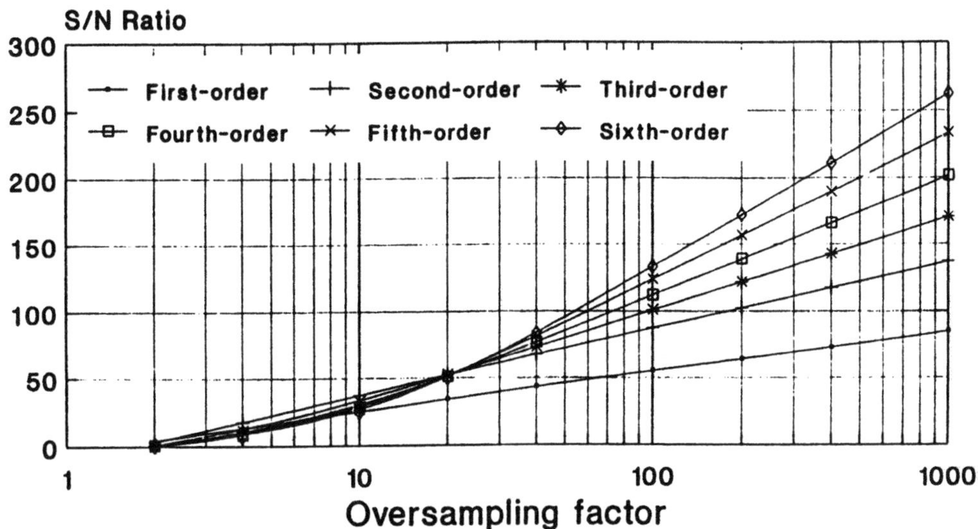

Figure 10.13 : Signal-to-noise improvement of a 1-bit system as a function of oversampling ratio and filter order

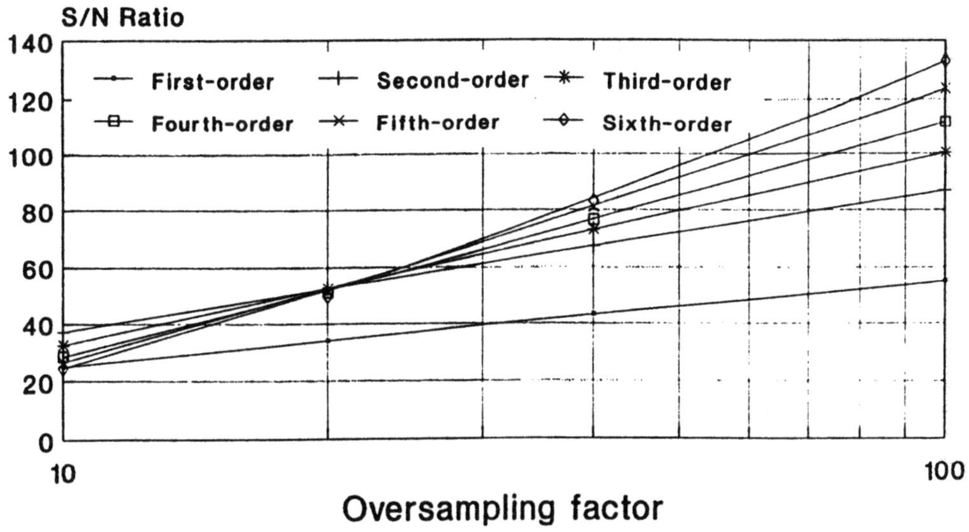

Figure 10.14 : Detail of signal-to-noise improvement of a 1-bit system with oversampling ratios between 10 and 100

will be given.

At this point it must be noted that at the moment the number of output bits is between 2 and 6 for systems having third- to sixth-order noise-shaping functions, the value of a must be determined and the increase in dynamic range according to Figure 10.12 can be used to determine the improvement in signal-to-noise ratio of such a system. Mostly designers use simulations to come to the required results.

10.3.5   Largely oversampled noise-shaper

Assume that the filter function in the feedback loop is given by:

$$1 - J(z) = (1 - z^{-1})^n. \tag{10.39}$$

Using this filter in an nth order noise-shaper, then equation 10.12 changes into:

$$F_n = \sqrt{\frac{\int_0^{\theta_1} [2(1 - \cos\theta)]^n \, d\theta}{\theta_1}}. \tag{10.40}$$

When the sampling frequency f_s in a system is much larger than the signal bandwidth f_b, then $\cos\theta_1$ can be approximated by:

$$\cos\theta_1 \approx 1 - \frac{\theta_1^2}{2!}. \tag{10.41}$$

Inserting 10.41 into 10.40, this equation simplifies into:

$$F_n \approx \sqrt{\frac{\int_0^{\theta_1} \theta^{2n} \, d\theta}{\theta_1}}. \tag{10.42}$$

After integration we obtain:

$$F_n \approx \frac{\theta_1^n}{\sqrt{(2n + 1)}}. \tag{10.43}$$

Because of the oversampling the additional factor G_n is equal to:

$$G_n = \sqrt{\frac{\theta_1}{\pi}}. \tag{10.44}$$

The total improvement of a largely oversampled nth order noise-shaper becomes:

$$H_n \approx \theta_1^{-(n+\frac{1}{2})} \sqrt{\pi(2n + 1)}. \tag{10.45}$$

When the oversampling ratio being ($\frac{\pi}{\theta_1}$ or $\frac{f_s}{2f_b}$) is increased with a factor two, the signal-to-noise ratio increases according to equation 10.45 with a factor: $2^{n+\frac{1}{2}}$.

When a limited oversampling ratio is used, then the exact solution of the integrals 10.40 must be used.

10.4 Multi-bit largely oversampled noise-shaper

The signal-to-noise ratio at the output of a multi-bit system can easily be calculated. When the amount of output bits is small, then equation 6.26 must be applied. Suppose that m output bits are used, then the signal-to-noise ratio becomes:

$$S/N_{quantizer} = (2^m - 2 + \frac{4}{\pi})\sqrt{1.5}H_n. \tag{10.46}$$

m is the number of output bits.
In equation 10.46 H_n represents the improvement in dynamic range due to the nth order noise-shaping filter. When the oversampling ratio is large then equation 10.46 can be expressed as:

$$S/N_{quantizer} = (2^m - 2 + \frac{4}{\pi})\sqrt{\frac{1.5f_s}{2f_b}(\frac{f_s}{2\pi f_b})^n}\sqrt{2n+1}. \tag{10.47}$$

It must be noted that the calculated results of equations 10.46 and 10.47 are *only* valid for multi-bit systems or systems that have $a = 0$ and are stable.

At the moment 1-bit systems are used, a correction for the stability term in the noise-shaping function is required that results in deviations from the given equation. The correction factor must be introduced at the moment third-order or higher-order 1-bits systems are analyzed. Use Figure 10.12 to obtain the correction factor depending on the stabilization variable a.

Equation 10.47 predicts that the signal-to-noise ratio increases with more than 6 dB per bit when the number of output bits is increased from a 1-bit signal into a multi-bit signal. The nonlinear relation of the signal-to-noise ratio on the number of bits with a small number of used bits causes this phenomenon. In Table 10.2 calculation results are shown.
In this table n is the filter order and m is the number of bits of the D/A

	S/N dB ($f_s/2f_b = 100$)				
n	m=1	m=2	m=3	m=4	m=5
1	58.7	66.9	73.8	80.3	86.5
2	91.0	99.2	106.1	112.5	118.8
3	N.A.	130.7	137.6	144.1	150.3
4	N.A.	N.A.	168.8	175.2	181.5
5	N.A.	N.A.	N.A.	206.2	212.4

Table 10.2 : S/N ratio of a multi-bit quantizer

converter. (N.A. = Not Applicable).

Furthermore when these quantizers are included in a system with a finite number of input bits, then the dynamic range of the total system is calculated from:

$$S/N_{system} = \frac{S/N_{input}}{\sqrt{1 + (S/N_{input}/S/N_{quantizer})^2}}. \tag{10.48}$$

The signal-to-noise ratio of the input is determined by the number of input bits applied to the system, while the signal-to-noise ratio of the quantizer is determined by the construction of the quantizer. Equation 10.47 determines in this case the S/N of the quantizer.

10.5 Stability analysis of noise-shapers

A noise-shaping system shows an idling pattern with the error signal being noise like or consisting of limit cycles that can be disturbed by the input signal and occur at sub-multiples of the sampling frequency such as $\frac{f_s}{2}$, $\frac{f_s}{3}$ and so on. These limit cycles are part of the error signal and must be distinguished from unstable behavior. Instability of the noise-shaping feedback loop gives rise to limit cycles at relatively low frequencies ($\frac{f_s}{25}$). These limit cycles are usually not effected by the input signal, and if such a limit cycle is present, it persists even if the input signal is removed. Hence the stability of a noise-shaper cannot be related to the stability of conventional linear systems.

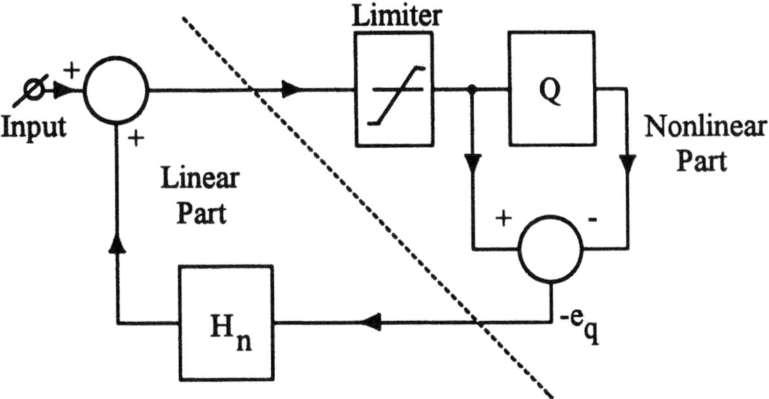

Figure 10.15 : Partitioning of a noise-shaper for stability analysis

Therefore, the criterion that will be used in this book is that a noise-shaper is unstable if a limit cycle can be present that disturbs the noise-shaping mechanism and that remains present even when the input signal is removed from the system.

10.5.1 Noise-shaper stability model

To perform the stability analysis of noise-shapers, a simple model will be used. In Figure 10.15 a partitioning of the noise-shaper into a linear part and a nonlinear part is shown. In the linear part of the system comprises the loop filter and all additional loop delays. The nonlinear part consists of the quantizer Q and the subtracter from which the quantization error is obtained. The limiter represents a limitation of the signal caused by overload of the system by large input signals. As a result of the nonlinear operation of the quantizer this function will be replaced by a global transfer λ. The global transfer λ represents the transfer of the fundamental waveform that is assumed to be present in the loop. In Figure 10.16 the stability model to be used during the analysis is given. In the case of a multi-level quantization, the quantizing steps are small when related to the input of the quantizer. As a result, the quantization error that is applied to the noise-shaping filter is small related to the input signal. Thus the global transfer λ is small, requiring a large value for $| H_N(\theta) |$ of the loop filter to sustain a stable limit cycle. Moreover, since there is no or very limited correlation between the input signal and the error signal, an unwanted limit cycle is not to be expected in this case.

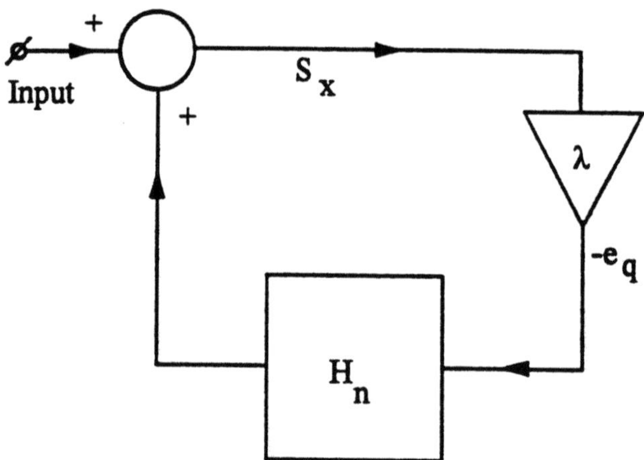

Figure 10.16 : Stability analysis model for noise-shapers

10.5.2   Root Locus stability analysis method

The Root Locus method is a simple and easy to apply method to analyze
the stability of a system. In case of a sampled data system, a stable region
of the system is obtained when this region is inside the unit circle. The
stability criterion using Figure 10.16 we have to analyze is:

$$\lambda H(z) = 1. \tag{10.49}$$

A system is stable when for all values of λ the roots of 10.49 are within the
unit circle $|z| = 1$.

The following rules apply for the construction of a root locus.

1. Loci originate on poles and terminate on zero of $H(z)$.

2. The root locus on the real axis always lies in a section of the real axis
   to the left of an odd number of poles and zeros on the real axis.

3. The root locus is symmetrical with respect to the real axis.

4. The number of asymptotes is equal to the number of poles of $H(z)$,
   n_p, minus the number of zeros n_z.

5. The breakaway points are given by the roots of:
   $$\frac{d(H(z))}{dz} = 0$$

or by: $D(z)\frac{dN(z)}{dz} - N(z)\frac{dD(z)}{dz} = 0.$
with $H(z) = \frac{N(z)}{D(z)}.$

10.5.3 First-order system

In a first-order system $J(z) = z^{-1}$. The root locus starts at $z = 0$ and ends at $z = \infty$. Because of the feedback loop λ cannot be positive. In Figure 10.17 the root locus of the first order system is shown. The figure shows

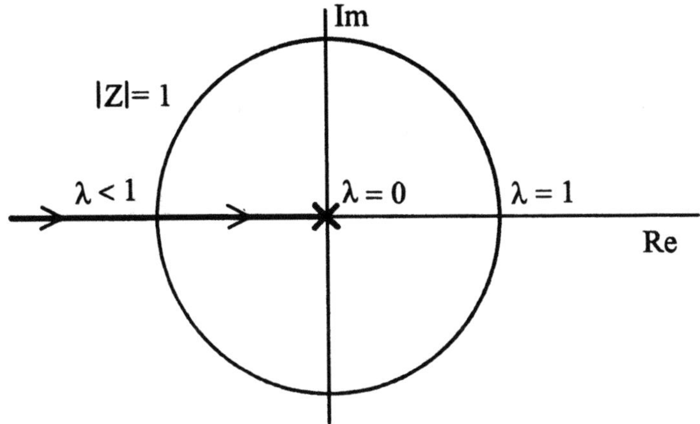

Figure 10.17 : Root locus of a first-order system

that for $z = -1$ $(f = \frac{1}{2}f_s)$ and $\lambda = -1$ the system becomes instable. This instability, however, is required for the idle pattern of a noise-shaper.

10.5.4 Second-order system

In a second-order coder the filter function $J(z)$ is implemented by:

$$J(z) = z^{-1}(2 - z^{-1}) = \frac{2z - 1}{z^2}. \tag{10.50}$$

The root locus starts at the double pole $z = 0$ and ends at $z = 0.5$ and $z = \infty$. The breakaway points are obtained from:

$$z(1 - z) = 0, \tag{10.51}$$

resulting in $z = 0$ and $z = 1$. The root locus of this system is shown in Figure 10.18. The circle is centered at $z = (0.5, 0)$ and has a radius of 0.5. Again this system becomes unstable for $z = -1$ and $\lambda = -\frac{1}{3}$. However, this instability is required and gives the idle pattern at half the sampling frequency.

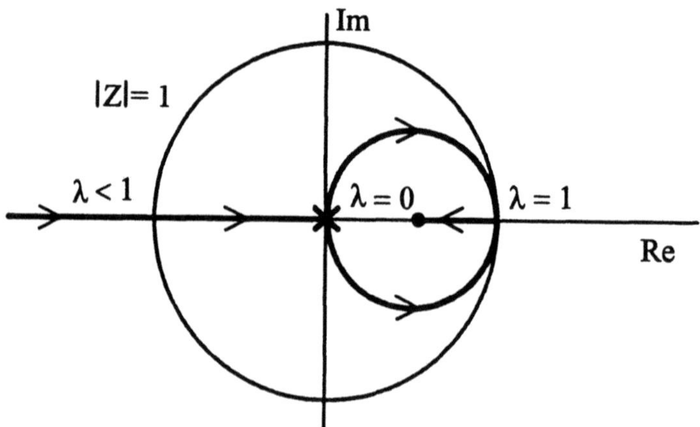

Figure 10.18 : Root locus of a second-order system

10.5.5   Third-order system

A split-up between a 1-bit coder and a multi-bit coder is needed.

1-bit Coder implementation

A general implementation of a third-order coder filter can be written as:

$$J(z) = \frac{3(1-a)z^2 - 3(1-a^2)z + (1-a^3)}{(z-a)^3}. \tag{10.52}$$

Here a determines the value of the stabilization function and $b = 1$. Inserting $a = 0.5$ in equation 10.52 we obtain:

$$J(z) = \frac{1.5z^2 - 2.25z + 0.875}{(z-0.5)^3}. \tag{10.53}$$

The root locus of 10.53 starts at the three-fold pole $z = 0.5$ and ends at $z = \infty$ and

$$z_{1,2} = \frac{3}{4} \pm j\frac{1}{12}\sqrt{3}. \tag{10.54}$$

The breakaway points are found from:

$$(z-0.5)^2(z-1)^2 = 0. \tag{10.55}$$

The results are double points at $z = 0.5$ and $z = 1$. The circles are centered at $z = (\frac{3}{4}, \pm\frac{1}{12}\sqrt{3})$. The root locus of this system is shown in Figure 10.19.

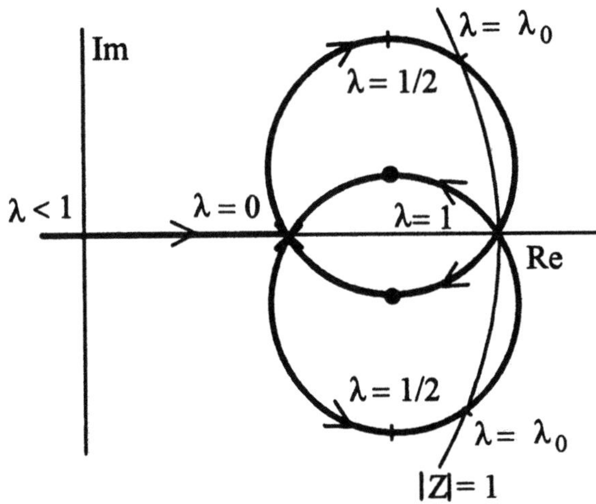

Figure 10.19 : Root locus of a 1-bit third-order system

From the root locus figure it can be seen that the system becomes unstable with increasing λ. The value of λ for which the unit circle is crossed is determined by:

$$\lambda_0 = \frac{1}{1 + \left(\frac{2\cos(\alpha)}{1+a}\right)^n}. \tag{10.56}$$

n determines the filter order ($n = 3, 4, 5, 6$, etc.) and $\alpha = \frac{\pi}{n}$.
The derivation of equation 10.56 will be given in subsection 10.5.7.
Inserting $n = 3$ and $a = 0.5$, we obtain for $\lambda_0 = 0.771$.
The value for λ_0 can also be obtained from the root locus plot. By graphically measuring the distance between the unity circle cross point and the zero and pole positions we obtain λ_0 from:

$$\frac{1}{\lambda_0} = \frac{1.5 z_1 z_2}{p_1^3}. \tag{10.57}$$

From the root locus plot we obtain: $z_1 = 0.29$, $z_2 = 0.55$, and $p_1 = 0.57$, resulting in $\lambda_0 \approx 0.774$. This graphically determined value is close to the exactly calculated value of 0.771. This means that the maximum gain of the quantizer in the loop may not exceed the value 0.771 to avoid unwanted limit cycles. A proper choice of input and output levels of the quantizer is required. Finally with $z = -1$ λ becomes -0.73.

Multi-bit third-order system

In a multi-bit third-order system the value of a can be chosen 0. Then the filter function $J(z)$ changes into:

$$J(z) = z^{-1}(3 - 3z^{-1} + z^{-2}).\tag{10.58}$$

The root locus of this system starts from a triple pole at $z = 0$ and ends at $z_1 = \infty$ and

$$z_{2,3} = \frac{1}{2} \pm j\frac{1}{6}\sqrt{3}.\tag{10.59}$$

The circles are centered around $z = (\frac{1}{2}, \pm\frac{1}{6}\sqrt{3})$. The root locus of this system is shown in Figure 10.20. As can be seen from Figure 10.20, this system

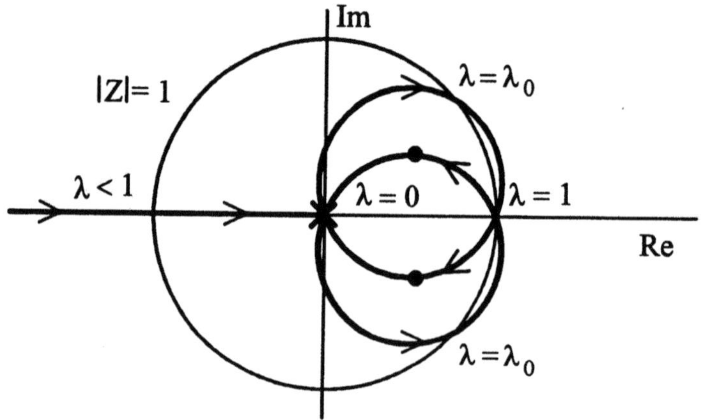

Figure 10.20 : Root locus of a third-order multi-bit system

is unstable for $z = -1$. However, this instability is required as idle pattern at half the sampling frequency.

Using equation 10.56 the value of λ_0 for which the unit circle is crossed with $a = 0$ is found to be 0.5. When a multi-bit system is used, then the error is $(2^N - 1)$ times smaller than in a 1-bit system. As a result the value of

$$\frac{\lambda_0}{2^N - 1}\tag{10.60}$$

is smaller than 1 and the system will be stable. As a result, the third-order system becomes stable with $N \geq 2$, because the quantizer gain never exceeds 1.

10.5.6 Fourth-order system

A 1-bit implementation and a multi-bit system will be discussed. It must be noted here that multi-bit means $N \geq 4$. When systems with 2 and 3 output bits are used, then a specific analysis for this problem must be performed. The value for a in the stabilization term will be between $a = 0$ and $a = \frac{2}{3}$.

1-bit fourth-order system

The transfer function of this system becomes:

$$J(z) = \frac{4z^3(a-1) - 6z^2(a^2-1) + 4z(a^3-1) - (a^4-1)}{(z-a)^4}.$$ (10.61)

The root locus starts at the quadruple pole on $z = a$ and ends on:

$$z_1 = \frac{a+1}{2}$$ (10.62)

$$z_{2,3} = \frac{(a+1)}{2} \pm j\frac{|a-1|}{2}$$ (10.63)

$$z_4 = \infty.$$ (10.64)

The breakaway points are $z = a$ and $z = 1$. The root locus of this system is shown in Figure 10.21. Using equation 10.56, the value of λ for which

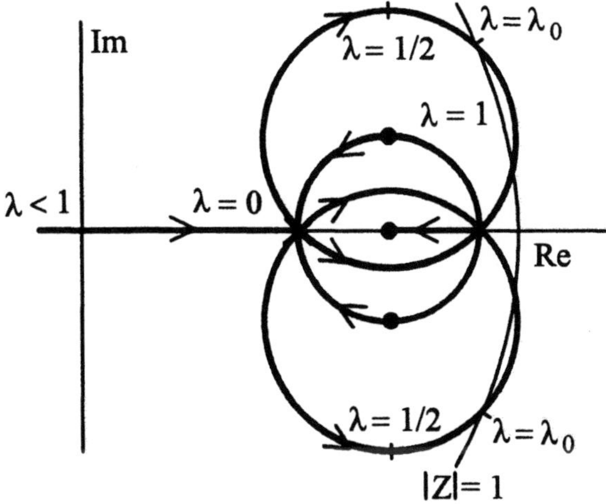

Figure 10.21 : Root locus of 1-bit fourth-order system

the root locus crosses the unit circle substitution of $a = 0.66$ gives $\lambda_0 = 0.65$.

The global gain of the quantizer must be chosen smaller then 0.65 to obtain a stable system. With $z = -1$ λ becomes -0.9.

Multi-bit fourth-order system

In a stable multi-bit system the largest improvement in dynamic range is obtained with $a = 0$. $J(z)$ then becomes:

$$J(z) = \frac{4z^3 - 6z^2 + 4z - 1}{z^4}.\tag{10.65}$$

The root locus starts at the quadruple pole $z = 0$ and ends at $z_1 = \frac{1}{2}$, $z_{2,3} = \frac{1}{2} \pm j\frac{1}{2}$ and $z = \infty$. The breakaway points are $z = 0$ and $z = 1$. The root locus of this system is similar to Figure 10.21 except the shift from a to $\frac{1}{2}$. The value of λ determined with $a = 0$ from equation 10.56 results in: $\lambda_0 = 0.2$. In multi-bit systems with output bit-numbers between 1 and 4 the value of a needed in a stable system changes from 0.66 to 0. A lower value for a can be used because with increasing bit number the error is reduced with $(2^N - 1)$.

10.5.7   Maximum global gain calculation

The maximum global gain of higher-order system starting with a third-order system giving the edge of stability can be calculated.
Starting from the stability criterion:

$$\lambda J(z) = 1.\tag{10.66}$$

and inserting for $J(z)$:

$$J(z) = 1 - \frac{(z-b)^n}{(z-a)^n}.\tag{10.67}$$

we obtain:

$$\frac{(z-b)^n}{(z-a)^n} = 1 - \frac{1}{\lambda}.\tag{10.68}$$

Introducing the variable Λ with $\pm\Lambda^n = 1 - \frac{1}{\lambda}$ equation 10.68 changes into:

$$\frac{(z-b)^n}{(z-a)^n} = \pm\Lambda^n.\tag{10.69}$$

Taking the nth root of equation 10.69 we obtain:

$$\Lambda = \frac{(z-b)}{(z-a)}e^{-j\alpha},$$ (10.70)

with $\alpha = k\frac{\pi}{n}$ and $k = 0, 1, .. , n - 1$. Solving equation 10.70 for z we obtain:

$$z = \frac{\Lambda a e^{j\alpha} - b}{\Lambda e^{j\alpha} - 1}.$$ (10.71)

At the moment the root locus touches the unit circle we obtain: $|z^2| = 1$. Inserting this value in equation 10.71, we obtain:

$$1 = |z^2| = \frac{b^2 - 2ab\Lambda\cos(\alpha) + a^2\Lambda^2}{1 - 2\Lambda\cos(\alpha) + \Lambda^2}.$$ (10.72)

Solving Λ from equation 10.72 in case $b = 1$ we obtain:

$$\Lambda = \frac{2\cos(\alpha)}{(1+a)}.$$ (10.73)

The value of λ_0 can be found from $-\Lambda^n = 1 - \frac{1}{\lambda}$ and we obtain:

$$\lambda_0 = \frac{1}{1 + \left(\frac{2\cos(\alpha)}{(1+a)}\right)^n},$$ (10.74)

with $\alpha = \frac{\pi}{n}$ and n is the filter order.

Note that only the negative value of Λ is used in equation 10.74 because the smallest value of λ_0 is restrictive for the stability.

The value of λ_0 is valid for values of a starting from 0 to nearly 1.

In Figure 10.22 the values of λ_0 as a function of a and the order of the loop filter are shown. When the maximum global gain is determined in a system, then the filter order and the stabilizing factor can be determined using Figure 10.22. Having determined the filter order and the stabilizing factor a, then the improvement in signal-to-noise ratio can be obtained using the curves from Figure 10.10 and Figure 10.12. The filter order can be optimally chosen in this way.

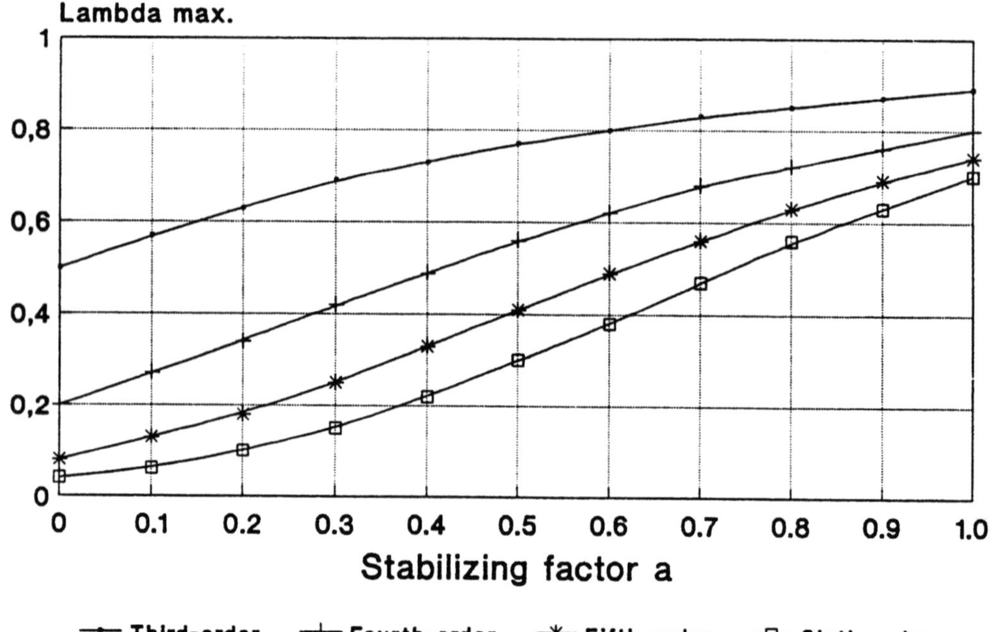

Figure 10.22 : Global gain as a function of a and the loop filter order

10.6   Practical noise-shaping D/A converters

In this section applications of the given theoretical treatment will be shown.
Most of the practical examples are intended for digital audio system applications.

10.6.1   16-bit D/A converter system

In this section an example of a 16-bit noise-shaping D/A converter system
will be described.

When an oversampling ratio of about 100 is used with a second-order noise-
shaper, a 16-bit D/A converter for applications in digital audio systems can
be designed.
In Figure 10.23 a block diagram of the total D/A converter is shown. At
the input of the converter the 16-bits serial data are converted into a 16-bits
parallel data word with a sampling frequency f_s. This conversion is performed by the block called SIPO. The 16-bit parallel data is then increased
in sample rate with a factor 4 using the first oversampling filter. This filter

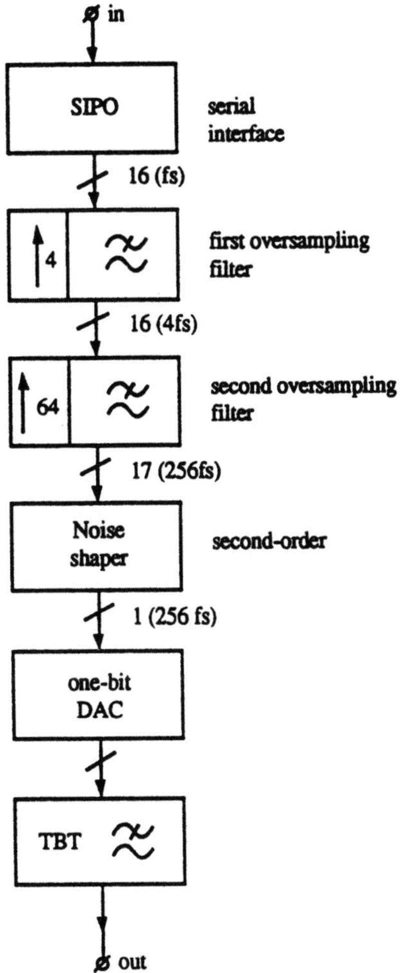

Figure 10.23 : Block diagram of an oversampled D/A converter system

can be of the type shown in Figure 10.2. The output sampling rate is $4f_s$. Then a second oversampling with a factor 64 is introduced. In this filter a linear interpolation with a factor 64 is used between two data points. As a result of this linear interpolation, a filtering with an amplitude response of $(\frac{\sin x}{x})^2$ is obtained. The output sample rate is now increased to $256f_s$. The second oversampling filter is followed by a second-order noise-shaper. The output of this noise-shaper is a 1-bit signal with a sample rate of $256f_s$. This signal is applied to a 1-bit D/A converter that reproduces the analog signal. The output of the 1-bit D/A converter is followed by a linear phase low-pass filter to remove the high-frequency components from the audio signal.

In Figure 10.24 the frequency spectrum of the oversampling filters is shown. In Figure 10.25 the amplitude response of the pass band of the total filter is shown. The amplitude decrease of the analog low-pass filter (Thomson Butterworth Thomson) is compensated for by an amplitude increase of the digital filter. As a result, a very accurate filtering characteristic is obtained with a very small pass-band ripple (less than 0.01 dB). As is shown in Figure

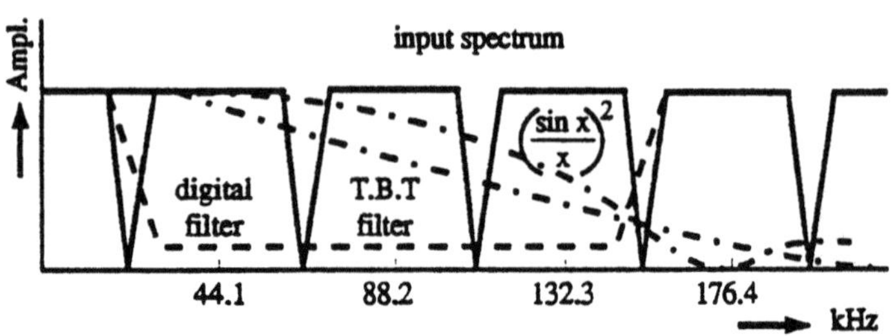

Figure 10.24 : D/A converter output signal frequency spectrum

10.24, a sharp filtering in the first four times oversampling filter is needed to reject the repeated input spectra of the digital signal with frequency f_s = 44.1 kHz. A block diagram of the second order noise-shaper is shown in Figure 10.27. The quantizer Q in Figure 10.26 generates the 1-bit signal. In higher-order noise-shapers it is necessary to implement a limiter function to avoid overloading of the system that may lead to unwanted limit cycles. In normal operation the limiter is transferring the input signal to the subtracter. The output of the subtracter is applied to the noise-shaping filter $H(\omega)$. In a first-order noise-shaper $H(\omega)$ is implemented as a delay. In a second-order system a more complex transfer function for $H(\omega)$ is needed. In Figure 10.27 a switched capacitor implementation of a 1-bit D/A converter using a single reference voltage V_1 is shown. The system consists of a set of switches driven by the two phases of the clock signal CL. A charge "bucket" is stored in capacitors C_1 and C_2, respectively. The two switches driven by the signals P^+ and P^- determine if a charge packet ($Q = C_1V_1 = C_2V_1$) is added or subtracted from the output signal. The output signal is generated and filtered by C_3 and R_1 using an operational amplifier to keep the voltage swing V_2 very small. The filter constant C_3R_1 is part of the analog low-pass output filter which suppresses the high frequency spectra of the signal. In Figure 10.28 the frequency spectrum of the 1-bit D/A converter signal without analog low-pass filtering is shown. From the figure it is seen that the noise is

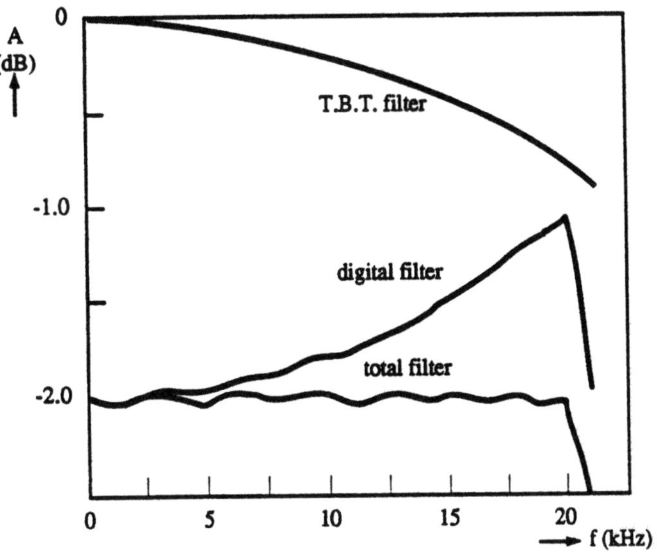

Figure 10.25 : Pass band amplitude response

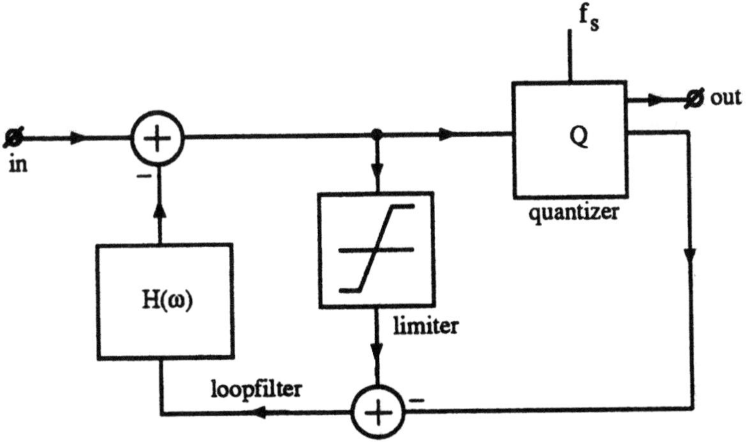

Figure 10.26 : Block diagram of a noise-shaper

removed from the pass-band and entered in the system at higher frequencies. As a result of this filtering effect the dynamic range of the system increases over the (small) signal band of interest. The figure visualizes the noise-shaping operation.

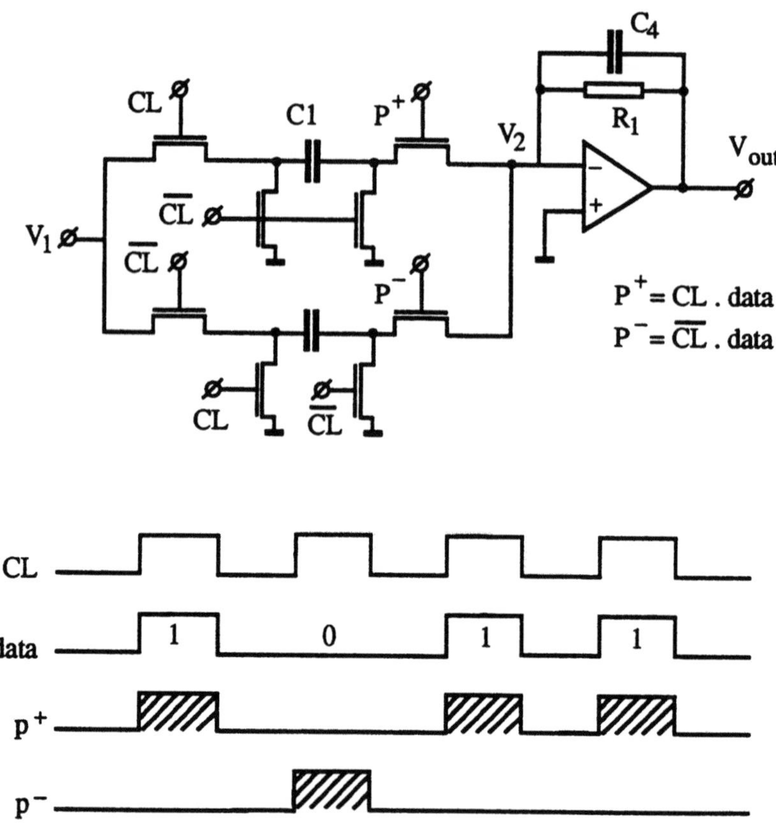

Figure 10.27 : Switched capacitor 1-bit D/A converter

10.6.2 18-bit D/A converter system

The limiting factor in high resolution D/A converters based on 1-bit D/A
converters with a large oversampling factor and a higher-order noise-shaper
is found to be the analog switching part with the output operational ampli-
fier(s) and analog output filters. In a practical design much attention must
be paid to the analog signal reconstruction part of the converter [94]. The
18-bit example described in this section uses basically the same digital signal processing part as the system of the previous section. In a 5 V CMOS
process the maximum analog signal swing is limited. The noise floor present
in a system determines the maximum dynamic range of such a system. In
a full differential system implementation the signal swing can be doubled
without increasing the supply voltage. A maximum increase in dynamic
range with 3-6 dB can be obtained, depending on the construction of the
output amplifiers. The differential circuit implementation compensates for

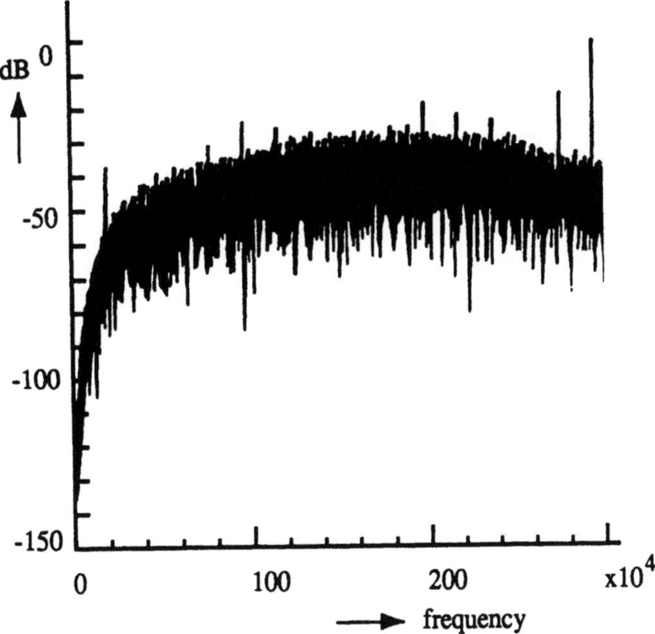

Figure 10.28 : 1-bit D/A converter spectrum

imperfections of, for example, switches and reduces the sensitivity of a system for supply voltage variations. In Figure 10.29 a fully differential circuit implementation of a dual 1-bit D/A converter is shown. The 1-bit D/A converter part consists per channel out of a logic part with switch driver circuitry, a switched capacitor differential D/A network and two operational amplifiers with highly linear feedback elements and a differential-to-single ended output amplifier. The external feedback elements are used to obtain the extreme linearity required for 18-bits resolution and linearity. At the output of the two operational amplifiers driven by the differential switched capacitor network two counter phase analog signals appear. These two signals are subtracted to obtain the analog output signal of the converter. In Figure 10.30 a block diagram of the system per channel is shown. Note that the output of the differential switched capacitor D/A converter is a charge. This charge is converted into a voltage by the two operational amplifiers with the external feedback resistors.

In Figure 10.31 the circuit diagram of the differential switched capacitor D/A converter is shown. Note that in this system again a single reference voltage is used for charging and discharging. Non overlapping clocks are needed to drive this system. The digital data are entered with the switches marked D

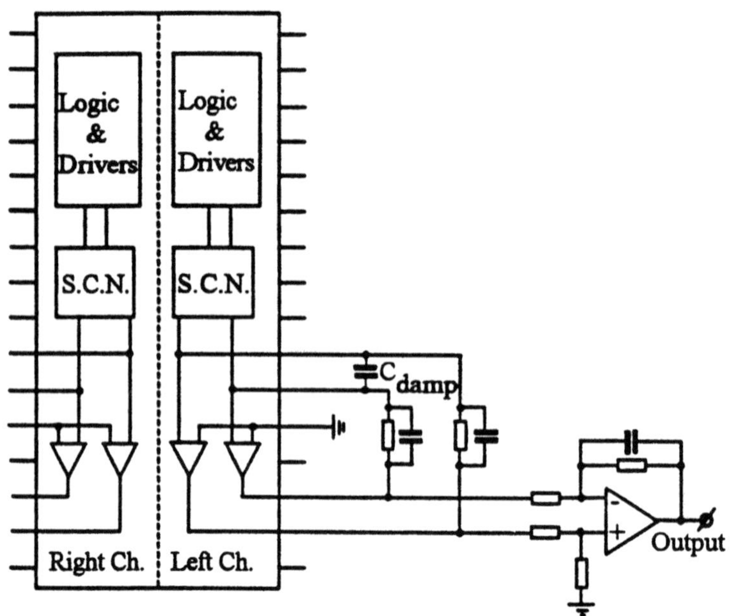

Figure 10.29 : Differential 1-bit D/A converter implementation

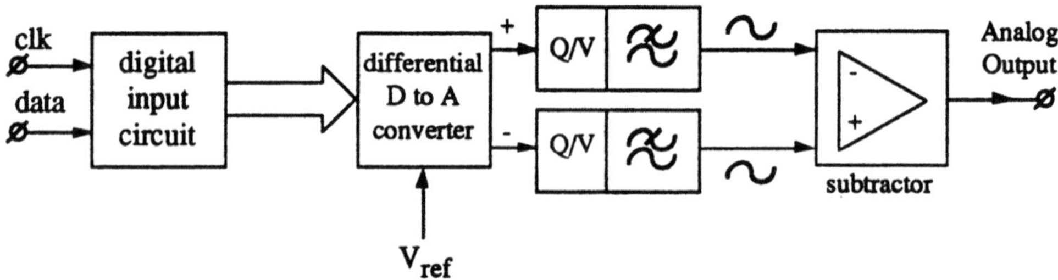

Figure 10.30 : Differential system topology

and $\bar{D}$. In Figure 10.32 the timing diagram of the non-overlapping clocks is shown. In a BiCMOS technology high-performance operational amplifiers can be designed by using a two stage system configuration with a Miller integrator frequency compensation. In Figure 10.33 the basic circuit diagram of such an amplifier is shown. In the input stage PMOS devices are used to perform the necessary level shift. Large device sizes are used to minimize the noise. The dynamic performance of the system is shown in Figure 10.34 and Figure 10.35. The measurements include quantization errors and distortion. The distortion in the system shows that with a full input signal the maximum dynamic range is limited to 104 dB.

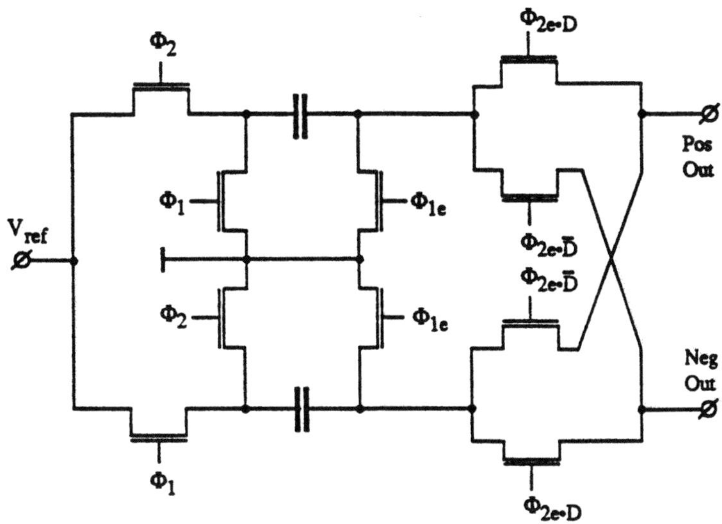

Figure 10.31 : Differential switched capacitor D/A converter

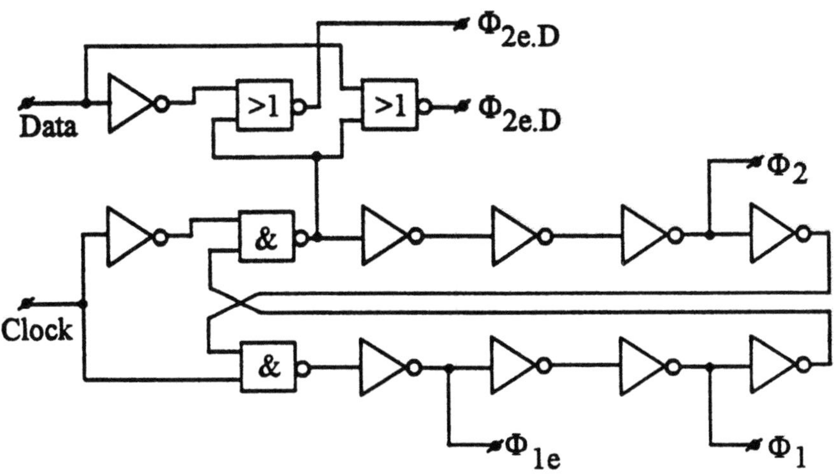

Figure 10.32 : Timing diagram of non-overlapping clock signals

The specifications of the 1-bit D/A converter are given in Table 10.3. These measurements are performed with a third-order noise-shaper operating at a frequency of 192 times f_s. An 18-bit input data word with a sampling frequency of 44.1 kHz is applied to the system.

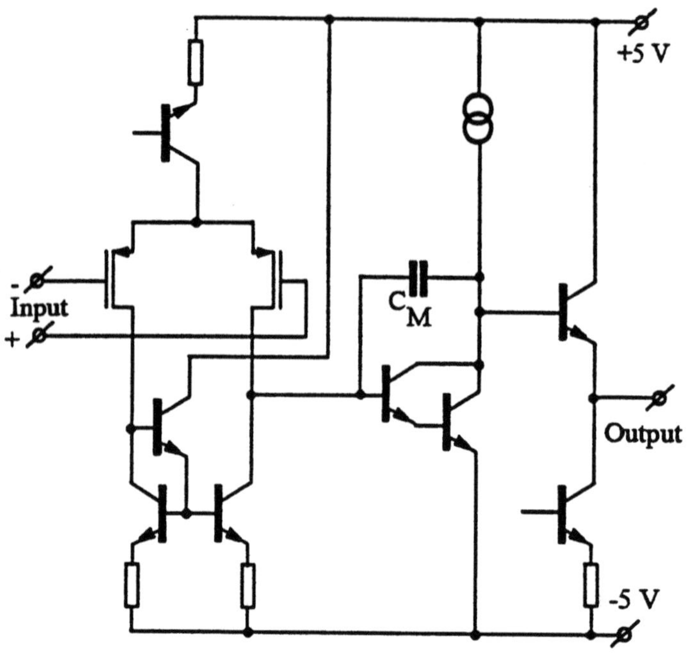

Figure 10.33 : BiMOS operational amplifier

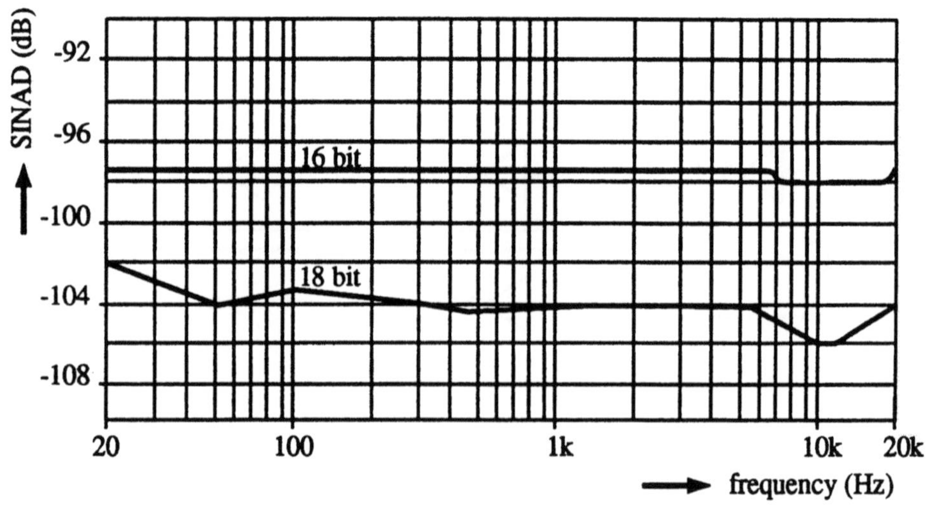

Figure 10.34 : THD plus noise as a function of frequency

10.7   Multi-bit noise-shaping D/A converter

The only way to increase the dynamic range of an oversampling converter
without increasing the sampling frequency or the order of the filter function is

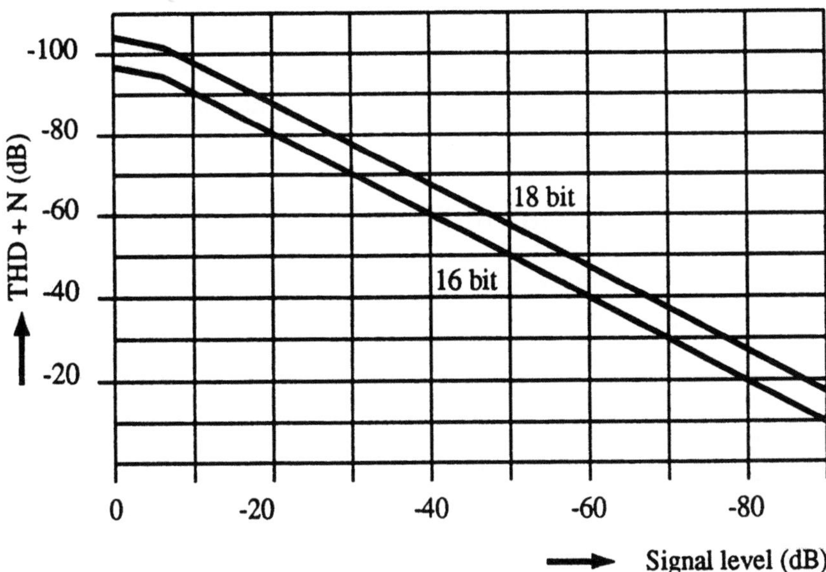

Figure 10.35 : THD plus noise as a function of amplitude

THD + Noise (Input signal 0 dB)	-104 dB
Dynamic range (Input signal -24 dB)	108 dB
Signal to silence (idle pattern)	112 dB
Crosstalk between channels ($f_{in} = 1$ kHz)	110 dB
Power dissipation (+5 V and -5 V supply)	750 mW
Maximum clock frequency	10 MHz
Chip size	2.31 × 3.16 mm

Table 10.3 : 1-bit D/A converter data

obtained by using a multi-bit D/A converter function. The advantage of the
1-bit converter over the multi-bit converter is found in the inherent linearity
of such a converter. A disadvantage of the 1-bit approach is the shaping
of the noise to the high-frequency band. Such large signal amplitude may
cause inter-modulation errors in the analog part of the D/A converter. These
inter-modulation products often appear in the audio band giving, audible
non-harmonic distortion. The multi-bit approach reduces the level of the
out-of band quantization signals at high output levels, while at small signal
levels the well-known 1-bit operation is obtained. However, the multi-bit

approach requires a D/A linearity equal to the full resolution of the system. Such a linearity is difficult to obtain without using calibration techniques. In this section an example of a multi-bit noise-shaping converter will be described.

10.7.1 Multi-bit system configuration

The architecture of the multi-bit D/A converter is shown in Figure 10.36 [95]. It consists of an oversampling filter which converts the input data with

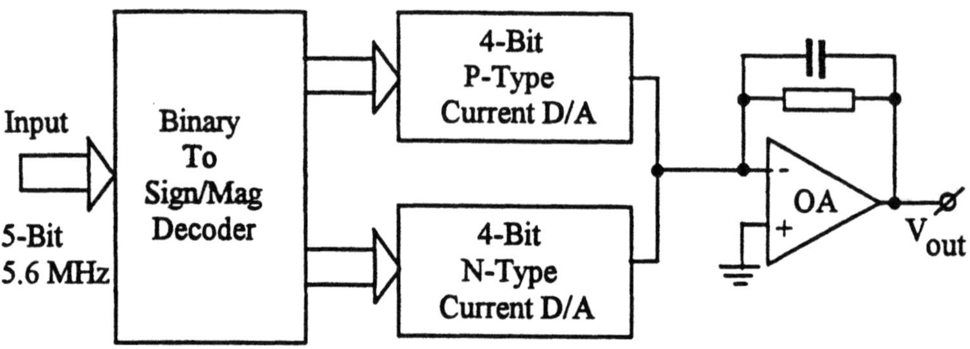

Figure 10.36 : multi-bit D/A converter architecture

a sampling rate of 44.1 kHz into a 5.6 MHz sampling rate. A third-order noise-shaper converts the data of the oversampling filter into a 5-bit noise-shaped code which is applied to the 5-bit D/A converter to obtain the analog output signal. When in a D/A converter the dynamic range is increased, then the noise present on the individual bit-values may become in the same order of magnitude as the quantization noise of the system. This is especially true when offset binary coding is used and small signals are generated. The noise present on the most significant bit values in such a case may have the same order of magnitude as the quantization noise. To overcome this problem a sign-magnitude coding is used to drive the multi-bit D/A converter. In the sign-magnitude coding the bit values in case of a bipolar signal depend linearly on the signal amplitude. As a result, the noise depends on these bit-values, thus increasing the dynamic range of the D/A converter. The sign-magnitude D/A converter consists of a 4-bit P-type D/A converter and a 4-bit N-type D/A converter implemented in a CMOS technology. The output currents of both D/A converters are converted into a voltage using an operational amplifier with a feedback network. To obtain the required linearity of the P- and N-type D/A converters a self-calibration procedure is used. The combination of the two 4-bit D/A converters results in an overall resolution of 5-bits.

10.7.2   Detail of sign-magnitude converter

In Figure 10.37 a more detailed system diagram of the sign-magnitude D/A converter is shown. In a 4-bit D/A converter system 15 equal current sources

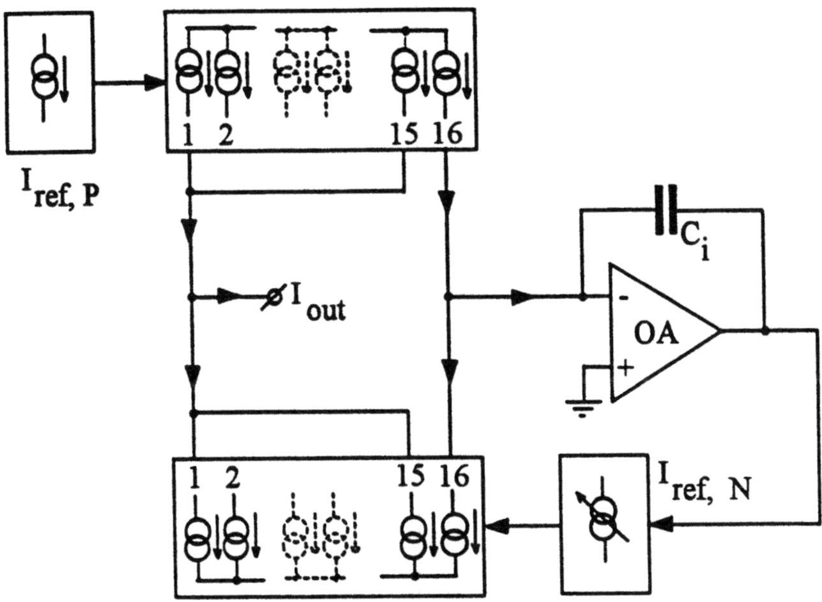

Figure 10.37 : Detailed sign-magnitude D/A converter system

can be used to obtain the required resolution. In this system an extra current source is used to calibrate the scale factor of the P-type D/A converter current to the N-type converter current. An operational amplifier is used to control the reference current source I_{refN} of the N-type D/A converter. In this way an accurate calibration of the N-type scale factor is obtained. The operational amplifier is configured as an integrator. In this way an averaging of the signals is obtained.

10.7.3   Bipolar self-calibration system

In Figure 10.38 a detail of the self-calibration system is shown. The part inside the dashed line and marked *common* is used for all of the current sources of the converter. Concentrating on the P-type converter part the transistor M_{P1} is biased by the voltage V_{refP} and conducts about 90% of the

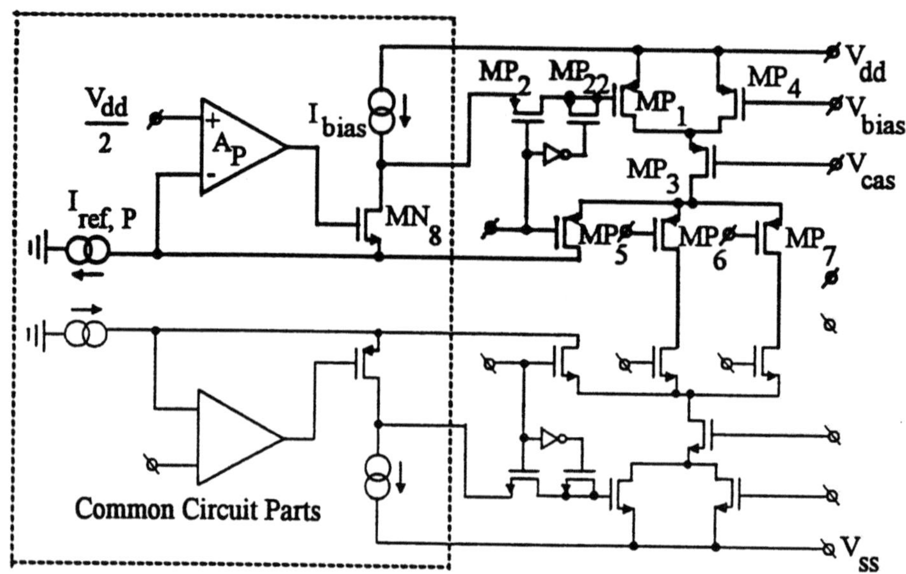

Figure 10.38 : Detail of a bipolar self-calibration system

reference current. Transistor M_{P2} conducts the error current necessary to calibrate the total current of M_{P1} plus M_{P2} to I_{refP}. During the calibration cycle transistor M_{P4} is active. At the end of the calibration cycle the current is available to the converter. Depending on the input data word, either transistor M_{P5} or transistor M_{P6} is active. The current in either case is applied to the *dump* line or the output line I_{out}. The operational amplifier A_P controls the drain voltage of transistor M_{P5} at the level V_{refOUT} resulting in equal biasing conditions during calibration and during output current generation. An identical explanation is valid for the N-type D/A converter. A more detailed explanation of the self-calibration system has been given in Chapter 7.

10.7.4   Total system implementation

The total analog subsystem has been implemented in a 1.6μm CMOS technology. The integral non-linearity of the system is limited to 90 dB at full scale. No special attention has been paid to push this linearity to 120 dB. At smaller signal levels the dynamic range increases while distortion remains at 90 dB. In Figure 10.39 the measured spectrum of a 1 kHz output signal attenuated -84 dB with respect to full scale is shown. The spectral component at 2.75 kHz introduced by the self-calibration system is found at a level of -130 dB referred to full scale. In Table 10.4 the specifications of the

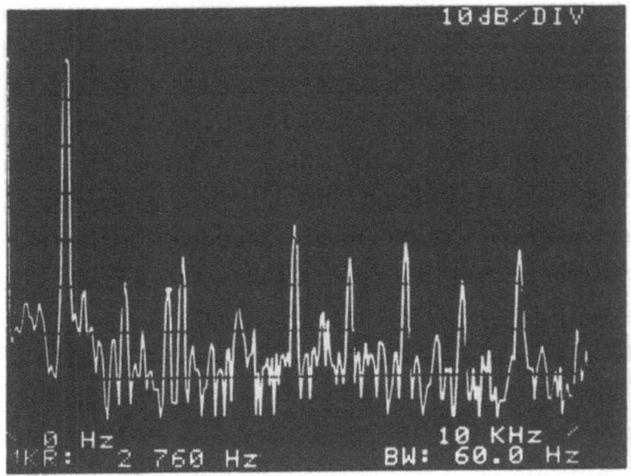

Figure 10.39 : Output spectrum of a 1 kHz signal attenuated 84 dB

analog D/A converter system are given. Using the multi-bit system config-

Output current	$\pm$ 1 mA
S/(N + THD) at 0 dB	90 dB
at -60 dB	55 dB
Sample frequency	5.6 MHz
Supply voltage	5 V
Power dissipation	100 mW
Temperature range	-10 to 70^0 C
Process	1.6 μm CMOS
Active chip area	7.5 mm^2

Table 10.4 : multi-bit D/A converter specification

uration a high performance D/A converter can be designed with a limited oversampling frequency.

10.8 Conclusion

Noise-shaping coders are excellent alternatives to conventional digital-to-analog converters for applications in largely oversampled digital audio and digital video systems. The 1-bit digital-to analog converter needs only two, or in special configurations one, reference voltage to perform the conversion. In these systems an excellent differential linearity with small input signals is obtained. The noise-shaping action randomizes the quantization noise rather well. Especially at small input signals the performance of the total conversion system is improved. The combination of the noise-shaping oversampling filter with the output filter function minimizes the component count, introduces a nearly linear phase response, and seems to be a very attractive alternative for digital audio signal conversion. When the maximum clock rate is limited, an increase in dynamic range can be obtained by combining a multi-bit D/A converter function with a noise-shaping function. This multilevel system optimizes the performance of a D/A converter with respect to dynamic range, filtering capability, and system complexity. Especially for high-resolution D/A converters, the noise-shaping functions results in a maximum obtainable dynamic range. The dynamic range calculations performed in this chapter are only valid for ideally operating noise-shaping converters. Some care must be used in handling these results. A real-time simulation of the practically designed converter gives the final result of the system. However, the given data give a good idea about the performance capabilities of noise-shaping coders. Furthermore, the stability analysis given proves to be useful for small input signal conditions. When input signal levels close to maximum are applied, then the nonlinearity of the quantizer influences the stability margin which usually results in an instable coder. Again,system simulations are needed to prove the overall stability.

Chapter 11

Sigma-delta converters

11.1   Introduction

In this chapter noise-shaping techniques applied to analog-to-digital converter systems will be described. Noise-shaping is very useful when speed can be exchanged with accuracy. In reference [96] an overview of theoretical and practical aspects of oversampling converters is given. The quantization errors in a noise-shaping system are removed from the signal band of interest. Furthermore, in an analog-to-digital converter system the input noise is filtered out by the input noise-shaping function. As a result, a reduced bandwidth can be used compared to, for example, successive approximation conversion methods. Again, the removed quantization errors appear with larger amplitudes as out-of-band noise in the system. With a digital filter these errors are removed. An increased dynamic range of the system is obtained. An example of such an operation is sigma-delta analog-to-digital conversion using single bit word-lengths [93,98,97]. The advantage of a 1-bit converter is the extreme linearity of such a device. A very good differential linearity is obtained with these converters. The most important design criteria will be given. The dynamic range performance is related to the noise-shaping coders described in Chapter 10. At the moment the dynamic range of a system must be enlarged, but the maximum clock rate of the system cannot be increased because of technology limitations, then a multi-bit digital-to-analog converter can be used in the feedback loop. At that moment, however, the linearity of the digital-to-analog converter determines the linearity and the distortion in the system. To overcome this problem *Dynamic Element Matching* or *Continuous Current Calibration* techniques can be used to obtain the extreme linearity of the D/A converter without needing extra trimming steps. In the stability analysis of the sigma-delta converters the root-locus method can be successfully applied. Examples for first-, second-, and third-order coders will be treated. A coder is said to be stable when idling occurs at frequencies close to half the sampling frequency. In this case a maximum of oversampling ratio is obtained, resulting in a maximum dynamic range. In most cases an idle pattern close to half the sampling frequency gives the best result. However, idle patterns at other frequencies, depending on the input signal randomly generated, are also allowed. In this case the quantization errors are randomized and appear as noise. Different architectures to implement higher-order sigma-delta converters will be introduced. These converters use special architectures to avoid stability problems. A special system that uses a signal-level-dependent filtering order introduces a feed-forward like filter coupling. At low signal levels the highest filter order is used, while with increasing signals the filter order is reduced. The reduced filter order results in a stable system with reduced dynamic range. Usually this reduction in dynamic range at large input signals is not a problem. A large dynamic signal range can be covered. A good optimum between filter order and stable system operation is found. At the end of this chapter a first-order implementation to be used in a 5-digit digital voltmeter with automatic offset compensation will be given.

11.2   General form of Sigma-delta A/D converters

In Figure 11.1 a general form of a sigma-delta A/D converter system is shown. The system uses a multi-bit quantizer (analog-to-digital converter) and a multi-bit digital-to-analog converter to reconstruct the analog signal. When multi-bit D/A converters are used to reconstruct the analog signal, then the linearity of such a converter is important. In case of high-resolution converters an accuracy problem in the D/A system is encountered. To overcome this accuracy problem a 1-bit system is used. In a 1-bit D/A converter the linearity is determined by the accuracy of switching between the reference signals. If a high switching accuracy can be guaranteed, then a very linear system is obtained. In the explanation a 1-bit system will be discussed. From the input signal the output signal of the 1-bit D/A converter is subtracted. The difference of these two signals is filtered by the loop filter, and the output signal of the loop filter is applied to the 1-bit quantizer or

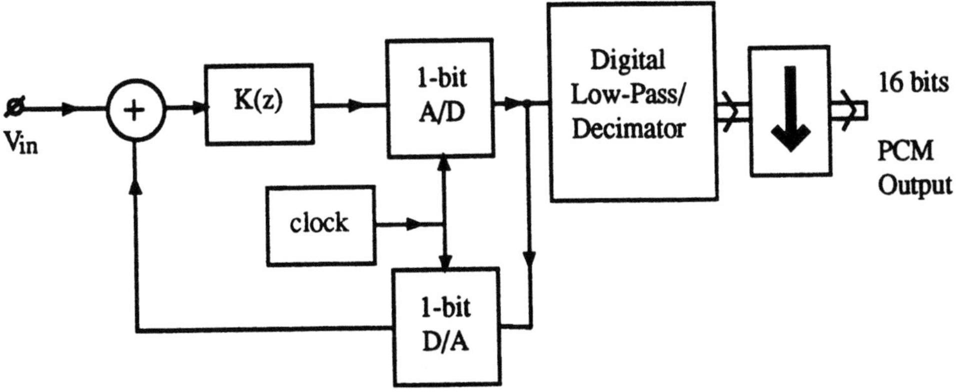

Figure 11.1 : Sigma-delta A/D converter system

A/D converter. The clock frequency of the system is high compared to the maximum analog input frequency while the order of the loop filter determines the dynamic range of the system. Equivalent equations apply for this system as given in the noise-shaping D/A converter chapter. Noise-shaping filters can be recalculated into sigma-delta filters showing nearly identical performance. In this way an identical analysis can be performed. The output of the 1-bit A/D converter is usually applied to a digital low-pass which rejects signals above the signal band of interest. Then sub-sampling or decimation is applied to obtain a multi-bit output code. The whole operation results in a binary-weighted digital output signal that can have a minimum sampling ratio equal to twice the signal bandwidth.

When the loop filter that is applied in this system consists of a continuous-time filter, then the analog signal band is filtered with the same filtering characteristic as is applied for noise-shaping. A cost-effective solution is obtained in this way. In the case of a discrete-time loop filter the anti-alias filtering must be performed before the analog signal enters the A/D converter. Discrete-time filters are mixing the high frequency input signals with the sampling clock, resulting in aliasing of signals which is not allowed.

The relationship between the sigma-delta modulator and the noise-shaper can be revealed by a transformation method shown in Figure 11.2 The transformation starts with the noise-shaper shown in Figure11.2(a). First the limiter is relocated according to Figure 11.2(b). This relocation is allowed because the output of the quantizer is not influenced by the omission of the

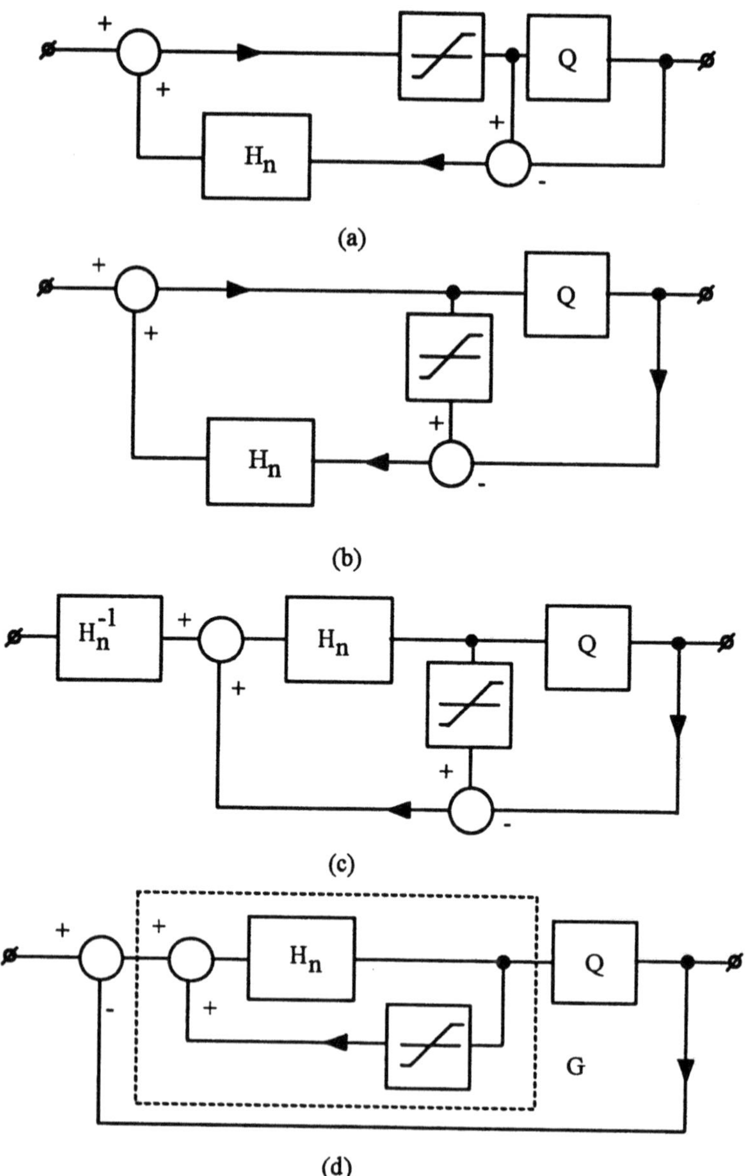

Figure 11.2 : Transformation of a noise-shaper into a sigma-delta modulator

limiter in its input. In the second step, shown in Figure 11.2(c), the loop filter is relocated. This operation does not change the output noise spectrum, and the influence on the signal transfer is canceled by an inverse filtering operation performed on the input signal. In the third step the two addition points in the loop are interchanged. The resulting system contains two nested feedback loops. The minor loop contains the noise-shaper loop filter and the limiter whereas the major loop contains the quantizer, the addition function, and the minor loop function as shown in Figure 11.2(d).

When the minor loop filter is considered as $K(z)$ and the limiter function is replaced by one, then the relationship between the noise-shaper filter $J(z)$ shown in Figure 10.4 and $K(z)$ is given by:

$$K(z) = \frac{J(z)}{1 - J(z)}. \tag{11.1}$$

Solving equation 11.1 for $J(z)$ we obtain:

$$J(z) = \frac{K(z)}{1 + K(z)}. \tag{11.2}$$

Equation 11.1 or 11.2 can be used to convert a noise-shaping filter $J(z)$ into a sigma-delta filter $K(z)$ or vice versa. The transformation is valid as long as the limiter is not active.

From Figure 11.1 the general transfer function of a sigma-delta A/D converter is obtained. Here N_q is the quantization error introduced by the analog quantizer.

$$\frac{Y(z)}{X(z)} = \frac{C_g K(z)}{1 + C_g K(z)} + \frac{N_q}{X(z)} \frac{C_g}{1 + C_g K(z)}. \tag{11.3}$$

In this equation C_g is the (non)linear gain of the quantizer. This constant is used for calculation purposes and depends on the signal level and the loop filter characteristic of the system. The root locus method that has been used for $J(z)$ in case of a noise-shaping coder can be used in the same manner for the sigma-delta converter. The characteristic equation in this system is:

$$1 + C_g K(z) = 0. \tag{11.4}$$

The root locus of $C_g K(z)$ as a function of C_g will be examined. Furthermore, it is possible to transform $K(z)$ into $J(z)$ according to equation 11.2 and then perform the root locus analysis. In practice this means that first- and second-order systems are mostly stable. Higher order systems require special measures for stabilization. The nonlinearity results in a signal dependent stability of the system. This is especially true when higher-order filter structures are used and the input signal increases towards the maximum input signal level.

11.2.1 Dynamic range

Formula 11.3 giving the transfer function of the system can be split-up into a signal transfer part and a noise transfer part what is very useful in case of stability and dynamic range (S/N) analysis.

Then the dynamic range of the system is found from:

$$S/N \approx \frac{X(z)}{N_q} K(z). \tag{11.5}$$

From equation 11.5 it is expected that the signal-to-noise ratio of the system is independent of the input signal amplitude. This is not true because with an increase in input signal the quantizer gain C_g decreases and therefore the efficiency of the filter operation is reduced.

To obtain an estimate of the signal-to-noise ratio as a function of the input signal the term $K(z)$ is analyzed. When the sigma-delta converter operates with a small input signal, the idle pattern is obtained when:

$$1 + C_g K(z) = 0 \tag{11.6}$$

or

$$C_g = \frac{-1}{K(z)}. \tag{11.7}$$

When the input signal increases, the value of C_g will decrease.

Putting $C_g = C_{g0}$ as the gain value for $A_{in} = 0$, with A_{in} is the amplitude of the input signal; then an estimate for C_g as a function of input signal amplitude A_{in} can be found from the describing function of a quantizer:

$$C_g = C_{g0}(1 - \frac{A_{in}}{A_{ref}}). \tag{11.8}$$

Here, A_{ref} is a reference-fitting amplitude.

The input amplitude dependent dynamic range can now be estimated as:

$$S/N \approx \frac{A_{in}}{N_q} K_0(z)(1 - \frac{A_{in}}{A_{ref}}). \tag{11.9}$$

$K_0(z)$ is the noise transfer function of the coder for $A_{in} = 0$.

It must be noted that equation 11.9 is a simplified model of a general decrease in dynamic range observed in these coders. In Figure 11.3 the normalized result of equation 11.9 is shown. In this figure $A_{ref} = 1.4$, $N_q = 1$ and $K_0(z)$

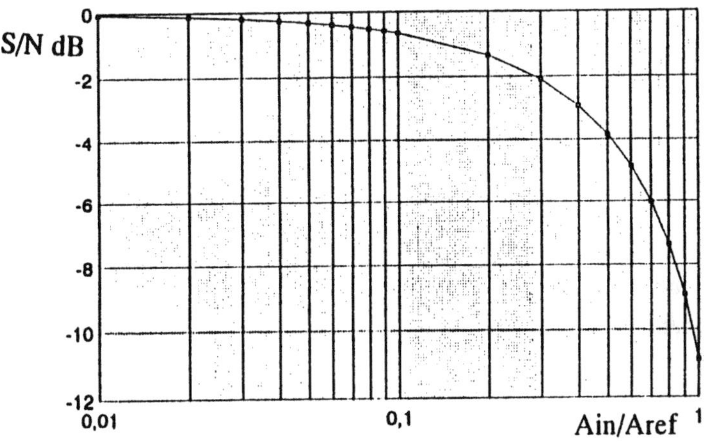

Figure 11.3 : Normalized S/N as a function of amplitude

$= 1$.

11.2.2   Phase uncertainty of a sampled quantizer

The stability of a sigma-delta analog-to-digital converter can be determined in two ways:

1. Root locus method

2. Phase characteristic of the feedback loop

In this section we will use the phase relation of the loop filter. In designing the loop filters of sigma-delta A/D converters, the phase characteristic of the loop filter is very important. As was shown in the description of the (analog) quantizer, there exists a phase uncertainty of $\pm\frac{\pi}{2}$ in such a quantizer at half the sampling frequency. The phase characteristic of the loop filter with the phase uncertainty of the quantizer can be used to design an optimum converter [90]. An important criterion is the generation of the idle pattern at half the sampling frequency with small input signals. Other frequencies are only allowed at the moment input signals are applied. In fig. 11.4 the phase uncertainty of the analog 1-bit quantizer as a function of frequency is repeated from Chapter 1. To obtain a stable idle pattern with a preference of oscillation at half the sampling frequency a feedback with a suitable

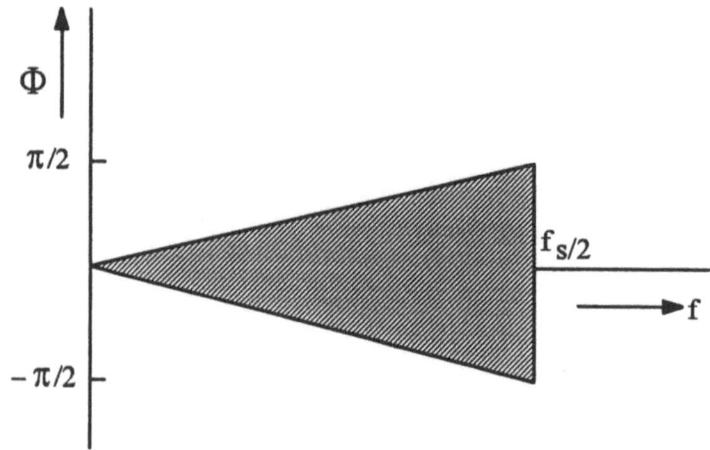

Figure 11.4 : Phase uncertainty of a 1-bit sampled quantizer

gain is needed in the sigma-delta modulator system. As a result, the phase characteristic of the loop filter can be nearly equal to the part necessary to obtain a total phase equal to π over the frequency band from zero to $\frac{f_s}{2}$. At $\frac{f_s}{2}$ the total phase must be π to obtain the preferred idle pattern at that frequency. To show a design example in Figure 11.5, the phase characteristic of the loop filter and the phase uncertainty of quantizer are given. The phase

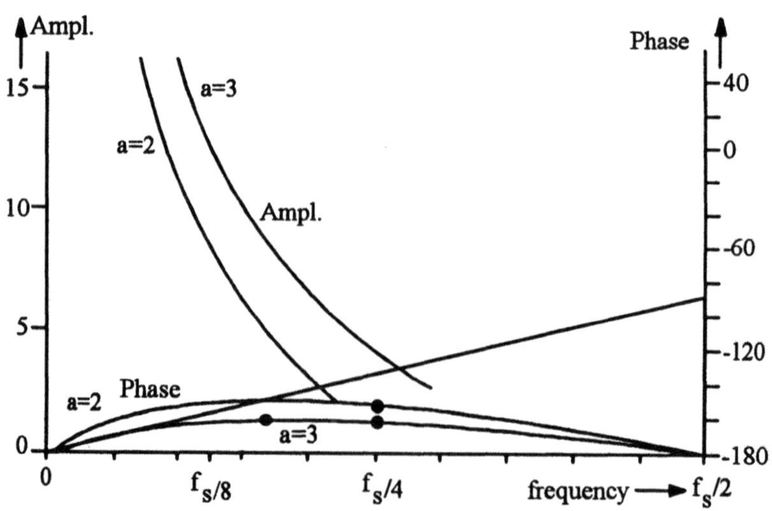

Figure 11.5 : Example of phase design criterion

curve of the loop filter may not enter the triangle introduced by the phase uncertainty of the quantizer. When the phase curve is inside this triangle, there exits always a possibility that the total phase equals π. However, only oscillations occur when the loop gain is equal or larger than 1. As long as the gain condition is not satisfied, no idle pattern at this frequency is obtained. Such a condition is obtained at low input frequencies where the phase curve crosses the uncertainty region. A conditional stable system is obtained in this way. The application of this criterion is valid in systems having a linear loop filter. When a limiter is used then this criterion does not give usable results. This procedure is an alternative method for the root locus stability analysis method.

11.2.3 Sigma-delta signal examples

If no analog input signal is applied to the A/D converter, then an idle pattern consisting of "1" and "0" signals at half the sampling frequency is obtained. The "1"-"0" pattern changes when an analog input signal is applied to the converter. The number of ones or zeros changes in such away that the analog input signal is nearly equal to the average output signal generated by the 1-bit D/A converter. The maximum output signal can be nearly equal to the positive or the negative reference voltage. As long as the switches in the D/A converter show a nearly ideal performance, the linearity of the converter is not determined by the reference voltages, because by changing the one-zero pattern, the output signal of the D/A converter varies according to a straight line between the minus reference voltage and the plus reference voltage.

In Figure 11.6 an example of the output signal of the A/D converter with a sine wave analog input signal is shown. To make the figure understandable the one-zero idle pattern at half the sampling frequency is removed with a simple digital filter. Note that in the top of the sine wave a large amount of "ones" is generated while in the bottom part of the sine wave a large amount of "zeros" is obtained. An example of an A/D converter using a continuous-time first-order filter with a time-discrete 1-bit D/A converter is shown in Figure 11.7. In the D/A converter a switched capacitor D/A converter approach is used. This construction is less sensitive to sampling clock uncertainties because a charge "bucket' is transferred to the integrator at every clock moment. The amount of charge is determined by the reference voltage V_{ref} and the capacitors C_1. Furthermore, the switches in the D/A converter are arranged in such a way that only a single reference voltage V_{ref} is needed to transfer a charge or discharge packet into integrator. Note that

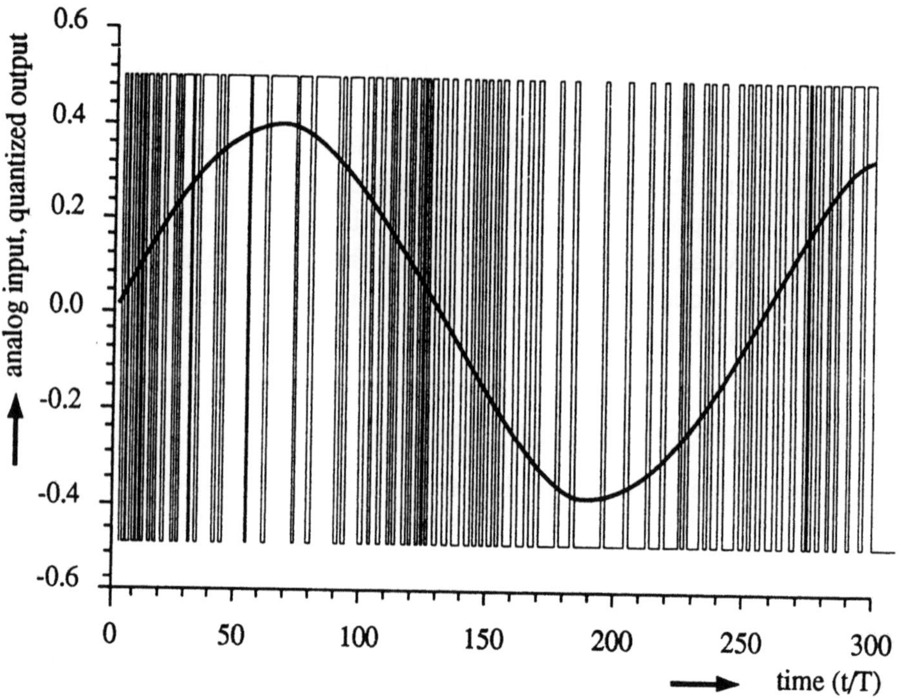

Figure 11.6 : Sigma-delta A/D converter input and output signals

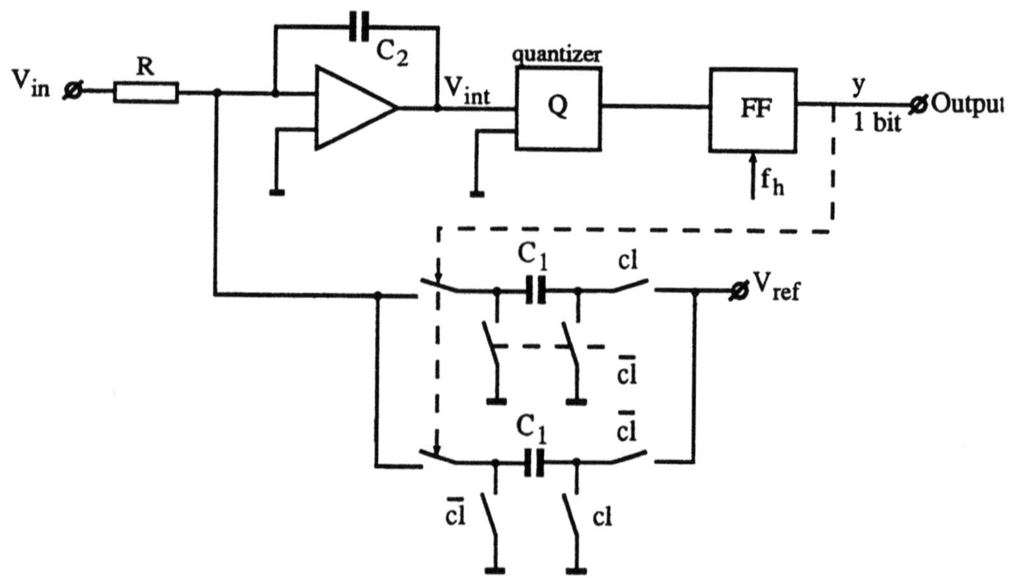

Figure 11.7 : A/D converter with continuous-time loop filter

in this system the analog input signal is filtered by the integrator as well. Although the filtering attenuation is small it helps to simplify the anti-alias filter at the analog input (this filter is not shown in the figure).

An example of a third-order noise-shaping A/D converter is shown in Figure 11.8. In this system a switched capacitor technique is used in the loop filter to obtain a high filtering accuracy. Note that in this system the high

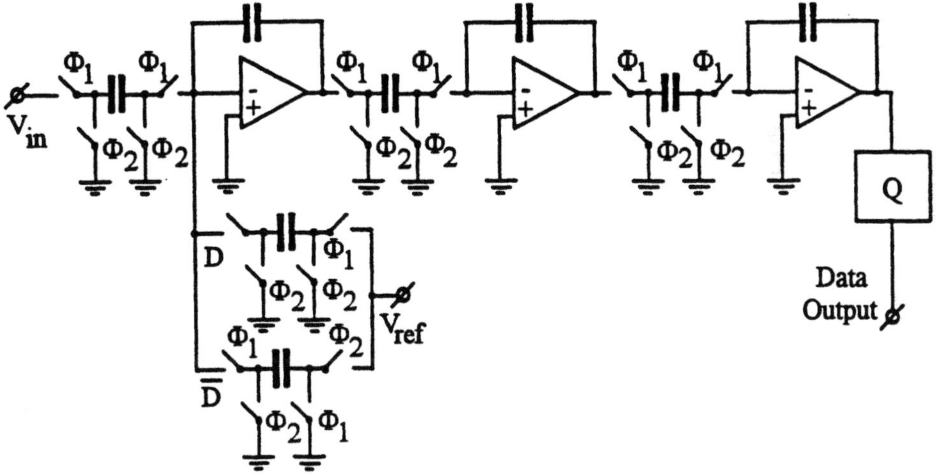

Figure 11.8 : Third order switched capacitor noise-shaping A/D converter

frequency components must be filtered out before an analog signal can be applied to the system to avoid aliasing. A continuous-time analog filter is needed to perform the anti-alias input filtering.

11.3   First-order A/D converter

In Figure 11.9 the basic circuit diagram of a first-order analog-to-digital converter is shown. The signal transfer function for this system using C_g as the gain of the quantizer becomes:

$$\frac{Y(z)}{X(z)} = \frac{C_g G(z)}{1 + C_g G(z)}. \tag{11.10}$$

The error signal E(z) becomes:

$$E(z) = N_q \frac{C_g}{1 + C_g G(z)}. \tag{11.11}$$

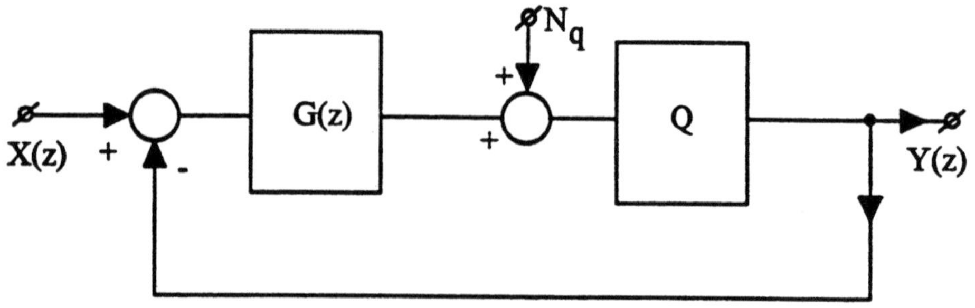

Figure 11.9 : First-order A/D converter

With

$$G(z) = \frac{z^{-1}}{1 - z^{-1}} \tag{11.12}$$

we obtain:

$$\frac{Y(z)}{X(z)} = \frac{C_g z^{-1}}{1 + (C_g - 1)z^{-1}} \tag{11.13}$$

or

$$\frac{Y(z)}{X(z)} = \frac{C_g}{z - (1 - C_g)}. \tag{11.14}$$

The error of the system can be calculated by substituting $G(z)$ into equation 11.11. The result is:

$$E(z) = N_q \frac{C_g(1 - z^{-1})}{1 + (C_g - 1)z^{-1}} = N_q \frac{C_g(z - 1)}{z - (1 - C_g)}. \tag{11.15}$$

With $z \approx 1$, what corresponds with $f_{signal} \approx 0$, the error signal $E(z)$ from equation 11.15 is nearly zero. This corresponds to the expected noise-shaping operation of the system.

The stability is analyzed by means of the root locus method using C_g as the parameter. In the first order system we have to analyze:

$$C_g G(z) = \frac{C_g z^{-1}}{1 - z^{-1}}. \tag{11.16}$$

The root locus starts at $z = 1$ and ends at $z = -\infty$. In Figure 11.10 the root locus is shown. The system is stable and gives the expected idle pattern at $\frac{f_s}{2}$ for $z = -1$.

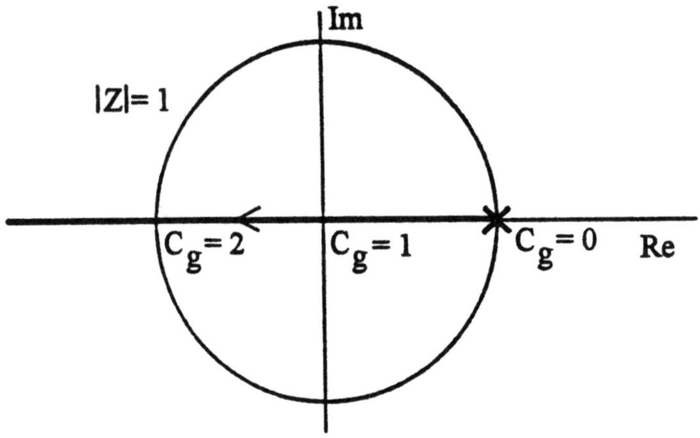

Figure 11.10 : Root locus of a first-order A/D system

11.4 Second-order A/D converter

In Figure 11.11 a second-order sigma-delta modulator is shown [91]. As the figure shows, the reproduced analog signal is subtracted after each integrator. The coefficient a_0 that determines the amount of analog signal to be subtracted from the output signal of the first integrator determines the total loop filter characteristic. The signal transfer function of the second-order

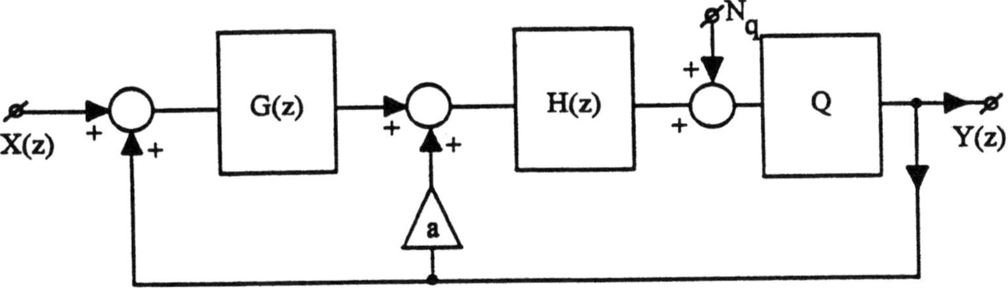

Figure 11.11 : Second-order sigma-delta modulator

coder becomes:

$$\frac{Y(z)}{X(z)} = \frac{C_g G(z) H(z)}{1 + C_g H(z)[a_0 + G(z)]}. \tag{11.17}$$

The error transfer function becomes:

$$E(z) = N_q \frac{C_g}{1 + C_g H(z)[a_0 + G(z)]}. \tag{11.18}$$

The coefficient a_0 can be chosen in such a way that an optimum performance of the converter is obtained.

Substituting for $G(z) = H(z) = \frac{z^{-1}}{1-z^{-1}}$ and simplifying this function, we obtain for:

$$K(z) = H(z)(a_0 + G(z)),\tag{11.19}$$

with $K(z)$ is the equivalent transfer function as shown in the general sigma-delta modulator Figure 11.1,

$$K(z) = \frac{z^{-1}(a_0 - z^{-1}(a_0 - 1))}{(1 - z^{-1})^2}.\tag{11.20}$$

Now $K(z)$ can be converted into an equivalent noise-shaper function $J(z)$ resulting in:

$$J(z) = \frac{z^{-1}(a_0 - (a_0 - 1)z^{-1})}{1 + (a_0 - 2)z^{-1} - (a_0 - 2)z^{-2}}.\tag{11.21}$$

By making $a_0 = 2$, equation 11.21 reduces into:

$$J(z) = z^{-1}(2 - z^{-1}),\tag{11.22}$$

which is equivalent to the second-order noise-shaper loop filter shown in Figure 10.18. $a_0 = 2$ is a good choice for a second-order system.

Stability analysis is performed on $C_g K(z)$. With $a_0 = 2$ we obtain:

$$C_g K(z) = C_g \frac{2z - 1}{(z - 1)^2}.\tag{11.23}$$

The root locus starts at the double pole $z = 1$ and ends at $z = 0.5$ and $z = -\infty$ respectively. In Figure 11.12 the root locus of the system is shown. The root locus of this system is nearly identical to the root locus of the second-order noise-shaper shown in Figure 10.18.

Assuming that a stable oscillating pattern with small input signals at $\frac{f_s}{2}$, then the gain C_g of the quantizer can be calculated using equation 11.23. Inserting z = -1 results in $C_g = \frac{4}{3}$. The nonlinearity of the analog quantizer results in a linearized gain of $\frac{4}{3}$ with an idle noise pattern at $\frac{f_s}{2}$.

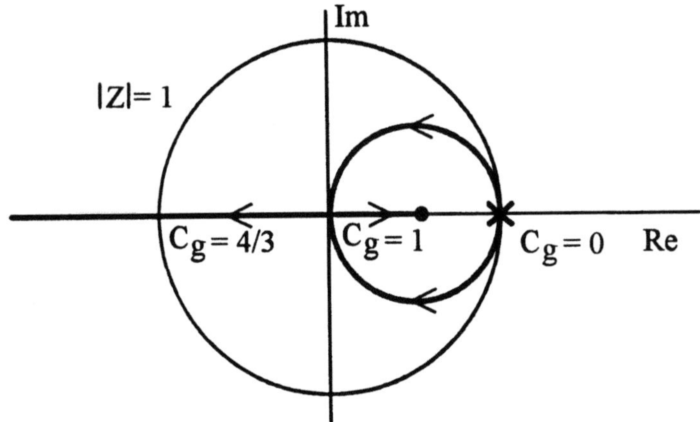

Figure 11.12 : Root locus of second-order system

11.5 Third-order A/D converter

An example of a third-order sigma-delta converter is shown in Figure 11.13. Again after each integrator the reconstructed analog value is subtracted from the quantized signal. Putting C_g for the quantizer gain again and having coefficients a_0, a_1, and a_2, the transfer function of the system can be calculated. The coefficient a_0 is added to obtain an extra variable in converting the sigma-delta coder into a noise-shaping coder.

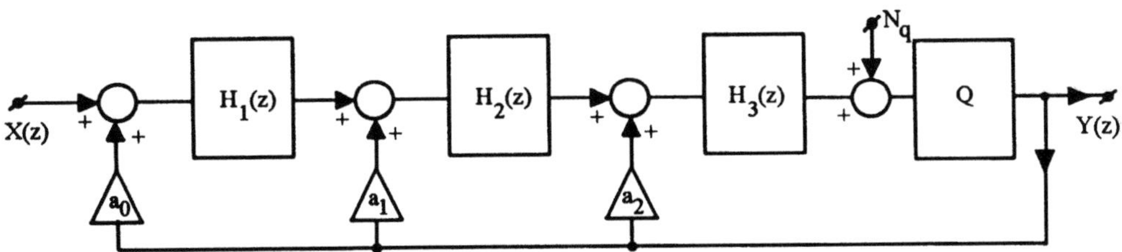

Figure 11.13 : Third-order noise-shaping coder

$$\frac{Y}{X} = \frac{C_g H_1(z) H_2(z) H_3(z)}{1 + C_g H_3(z)[a_2 + H_2(z)(a_1 + a_0 H_1(z))]}. \tag{11.24}$$

By inserting $H_1(z) = H_2(z) = H_3(z) = \frac{z^{-1}}{1-z^{-1}}$, the equivalent transfer function $K(z)$ becomes:

$$K(z) = \frac{z^{-1}}{(1 - z^{-1})^3}(a_2 + (a_1 - 2a_2)z^{-1} + (a_2 - a_1 + a_0)z^{-2}) \tag{11.25}$$

or

$$K(z) = \frac{(a_0 - a_1 + a_2) + (a_1 - 2a_2)z + a_2 z^2}{(z-1)^3}.$$

(11.26)

By converting $K(z)$ into the noise-shaping filter $J(z)$, we obtain:

$$J(z) = \frac{(a_0 - a_1 + a_2) + (a_1 - 2a_2)z + a_2 z^2}{(a_0 - a_1 + a_2 - 1) + (a_1 - 2a_2 + 3)z + (a_2 - 3)z^2 + z^3}.$$

(11.27)

A filter function already used as a third-order noise-shaper in Chapter 10 is:

$$J(z) = 1 - \left(\frac{z-1}{z-a}\right)^3$$

(11.28)

or

$$J(z) = \frac{(1 - a^3) - 3z(1 - a^2) + 3z^2(1 - a)}{(z-a)^3}.$$

(11.29)

The coefficients of equations 11.27 and 11.29 must be equal we obtain:

$$a_2 = 3(1 - a)$$

(11.30)

$$a_1 = 3a(1 - a)^2$$

(11.31)

$$a_0 = (1 - a)^3.$$

(11.32)

By inserting $a = 0.5$ as a good choice for an optimum third-order performance, we obtain:

$$a_2 = 1.500$$

(11.33)

$$a_1 = 0.375$$

(11.34)

$$a_0 = 0.125.$$

(11.35)

The value of coefficient a_0 can be made equal to 1 using equation 11.26 and
modifying the gain C_g; thus we obtain:

$$a_{20} = \frac{3}{(1-a)^2}$$

(11.36)

$$a_{10} = \frac{3a}{(1-a)}$$

(11.37)

$$C_{g0} = a_0 C_g.$$

(11.38)

Basically the small gain change is not a problem, and the quantizer does not
show a change in performance. 1-bit sigma-delta converters are not dependent on the quantizer gain.

The stability of the system is found from:

$$K(z) = \frac{(a_0 - a_1 + a_2) + (a_1 - 2a_2)z + a_2 z^2}{(z-1)^3}. \tag{11.39}$$

The root locus starts at $z = 1$ with a triple pole and ends at:

$$z_{1,2} = (1 - \frac{a_1}{2a_2}) \pm \sqrt{(\frac{a_1}{2a_2} - 1)^2 - (\frac{a_0 - a_1}{a_2} + 1)} \tag{11.40}$$

and $z_3 = -\infty$.

The minimum loop gain with $a = 0.5$ equals 0.78 for the given coefficient values. When $a_0 = 1$, this gain reduces to $C_g = 0.0975$.

The root locus of this system is shown in Figure 11.14. The minimum loop gain that gives a stable system is obtained when the root locus crosses $|z| = 1$. From the root-locus it can be seen that when the gain of the signal is

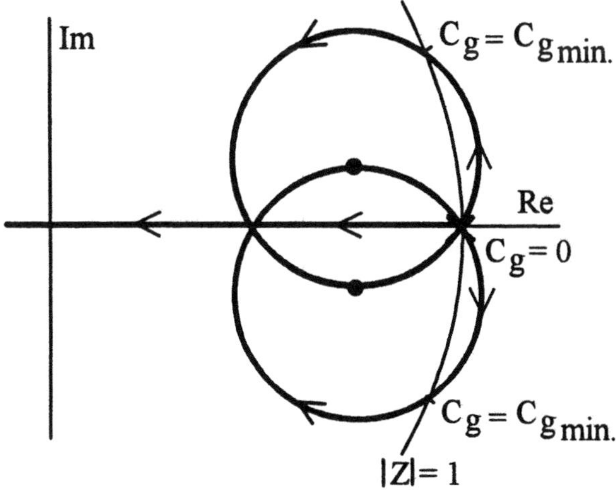

Figure 11.14 : Root locus of third-order A/D converter

reduced below a value of 0.0975 this system will become instable. A proper choice of quantizer limiting level and D/A output level avoids the instability problem of the system. The value of C_g can be calculated when a small input signal is present and the coder shows an idle pattern at $\frac{f_s}{2}$.

Inserting $z = -1$ into equation 11.39 results in $C_g = \frac{8}{37}$. From the third-order noise-shaper implementation it was shown that the coefficients as used in equation 11.39 results in a maximum dynamic range.

11.6 Multi-stage sigma-delta converter (MASH)

To avoid the stability problem of, for example, a third-order noise-shaping coder, a so-called "MASH" converter system consisting of three cascaded first order coders is used. In Figure 11.15 this system is shown [99]. From

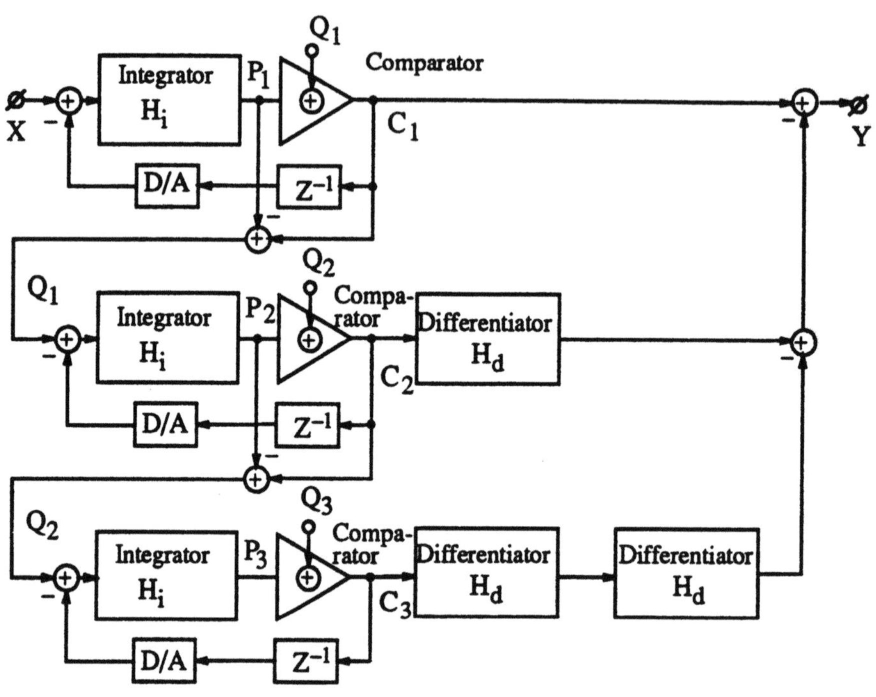

Figure 11.15 : MASH noise-shaping coder system

the figure it is seen that the error signal P_1 of the first first-order noise-shaping coder is applied to the second first-order coder. This second-order coder quantizes the error signal and after differentiation of this second output signal the total output signal is corrected. Then the error signal P_2 of the second coder is applied to the third coder which quantizes this signal again. After a two times differentiation, the signal of the third quantizer is added to the output signal again. As a result of this operation, the order of the noise-shaping coder is increased from a first order of the first encoder up to a third order. This increase in noise-shaping filtering performance is obtained without introducing a stability problem because every individual stage of the encoder is a first-order noise-shaper. The transfer function of this coder is given by:

$$K(z) = C_g(\frac{z^{-1}}{1 - z^{-1}})^3.$$

(11.41)

The transfer function of the integrators in the loop including the z^{-1} delay of the quantizer is given by:

$$I(z) = \frac{z^{-1}}{1 - z^{-1}}.$$

(11.42)

A problem in this system is the accuracy with which the output signal P_1 is encoded with respect to the full resolution of the total system. The accuracy of the 1-bit digital-to-analog converter recovering the analog output signal of the first encoder influences the accuracy of signal P_1 and so does the subtracter determining the input signal to the second encoder. In the second encoder an accuracy problem is again found but less stringent requirements are needed here. Therefore, accuracy problems may be found when it is simply increased into a fifth- or even higher-order noise-shaping system.

At the output of the MASH converter a multi-bit digital word is found due to the addition of the output signals of the successive 1-bit quantizers. By using an extra noise-shaper, this multi-bit output word can be reduced into a 1-bit largely oversampled output signal.

In reference [92] an example of MASH structure using a 1-bit "coarse" quantizer and a $N = 3$-bit "fine" quantizer is presented. The block diagram of this converter system is shown in Figure 11.16 In the first "coarse" quantizer a 1-bit system approach is used because of the very high accuracy that can be obtained in a 1-bit system without needing trimming. The second "fine" quantizer uses a multi-bit quantizer. Because the multi-bit operates on the "error" signal from the coarse quantizer, the accuracy requirements are relaxed. The advantage of the multi-bit "fine" quantizer is the increase in dynamic range compared to a 1-bit implementation and the reduction in out-of-band noise.

11.7 Feed-forward A/D converter system

A system that combines the higher-order filter function for small signals with a gliding filter order function depending on the signal amplitude tending to reduce into a first-order for large signals is shown in Figure 11.17. Basically

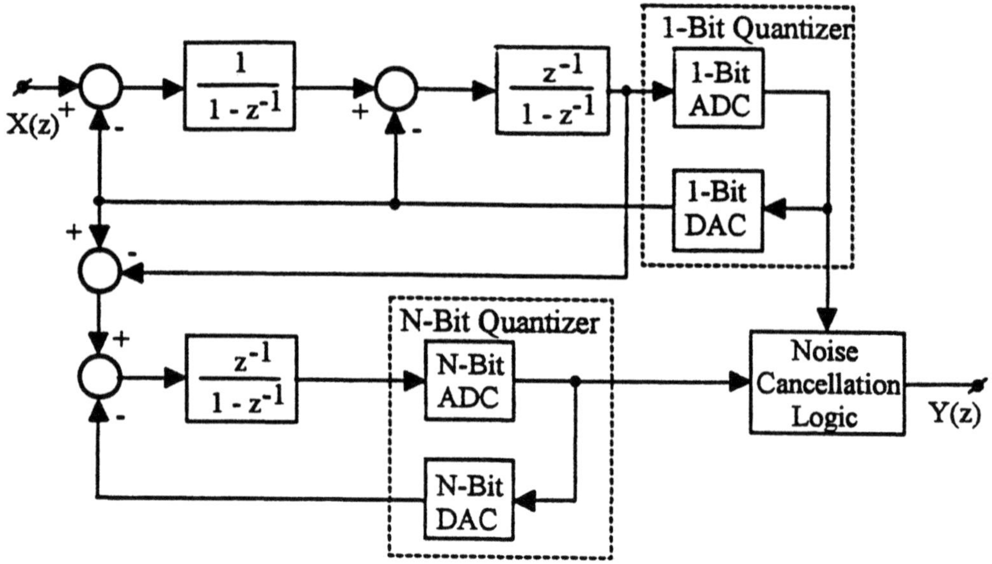

Figure 11.16 : Block diagram of cascaded multi-bit sigma-delta modulator

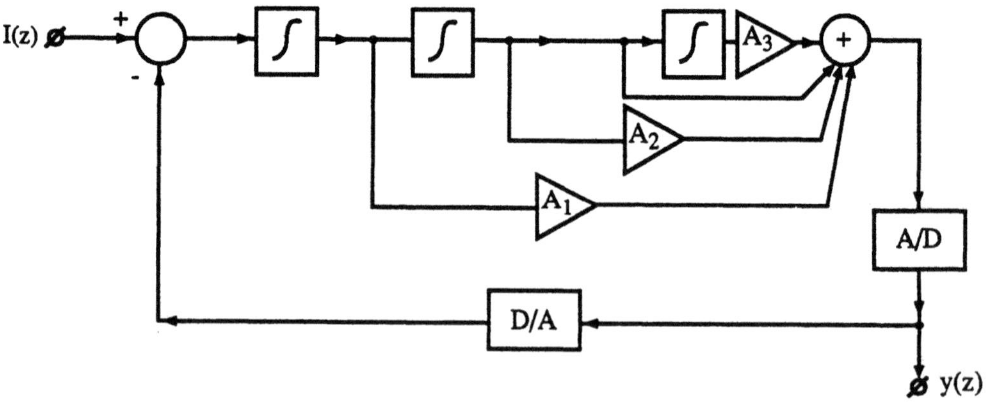

Figure 11.17 : Feed-forward A/D converter system

the system consists of a cascade of integrator sections. A feed-forward technique using coefficients A_N with N, depending on the number of stages, feeds signals directly to the comparator stage (1-bit A/D). When the input signal increases, then the signals are amplified so much that a clamping occurs. At that moment, the feed-forward operation takes over and a stable system is obtained. When the transfer function of an integrator is determined by:

$$I(z) = \frac{z^{-1}}{1 - z^{-1}}, \tag{11.43}$$

then the transfer function of a third-order filter implementation is given by:

$$K(z) = \frac{z^{-1}}{(1 - z^{-1})^3} A_1 \left(1 + \frac{A_2 - 2A_1}{A_1} z^{-1} + \frac{A_3 + A_1 - A_2}{A_1} z^{-2}\right). \quad (11.44)$$

In this equation the coefficients $A_{(n)}$ determine the weighting of the output signals of the integrators. Expressions for the coefficients $A_{(n)}$ can be obtained starting with a third-order noise-shaping filter:

$$1 - J(z) = \left(\frac{z - 1}{z - a}\right)^3 \quad (11.45)$$

Converting the noise-shaping filter $J(z)$ into the sigma-delta filter $K(z)$ results in expressions for the coefficients $A_{(n)}$, we obtain:

$$A_1 = 3(1 - a) \quad (11.46)$$
$$A_2 = 3(1 - a)^3 \quad (11.47)$$
$$A_3 = (1 - a)^3. \quad (11.48)$$

With $a = 0.5$ equation 11.44 becomes:

$$K(z) = \frac{z^{-1}}{(1 - z^{-1})^3} 1.5\left(1 - 1.5z^{-1} + \frac{7}{12}z^{-2}\right). \quad (11.49)$$

giving a stable coder which already was shown in a previous analysis.

Basically the gain A_1 can be incorporated into the quantizer gain in case of a 1-bit system. It is clear that other choices for filter coefficients are possible to obtain the required performance of the converter and give a sufficient stability. At this moment the special architecture of this coder will become active.

In practical solutions a filter order between three and five seems to be a good optimum. Due to the higher order filtering operation, a better randomizing of the quantization errors occurs at small input signals. Therefore, the small signal performance is extremely good, while with increasing signals the filter order decreases until in extreme circumstances the final first-order stage is reached. The outputs of the integrators that are closest to the quantizer have a predetermined maximum signal value before limiting occurs. During limiting no signals are transferred, and therefore this part of the filter function is eliminated.

The system shown in Figure 11.17. The outputs of the integrators are multiplied by the filtering coefficients (A_1, A_2, .. A_n) and when needed the output signals of the coefficient stages are clipped at predetermined levels. Basically the same idea can be used for application in D/A converters. In this case parts of the feedback filtering function are bypassed to reduce the filter order at high signal levels. A comparable system is used in operational amplifiers to reduce the amplifier order at high frequencies to obtain a first-order frequency response (see Chapter 8).

11.8   Nth-order sigma-delta architecture

A general form of an Nth-order sigma-delta converter is shown in Figure 11.18 The system consists of a feed-forward path of integrators determined

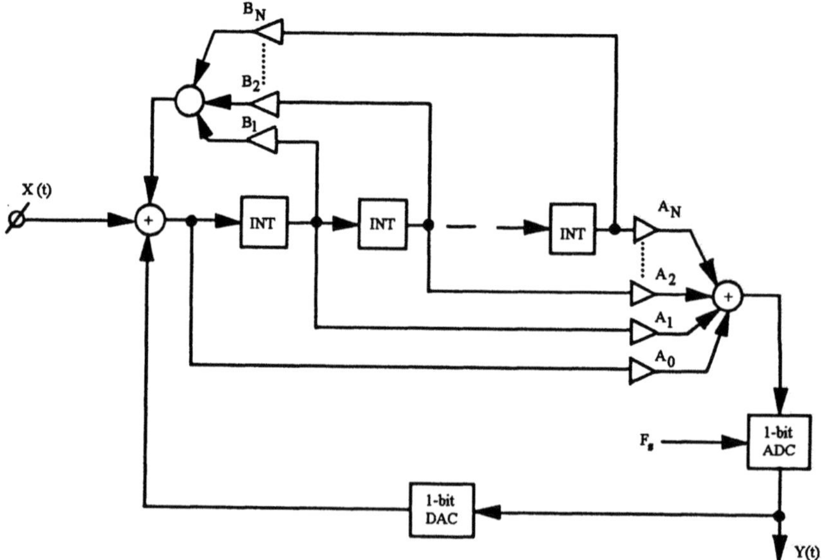

Figure 11.18 : Nth-order sigma-delta converter architecture

by the coefficients A_N and a feedback path determined by the coefficients B_N. When the coefficients B_N are not implemented, the converter of the previous section is obtained. At the moment the coefficients B_N are introduced, additional "zero" transmissions at different frequencies in the passband of the low-pass filter are introduced. These additional zeros reduce the quantization errors in the total signal band, resulting in an increase in dynamic range. In Figure 11.19 the noise transfer functions of a fourth-order sigma-delta converter according to Figure 11.18 are shown. Splitting

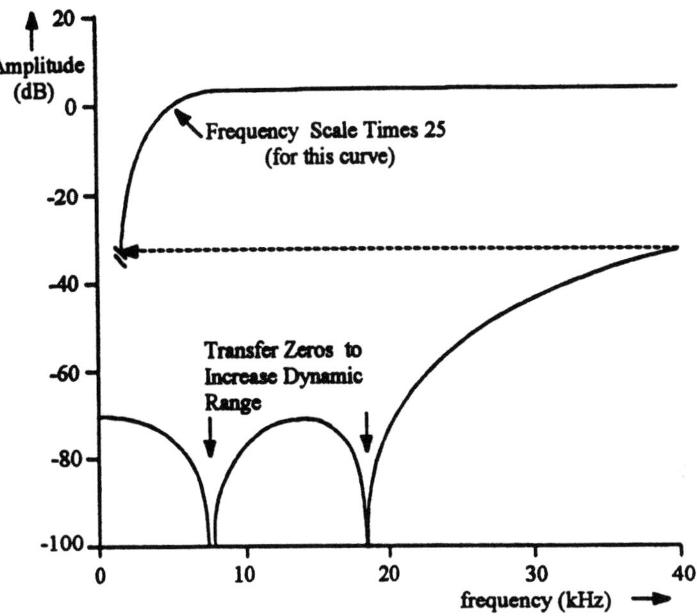

Figure 11.19 : Noise transfer function of a sigma-delta converter

up the transfer of the sigma-delta converter into a signal and an error term
we obtain:

$$Y(z) = C_g K_x(z) X(z) + C_g K_e(z) N_q. \tag{11.50}$$

Here $K_x(z)$ represents the signal transfer function, and $K_e(z)$ is the noise-transfer function. In case of Figure 11.18 we obtain:

$$K_x(z) = \frac{\sum_{i=0}^{N} A_i (z-1)^{N-i}}{z[(z-1)^N - \sum_{i=1}^{N} B_i (z-1)^{N-i}] + \sum_{i=0}^{N} A_i (z-1)^{N-i}} \tag{11.51}$$

$$K_e(z) = \frac{(z-1)^N - \sum_{i=0}^{N} A_i (z-1)^{N-i}}{z[(z-1)^N - \sum_{i=1}^{N} B_i (z-1)^{N-i}] + \sum_{i=0}^{N} A_i (z-1)^{N-i}}. \tag{11.52}$$

As can be seen from $K_e(z)$, the coefficients B_i result in zero transmissions
in the passband of the noise transfer.

In reference [100] a fourth-order converter design has been presented. Coefficients values for A_i are determined using a Butterworth filter design. The
results are given here; we obtain:

$$A_0 = 0.865300 \tag{11.53}$$

$$A_1 = 1.192000 \tag{11.54}$$

$$A_2 = 0.390600 \tag{11.55}$$

$$A_3 = 0.069260 \tag{11.56}$$

$$A_4 = 0.005395. \tag{11.57}$$

The B_i coefficients are obtained by using a Chebyshev equi-ripple design. The coefficients obtained in this way are:

$$B_1 = -3.54010^{-3} \tag{11.58}$$

$$B_2 = -3.54210^{-3} \tag{11.59}$$

$$B_3 = -3.13410^{-6} \tag{11.60}$$

$$B_4 = -1.56710^{-6}. \tag{11.61}$$

The sample frequency of the converter is $f_s = 2.1$ MHz and the audio bandwidth is $f_b = 20$ kHz. The filter order $N = 4$.

The curve of the simulated signal-to-noise ratio of this system is shown in Figure 11.20. The use of zeros in the pass-band improves the dynamic

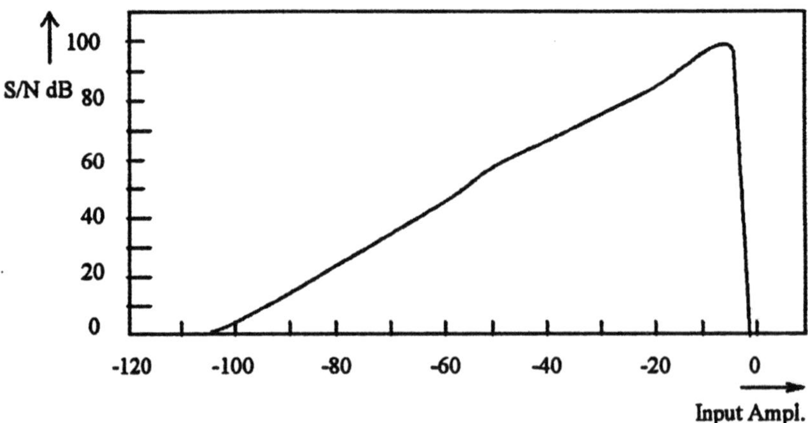

Figure 11.20 : Simulated S/N ratio of 4th order sigma-delta converter

range of the system at the cost of more complexity and sometimes a smaller maximum input signal range.

11.9   Idle pattern

The idle pattern generated in sigma-delta converters depends on the order of noise-shaping used in the coder. In Figure 11.21 an example of an idle

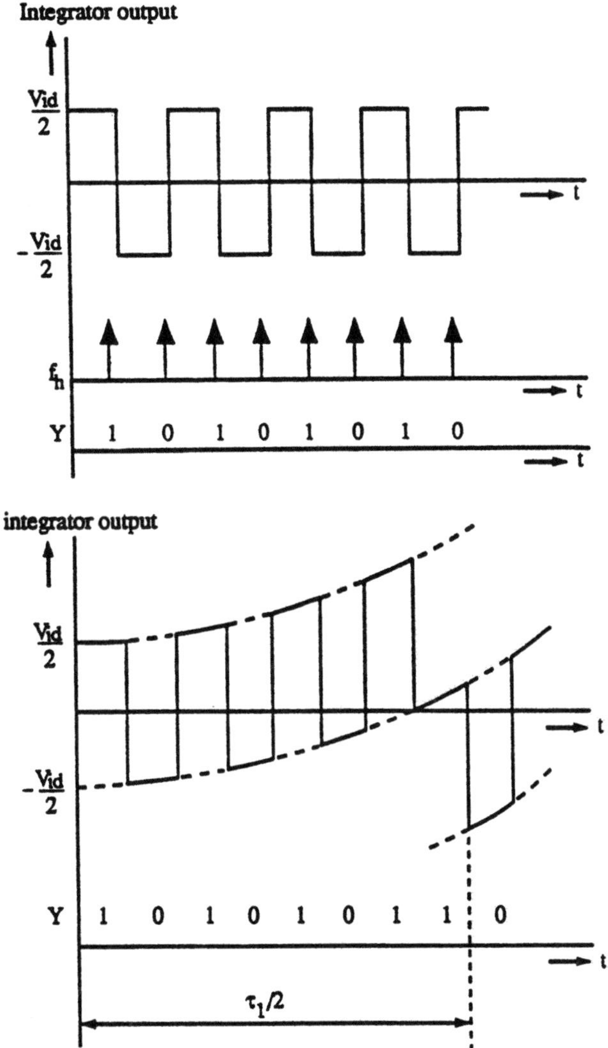

Figure 11.21 : (a) Idle pattern of a sigma-delta A/D converter with zero input signal and (b) with a sine wave input signal

pattern with zero input and with a sine wave input signal are shown. When well-determined dc input signals are applied to a first-order converter, some stable patterns can be found. The following analysis gives some of the most common patterns.

Suppose that during n_1 successive clock pulses the capacitor is charged and during n_2 clock pulses the capacitor is discharged. Suppose furthermore that after n_r repetitions of the previous pattern during one clock pulse an additional charge pulse is added; then we obtain for the total charge:

$$Q_{charge} = n_r \times n_1(Q_{ref} - Q_{analog}) + Q_{ref} - Q_{analog}. \tag{11.62}$$

The amount of charge discharged from the integration capacitor becomes:

$$Q_{discharge} = n_r \times n_2(Q_{ref} + Q_{analog}). \tag{11.63}$$

The operation of the A/D converter forces the total charge and discharge to cancel, which means:

$$Q_{charge} = Q_{discharge}. \tag{11.64}$$

After rearranging, we obtain:

$$\frac{Q_{analog}}{Q_{ref}} = \frac{n_r(n_1 - n_2) + 1}{n_r(n_1 + n_2) + 1}. \tag{11.65}$$

Note that if a positive input signal is applied, then the number of discharge pulses n_2 must be equal to 1 to obtain the required idle pattern at half the sampling frequency. At the same time n_1 becomes 1 for a negative input signal.

A special class of stable patterns is obtained when the repetition n_r becomes very large ($n_r \gg 1$). Equation 11.65 simplifies into:

$$\frac{Q_{analog}}{Q_{ref}} = \frac{n_1 - n_2}{n_1 + n_2} \tag{11.66}$$

Again, with a positive input signal n_2 is 1 and with negative input signals n_1 is 1.

By using $n_2 = 1$ and varying n_1 from 1, 2, 3, and so on, an analog input signal of 0, $\frac{1}{3}Q_{ref}$, $\frac{1}{2}Q_{ref}$, and so on, must be applied.

The drawback of stable patterns is found in the frequency distribution of the quantization error. At that moment correlations with the input signal are obtained. Such correlations may results in whistles and birdies in the quantized signal. In audio applications such a correlation is not allowed because the human ear is very sensitive for repetitive signals.

11.9.1 Threshold effect of a first-order converter

The idle pattern generated in a sigma delta A/D converter introduces a threshold effect below which no signal quantization is obtained. A signal is quantized at the moment the integrated input signal is equal to half the amplitude of the idle pattern. An example of such a quantization is shown in Figure 11.21(b). The figure shows that at the moment the amplitude of the integrated input sine wave crosses the zero line, an extra discharge pulse is needed. This extra pulse changes the 1 0 1 0 . . pattern into a 1 0 1 0 1 1 0 pattern. The extra 1 contains the quantization information.

The threshold effect of the A/D converter shown in Figure 11.7 can be calculated rather simply. Assume that no input signal is applied to the converter and suppose, furthermore, that all elements are ideal; then the amplitude of the idle pattern becomes:

$$V_{id} = \frac{C_1}{C_2} V_{ref}. \tag{11.67}$$

Here V_{id} is the peak-to-peak amplitude of the idle pattern and V_{ref} is the reference voltage applied to the converter. The discrete-time D/A converter used results in a gain ratio determined by the capacitors C_1 and C_2. The input signal is integrated using the continuous-time integrator with R and C_2.

If a sine wave $V_p \sin \omega_{in} t$ with a small amplitude is applied to the converter, this signal is integrated and added to the charge generated by the capacitive D/A converter circuit. Suppose that the amplitude of the input signal is so small that it takes a time τ_1 before the idle pattern is disturbed. The amplitude of the sine wave signal over the time τ_1 then becomes:

$$\int_0^{\tau_1} \frac{V_p}{R} \sin 2\pi f_{in} t\, dt = C_2 V_{out}. \tag{11.68}$$

By reworking equation 11.68 the output voltage V_{out} becomes:

$$V_{out} = \frac{1}{2\pi f_{in}} \frac{V_p}{C_2 R} [\cos(2\pi f_{in}\tau_1) - 1]. \tag{11.69}$$

The minimum value of equation 11.69 is obtained if:

$$2\pi f_{in}\tau_1 = \pi \ or \ \tau_1 = \frac{1}{2f_{in}}. \tag{11.70}$$

Inserting the value of 11.70 into equation 11.69 results in:

$$V_{out} = -\frac{1}{\pi f_{in}} \frac{V_p}{C_2 R}. \tag{11.71}$$

The maximum value of equation 11.69 is obtained when $\cos 2\pi f_{in}\tau_1 = 1$. Then a maximum output signal $V_{out} = 0$ is obtained. The peak-to-peak value of the output signal V_{out} is equal to the value given by equation 11.71. The idle pattern is disturbed when the output voltage $V_{out} = \frac{1}{2}V_{id}$. The following result is then obtained:

$$V_{out} = V_{id}. \tag{11.72}$$

To obtain a relation between the maximum input signal value V_{max} and the threshold effect of the converter, a relation between the reference voltage V_{ref} and the maximum signal level is needed. Using a charge balance equation, the following relation is valid:

$$\frac{V_{max}}{R} \frac{1}{f_s} = V_{ref}C_1. \tag{11.73}$$

By combining equations 11.67, 11.68, 11.71, and 11.73, the following result is obtained:

$$V_p = -\pi \frac{f_{in}}{f_s} V_{max}. \tag{11.74}$$

The peak threshold voltage can be defined as:

$$V_{th} = \mid V_p \mid. \tag{11.75}$$

Inserting equation 11.75 into equation 11.74 results in:

$$V_{th} = \frac{\pi f_{in}}{f_s} V_{max}. \tag{11.76}$$

From equation 11.76 it is seen that the threshold voltage linearly increases with the input signal frequency. The result of the threshold is the introduction of signal distortion. This distortion increases with increasing input frequency. The threshold voltage can be reduced by increasing the order of the loop filter.

11.9.2 Threshold effect of a second-order converter

In the second-order sigma-delta A/D converter an extra integrator is inserted in the loop. To obtain stability, which means an idle pattern around $\frac{f_s}{2}$, a zero must be added. A good practical approach is found when the frequency f_n of the zero is at about $\frac{1}{10}f_s$. Because of the increased loop gain, a reduction of the threshold voltage is obtained at lower frequencies. An estimate of the threshold voltage can be determined. The result becomes:

$$V_{th} = \frac{\pi f_{in}^2}{f_s f_n} V_{max}. \tag{11.77}$$

From equation 11.77 it is found that the threshold reduces with a factor $\frac{f_{in}}{f_n}$. With large oversampling ratios the improvement is obtained. However, a second-order system results in an increase in dynamic range. In practice this results in the fact that the frequency for which the threshold is reached is still in the signal band of interest.

At that moment the threshold determines the dynamic range and not the loop filter. To overcome this problem a dither signal is usually added. Such a dither signal linearizes and reduces the threshold, resulting in an improvement is dynamic range and a lower distortion.

11.9.3 Dither signals

To avoid correlation of idle patterns with the input signal in an A/D converter, a dither signal can be applied. Such a dither signal can be a small noise signal or a signal at half the sampling frequency. The amplitude of this dither signal is small to avoid a reduction in dynamic range of the total converter system. The purpose of the dither signal is to disturb the fixed signal patterns in the converter. In this way no correlation of the quantization noise and the input analog signal exists so a mere noise-like error signal is obtained. At the same time the dither signal reduces the threshold of the converter, resulting in a reduction in distortion of the signal and a coding of low-level, high-frequency input signals. In Figure 11.22 the frequency spectrum of the error signal of an A/D converter with a sine wave input and no dithering is shown. From the spectrum shown in Figure 11.22 a large number of correlated signal components can be distinguished. When a small dither signal is applied, the error spectrum is more randomized. The result is shown in Figure 11.23. As can be seen from the spectrum, the amplitude is reduced while the number of correlated components is reduced.

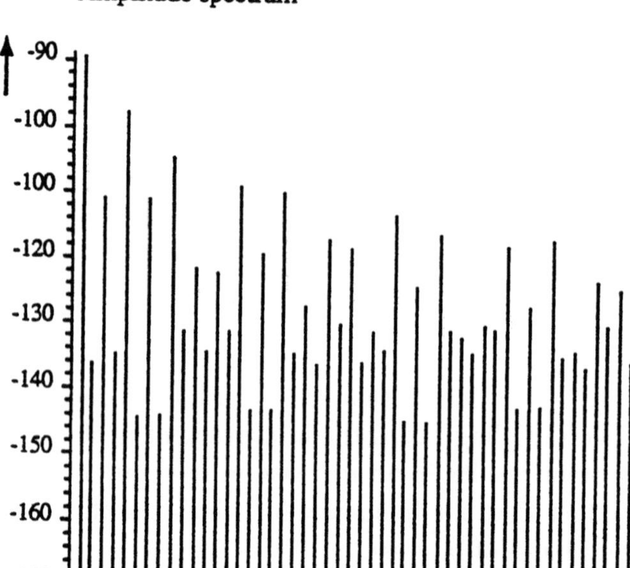

Figure 11.22 : Error signal spectrum of an A/D converter with an un-dithered sine wave

11.9.4   Threshold signal distortion

When a signal is applied to a transfer function with a threshold around zero, distortion is obtained. By applying a sine wave at the input of the converter, the output signal can be approximated by a sine wave minus a square wave with an amplitude equal to the threshold voltage and a frequency equal to the input sine wave.

By using equation 11.76, the amplitude of the square wave is found. This square wave is expanded in its Fourier series. This gives:

$$V_{sq} = \frac{4V_{th}}{\pi}\left(\sin f_{in} + \frac{\sin 3 f_{in}}{3} + \frac{\sin 5 f_{in}}{5} + ...\right) \qquad (11.78)$$

The series in 11.78 only contains odd harmonics. In Figure 11.24 the result of a simulation is shown. The simulation result shows the increase of the distortion and noise at the high-frequency band edge of the system. This is because of the noise-shaping. At the moment a dc offset is added to the system, the series expansion changes and even order terms will appear as

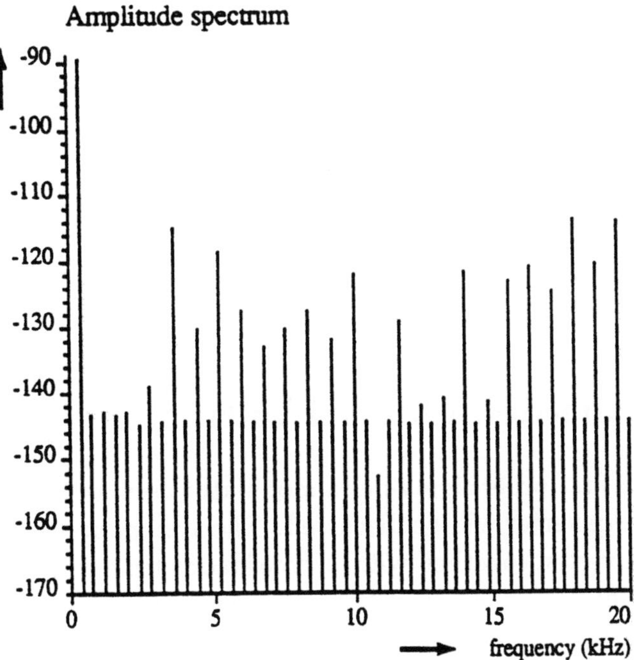

Figure 11.23 : Error spectrum of a dithered sine wave signal

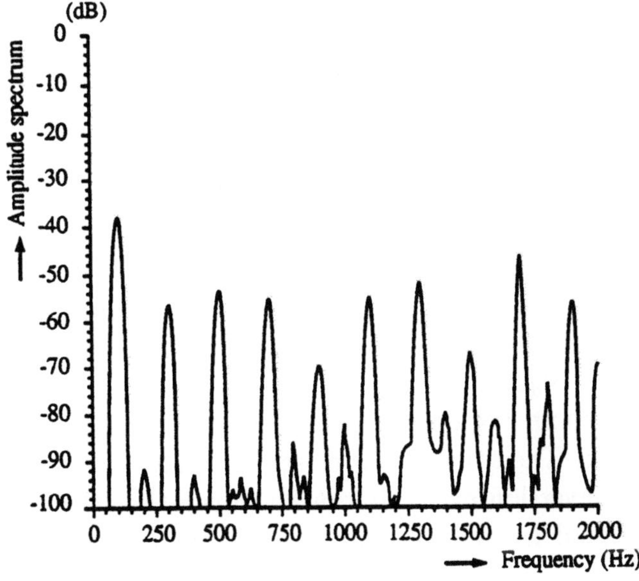

Figure 11.24 : Distortion simulation of a sigma-delta A/D converter

well. Equation 11.78 now changes into:

$$V_{sqdc} = \frac{Vth}{2}[k + \frac{2}{\pi}(\sin k\pi \cos f_{in} + \frac{1}{2} \sin 2k\pi . \cos 2f_{in} + .. \qquad (11.79)$$

In a practical situation k is between $\frac{1}{2}$ and $\frac{1}{4}$ for a dc offset between zero and about $\frac{V_{th}}{2}$. In equations 11.78 and 11.79 V_{sq} and V_{sqdc} are the amplitudes of the fundamental and the distortion products introduced by the threshold effect of the converter. The ratio between these components and the maximum signal level V_{max} results in the distortion of the converter.

11.10   Sigma-delta digital voltmeter

An example of a first-order sigma-delta modulator used as a five-digit digital voltmeter A/D converter will be discussed. The system uses full-time pulses in the D/A converter together with a continuous-time first-order filtering operation. The basic converter system is shown in Figure 11.25. The input signal is applied to the converter as a current I_{in}. The charge and discharge reference currents are called I. The signal is integrated across the capacitor C. The voltage across the capacitor is monitored by the comparator *comp* that drives the flip-flop FF. This flip-flop is toggled by the system clock f. The output of the flip-flop controls the switch S that switches the charge or discharge currents I to the capacitor C. When no input signal is applied to the system, then a triangular wave is generated across the capacitor C. The average value of the capacitor voltage is zero. When an input signal is applied, then the one-zero pattern changes. The output of the flip-flop called *Data* is applied to an up-down counter which performs the conversion of the 1-bit sigma-delta signal into a binary-weighted output signal. In Figure 11.26 the total digital voltmeter system is shown. The sigma-delta converter is enlarged with a timer $(Timer(N))$, an up/down counter and a gating function. The timer function counts the total amount of up and down pulses $N = n_{up} + n_{down}$. During this time the gate is opened and the difference between the up and down pulses $n = n_{up} - n_{down}$ are counted in the up-down counter. Then the gate is closed and the system must be reset to start the next conversion cycle. It can be shown that the ratio between the analog input signal I_{in} and the reference current I_{ref} is equal to:

$$\frac{I_{in}}{I_{ref}} = \frac{n_{up} - n_{down}}{n_{up} + n_{down}} = \frac{n}{N}. \qquad (11.80)$$

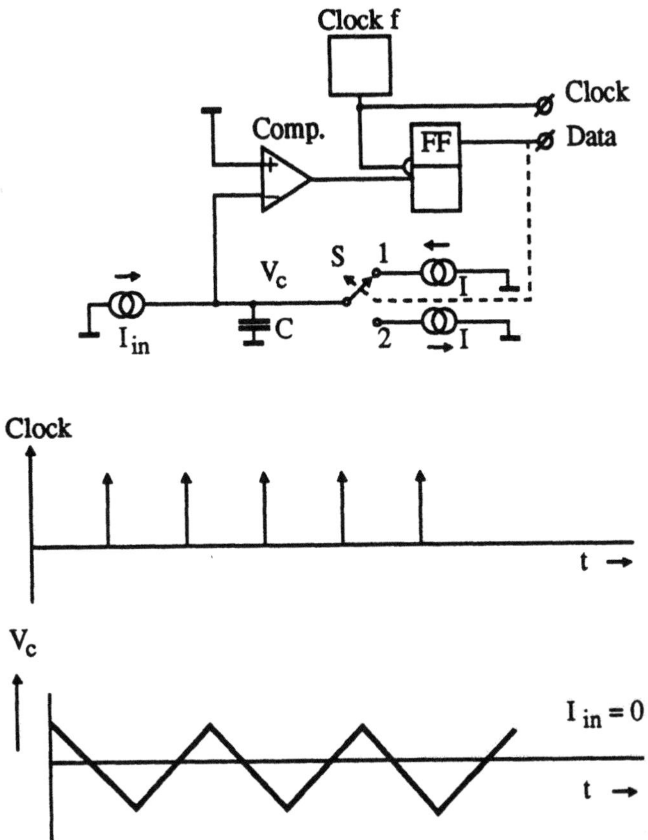

Figure 11.25 : Basic sigma-delta digital voltmeter system

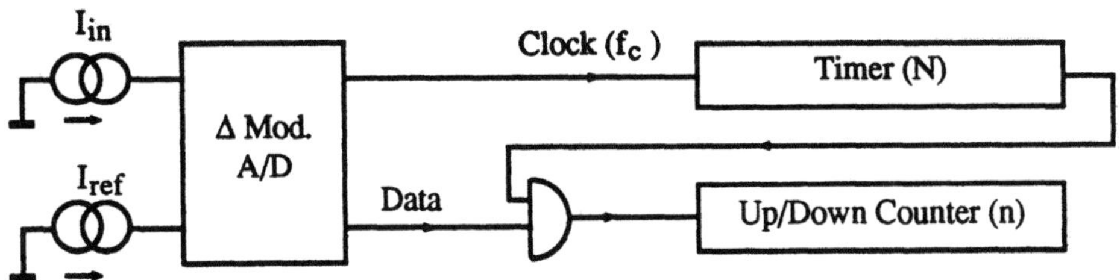

Figure 11.26 : Total digital voltmeter system

The resolution of the converter is determined by the length N of the timer circuit, while the conversion time T_c at a certain resolution N is determined by the clock frequency f_c, so

$$T_c = \frac{N}{f_c}. \tag{11.81}$$

When the conversion time T_c is related to the mains frequency, then due to the integration of a full- or a multiple period of the mains frequency, a very good rejection of this frequency is obtained.

11.10.1 Auto-zero circuit

In high-resolution analog systems dc offset is a problem because it adds or subtracts from small input signals. To overcome the offset problem in the analog circuit part of the converter, an auto-zero system is used. The basic configuration is shown in Figure 11.27 In Figure 11.27 the offset of

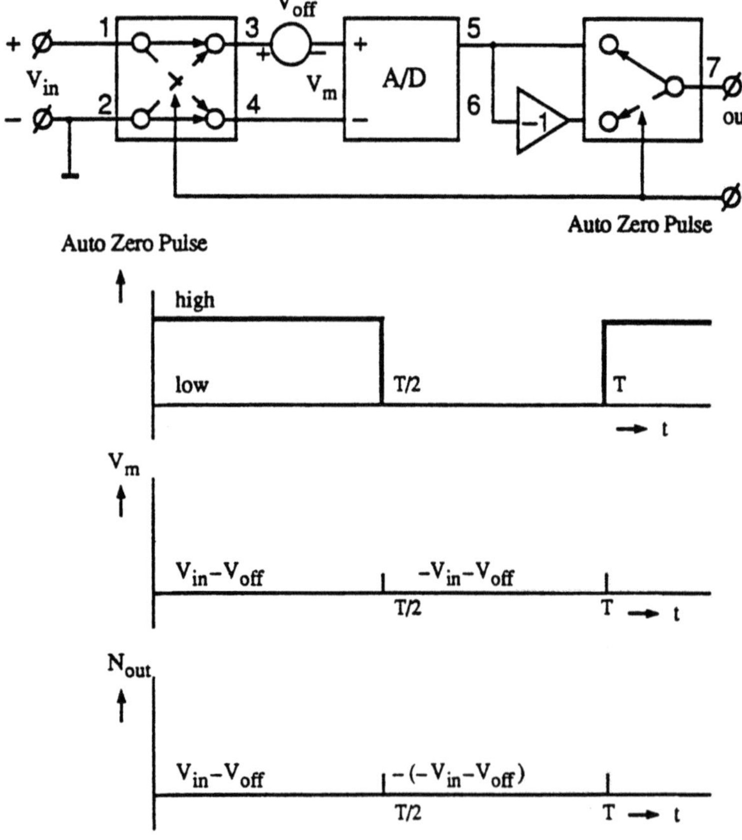

Figure 11.27 : Auto-zero system

the analog subsystem is given by V_{off}. At the input of the system a set of switches is added. These switches invert the analog input signal with respect to the A/D converter. At the output the data pulses are inverted, depending on the input switch settings. In this way the input "inversion" is removed from the data which are applied to the counter section. The system basically operates by dividing the conversion time T_c into two equal parts $T_c/2$. During the first part of the conversion time input terminals 1 and 3 are connected as is done with 2 and 4. In the second part terminals 3 and 4 are interchanged. During the first part of the measuring time, a voltage V_{m1} equal to $V_{in} - V_{off}$ is measured. During the second half of the measuring time a voltage V_{m2} equal to $-V_{in} - V_{off}$ is converted. The difference between V_{m1} and V_{m2} must be calculated. By inserting the data inverter during the second part of the measuring time, the data inversion is performed and the subtraction is transformed into an addition. An addition is very easy to implement by counting of pulses. As a result, the measured voltage is equal to:

$$V_m = \frac{1}{2}(V_{in} - V_{off} + V_{in} + V_{off}) = V_{in}. \tag{11.82}$$

As can be seen from equation 11.82, the auto-zero circuit ideally cancels the offset voltage without increasing the conversion time. A second advantage of this system is that the measured values for positive and negative signals is exactly equal because there is no difference between measuring a positive signal or a negative signal according to equation 11.82 except from the sign determining algorithm.

11.10.2 Analog subsystem implementation

A simplified diagram of the analog subsystem is shown in Figure 11.28. The system consists of a voltage-to-current converter V to I, the differential PNP transistor pair T_3 and T_4 controlled by the flip-flop FF, a comparator $comp$, the integration capacitor C, and the reference current source I_{ref}. An additional operational amplifier $Op.$ $Amp.$ controls the current sources I_1 and I_2 in the voltage-to-current converter. The voltage-to-current converter is drawn as the differential amplifier pair T_1 and T_2 with emitter degeneration resistor R_1.

The equal charge and discharge currents are generated by the reference current I_{ref} and the differential pair T_3 and T_4. As long as the base currents are very small then these charge and discharge currents are exactly equal. In

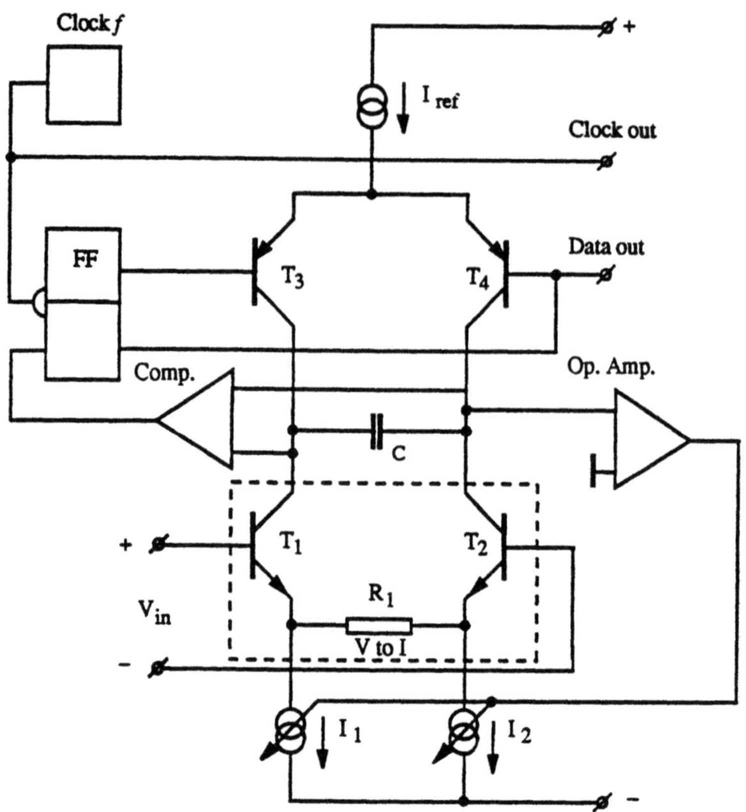

Figure 11.28 : Analog subsystem implementation

a BiCMOS technology the switches T_3 and T_4 will be replaced by MOS devices. No gate currents will flow in these devices. The operational amplifier *Op. Amp.* controls the currents I_1 and I_2 in such a way that the potential of the collectors of T_2 and T_4 is ate zero volt. The sigma-delta converter operation performs the same with the collector voltage of T_1 and T_3. The input system is a full-differential system that is required for the auto-zero function. Digital output signals of the conversion section are *clock out* and *data out*. These signals are applied to the digital counting section.

11.10.3   Basic voltage-to-current converter

In Figure 11.29 a circuit diagram of a basic voltage-to-current converter is shown. The basic voltage-to-current converter consists of the PNP input pair T_1, T_2, and two current mirrors (T_3, T_4, T_5 and T_6, T_7, T_8) with biasing current sources I_0. With $V_{in} = 0$, due to the current mirror action, a current

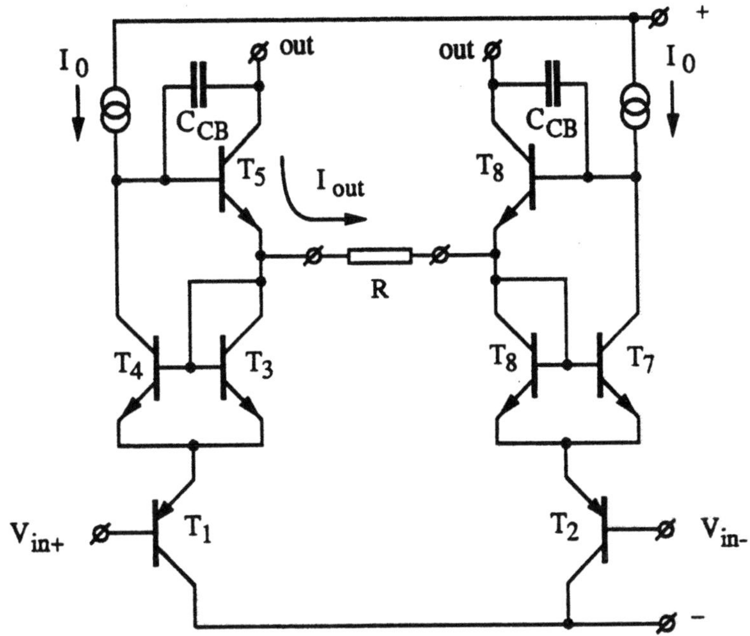

Figure 11.29 : Basic voltage-to-current converter

I_0 must flow through T_3 and T_4 as well as through T_6 and T_7. Therefore, the currents through T_1 and T_2 are constant and have a value $2I_0$. When an input voltage is applied, this voltage is exactly reproduced across the conversion resistor R, because the voltage drops across T_1, T_3 and T_2, T_6 remain constant. As a result, the converted output current I_{out} can only flow through transistors T_5 and T_8. As long as the base currents of T_5 and T_8 are small, a constant current will flow through the input devices T_1, T_2 and the diodes T_3, T_6, keeping the voltage drops constant. An accurate voltage-to-current conversion is obtained with this system. In the final system a more complicated circuit based on the same system is used.

11.10.4 Complete digital voltmeter system

The complete digital voltmeter system is implemented in a CMOS digital part for control and display function and a bipolar part that contains the complete sigma-delta A/D converter section with reference source and gain selection. In Table 11.1 the performance of the system is shown. The internal reference voltage is based on the band-gap principle of silicon.

Input voltage	100 mV and 1 V
	(bipolar)
Resolution	5 digits plus sign
	(f_{clock} = 200 kHz)
Linearity	$\pm$ 10^{-5} of full scale
Zero stability	< 10^{-6} per degree C
Temp. coefficient. of	5.10^{-6} per degree C
reference source	over Δ T = 80^{0} C
Supply voltage dependence	
of reference source	5.10^{-6} per V
Power supply	$\pm$ 7.5 V 20 mW

Table 11.1 : Digital voltmeter performance data

11.11   Conclusion

Sigma-delta converters are excellent alternatives for high-resolution analog-to-digital converters for applications in digital audio, digital video and digital voltmeter systems. The sigma-delta system needs only two or, in special configurations one reference voltage to perform the conversion. In these systems an excellent differential linearity with small input signals is obtained. The combination of the noise-shaping filter with the anti-alias function minimizes the components count and seems to be a very attractive alternative for digital audio signal conversion. Even a larger dynamic range can be obtained compared to direct conversion methods (e.g.,successive approximation algorithm). When the maximum clock rate is limited, an increase in dynamic range can be obtained by combining a multi-bit D/A converter function with a noise-shaping function. At the moment a filtering order above a second-order is used, then feed-forward techniques can be used to obtain a stable operation of the converter under all signal conditions. This multi-level system optimizes the performance of an A/D converter with respect to dynamic range, filtering capability, and system complexity. The presented dynamic range calculations are for idealized systems. Care must be taken in using the presented results. However, the results help to understand the operation of the sigma-delta converter. When input signal levels close to maximum are applied, then the nonlinearity of the quantizer influences the stability margin which usually results in an unstable coder. Again,system simulations are needed to prove the overall stability.

Problems

In this chapter problems per chapter will be given.

P 1.1

Explain why with analog-to-digital converters an analog prefiltering is required.

P 1.2

What is the minimum value of the stop-band attenuation of the analog prefilter when an N-bits A/D converter is used?

P 1.3

What is the advantage of a combination of an analog prefilter followed by a four times oversampled analog-to-digital converter followed by a digital filter with respect to a full analog prefilter?

P 1.4

Derive the formula giving a relation between the sample time uncertainty and the maximally sine input frequency for a $\frac{1}{2}$ LSB peak-to-peak error in case of a $N - bit$ converter.

Calculate the time uncertainty for a 50 MHz input sine wave and a 10-bit converter in case the amplitude error at maximum signal slope must be smaller than $\frac{1}{2}$ LSB.

P 2.1

What is the advantage of a serial input data register above a full parallel input in case a high-resolution current output digital-to-analog converter is used?

P 2.2

What does *monotonicity* mean for a converter? Does monotonicity automatically include the integral nonlinearity (INL) specification of $\frac{1}{2}$ LSB? If not, give the right definition.

P 2.3

During the measurement of the nonlinearity of a digital-to-analog converter the offset has not been corrected. By applying the digital code 000..0 an output voltage $V_{out} = 3.52$ mV has been measured.

The values of the bit weights at the output measured are: $B_0 = 2571{,}79$ mV, $B_1 = 1284{,}22$ mV, $B_2 = 645{,}73$ mV, $B_3 = 326{,}23$ mV, $B_4 = 158{,}7$ mV, $B_5 = 88{,}25$ mV, $B_6 = 45{,}85$ mV, $B_7 = 23{,}37$ mV, $B_8 = 14{,}82$ mV and $B_9 = 8{,}58$ mV.

1. Calculate the integral nonlinearity (INL) of this converter.

2. What can you tell about the converter after examining the result of question 1)

3. Calculate the differential nonlinearity (DNL) in case of the most significant bit carry (code = 100..0 to 011..1).

P 2.4

What are glitches in D/A converters and how do they occur?
What precautions have to be taken to minimize glitches in a full parallel data input digital-to-analog converter? The inputs drive immediately the bit switches.

P 2.5

Explain why in an analog-to-digital converter distortion occurs when a signal frequency close to half the sampling frequency is applied, especially when non-harmonic distortion is found.

When, with the same signal bandwidth of the system the sampling frequency is increased with a factor 4, does an improvement in distortion occur?

P 2.6

Why is a sample-and-hold amplifier used in front of an analog-to-digital converter. Do you know another application of such an amplifier?

P 2.7

Why is the differential non-linearity more important than the integral non-linearity when small signals are reproduced using a D/A converter?

P 2.8

Over what bandwidth with respect to the sampling frequency is the signal-to-noise ratio of a converter defined? What is the expression that defines the signal-to-noise ratio of an N-bits converter.

P 3.1

During the measurement of the signal-to-noise ratio of a digital-to-analog converter using a distortion analyzer the fundamental signal is rejected by the measurement set.

1. Give three (distortion) components that are measured by the distortion analyzer.

2. At high-signal frequencies what is the dominating factor?

P 3.2

What is the minimum amount of data needed to determine the integral non-linearity of an n-bits binary-weighted digital-to-analog converter?

P 3.3

What important dynamic parameters can be measured with the "zero-beat" measurement technique of a flash-type analog-to-digital converter?

P 3.4

What is the advantage of using the "zero-beat" dynamic test set-up in case two sample-and-hold amplifiers are cascaded and clocked with different clock frequencies?

P 4.1

Why doesn't a flash-type analog-to-digital converter need an input sample-and-hold amplifier?

P 4.2

Does a folding analog-to-digital converter require a sample-and-hold amplifier at the input?

P 4.3

Explain why a CMOS two-step analog-to-digital converter can operate without a sample-and-hold amplifier.

P 4.4

Explain how the bit error rate in a master-slave comparator flip-flop can be improved.

P 4.5

Give the advantages and disadvantages of a high-speed analog-to-digital converter using a direct Gray code circuit implementation.

P 4.6

What is the advantage of interpolation compared to direct fine flash amplitude conversion in folding analog-to-digital converters?

P 4.7

In a 10-bit two-step analog-to-digital converter is a 5-bit digital-to-analog converter used to convert the coarse digital information into a quantized analog signal. Explain what the integral nonlinearity of the 5-bit digital-to-analog converter must be compared to the 10-bit overall resolution.

P 5.1

What small signal ac parameter introduces third-order distortion in full-flash analog-to-digital conversion?

P 5.2

What is the reason for using a clock and/or signal wire tree in flash analog-to-digital converters?

P 5.3

What parameters determine the bit error rate of a master flip flop comparator stage.

P 5.4

Determine the minimum small signal bandwidth of the input circuit of a 10-bit flash type analog-to-digital converter when the third-order distortion must be smaller than 70 dB.

P 6.1

Why is the conversion rate of pulse-width digital-to-analog converters relatively low, and what can be done to improve the conversion speed when a high resolution is needed?

P 6.2

What is the advantage of an R-2R ladder-network compared to a binary-weighted resistor network in integrated circuits?

P 6.3

Explain why in a binary-weighted bipolar R-2R network emitter scaling is used. What is the disadvantage of emitter scaling in high-resolution digital-to-analog converters?

P 6.4

Give a circuit design of an R-2R bipolar digital-to-analog converter that provides an alternative solution to emitter scaling. What is the disadvantage of this system?

P 6.5

What is the advantage of binary-weighted current networks using device scaling techniques *only*.

P 6.6

What is the advantage of a converter design based on a *monotonic by design* criterion? Can the integral nonlinearity specification or the differential nonlinearity specification easily be obtained? Explain why.

P 6.7

The *dynamic element matching* design system results in a high division accuracy. Calculate the division accuracy of a divide by two stage when the passive divider error is 1 % and the timing error is 10^{-4}.

P 6.8

Can the *dynamic element matching* technique be applied when currents with a real division ratio (1 to 2.3, for example) are applied.

P 6.9

What is the size of the "ripple" current coming from a *dynamic element matching* current divider? Is this an advantage or a disadvantage.

P 6.10

In the practical *continuous current calibration* technique no storage capacitor is drawn in the circuit. How is analog storage performed in this system?

P 6.11

Give the disturbing sources that decrease the duplication accuracy in the *continuous current calibration* digital-to-analog converter. Give a circuit diagram that improves the duplication accuracy in a simple way.

P 6.12

Give a circuit diagram of a *continuous current calibration* system that does not need a compensation current smaller than the total output current.

P 7.1

In practical operational amplifiers dc offset is always found. Explain if this offset voltage influences the conversion result of a dual-slope integrating analog-to-digital converter.

P 7.2

Is the integral nonlinearity of a successive approximation analog-to-digital converter influenced by comparator offset? Explain what the influence of offset on the result is.

P 7.3

In self-calibrating analog-to-digital converters a high accuracy can be obtained. What is the disadvantage of the calibration cycle in continuous operating systems like digital audio?

P 7.4

What is the accuracy-determining element in algorithmic recycling analog-to-digital converters? Give a system that overcomes comparator offset problems.

P 8.1

What is the purpose of a sample-and-hold amplifier?

P 8.2

When a sample-and-hold amplifier is designed using a single MOS transistor as a switch in a non-inverting application, does the rise and fall time of the sampling clock have influence on the sampling accuracy of the system?

P 8.3

When the clock feed-through during switching from the sample mode into the hold mode is signal dependent, how is this feed-through found in the sampled output result?

P 8.4

When a diode bridge is used as a sampling switch, what value should have
the current through the diodes compared to the signal current flowing into
the hold capacitor? What kind of error is expected when the current through
the sampling bridge is in the same order as the hold capacitor current?

P 9.1

Calculate the temperature coefficient of a base-emitter current stabilizer if
$V_{BE} = 0.85$ V and $V_g = 1.205$ V.

P 9.2

A current stabilizer has been built using a PNP current independent current
mirror with ratio 1 and an NPN current dependent current mirror with emitter ratio of 5. Draw the current ratios as a function of the input currents.
What observation is obtained when the curves are crossing? What can be
done to obtain a well defined stabilized current?

P 9.3

As a voltage reference a physical constant of silicon is used. What voltage
is obtained? Give a circuit diagram of such a reference voltage source using
a 1 to 1 resistor ratio and a 1 to 5 emitter size of NPN devices.

P 9.4

Which two components as a function of temperature compensate to give a
stable output voltage depending on a physical constant of silicon?

P 9.5

When a transistor plus a resistor in the emitter of this transistor are used
as a current source, calculate the value of the voltage across the emitter
generating resistor to obtain a 0.1% accuracy of the output current when
the transistor base-emitter current i_0 has an accuracy of 1.0 % and the
resistor has an accuracy of 0.1 %.

P 10.1

Derive the formula giving the improvement in signal-to-noise ratio of a first-order, noise-shaping, digital-to-analog converter.

P 10.2

What is the advantage of a switched capacitor type digital-to-analog converter above the full-signal 1-bit digital-to-analog converter in a noise-shaping coder?

P 10.3

What is the advantage of a 1-bit digital-to-analog converter compared to a multi-bit digital-to-analog converter used in a noise-shaping coder?

P 10.4

What is the advantage of a noise-shaping coder compared to a binary-weighted converter when a small (bipolar) output signal is generated?

P 10.5

What is the advantage of a differential implementation of a noise-shaping coder?

P 10.6

What is the advantage of a multi-bit D/A converter incorporated in a noise-shaping coder?

P 11.1

Derive the relation between a sigma-delta filter function and a noise-shaping coder filter function.

P 11.2

What is the limiting factor that limits the number of stages in a MASH type sigma-delta converter?

P 11.3

Is the phase design stability method applicable to nonlinear (clamped) filter structures. Explain why.

P 11.4

What is the purpose of a dither signal applied to a sigma-delta converter? What is the disadvantage of the application of a dither signal?

P 11.5

What is the advantage of a switched capacitor filter implementation in a sigma-delta converter?

P 11.6

Can a switched capacitor filter function in a sigma-delta converter be used as an anti-alias filter? Explain why.

P 11.7

Can a small input signal close to half the sampling frequency be converted into a digital value using a sigma-delta converter? What problem occurs? What can be done about this problem.

Bibliography

[1] R.J. van de Plassche, D. Goedhart, *"A monolithic 14-bit D/A converter,"* IEEE Journal of Solid-State Circuits, vol. SC-14, pp. 552-556, June 1979.

[2] H.J. Schouwenaars, E.C. Dijkmans, B.M.J. Kup, E.J.M. van Tuijl, *"A monolithic dual 16-bit D/A converter,"* IEEE Journal of Solid-State Circuits, vol. SC-21, pp. 424-429, June 1986.

[3] R.J. van de Plassche, H.J. Schouwenaars, *"A monolithic 14-bit A/D converter,"* IEEE Journal of Solid-State Circuits, vol. SC-17, pp. 1112-1117, Dec. 1982.

[4] R.J. van de Plassche, H.J. Schouwenaars, *"A monolithic high-speed Sample-and- Hold amplifier for digital Audio,"* IEEE Journal of Solid-State Circuits, vol. SC-18, pp 716-722, Dec. 1983.

[5] R.J. van de Plassche, P. Baltus, *"An 8-bit 100 Mhz full Nyquist Analog-to-Digital Converter,"* IEEE Journal of Solid-State Circuits, vol. SC-23, pp. 1334-1344, Dec. 1988.

[6] C.J. van Valburg, R.J. van de Plassche, *"An 8-b 650 MHz Folding ADC,"* IEEE Journal of Solid-State Circuits, vol. 27, pp. 1662-1666, Dec. 1992.

[7] A.W.M. van den Enden, N.A.M. Verhoekx, *"Discrete-time signal processing,"* Prentice Hall, 1989.

[8] W.R. Bennett, *"Spectra of quantized signals"* Bell System Technical Journal, vol. 27 pp. 446-472, July 1948.

[9] M. Schwartz, *"Information transmission, modulation, and noise,"* McGraw-Hill, 1980.

[10] A.B. Carlson, "*Communication systems,*" McGraw-Hill 1975.

[11] E.F. Stikvoort, "*Some subjects in digital audio, noise-shaping, sample-rate conversion, dynamic range compression and testing,*" Ph.D. Thesis, University of Technology Eindhoven, September 1 1992.

[12] H.J.M. Veendrick, "*The behavior of flip-flops used as synchronizers and prediction of their failure rate,*" IEEE Journal-of Solid-State Circuits, vol. SC-15, pp. 169-176, April 1980.

[13] R.J. van de Plassche, E.C. Dijkmans, "*A monolithic 16-bit D/A conversion system for digital audio,*" Premier AES Conference, Rye, New York, pp. 54-60, June 1982.

[14] E.R. Hnatek, "*A user's handbook of D/A and A/D converters,*" John Wiley & Sons, 1976.

[15] D.H. Sheingold, "*Analog-digital conversion handbook,*" Prentice-Hall, 1986.

[16] M.V. Mahoney, "*DSP-based testing of analog and mixed signal circuits,*" The Computer Society of the IEEE.

[17] H. Kitayoshi, S. Sumida, K. Shirakawa, S. Takeshita, "*DSP synthesized signal source for analog testing stimulus and new test method,*" Proceedings of IEEE International Test Conference 1985, pp. 825-834.

[18] M. Corradetti, I. Olivia, "*MOS A/D and D/A converter circuits based on the stochastic principle,*" International Conference "Mikroelektronik" Munich, November 27-29, 1972.

[19] P. Holloway, M. Norton, "*A high yield, second generation 10-bit monolithic DAC,*" ISSCC Digest of Technical Papers, pp. 106-107, February 1976.

[20] A.P. Brokaw, "*A monolithic 10-bit A/D using I^2L and LWT thin film resisters*" IEEE Journal of Solid-State Circuits, vol. 13, pp. 736-745, December 1978.

[21] J.A. Schoeff, "*An inherently monotonic 12-bit DAC,*" IEEE Journal of Solid-State Circuits, vol. SC-14, pp. 904-911, Dec. 1979.

[22] P. Holloway, "*A trimless 16-bit digital potentiometer,*" ISSCC Digest of Technical Papers, pp. 66-67, February 1984.

[23] M.J.M. Pelgrom, *"A 10-b 50-MHz CMOS D/A converter with 75-Ω buffer"* IEEE Journal of Solid-State Circuits, vol. 25, pp. 1347-1352, December 1990.

[24] K. Maio, S-I. Hayashi, M. Hotta, T. Watanabe, S. Ueda, N. Yokozawa, *"A 500 MHz 8-bit D/A converter,"* IEEE Journal of Solid-State Circuits, vol. 20, pp. 1133-1136, December 1985.

[25] P.H. Saul, J.S. Urquhart, *"Techniques and technology for high-speed D-A conversion,"* IEEE Journal of Solid-State Circuits, vol. 19, pp. 62-68, February 1984.

[26] T. Kamoto, Y Akazawa, M. Shinagawa, *"An 8-bit 2-ns monolithic DAC,"* IEEE Journal of Solid-State Circuits, vol. 23, pp. 142-146, February 1988.

[27] K-C. Hsieh, Th.A. Knotts, G.L. Baldwin, T. Hornak, *"A 12-bit 1-Gword/s GaAs digital-to-analog converter system,"* IEEE Journal of Solid-State Circuits, vol. 22, pp. 1048-1055, Decemebr 1987.

[28] J.L. McCreary, P.R. Gray, *"All-MOS charge redistribution analog-to-digital conversion techniques - part I,"* IEEE Journal of Solid-State Circuits, vol. SC-10, pp. 371-379, Dec. 1975.

[29] J.L. McCreary, D.A. Sealer, *"Precision capacitor ratio measurement technique for integrated circuit capacitor arrays,"* IEEE Transactions on Instrumentation and Measurements, vol. IM-28, pp. 11-17, March 1979.

[30] H.B. Aasnaes, Th.J. Harrison, *"Triple play speeds A/D conversion,"* Electronics, pp. 69-72, April 1968

[31] J.L. McCreary, *"Matching properties and voltage and temperature dependence of MOS capactitors,"* IEEE Journal of Solid-State Circuits, vol. SC-16, pp. 608-616, December 1981.

[32] H.J. Schouwenaars, D.W.J. Groeneveld, H. Termeer, *"A stereo 16-bit CMOS D/A converter for digital audio,"* IEEE Journal of Solid-State Circuits, vol. SC-23, pp. 1290-1297, Dec. 1988.

[33] M.J.M. Pelgrom, A.C.J. Duinmaijer, A.P.G. Welbers, *"Matching properties of MOS transistors,"* IEEE Journal of Solid-State Circuits, vol. 24, pp. 1433-1439, October 1989.

[34] V. Valencic, P. Deval, "*A 750 ksample/s 8-bit low-power pipelined A/D converter*," Digest of Technical Papers, ESSCIRC 1988, Manchester UK, pp. 42-45, Sept. 1988.

[35] C.C. Shih, P.R. Gray, "*Reference refreshing cyclic analog-to-digital and digital-to-analog converters*," IEEE Journal of Solid-State Cicruits, vol. SC-21, pp. 544-554, August 1986.

[36] Product Note 5180A-2, Hewlett-Packard, "*Sine wave curve fitting*," in Dynamic Performance Testing of A to D converters.

[37] J. Doernberg, H.-S. Lee, D.A. Hodges, "*Full-Speed testing of A/D Converters*," IEEE Journal of Solid-State Circuits, vol. SC-19, pp. 820-827, Dec. 1984.

[38] J. Peterson, "*A monolithic video A/D converter*," IEEE Journal of Solid-State Circuits, vol. SC-14, pp. 932-937, Dec. 1979.

[39] M. Inoue, H. Sadamatsu, A. Matsuzawa, A. Kanda, T. Takemoto, "*A monolithic 8-bit A/D converter with 120 MHz conversion rate*," IEEE Journal of Solid-State Circuits, vol. SC-19, pp. 837-841, Dec. 1984.

[40] B. Zojer, B. Astegher, H. Jessner, R. Petschacher, "*A 10 b 75 MHz subranging A/D converter*," IEEE Journal of Solid-State Circuits, vol. 25, pp. 1339-1346, December 1990.

[41] Y. Fujita, E. Masuda, S. Sakamoto, T. Sakaue, Y. Sato, "*A bulk CMOS 20 Ms/s 7-b flash ADC*" ISSCC Digest of Technical Papers, pp. 56,57, February 1984.

[42] T. Takemoto, M. Inoue, H. Sadamatsu, A. Matsuzuwa and K. Tsuji, "*A fully parallel 10-bit A/D converter with video speed*," IEEE Journal of Solid-State Circuits, vol. SC-17, p.p 1133-1138 , Dec. 1982.

[43] B. Zojer, R. Petscharher, W. Luschnig, "*A 6-bit 200 MHz full Nyquist A/D converter*," IEEE Journal of Solid-State Circuits, vol. SC-20, pp. 780-786, June 1985.

[44] B. Peetz, B. Hamilton, J. Kang, "*An 8-bit 250 Megasample per second analog-to-digital converter: operation without a sample-and-hold*," IEEE Journal of Solid-State Circuits, vol. SC-21, pp. 997-1002, Dec. 1986.

[45] Y. Yoshi, M. Nakamura, K. Hirasawa, A. Kayanuma, K. Asano, "*An 8-bit 350 MHz flash ADC*," ISSCC Digest of Technical Papers, pp. 96-97, February 1987.

[46] Y. Akazawa, A. Iawata, T. Wakimoto, H. Nakamura, H. Ikawa, "*A 400 Mspsec 8-bit flash AD conversion LSI*," ISSCC Digest of Technical Papers, pp. 98-99, February 1987.

[47] T. Wakimoto, S. Konaka, "*Si bipolar 2 GHz 6-bit flash A/D conversion LSI*", IEEE Journal of Solid-State Circuits, vol. SC-23, pp. 1345-1350, Dec. 1988.

[48] D. Daniel, U. Langmann, B.G. Bosch, "*A silicon bipolar 4-bit 1 Gsample/sec full Nyquist A/D converter*", IEEE Journal of Solid-State Circuits, Vol. SC-23, pp. 742-749, June 1988.

[49] A. Dingwall, "*Monolithic expandable 6-bit 20 MHz CMOS/SOS A/D converter*", IEEE Journal of Solid-State Circuits, vol. SC-14, pp. 926-932, Dec. 1979.

[50] U. Fiedler, D. Seitzer, "*A high-speed 8 bit A/D converter based on a Gray-code multiple folding circuit*", IEEE Journal of Solid-State Circuits, vol. SC-14, pp. 547-551, June 1979.

[51] R.J. van de Plassche, R.E.J. van der Grift, "*A high-speed 7-bit A/D converter*", IEEE Journal of Solid-State Circuits, vol. SC-14, pp. 938-943, Dec. 1979.

[52] R.E.J. van der Grift, R.J. van de Plassche, "*A monolithic 8-bit video A/D converter*", IEEE Journal of Solid-State Circuits, vol. SC-19, pp. 374-378, Dec. 1984.

[53] R.E.J. van der Grift, W.J.M. Rutten, M. van der Veen, "*An 8-bit video ADC incorporating folding and interpolation techniques*", IEEE Journal of Solid-State Circuits, vol. SC-22, pp. 944-953, Dec. 1987.

[54] K. Poulton, J.J. Corcoran, T. Hornak, "*A 1 GHz 6-bit ADC system*", IEEE Journal of Solid-State Circuits, vol. SC-22, pp. 962-970, Dec. 1987.

[55] N. Fukushima, T. Yamada, N. Kumazawa, Y. Hasegawa, M. Soneda, "*A CMOS 8-bit 40 MHz 2-step parallel A/D converter with 105 mW*

power consumption", ISSCC Digest of Technical Papers, pp. 14-15, February 1989.

[56] M. Ishikawa, T. Tsukahara, *"An 8-bit 40 MHz CMOS subranging ADC with pipelined wideband S/H"*, ISSCC Digest of Technical papers, pp. 12-13, February 1989.

[57] R.J. van de Plassche, *"An autozero MOS comparator"*, Dutch Patent nr. 89.000.90.

[58] A. Dingwall, V. Zazzu, *"An 8 MHz CMOS subranging 8-bit A/D converter"*, IEEE Journal of Solid-State Circuits, vol. SC-20, pp. 1138-1143, Dec. 1985.

[59] T. Shimizu, M. Hotta, K. Maio, *"A 10-bit 20 MHz two-step parallel ADC with internal S/H"*, ISSCC Digest of Technical Papers, pp. 224-225, February 1988.

[60] R.J. van de Plassche, P. Baltus, *"The design of an 8-bit folding Analog-to- Digital converter"*, Philips Journal of Research, vol. 42, pp. 482-509, Dec. 1987.

[61] J. Neal, J. Surber, *"Track-and-holds take flash converters to their limits"*, Electronic Design, pp. 381-388, May 3, 1984.

[62] W.D. Mack, M. Horowitz, R.A. Blauschild, *"A 14b PCM DAC"*, ISSCC Digest of Technical Papers, pp. 85,86, February 1982.

[63] R.J. van de Plassche, *"Dynamic element matching for high-accuracy monolithid D/A converters"*, IEEE Journal of Solid-State Circuits, vol SC-11, pp. 795-800, Dec. 1976.

[64] G. Kelson, H.H. Stellrecht, D.S. Perloff, *"A monolithic 10-bit digital-to-analog converter using ion implantation"*, IEEE Journal of Solid-State Circuits, vol. SC-8, Dec. 1973.

[65] J.R. Naylor, *"A complete high-speed voltage output 16-bit monolithic DAC"*, IEEE Journal of Solid-State Circuits, vol. SC-18, pp. 729-735, Dec. 1983.

[66] M.P. van Alphen, R.E.J. van der Grift, J.M. Pieper, R.J. van de Plassche, *"The PM 2517 automatic digital multimeter"*, Philips Technical Review, vol. 38, pp. 181-194, 1978/79 No. 7/8

[67] D.W.J. Groeneveld, H.J. Schouwenaars, H. Termeer, *"A self calibration technique for monolithic high-resolution D/A converters"*, IEEE Journal of Solid-State Circuits, vol. SC-24, pp. 1517-1522, Dec. 1989.

[68] D. Vallencourt, Y.P. Tsividis, S.J. Daubert, *"Sampled current sources"*, ISCAS Proc. pp. 1592-1595, Portland 1989.

[69] G. Wegmann, E.A. Vittoz, *"Very accurate dynamic current mirrors"*, Electronic Letters, May 11 1989, vol. 25, No. 10, p. 644.

[70] G. Wegmann, E.A. Vittoz, *"Analysis and improvements of accurate dynamic current mirrors"*, IEEE Journal of Solid-State Circuits, vol. 25, pp. 699-706, June 1990.

[71] H.J. Schouwenaars, *"Low-power digital-to-analog conversion"*, M. Phil Thesis, University of Wales, Swansea, April 10 1989.

[72] B. Ginetti, P Jespers, A. Vandemeulebroecke, *"A CMOS 13 bits cyclic RSD A/D converter"*, Proceedings ESSIRC 1991, Milan, pp. 345-348, September 1991.

[73] K. Maio, M. Hotta, N. Yokozawa, M. Nagata, K. Kaneko, T. Iwasaki, *"An untrimmed D/A converter with 14-bit resolution"*, IEEE Journal of Solid-State Circuits, vol. 16, pp. 616-621, December 1981.

[74] H.S. Lee, D.A. Hodges, *"A self-calibrating 15-bit CMOS A/D converter"*, IEEE Journal of Solid-State Circuits, vol. SC-19, pp. 813-819, Dec. 1984.

[75] G. Erdi, P.R. Henneuse, *"A precision FET-less sample-and-hold with high charge-to-droop current ratio"*, IEEE Journal of Solid-State Circuits, vol. SC-14, pp. 864-873, Dec. 1978.

[76] R.J. Widlar, *"New developments in ic voltage regulators"*, IEEE Journal of Solid-State Circuits, vol. SC-6, pp. 2-7, February 1971.

[77] K.E. Kuijk, *"A reference voltage source"*, IEEE Journal of Solid-State Circuits, vol. SC-8, pp. 222-226, June 1973.

[78] A.P. Brokaw, *"A simple three-terminal IC bandgap reference"*, IEEE Journal of Solid-State Circuits, vol. 9, pp. 388-393, December 1974.

[79] G.C.M. Meijer, *"A new configuration for temperature transducers and bandgap references"*, Digest of Technical Papers, pp. 142-145, European Solid-State Circuits Conference, Amsterdam, 1978.

[80] R.J. van de Plassche, *"Stabilized current and voltage reference sources"*, US Patent 4,816,742, March 28, 1989.

[81] M. Ishikawa, T. Tsukahara, *"An 8-bit 50 MHz CMOS subranging A/D converter with pipelined wide-band S/H"*, IEEE Journal of Solid-State Circuits, vol. SC-24, pp. 1485-1491, December 1989.

[82] M.V. Kolluri, *"A 12-bit 500 nsec subranging ADC"*, IEEE Journal of Solid-State Circuits, vol SC-24, pp. 1498-1505, December 1989.

[83] R. Jewett, J. Corcoran, G. Steinbruch, *"A 12b 20 Ms/s ripple through ADC"*, ISSCC Digest of Technical Papers, pp. 34,35, February 1992.

[84] S.H. Lewis, H.S. Fetterman, G.F. Gross Jr., R. Ramachandran, T.R. Viswanathan, *"A 10-b 20-Msample/s analog-to-digital converter"* IEEE Journal of Solid-State Circuits, vol. 27, pp. 351-358, March 1992.

[85] A. Matuszawa, M. Kagawa, M. Kanok et al., *"A 10-bit 30 MHz two-step parallel BiCMOS ADC with internal S/H"*, ISSCC Digest of Technical Papers, pp. 162,163, Februari 1990.

[86] Bang-Sup Song, S-H. Lee, M.F. Tompsett, *"A 10-bit 15 MHz recycling two-step A/D converter"*, IEEE Journal of Solid-State Circuits, vol. 25, pp. 1328-1337, December 1990.

[87] P.W. Li, M.J. Chin, P.R. Gray, R. Castello, *"Ratio-independent algorithmic analog-to-digital conversion technique"*, IEEE Journal of Solid-State Circuits, vol. 19. pp. 828-835, December 1984.

[88] H. Onodera, T. Tateishi, K. Tamaru, *"A cyclic A/D converter that does not require ratio-matched componenets"*, IEEE Journal of Solid-State Circuits, vol. 23, pp. 152-158, February 1988.

[89] S. Wolfram, *" "Mathematica" A system for doing mathematics by computer"*, Addison-Wesley Publishing Company, Redwood City 1991.

[90] M.H.H. Höfelt. *"On the stability of a one-bit quantized feedback system"*, Proceedings of the International Conference on ASSP (ICASSP), Washington, pp. 844-848, 1979.

[91] J.C. Candy, "*A use of doouble integration in sigma-delta modulation*", IEEE Transactions on Communications, vol. COM-35, pp. 481-489, May 1987.

[92] Brian P. Brandt, Bruce A. Wooley, "*A 50-MHz multibit sigma-delta modulator for 12-b 2-MHz A/D conversion*", IEEE Journal of Solid-State Circuits, vol. SC 26, pp. 1746-1756, December 1991.

[93] P. Ferguson, A. Ganesan, R. Adams, S. Vincelette, R. Libert, A. Volpe, D. Andreas, A. Charpentier, J. Dattorro, "*An 18-bit 20 kHz Dual sigma-delta A/D converter*", ISSCC Digest of Technical Papers, pp. 890-891, February 1991.

[94] B.M.J. Kup, E.C. Dijkmans, P.J.A. Naus, J. Sneep, "*A bit-stream digital-to-analog converter with 18-b resolution*", IEEE Journal of Solid-State Circuits, vol. 26, pp. 1757-1763, December 1991.

[95] H.J. Schouwenaars, D.W. Groeneveld, C.A.A. Bastiaansen, H.A. Termeer, "*An oversampled multibit CMOS D/A converter for digital audio with 115-dB dynamic range*", IEEE Journal of Solid-State Circuits, vol. 26, pp. 1775-1780, December 1991.

[96] J.C. Candy, G.C. Temes, "*Oversampling delta-sigma data converters*", IEEE Press, 1992.

[97] D.B. Ribner, R.D. Baertsch, S.L. Garverick, D.T. McGrath, J.E. Krisciunas, T. Fuji, "*16 b third-order sigma-delta modulator with reduced sensitivity to non-idealities*", ISSCC Digest of Technical Papers, pp. 66,67, February 1991.

[98] F. Op'tEynde, G. Ming, W. Sansen, "*A CMOS fourth-order 14 b 500 k-samples/s sigma delta ADC converter*", ISSCC Digest of Technical Papers, pp. 62,63, February 1991.

[99] Y. Matsuya, K. Uchimura, A. Iwata, T. Kobayashi, M. Ishikawa, T. Yoshitome, "*A 16-bit oversampling A-to-D conversion technology using triple-integration noise-shaping*", IEEE Journal of Solid-State Circuits, vol. 22, pp. 921-929, December 1987.

[100] W.L. Lee, C.G. Sodini, "*A topology for higher order interpolative coders*", Proceedings 1987 International Symposium on Circuits and System, pp. 459-462, May 1987.

[101] P.J.A. Naus, *"Bitstream D/A conversion for digital audio"*, M.Phil thesis, University of Wales, September 1991.

[102] B. Astegher, A. Lechner, H. Jessner, *"A novel all-NPN sample-and-hold circuit"*, ESSCIRC Digest of Technical Papers, pp. 88-91, Sept. 1989.

[103] P. Vorenkamp, J.P.M. Verdaasdonk, *"Fully bipolar 120-Msample/s 10-b track-and-hold circuit"*, IEEE Journal-of-Solid State Circuits, vol. SC 27, pp. 988-992, July 1992.

[104] P.R. Gray, R.G. Meyer, *"Analysis and design of analog integrated circuits"*, 2nd ed., John Wiley and Sons, 1984.

[105] P.R. Gray, *"High-speed CMOS A/D converters"*, Summer course EPFL, Lausanne, Switzerland, July 2-6, 1990.

[106] S. Sutarja, P.R. Gray, *"A pipelined 13-bit 250 ks/s 5-V analog-to-digital converter"*, IEEE Journal of Solid-State Circuits, vol. 23, pp. 1316-1323, December 1988.

[107] A. Matsuzawa, M. Kagawa, M. Kanoh, K. Tatehara, T. Yamaoka, K. Shimizu, *"A 10b 30 MHz two-step parallel Bi-CMOS ADC with internal S/H"*, ISSCC Digest of Technical Papers, pp. 162,163, February 1990.

[108] P.J. Lim, B.A. Wooley, *"A high-speed sample-and-hold technique using a Miller hold capacitance"*, IEEE Journal of Solid-State Circuits, vol. 26, pp. 643-651, April 1991.

[109] G. Wegmann, E.A. Vittoz, F. Rahali, *"Charge injection in analog MOS switches"*, IEEE Journal-of-Solid State Circuits, vol. SC 22, pp. 1091-1097, December 1987.

[110] E.A. Vittoz, *"Microwatt switched capacitor circuit design"*, Electro-components Science and Technology, vol. 9, pp. 263-273, 1982.

Index

resolution
  A/D converter 445
  effective 15
  increase 31
    oversampling 154
  loss of 149
reverse bias
  maximum 151
reverse zener voltage 151
ripple current 240, 255
ripple frequency 252
rise time 9
roll-off
  analog signal
    coarse divider 234
  first-order 315, 326
  second-order 314
  second order 325
ROM 115, 132, 172
  decoder 112
  encoder circuit 180
root locus 397
  alternative method
    stability analysis 421
  method
    stability analysis 390
rounding
  folding signals 161
    high frequencies 155
rounding-off
  tips of folded signals 149

S

S/H
  basic configuration 298
  bipolar inverting 324
  differential 316
    low-power 319
  differential MOS 322

  inverting architectures 312
  non-inverting BiCMOS 311
  non-inverting mode 298
  switched capacitor 310
S/N improvement
  oversampling factor
    filter order 384
sample-and-hold
  amplifier 3, 5, 25
    reconstruction 28
  built-in
    function 124
  parameter overview 103
sample-and-hold amplifier 88, 107, 133, 135, 137, 148, 190, 211, 274, 277
sample-and-hold amplifiers 144, 297
  inverting, non-inverting 100
  tests 100
sample-to-hold step 71
sample charge
  accurate transfer 324
sampling
  digital 148
  Nyquist 30
  signal dependent
    S/H 300
  signal plus offset 123
sampling bridge 312
sampling clock
  doubling 125
  rise, fall time 111
  uncertainty 421
sampling frequency
  maximum 153
sampling moment
  frequency dependent 208
sampling oscilloscope 312
sampling rate

If you have any concerns about our products,
you can contact us on
ProductSafety@springernature.com

In case Publisher is established outside the EU,
the EU authorized representative is:
Springer Nature Customer Service Center GmbH
Europaplatz 3, 69115 Heidelberg, Germany

Printed by Libri Plureos GmbH
in Hamburg, Germany